551.362/CAS

WOODSIDE HOUSTON LIBRARY
0116448

AF825540

HOUSTON LIBRARY COPY
(Woodside Energy (USA) Inc.)
DO NOT REMOVE
FROM THIS OFFICE

BC: 0116448
Call # 551.362 CAS

Investigations in Geophysics Series, Volume 8
Michael R. Cooper, Series Editor

Offset-Dependent Reflectivity—

Theory and Practice of AVO Analysis

HOUSTON LIBRARY COPY
(Woodside Energy (USA) Inc.)
DO NOT REMOVE
FROM THIS OFFICE

Edited by
John P. Castagna and Milo M. Backus

Society of Exploration Geophysicists
P.O. Box 702740, Tulsa, OK 74170-2740

Library of Congress Cataloging-in-Publication Data
Offset-dependent reflectivity : theory and practice of AVO analysis /
edited by John P. Castagna and Milo M. Backus.
p. cm.—(Investigations in geophysics series : v. 8)
Includes bibliographical references and index.
ISBN 0-931830-46-X : $119.00
1. Petroleum--Prospecting. 2. Natural gas--Prospecting.
3. Amplitude variation with offset analysis. I. Castagna, John
Patrick. II. Backus, Milo M., 1932 III. Series
TN271.P4034 1993
622′.1828--dc20 93-14279
CIP

ISBN 0931830-46-X (Series)
ISBN 1-56080-059-3 (Volume)

© 1993 by Society of Exploration Geophysicists
All rights reserved. This book or parts hereof
may not be reproduced in any form without permission in writing from the publisher.

Published 1993
Second Printing 1993
Third Printing 1995
Fourth Printing 1997
Fifth Printing 1999
Printed in the United States of America

Contents

Preface

This volume is intended to provide a comprehensive review of amplitude-variation-with-offset (AVO) analysis as applied to hydrocarbon exploration. From conception to fruition, it has taken over six years to finish the job.

Failures of flesh and bone, editorial changes, job reassignments and relocations, and the more ordinary obstacles, conspired to delay completion. In 1987 we had a core group of six papers (Thomsen; Castagna et al. a and b; Chiburis; Wrolstad; and Estill and Wrolstad). The remaining papers were added in 1989 and 1990 with the exception of the tutorial which was added in 1991. Many of the papers have been revised "organically" over the years. Perhaps due to the quality of the original contributions, or possibly as a consequence of the important and controversial subject matter, we feel that the delay did not materially decrease the significance of these papers.

We are deeply grateful to the thirty-five authors whose patience, cooperation, and encouragement enabled us to complete the task. Thanks also to Sabra Gronemeier whose contributions are too many to enumerate. Special acknowledgments are owed to Lynn Griffin without whose remarkable skill and dedication this volume could not have been completed.

J. P. Castagna
ARCO Oil and Gas Co.

M. M. Backus
Univ. of Texas at Austin

Introduction

"The *P*-wave reflection coefficient at an interface separating two media is known to vary with angle of incidence. The manner in which it varies is strongly affected by the relative values of Poisson's ratio of the two media. For moderate angles of incidence, the relative change in reflection coefficient is particularly significant when Poisson's ratio differs greatly between the two media.

Theory and laboratory measurements indicate that gas sands tend to exhibit abnormally low Poisson's ratios. Embedding a low velocity gas sand into sediments having 'normal' Poisson's ratios should result in an increase in reflected *P*-wave energy with angle of incidence. This phenomenon has been observed on conventional seismic data recorded over known gas sands."

With these words, W. J. Ostrander ushered in a new era in seismic interpretation. Although many workers were aggressively moving forward with amplitude variation with offset (AVO) analysis before 1982, Ostrander's presentation at the 52nd Annual International Meeting of the Society of Exploration Geophysicists popularized the concept as an exploration tool. The presentation proved to be a precursor to the avalanche of literature on the subject which has appeared during the past decade.

The tremendous promise of the isolation and analysis of offset-dependent reflectivity lies in the theoretical relationship between reflection coefficient, incident angle, and the variation in compressional-wave velocity (V_P), shear-wave velocity (V_S), and density (ρ) across an interface. These contrasts are, in turn, dependent on rock properties variations. In particular, an interface between gas-saturated and water-saturated unconsolidated or poorly consolidated sediments exhibits distinctive relationships between these parameters. Thus, bright spots caused by gas reservoirs potentially can be distinguished from high amplitude events of the same polarity caused by coal or other low impedance units. In addition to direct hydrocarbon indication for gas and high GOR oils, other potential applications include lithology or porosity indication, extraction or "true" zero-offset reflectivity, and improved imaging resulting from proper treatment and exploitation of waveform variations with offset.

All too often, however, the potential of AVO analysis has not been realized. Under proper (sometimes fortuitous) circumstances, AVO analysis has proved to be a useful prospecting tool (success in the Sacramento Valley is particularly notable). In other situations, the technique has simply failed (although the literature seems to be strongly biased toward successful examples). Table 1 shows questions concerning AVO analysis raised at the AVO workshop conducted at the 1983 SEG convention. These questions remain relevant 10 years later.

The enormous potential value of the method and the corresponding need for education and further research on the subject are the motivation for this volume. We make no attempt to present a robust methodology for AVO analysis. Such a methodology requires special data acquisition and processing considerations as well as an integrated approach to interpretation, and is certain to remain proprietary within the confines of those organizations which are successfully exploiting AVO analysis in the quest for hydrocarbons. We also make no claim to exhaustively treat the subject. It is our intention simply to (1) provide a general introduction to the subject and review the current state-of-the-art, (2) provide useful reference materials and data, and (3) present recent original contributions at the leading edge of AVO technology. It is hoped that this volume will be useful to explorationists and students, as well as to researchers in the field. Thus, we have solicited a wide variety of original contributions. Some papers are at a basic tutorial level with broad appeal, while others will attract only a select audience. Some are applied or empirical while others are highly theoretical. A diversity of approaches and orientations will become apparent to the reader who ultimately will have to synthesize this information and utilize it in a practical way.

This volume is divided into six chapters. Chapter 1 covers the basic principles of offset-dependent reflectivity as applied to hydrocarbon exploration and development. Two papers, (1) Spratt, Goins, and Fitch and (2) Dey-Sarkar and Svatek present integrated approaches to practical AVO analysis as applied by two major oil companies. The properties of AVO indicators, like those presented throughout this volume, are then investigated by Swan. Kan and Young discuss lessons learned from full-waveform elastic modeling while Thomsen describes the complicating effects of realistic earth media which exhibit weak anisotropy.

Chapter II covers the fundamental petrophysical

Table 1. Questions raised at the 1983 AVO workshop, 53rd Annual International Meeting, Society of Exploration Geophysicists

—What is the current state of our knowledge of operative in-situ rock properties?

—Under what plausible real earth conditions do we expect a significant change in reflection coefficients with angle?

—What are the important rock property variables and what is the sensitivity of angle dependence to these variables?

—What are the available methods of seismogram synthesis and how do they compare?

—What have we learned about the generalized reflectivity of layers, layer sequences, and transition zones?

—What are some of the important characteristics of solid earth synthetic seismograms obtained for realistic earth models?

—What are the effects of anisotropy?

—In what circumstances have we been able to exploit angle dependent reflectivity effects in exploration?

—Are special data acquisition, processing, and display techniques helpful in isolating the derived target characteristics?

—What are the most promising likely applications in the short term?

—How significant are the changes in waveform and amplitude observed in practice?

—Should we recast our basic model for certain environments?

—What is the relative potential merit of shear impedance estimates made from (a) P-P AVO, (b) P-S-P conversions, (c) P-S-S conversions, (d) S-S AVO?

—What is the likelihood of providing multiparameter subsurface images for exploration use?

—What are the most significant gaps in our knowledge, understanding, and capabilities with regard to the change in reflectivity with offset?

principles which form the basis for AVO analysis. The first paper (Castagna, Batzle, Tubman, Gaiser, and Burnett) describes various techniques for measuring or estimating shear-wave velocities, while the second (Castagna, Batzle, and Kan) describes the relationships between rock properties and wave propagation parameters such as velocity and attenuation.

Chapter III contains three papers relating to seismic data processing for AVO analysis. Resnick illustrates the special processing necessary for AVO analysis, Luh shows the importance of Q compensation, and Shang, McDonald, and Gardner present an automated method for amplitude-versus-angle extraction in structurally complex areas.

Five AVO case studies with real data and real exploration objectives constitute Chapter IV. The authors (1) Chiburis, (2) Hong, Castagna, and Sicking, (3) Regueiro, (4) Wrolstad, and (5) Estill and Wrolstad provide examples of AVO interpretation from four continents.

Chapters V and VI, covering elastic inversion and multicomponent AVO, have a distinct eye on the future. Demirbag, Coruh, and Costain present an inversion approach which is computationally practical, while Carrazone and Srnka provide a real data example of AVO inversion. Lörtzer and Berkhout study multicomponent inversion, while Zaengle and Frasier provide a case study of multicomponent AVO analysis.

Chapter I
Principles

AVO Analysis—Tutorial and Review

*John P. Castagna**

INTRODUCTION

In conventional utilization of the seismic reflection method, it has traditionally been assumed that seismic signals can be viewed as a band-limited normal incidence reflection coefficient series with appropriate traveltime and amplitude variation due to propagation through an overburden. Ostrander (1982) demonstrated that gas sand reflection coefficients vary in an anomalous fashion with increasing offset and showed how to utilize this anomalous behavior as a direct hydrocarbon indicator on real data. This work popularized the methodology which has come to be known as amplitude variation with offset analysis (AVO). In addition to being a "seismic lithology" tool, AVO provides an improved model of the reflection seismogram which allows us to better estimate both normal incidence reflection coefficients and "background" velocities.

Exploration geophysics is, to a large extent, a science of anomalies. It is probably safe to assume that most hydrocarbons found in the past fifty years have been associated with some kind of geophysical anomaly. Explorationists routinely utilize deviations from expected gravity, seismic traveltime, and seismic amplitude, without recovering uniquely determined absolute density, depth, or reflectivity. Consequently, significant risk is associated with drilling a well, even when all the technology and analysis available have been thrown at the problem. Experience has shown that geophysical anomalies can be used to reduce risk, and consequently, to identify new prospects.

Thus, one would expect the exploration geophysicist, who is concerned primarily with finding hydrocarbons, to have somewhat different expectations from AVO analysis, than say a research physicist, who may be primarily concerned with getting the right answer. Taking the latter approach, if one makes very simplistic assumptions, methods can be developed which work well on synthetic data. However, the more one appreciates the complexities of the real world with real geology, the more it becomes apparent that AVO analysis doesn't work. The only fault with this conclusion is that explorationists are successfully using AVO anomalies to find hydrocarbons throughout the world. The explorationist does not require answers that are correct in absolute terms. The presence of a deviation from some background trend may be sufficient; the magnitude of the deviation in absolute units may not even be required. An appropriate analogy is the spontaneous potential (SP) log. If I am told that an SP log reading is −30 millivolts at some depth, I cannot interpret the rock properties, even with the most sophisticated modeling and inversion schemes. Of course, the SP log is an anomaly log without a meaningful absolute scale. When I see that −30 millivolts is a significant deviation from the background trend produced by surrounding shales, I realize that the SP log is very valuable indeed in the context of my particular problem—that of identifying sands.

I am not advocating a know-nothing approach to AVO analysis. To the contrary, it is extremely important to understand the physics of wave propagation in a real earth to the best of our ability. This is required to correctly process and appropriately interpret the data, with full recognition of potential pitfalls and the expected applicability of the method to the exploration problem at hand. What I am advocating, is a faith in the inherent robustness of geophysical methods when employed to detect and map anomalies, which should also be applied to AVO analysis.

The purpose of this paper is to provide a tutorial introduction to AVO analysis, in order to lay the foundation for, and to place in context, the more advanced papers which constitute this volume.

OFFSET-DEPENDENT REFLECTIVITY

In exploration geophysics, we only rarely deal with simple isolated interfaces. Nevertheless, we must begin our understanding of offset-dependent reflectivity with the partitioning of energy at just such an inter-

*ARCO Oil and Gas Company

face. We must be careful, however, to remember that most reflections we see are a superposition of events from a series of layers, and will have more complex AVO behavior than will be shown here.

Consider two semi-infinite isotropic homogeneous elastic media in contact at a plane interface. Further, consider an incident compressional plane wave impinging upon this interface. The plane-wave assumption is valid at source-to-receiver distances which are much longer than the wavelength of the incident wave and is generally acceptable for precritical reflection data at exploration depths and frequencies. As shown in Figure 1, reflection at an interface involves energy partition from an incident P-wave to (1) a reflected P-wave, (2) a transmitted P-wave, (3) a reflected S-wave, and (4) a transmitted S-wave. The angles for incident, reflected, and transmitted rays synchronous at the boundary are related according to Snell's law,

$$p = \frac{\sin \Theta_1}{V_{P1}} = \frac{\sin \Theta_2}{V_{P2}} = \frac{\sin \Phi_1}{V_{S1}} = \frac{\sin \Phi_2}{V_{S2}} \tag{1}$$

where,

V_{P1} = P-wave velocity in medium 1,
V_{P2} = P-wave velocity in medium 2,
V_{S1} = S-wave velocity in medium 1,
V_{S2} = S-wave velocity in medium 2,
Θ_1 = incident P-wave angle,
Θ_2 = transmitted P-wave angle,
Φ_1 = reflected S-wave angle,

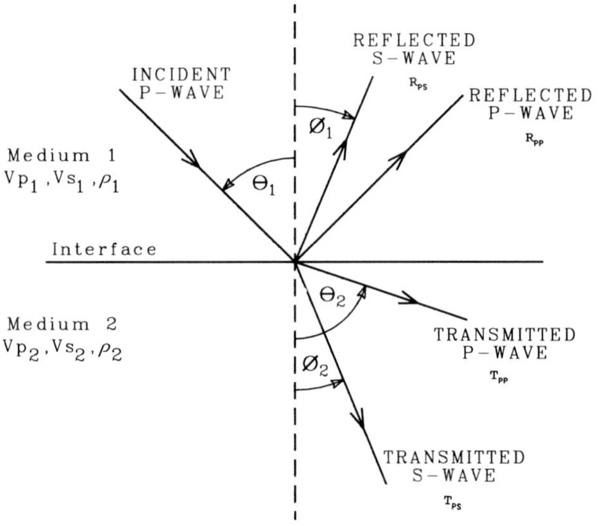

Fig. 1. Reflection and transmission at an interface between two infinite elastic half-spaces for an incident P-wave.

Φ_2 = transmitted S-wave angle,

and p is the ray parameter.

The P-wave reflection coefficient as a function of incidence angle $R_{PP}(\Theta_1)$ is defined as the ratio of the amplitude of the reflected P-wave to that of the incident P-wave. Similarly, the P-wave transmission coefficient $T_{PP}(\Theta_1)$ is the ratio of the amplitude of the transmitted P-wave to that of the incident P-wave. Also, $R_{PS}(\Theta_1)$ is the ratio of the amplitudes of reflected S-wave and incident P-waves, and $T_{PS}(\Theta_1)$ is the ratio of the transmitted S-wave and incident P-wave amplitudes. The following discussion is for particle displacement amplitudes, with polarity assigned according to the horizontal component of displacement.

At normal incidence, there are no converted S-waves and the P-wave reflection coefficient R_P is given by

$$R_P = \frac{I_{P2} - I_{P1}}{I_{P2} + I_{P1}} = \frac{1}{2}\frac{\Delta I_P}{I_{PA}} \approx \frac{1}{2} \ln (I_{P2}/I_{P1}) \tag{2}$$

where I_P is the continuous P-wave impedance profile,

I_{P2} = impedance of medium 2 = $\rho_2 V_{P2}$,
ρ_2 = density of medium 2,
I_{P1} = impedance of medium 1 = $\rho_1 V_{P1}$,
ρ_1 = density of medium 1,
I_{PA} = average impedance across the interface = $(I_{P2} + I_{P1})/2$, and,
$\Delta I_P = I_{P2} - I_{P1}$.

The logarithmic approximation is acceptable for reflection coefficients smaller than about $\pm$ 0.5.

The P-wave transmission coefficient at normal incidence T_P is given by

$$T_P = 1 - R_P \tag{3}$$

The variation of reflection and transmission coefficients with incident angle (and corresponding increasing offset) is referred to as offset-dependent-reflectivity and is the fundamental basis for amplitude-versus-offset analysis.

Knott (1899) and Zoeppritz (1919) invoked continuity of displacement and stress at the reflecting interface as boundary conditions to solve for the reflection and transmission coefficients as a function of incident angle and the media elastic properties (densities, bulk and shear moduli). By continuity of displacement, it is meant that the interface responds to the incident wave essentially as if the two media were welded together. That is, there is no cavitation (pulling apart) or slippage along the interface. Continuity of normal and tangential stress is also required to avoid infinite acceleration at the discontinuity.

The resulting Knott and Zoeppritz equations are notoriously complex and subject to troublesome sign, convention, or typographical errors (Hales and Roberts, 1974). Their complexity is such that they yield little physical insight, and the use of digital computers was required before they could be used routinely for exploration applications. Young and Braile (1976) modified expressions from Cerveny and Ravindra (1971) and provided a simple computer subroutine for coefficient calculation. Aki and Richards (1980) and Waters (1981; with obligatory sign error) gave the equations in easily solved matrix form

$$\mathbf{Q} = \mathbf{P}^{-1}\mathbf{R} \tag{4}$$

where **Q**, **P**, and **R** are given in Appendix A.

The coefficients R_{PP}, R_{PS}, T_{PP}, and T_{PS}, at any given incident angle Θ_1 are completely determined by the density and the P-wave and S-wave velocities of each medium. These parameters are, in turn, dependent on medium physical properties such as lithology, porosity, and pore fluid content.

Figure 2 (after Richards, 1961) illustrates the reflection coefficient variation for a variety of impedance ratios with a constant V_P/V_S of 2 in each layer (see Table 1). Keep in mind that Figure 2 shows the entire range of possible incidence angles, whereas common exploration incidence angles are about 30 degrees or less. Some observations follow:

(1) Reflected P-wave amplitude local maxima may occur at normal incidence, the first critical angle, and possibly near the S-wave critical angle. Total internal reflection always occurs at grazing incidence (90 degrees incident angle).

(2) The change of the reflection coefficient with respect to angle of incidence is small at low angles.

(3) The first critical angle Θ_{C1} is given by

$$\sin \Theta_{C1} = \frac{V_{P1}}{V_{P2}}. \tag{5}$$

Readers familiar with the principle of total internal reflection in optics may recall that beyond the critical angle all energy is reflected. This is also true for an interface between two fluid layers. However, for elastic layers, reflected P-wave energy decreases after the first critical angle due to increased conversion to reflected and transmitted S-waves. Consequently, there can be a second critical angle Θ_{C2} given by

$$\sin \Theta_{C2} = \frac{V_{P1}}{V_{S2}}. \tag{6}$$

Beyond this critical angle there are no transmitted S-waves. Note that when V_{P1} is greater than V_{S2} there is no second critical angle, and converted S-waves will be transmitted at all angles below 90 degrees. Similarly, when V_{P1} is greater than V_{P2}, there is no first critical angle, and P-waves can always be transmitted.

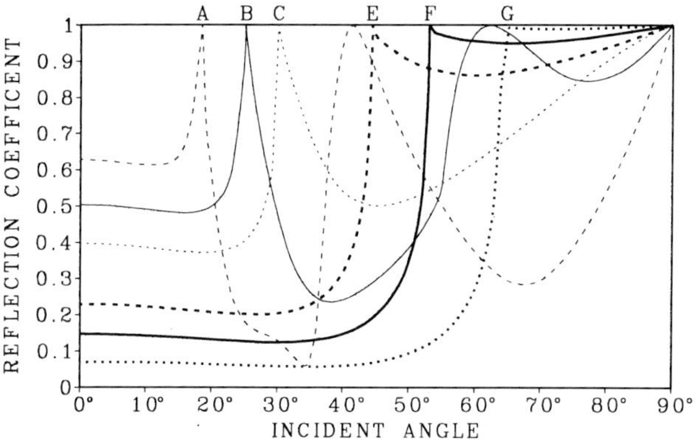

Fig. 2. P-wave reflection coefficient versus incident angle for six models given in Table 1 (after Richards, 1961).

Table 1. Parameters of first layer for six Zoeppritz models from Richards (1961) for Paleozoic limestone structures in Western Canada. These models correspond to reflection coefficient curves given in Figure 2 when the second layer has the parameters: V_P = 20 000 ft/s (6096 m/s), V_S = 10 000 ft/s (3048 m/s), and ρ is 2.65 gm/cc. Note that V_P/V_S = 2 for all layers.

Overburden Model	V_P (ft/s)	V_P (m/s)	V_S (ft/s)	V_S (m/s)	ρ (gm/cc)
A	6000	1829	3000	914	2.02
B	8270	2521	4135	1260	2.12
C	10000	3048	5000	1524	2.20
E	14000	4267	7000	2133	2.38
F	16000	4877	8000	2438	2.47
G	18000	5486	9000	2743	2.56

(4) At near normal incidence, R_{PP} initially decreases very slightly with increasing angle. Deviations from this general behavior can result from changes in V_P/V_S across the interface, and form the basis for seismic lithology analysis.

Some additional features of the Zoeppritz equations are:

(1) S-wave conversions, both reflected and transmitted, are generally strongest between the first critical angle and grazing incidence. S-wave conversions can be very strong at offsets commonly encountered in seismic prospecting.

(2) The reflection coefficient becomes complex beyond the critical angle, and a corresponding phase shift is introduced.

(3) The reflection and transmission coefficients are frequency independent.

(4) The reflection coefficient for a downgoing P-wave incident in medium 1 is generally not equal and opposite to that of an upgoing P-wave incident on the same interface in medium 2. Thus, the reflection coefficient behavior is asymmetrical.

Some exceptions must be considered for spherical waves which exhibit frequency-dependent reflection coefficients and for which the reflection coefficient maximum occurs beyond the critical angle. For most exploration applications, however, the plane-wave assumption is considered adequate.

Muskat and Meres (1940), without the benefit of a computer, calculated reflection coefficient versus angle of incidence for a large number of cases. However, partially due to the lack of information about V_P/V_S ratios in sedimentary rocks available at the time, and perhaps more importantly, due to the very laborious computations required, a constant V_P/V_S ratio of $\sqrt{3}$ (a classical Poisson solid) was assumed for each medium. This assumption colored the exploration community's thinking for many years. By assuming constant V_P/V_S, one is led to the conclusion that, for exploration geometries, amplitude decreases with increasing angle of incidence, and the overall effect is rather small. In fact, this conclusion provided theoretical justification for such procedures as stacking, convolutional modeling, bright spot analysis, and one dimensional (1-D) seismic inversion. The assumption that stacked reflection amplitudes mimic normal incidence reflection coefficients has been very effectively exploited by seismic interpreters. However, the potential of AVO analysis was generally overlooked. Application of AVO was limited to a few special cases (Anstey, 1977): (1) A weak target reflector can be intentionally illuminated at the critical angle to obtain a stronger signal. (2) As previously noted, if V_{P1} is greater than V_{P2}, no critical angle amplitude maximum will be observed. Thus, the polarity of the reflection may be identified if long offsets are available. (3) In some cases, salt reflections can be identified by a characteristic increase of amplitude with offset caused by an increase in velocity accompanied by a decrease in density across the interface.

Now let us turn our attention to the precritical angle of incidence range which is most important for exploration applications. Koefoed (1955) first pointed out the practical possibilities of using AVO analysis as an indicator of V_P/V_S variations. Figure 3 shows some of his reflection coefficient versus angle curves. Note that the calculations were restricted only to angles of incidence likely to be encountered in seismic exploration, and that the Poisson's ratio was allowed to vary across the interface (see Table 2 for model parameters). Poisson's ratio σ is related to V_P/V_S by

$$\sigma = \frac{\frac{1}{2}\left(\frac{V_P}{V_S}\right)^2 - 1}{\left(\frac{V_P}{V_S}\right)^2 - 1}, \tag{7a}$$

and

$$\frac{V_S}{V_P} = \sqrt{\frac{\frac{1}{2} - \sigma}{1 - \sigma}}. \tag{7b}$$

Figure 3 forms the empirical basis for some very general conclusions (Koefoed's rules) concerning the information content of reflection coefficients which will be discussed in a later section. Some additional observations are: (1) Figure 3a, 3b, and 3c illustrate the basic principle exploited when AVO is used for hydrocarbon detection; as σ_2 decreases (as would occur when gas replaces brine), the reflection coefficient becomes more negative with increasing offset. Contrast this to the often abused rule of thumb that "gas causes amplitude to increase with offset" which is only correct for near zero or negative R_P (keep in mind that increased "amplitude" results from increasing the magnitude of the reflection coefficient in a positive or negative direction). (2) For $\Delta\sigma = 0$ (Figures 3d and 3f), lowering the average Poisson's ratio, $(\sigma_1 + \sigma_2)/2$, also causes the reflection coefficient to become more negative with increasing offset. (3) Comparison of Figures 3b and 3e shows that the effect of lowering σ_2 occurs for all V_{P2}/V_{P1} ratios. (4) If V_{P2}/V_{P1} is varied while density and Poisson's ratio remain constant, the magnitude of the AVO effect is essentially unchanged (the AVO curves are shifted by the normal incidence reflection coefficient R_P).

Koefoed (1955) suggested what is today a somewhat overlooked application, that of using characteristic AVO responses to help correlate across faults. More prophetically he stated:

"... in a more remote future it may become possible to draw conclusions concerning the lithological nature of rock strata from the shapes of the reflection coefficient curves of the interfaces by which they were bounded."

APPROXIMATIONS TO THE KNOTT AND ZOEPPRITZ EQUATIONS

The unwieldy nature of the Knott-Zoeppritz equations defies physical insight and makes visualizing how the variation of a particular parameter will effect the reflection coefficient curve very difficult. Approximations are extremely useful for practical applications as they more readily reveal the information content contained in the amplitude behavior, do not require a computer to evaluate, and provide the basis for certain AVO processing techniques.

Bortfeld (1961) linearized the Zoeppritz equations by assuming small changes in layer properties and obtained

$$R_{PP}(\Theta_1) \approx \frac{1}{2} \ln \left[\frac{V_{P2}\rho_2 \cos \Theta_1}{V_{P1}\rho_1 \cos \Theta_2} \right] + \left(\frac{\sin \Theta_1}{V_{P1}} \right)^2 (V_{S1}^2 - V_{S2}^2) \times \left[2 + \frac{\ln \left(\frac{\rho_2}{\rho_1} \right)}{\ln \left(\frac{V_{S2}}{V_{S1}} \right)} \right]. \quad (8)$$

This approach was also followed in Richards and Frasier (1976) and Aki and Richards (1980) who derived a form simply parameterized in terms of the changes in density, P-wave velocity, and S-wave velocity across the interface:

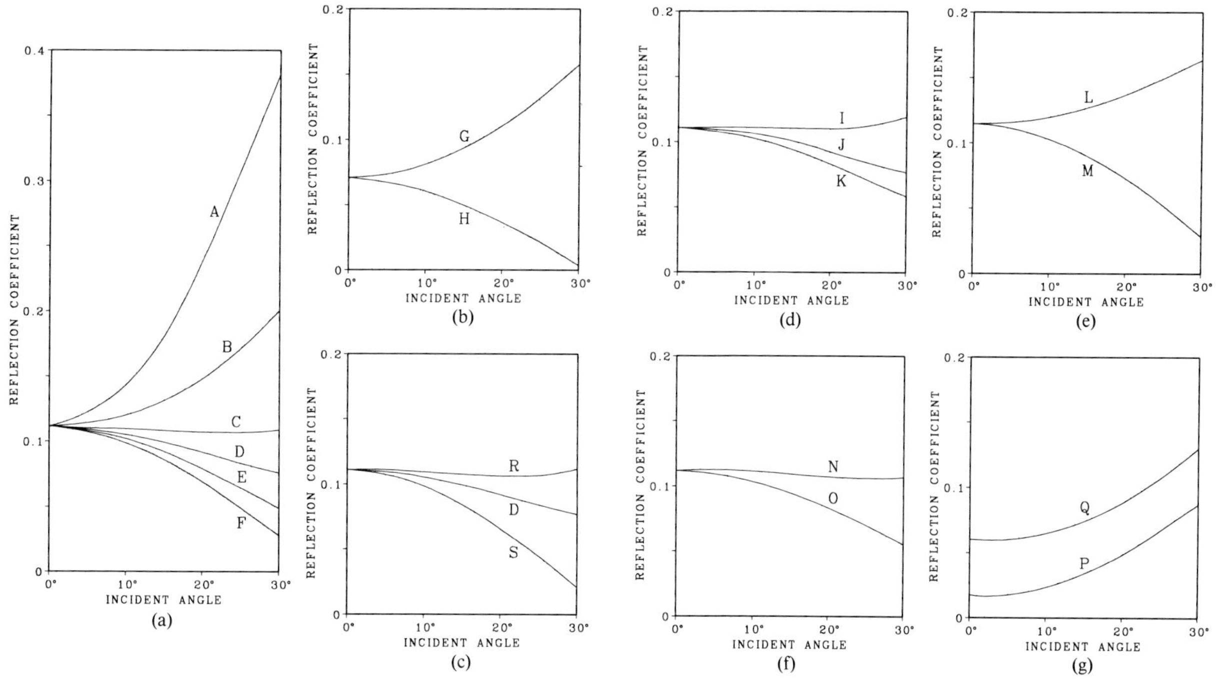

Fig. 3. P-wave reflection coefficient versus incident angle for model parameters given in Table 2 (from Koefoed, 1955). (a) and (b) show varying Poisson's ratio in the second medium, (c) and (e) show varying Poisson's ratio in the first medium, (d) shows varying values of Poisson's ratio which is equal in both half-spaces, and (f) shows varying values of the P-wave velocity contrast.

$$R_{PP}(\Theta) \approx \frac{1}{2}(1 - 4p^2V_{Sa}^2)\left(\frac{\Delta\rho}{\rho_a}\right) + \frac{1}{2\cos^2(\Theta)} \times \frac{\Delta V_P}{V_{Pa}} - 4p^2V_{Sa}^2\frac{\Delta V_S}{V_{Sa}} \quad (9)$$

where

$$\begin{aligned}
\Delta_\rho &= \rho_2 - \rho_1,\\
\Delta V_P &= V_{P2} - V_{P1},\\
\Delta V_S &= V_{S2} - V_{S1},\\
\rho_a &= (\rho_2 + \rho_1)/2,\\
V_{Pa} &= (V_{P2} + V_{P1})/2,\\
V_{Sa} &= (V_{S2} + V_{S1})/2,\\
\Theta &= (\Theta_1 + \Theta_2)/2
\end{aligned}$$

and p is the ray parameter as defined by equation (1). Because interchanging the medium properties simply changes the sign of $\Delta\rho$, ΔV_P, and ΔV_S, we see that equation (9) is fundamentally different from the Zoeppritz equations in that the reflection coefficient is symmetrical. According to Backus (1992; personal communication) for reflections from layers rather than from isolated interfaces, superposition of the P-wave and the locally converted S-wave may roughly cancel the asymmetry predicted by the Zoeppritz equations for primaries only. Thus, for some uses, the Aki and Richards approximation is more appropriate than the exact Zoeppritz equations. As a further approximation, Θ_1 is usually used rather than Θ.

For the special case of fluid layers, the reflection coefficient is given exactly by

$$R_{PP}(\Theta_1) = \frac{V_{P2}\rho_2 \cos\Theta_1 - V_{P1}\rho_1 \cos\Theta_2}{V_{P2}\rho_2 \cos\Theta_1 + V_{P1}\rho_1 \cos\Theta_2}. \quad (10)$$

Equation 10 is referred to as the "acoustic" reflection coefficient, $R_f(\Theta_1)$, rather than "fluid factor" as originally used by Hilterman and later used in Smith and Gidlow (1987) to mean something entirely different (see following discussion). Hilterman's course notes pointed out the physical insight to be gained by separating the Bortfeld approximation into "fluid" and "shear" terms. He recognized that the first term of the Bortfeld equation is about equal to equation (10) and modified Bortfeld's equation such that

$$R_{PP}(\Theta_1) = R_f(\Theta_1) + R_{sf}(\Theta_1) \quad (11)$$

where, $R_{sf}(\Theta_1)$, is the "shear factor" given by

$$R_{sf}(\Theta_1) = \left(\frac{\sin\Theta_1}{V_{P1}}\right)^2 (V_{S1} + V_{S2})\left[3(V_{S1} - V_{S2}) + 2\,\frac{V_{S2}\rho_1 - V_{S1}\rho_2}{\rho_2 + \rho_1}\right]. \quad (12)$$

Table 2. Parameters for Zoeppritz models from Figure 3 (Koefoed, 1955).

Model	ρ_2/ρ_1	V_{P2}/V_{P1}	σ_1	V_{P1}/V_{S1}	σ_2	V_{P2}/V_{S2}
A	1.0	1.25	.25	1.73	.50	∞
B	1.0	1.25	.25	1.73	.40	2.45
C	1.0	1.25	.25	1.73	.30	1.87
D	1.0	1.25	.25	1.73	.25	1.73
E	1.0	1.25	.25	1.73	.20	1.63
F	1.0	1.25	.25	1.73	.15	1.56
G	1.0	1.15	.25	1.73	.40	2.45
H	1.0	1.15	.25	1.73	.15	1.56
I	1.0	1.25	.40	2.45	.40	2.45
J	1.0	1.25	.25	1.73	.25	1.73
K	1.0	1.25	.15	1.56	.15	1.56
L	1.0	.80	.40	2.45	.25	1.73
M	1.0	.80	.15	1.56	.25	1.73
N	1.0	.80	.40	2.45	.40	2.45
O	1.0	.80	.15	1.56	.15	1.56
Q	.9	1.25	.25	1.73	.36	2.14
P	.9	1.15	.25	1.73	.36	2.14
R	1.0	1.25	.15	1.56	.25	1.73
S	1.0	1.25	.40	2.45	.25	1.73

For larger contrasts between layer properties, R_{sf} can be multiplied by $[1 - R_f(\Theta_1)^2]$ to better approximate the full Zoeppritz equations (Banik, personal communication).

Separation of the Bortfeld equations into "acoustic or fluid" and "shear" terms can also be achieved by assuming constant density (Hilterman, unpublished course notes, 1983) with

$$R_{PP}(\Theta_1) \approx \frac{V_{P2} \cos \Theta_1 - V_{P1} \cos \Theta_2}{V_{P2} \cos \Theta_1 + V_{P1} \cos \Theta_2} + 2\left(\frac{\sin \Theta_1}{V_{P1}}\right)^2 (V_{S1}^2 - V_{S2}^2) \qquad (13)$$

Figure 4, after Hilterman (unpublished course notes, 1983), illustrates the effect of the shear factor. Note that varying the shear factor may cause $R_{PP}(\Theta_1)$ to increase or decrease with increasing offset.

Shuey (1985) presented another form of the Aki and Richards (1980) approximation

$$R_{PP}(\Theta_1) \approx R_P + \left(A_0 R_P + \frac{\Delta\sigma}{(1-\sigma)^2}\right) \sin^2 \Theta_1 + \frac{1}{2}\frac{\Delta V_P}{V_{Pa}} (\tan^2 \Theta_1 - \sin^2 \Theta_1) \qquad (14)$$

where, R_P, is the normal incidence reflection coefficient, A_0 is given by

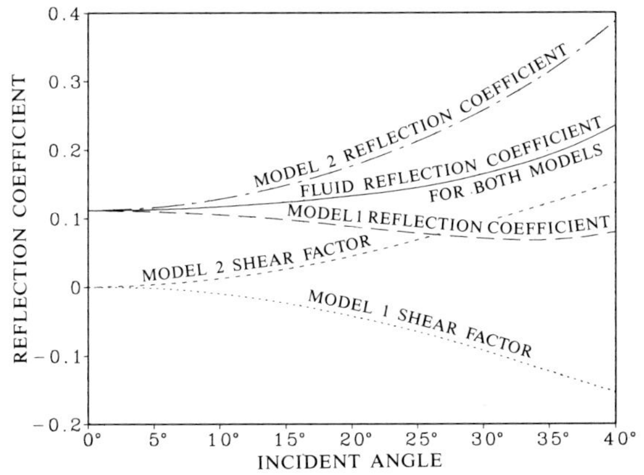

Fig. 4. The effect of the shear term, R_{sf}, on the P-wave reflection coefficient for two models. Model 1 has V_{P1} = 10 000 ft/s (3048 m/s), V_{P2} = 12 500 ft/s (3810 m/s), Poisson's ratio equals .25 for both media and density is constant. Model 2 is the same except Poisson's ratio is .1 for medium 1 and .4 for medium 2 (after Hilterman, unpublished course notes, 1983).

$$A_0 = B_0 - 2(1 + B_0)\left(\frac{1-2\sigma}{1-\sigma}\right) \qquad (15)$$

and

$$B_0 = \frac{\dfrac{\Delta V_P}{V_{Pa}}}{\dfrac{\Delta V_P}{\Delta V_{Pa}} + \dfrac{\Delta\rho}{\Delta\rho_a}}. \qquad (16)$$

The advantage of this form is that each term simply describes a different angular range of the offset curve. The first term is the normal incidence reflection coefficient, the second term predominates at intermediate angles, and the third term is dominant as the critical angle is approached. Thus, for restricted angles of incidence, we have an equation linear in $\sin^2 \Theta_1$.

$$R_{PP}(\Theta_1) \approx R_P + B \sin^2 \Theta_1. \qquad (17)$$

One often sees "A" used to represent the normal incidence reflection coefficient R_P. B is called the "AVO gradient" and is given by

$$B = A_0 R_P + \frac{\Delta\sigma}{(1-\sigma)^2}. \qquad (18)$$

Wiggins et al. (1983) showed that if V_P/V_S is about 2, for small angles then

$$B = R_P - 2R_S. \qquad (19)$$

This approximation greatly simplifies AVO interpretations.

The coefficients of the Wiggins and Shuey approximations are easily obtained by linear regression and form the basis of various "weighted stacking" procedures (see for example Smith and Gidlow, 1987).

Hilterman (1989) derived another convenient approximation:

$$R_{PP}(\Theta_1) \approx R_P \cos^2 \Theta_1 + 2.25\Delta\sigma \sin^2 \Theta_1. \qquad (20)$$

Thus, at small angles R_P dominates the reflection coefficient, whereas $\Delta\sigma$ dominates at larger angles. Thus, one can think of a near offset stack as imaging P-wave impedance contrasts, while the far offset stack images Poisson's ratio contrasts.

INFORMATION CONTENT

Koefoed (1955; see Figure 3) empirically established five rules which were later verified in Shuey (1985) for moderate angles of incidence:

(1) "When the underlying medium has the greater

longitudinal (P-wave) velocity and other relevant properties of the two strata are equal to each other, an increase of Poisson's ratio for the underlying medium causes an increase of the reflection coefficient at the larger angles of incidence.'' Shuey (1985) points out that the qualifications concerning the medium properties are not necessary for this and the next rule.

(2) ''When, in the above case, Poisson's ratio for the incident medium is increased, the reflection coefficient at the larger angles of incidence is thereby decreased.''

(3) ''When, in the above case, Poisson's ratios for both media are increased and kept equal to each other, the reflection coefficient at the larger angles of incidence is thereby decreased.'' This is strictly true only when Shuey's parameter B is greater than −1.

(4) ''The effect mentioned in (1) becomes more pronounced as the velocity contrast becomes smaller.''

(5) ''Interchange of the incident and the underlying medium affects the shape of the curves only slightly, at least up to the values of the angle of incidence of about 30 degrees.'' In other words, the reflection coefficient is nearly, but not quite, symmetrical at moderate angles.

For constant P-wave velocity and density, rules (1) and (3) imply that various combinations of Poisson's ratio can result in the same variation of the reflection coefficient with offset. This means that the inversion of the Zoeppritz equations for Poisson's ratio is fundamentally nonunique. Rosa (1976) demonstrated this analytically by differentiating the Zoeppritz equations with respect to the various model parameters. Figure 5 implies that many combinations of σ_1 and σ_2 can have the same reflection coefficient.

This was rediscovered and reported in Morley (1984) and de Haas and Berkhout (1988) among others. Figure 6, from the latter paper, shows the error in predicted reflection coefficient of a particular model as σ_1 and σ_2 are varied. Morley (1984) goes on to state that little information over post-stack inversion is obtained if amplitudes are only measured out to half of the critical angle. This has led some workers to suggest that multicomponent data are required for unambiguous inversion (see following discussion). Of course, if inversion for absolute Poisson's ratio is not required, the outlook for successful utilization of conventional AVO data is brighter.

Aki and Richards (1980), Wiggins (1983), and Shuey (1985) show that the additional information obtainable by AVO analysis is $\Delta\sigma$ (or equivalently R_S). Fortunately, this may be sufficient for many applications.

As the incident angle aperture increases, the third term in Shuey's approximation becomes significant

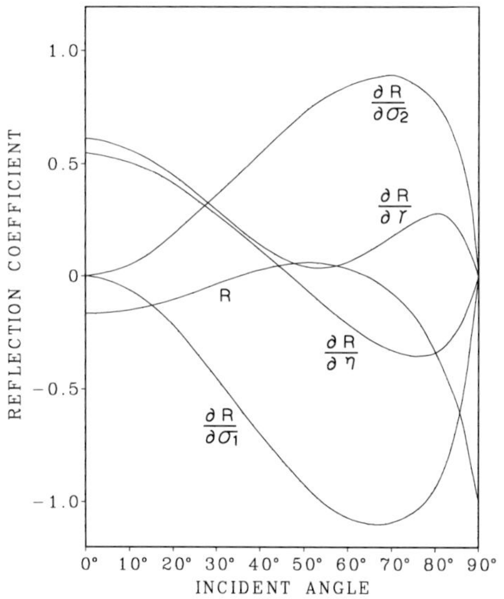

Fig. 5. The reflection coefficient for P-waves as a function of incidence angle and its partial derivatives with respect to the elastic parameters (from Rosa, 1976). $\gamma = V_{P2}/V_{P1} \cdot \zeta = \rho_2/\rho_1$.

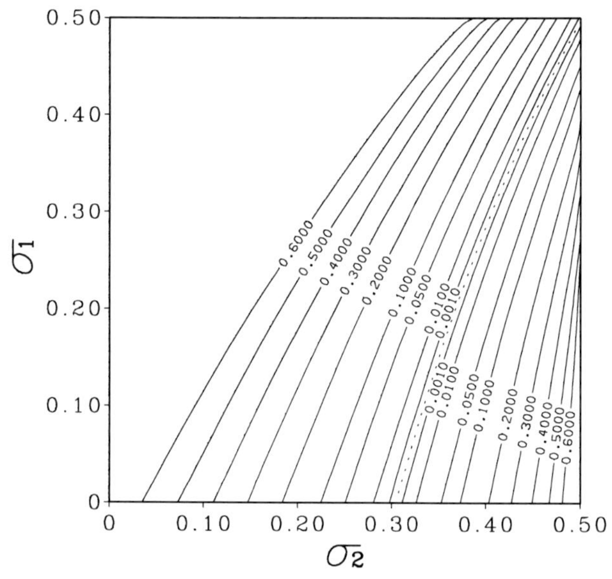

Fig. 6. Error in predicted reflection coefficient as Poisson's ratio in each medium is varied from the ''true'' Poisson's ratio of .1014 in medium 1 and .3333 in medium 2. The P-wave velocity ratio between the two media is .8, and the density ratio is .9 (from de Haas and Berkhout; 1988).

(e.g., Dey-Sarkar and James, 1986; Balogh and Snyder, 1986) and provides new information (P-wave velocity contrast from which density contrast can be unraveled). Even with an incident angle of 45 degrees, recovery of the third term is less robust than second term recovery. Nevertheless, Parson (1986) showed promising results for three-parameter gradient separation. The three parameter fit to $R_{PP}(\Theta_1)$ is important in direct detection since it can potentially provide a distinction between bright spots caused by low or high gas saturations.

PETROPHYSICAL BASIS

Gassmann's (1951) equations provide the fundamental basis for direct hydrocarbon indication. These equations predict a large drop in P-wave velocity and a small increase in S-wave velocity when even a small amount of gas is introduced into the pore space of a compressible brine-saturated sand. This drop (along with the corresponding density change) changes the P-wave reflection coefficient (resulting in "bright" or "dim" spots) and causes a drop in V_P/V_S (which causes AVO anomalies).

Ultrasonic laboratory measurements (such as Gregory, 1976; Domenico, 1976) do not quantitatively agree with Gassmann predictions. However, low-frequency laboratory measurements (Murphy, 1982) show excellent agreement with Gassmann's equations (see Figure 7). For seismic frequencies, we assume the Gassmann's equations are generally valid for sandstones. Gassmann's equations are strictly valid only when the following assumptions are applicable: (1) The porous rock framework (skeleton) is macroscopically isotropic and homogeneous. (2) The skeleton, grains, fluids, and saturated rock obey Hooke's law (stress is proportional to strain), and (3) The pore space is interconnected, fluid pressure is uniform, no fluid enters or leaves any volume of the system, and no cavitation occurs. It would seem that the shalier the rock, the more likely it is these assumptions will be violated.

Oil and water are often assumed to have similar acoustic properties, and to be indistinguishable with seismic methods. However, recent work (Rafipour, 1987; Wang et al., 1990) suggests that compressible oils with high gas-oil-ratio (GOR) may be detectable. Figure 8 shows predicted sonic log transit-times in an offshore Gulf of Mexico oil sand for various pore fluids. It is important to distinguish "live" oil from "dead" oil properties. Under in-situ pressure-temperature conditions, oils contain some amount of dissolved gas and are said to be "live". When the oil is

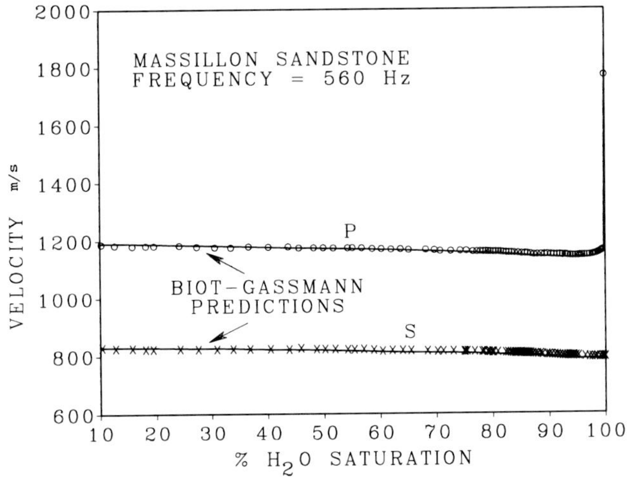

Fig. 7. Bar resonance data of Murphy (1982) for Massillon sandstone. The data are in agreement with Gassmann's equations (solid lines).

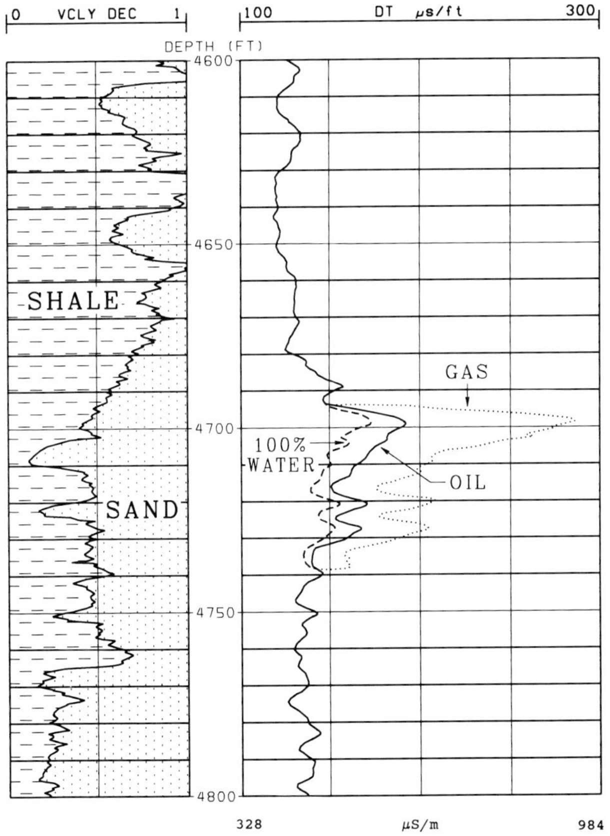

Fig. 8. Sonic log through an oil sand and calculated sonic logs for various Gassmann fluid substitutions.

produced, gas comes out of solution as the oil equilibrates to surface conditions leaving the oil "dead". The amount of exsolved gas is used to determine the GOR. It is important to note, as seen in Figure 8, that "live" oil-saturated sandstone can be significantly slower than water-saturated rock.

Angerer and Richgels (1990) describe various factors affecting P-wave and S-wave velocity: mineralogy, porosity, pore fluids, depth of burial, pressure, and temperature. Pore shape distribution and degree of lithification are also important factors. To first order, porosity, depth of burial, pressure, and lithification generally affect V_P and V_S in the same way. Thus, excepting unusual pore shapes, a given lithology will have a well defined V_P-V_S relationship. Given V_P and lithology, Greenberg and Castagna (1992) showed that S-wave velocity prediction can be quite robust, however, the prediction error (about 5 percent) may be inadequate when viewed in terms of reflection coefficients.

Locally calibrated V_P-V_S and V_P-ρ trends may provide essential a priori information for AVO analysis and may transform a nonunique seismic inversion problem into a well-constrained hydrocarbon detection scheme. Two parameters extractable by AVO analysis, R_P and R_S, can be crossplotted to identify gas sands in a clastic stratigraphic section. We use V_P-V_S and V_P-ρ trend curves (Castagna et al., 1993 this volume, pages 135–171.) for shale, brine sand, and gas sand. We also use the following empirical equation for gas sands:

$$\rho^{gs} = .199 V_P^{gs} + 1.53 \tag{21}$$

where ρ^{gs} is gas-sand density and V_P^{gs} is gas-sand velocity in km/s. The reader is cautioned that local calibration of trends (particularly V_P-ρ) is strongly recommended.

Figure 9 shows shale over brine sand and shale over gas sand reflection coefficients for a range of shale V_P ranging from 1.5 to 5.5 km/s. (A similar plot for shale over shale reflection coefficients is very similar to the shale over brine sand plot). Some very significant observations can be made:

(1) Large negative P-wave reflection coefficients (R_P less than $-.1$): Gas sand and brine sand reflections are unambiguously separated. The more negative the P-wave reflection coefficient or the lower the shale velocity, the greater the separation between gas and brine sands.

(2) Near-zero P-wave reflection coefficients (R_P between $-.1$ and $.1$): Gas sand and brine sand reflection coefficients are unambiguously separated only for low shale velocities (less than about 3.5 km/s) and only if shale velocity is approximately known.

(3) Large positive P-wave reflection coefficients (R_P greater than $.1$): Gas sand and brine sand reflection coefficients are unambiguously separated only for the lowest shale velocities (less than about 2.5 km/s) and only if the shale velocity is approximately known. The separation decreases with increasing R_P.

Similarly, comparison of Figure 9 with Figure 10 shows that the gas-sand over brine-sand (or shale) reflection can be distinguished from a shale over brine-sand reflection providing the expected gas-sand

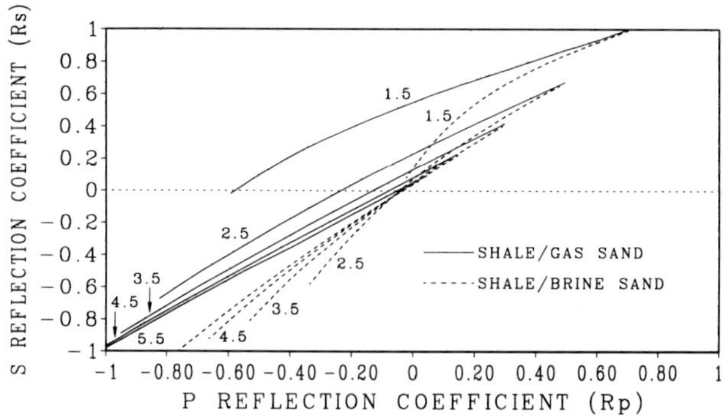

Fig. 9a. Reflection coefficients for shale over gas sand interface (solid) and shale over brine sand interface (dashed) for shale velocities ranging from 1.5 to 5.5 km/s.

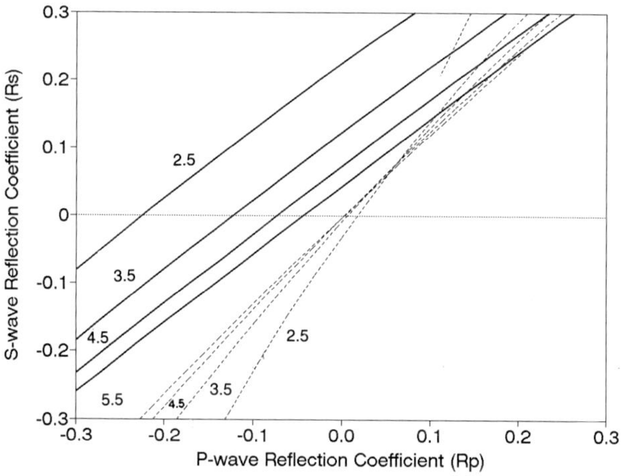

Fig. 9b. Expanded view of Figure 9a to emphasize the reflection coefficient range which is most appropriate for exploration applications.

velocity is less than about 4.5 km/s. Furthermore, if we assume equal porosity above and below the gas-water contact, R_S is about zero.

For completeness, note that gas sand over gas sand reflections are characterized by $R_P = R_S$ as V_P/V_S is approximately constant.

These R_P-R_S crossplots cover a very broad range of velocities. In practice, the range of expected velocities is much more limited. In general, velocity-depth functions for sands and shales can be constructed from well logs, or velocity analyses and petrophysical concepts. For example, a particular offshore Louisiana region has the following average properties above geopressure:

$$V_P^{sh} = .394\ Z + 1.73 \qquad (22a)$$

and

$$V_P^{ss} = 1.12\ V_P^{sh} - .24 \qquad (22b)$$

where

V_P^{sh} = shale P-wave velocity in km/s,
V_P^{ss} = sandstone P-wave velocity in km/s, and
Z = depth (km).

Combining equations (22a) and (22b) with V_P-V_S and V_P-ρ trend curves, yields R_P/R_S as a function of depth for shale over brine sand and shale over gas sand reflections. As shown in Figure 11, R_P/R_S is an excellent gas indicator at shallow depths but becomes less discriminating as depth increases. By carrying error bars with the R_P/R_S versus depth trends, and knowing the error bars for the AVO analysis, one can compute a confidence in the hydrocarbon indication. As R_P/R_S becomes unstable at small R_S, the difference R_P minus R_S may be a preferable indicator in some instances. A similar approach can be used to investigate the effects of coal or other lithologies.

An important word of caution: errors in the extraction of R_P and R_S tend to be biased (by anisotropy for example; see the following section) and the results may not conform to our petrophysical expectations. When this is the case, one should be willing to move the non-pay "baseline", as is commonly done in well log analysis, to empirically correct these biases (see also Smith and Gidlow, 1987; Spratt et al., 1993 this volume, pages 37–56.). With this empirical approach, the role of petrophysics differs from the forward modeling and inversion approaches. Petrophysical analysis provides: (1) The expected systematics of variation (for example; gas sands should plot to the upper left on Figure 9). (2) The expected magnitude of the anomaly and the petrophysical S/N. (3) The expectation of petrophysical uniqueness (or lack thereof) of the indicator, and (4) The basis for exploration where sufficient empirical data are not available.

COMPLICATING FACTORS, PROBLEMS, AND PITFALLS

All too often, the tremendous potential of AVO analysis as a prospecting tool has not been realized partially because of the wide variety of complications, problems, and pitfalls involved in isolating and interpreting the offset-dependent-reflectivity (ODR). Table 3 shows the myriad of factors affecting seismic amplitudes. For most AVO analysis techniques, all factors other than reflection coefficient variation with offset must be considered undesired noise (which should be attacked by processing or properly accounted for) or

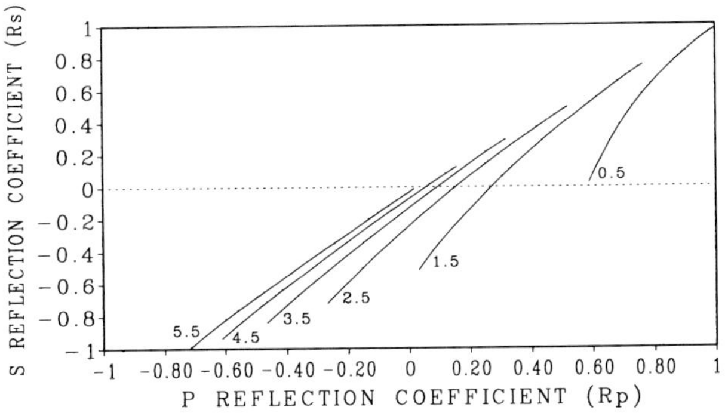

Fig. 10. Reflection coefficients for gas sand over shale interface for gas sand velocities ranging from 1.5 to 5.5 km/s. R_P/R_S is always less than unity. Comparison to Figure 9 shows for gas sand velocities less than 4.5 km/s that there is no overlap with shale over brine sand (or brine sand over shale) reflection coefficients.

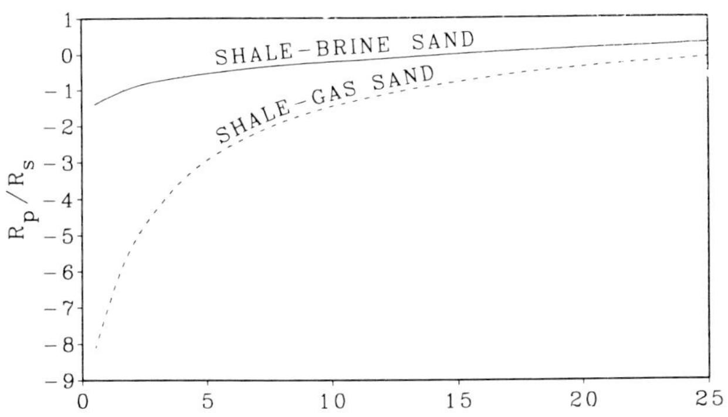

Fig. 11. AVO sensitivity for a particular Louisiana area. R_P/R_S versus depth for shale over brine sand and shale over gas sand interfaces.

must be appropriately comprehended as signal by the processing/analysis techniques. We must always be aware of the angle-dependent total system response/wavelet variation when attempting to isolate ODR.

Composite Reflections

According to Sherwood et al (1983): "The simple interface reflection coefficient is only a starting point toward understanding offset-dependent-reflectivity. More often we deal with reflections from layers, transition zones, or complex layered sequences." Modeling and inversion are required to resolve the effects of superposition of reflections from more than one interface.

The classical bed tuning phenomenon (Widess, 1973) becomes more complex when considered as a function of offset. Ostrander (1984), Hindlet and McDonald (1986), Ball (1988), and Swan (1988) among others considered this phenomenon. As shown in Ostrander (1984), the apparent change in time thickness of a layer is approximately obtained by differentiating the normal moveout equation. Also

$$\frac{\Delta t_\Theta}{\Delta t_0} \approx \cos\Theta \tag{23}$$

where t_0 is the normal incidence arrival time and t_θ is the arrival time for angle θ in the layer. Figure 12 from Ostrander (1983 unpublished course notes) shows that tuning can cause amplitude to increase or decrease with offset depending on the normal incidence time thickness-frequency product. With increasing offset, the time thickness decreases and the amplitude varies according to Figure 12.

Table 3. Factors affecting seismic amplitudes.

A. Desired information (signal)
   1. reflection coefficients versus angle of incidence
B. Potential information (considered noise for some methods; signal for others)
   1. composite reflections from multiple interfaces
   2. tuning caused by NMO convergence
   3. mode conversions
C. Factors without offset dependence (noise)
   1. random noise
   2. instrumentation
   3. source/receiver coupling
D. Factors with offset dependence (noise)
   1. source/receiver directivity including ghosting and array response
   2. emergence angles
   3. coherent noise, multiples
   4. spherical spreading
   5. processing distortions, NMO errors and stretch
   6. inelastic attenuation and anisotropy
   7. transmission coefficients and scattering above target
   8. structural complexity

Near Surface Considerations

There are a variety of near surface effects which perturb seismic amplitudes. Source strength and receiver coupling variations, particularly when related to slowly varying near-surface impedance or surface conditions, can modify AVO. Surface-consistent processing can be used to combat this. Other obvious emergent angle dependent effects include, source radiation pattern, geophone response, and array response. These effects are well known and readily corrected. Energy partitioning due to free surface mode conversions can also be corrected (Cerveny and Ravindra, 1971). In the marine case, angle dependent ghosting is considered a second-order problem. Near-surface transmission effects may also be important (see following discussion).

General Data Quality Issues

Seismic data quality is the most fundamental requirement for successful AVO analysis. Sufficient signal-to-noise (S/N), bandwidth, aperture (offset),

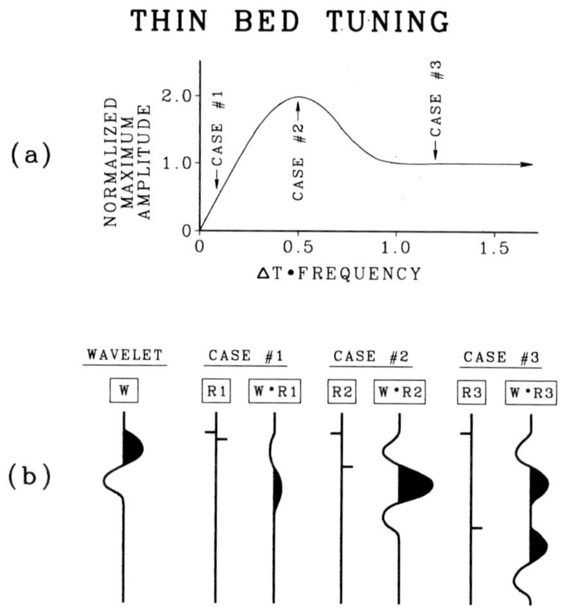

Fig. 12. (a) Normalized maximum amplitude versus frequency-time thickness product showing effect of thin bed tuning. Cases 1, 2, and 3 are indicated. (b) Convolution of wavelet with three reflectivity series R1, R2, and R3 which are below, at, and above tuning, respectively (from Ostrander, unpublished course notes, 1983).

and fold are all basic requirements. Channels, if not amplitude calibrated, must at least be consistent throughout the survey. Channel balancing may be problematical. As pointed out in Frasier (1988), individual traces have lower S/N than stacked traces, so channel balancing attempts may only balance the noise.

Source-generated coherent noise in the form of multiples, converted waves, diffractions, etc. is particularly troublesome. As shown in Pan et al. (1990) incoherent noise is less of a problem.

To suppress coherent noise *f-k* filtering is often applied. However, coherent noise, such as multiples, with varying spatial frequency (moveout) across a gather, may be unevenly attenuated, thereby biasing the AVO estimate. Multiples can also be attacked by ray parameter filtering or by predictive deconvolution in various domains. Multiples with small residual moveout (1/2 to 1 cycle) create a false amplitude variation with offset and are extremely difficult to remove. This is probably the single most important data quality issue for marine AVO.

A great promise of full-waveform inversion is that certain source-generated noise may become signal.

Curvature and Structural Complexity

By analogy with optics, we expect curved interfaces to act as lenses and curved mirrors, concentrating or dispersing seismic energy. Hilterman (1975) defined the curvature effect (CE) for a reflecting interface as the ratio of the reflection amplitudes from a curved interface to that of a flat interface. For normal incidence, Hilterman (1975) gives:

$$CE = (1 + Z/A)^{-1/2} \tag{24}$$

where A is the radius of curvature of the bed and Z is the depth.

Shuey et al. (1984) extended this equation to the non-normal incidence case

$$CE(\Theta_1) = \left(1 + \frac{Z}{A_x \cos^2 \Theta_1}\right)^{-1/2} \left(1 + \frac{Z}{A_y}\right)^{-1/2} \tag{25}$$

where,

A_x = radius of curvature in x direction,
A_y = radius of curvature in y direction, and
Z = depth of crest of anticline or trough of syncline.

Note that the radii are negative for buried focus synclines and that Shuey's equation reduces to Hilterman's at normal incidence. Equation (25) indicates that the curvature effect (1) decreases with offset for an anticline, (2) increases with offset for a syncline with a focus above the surface, and (3) decreases with offset for a syncline with a focus below the surface. Macleod and Martin (1988) showed that DMO corrects the curvature effect.

Bernitsas (1990) extended Shuey's equation to the general case of three-dimensional curvature:

$$CE(\Theta_1) = (1 + Z/A_x)^{-1/2}(1 + Z/A_y)^{-1/2} \times \left[\frac{X^2}{Z(A_x + Z)} + \frac{Y^2}{Z(A_y + Z)} + 1\right]^{-1/2} \tag{26}$$

where X, Y, and Z are the spatial coordinates.

The effects of complex surface roughness have not been adequately investigated. Herman and Blonk (1990) suggest that surface roughness can be emulated by an appropriate transition layer.

As elucidated in Resnick et al. (1987), problems caused by dip include (1) errors in angle of incidence, (2) mixing information from different subsurface locations, (3) incorrect normal moveout, and (4) interference of mispositioned events.

The lensing effect of curved layers above the target are best compensated for by amplitude preserving prestack migration. Complex structure also commonly results in lateral overburden velocity gradients and associated AVO errors.

Geometrical Spreading

Newman (1973) gives the equations needed to correct for geometrical spreading. At normal incidence

$$D_o = tV_a^2/V_1, \tag{27a}$$

$$t_o = \sum t_i, \tag{27b}$$

and

$$V_a^2 = \sum t_i V_i^2/t_o \tag{27c}$$

where,

D_o = normal incidence geometric divergence,
V_1 = velocity in first layer,
t_o = zero-offset two-way reflection time,
t_i = interval two-way transit time of the ith layer,
V_a = time-weighted rms velocity, and
n = the number of layers i.

At non-normal incidence

$$D(\Theta_1) = \frac{(X^2 + 2X \sum d_i \tan^3 \Theta_i)^{1/2}}{\tan \Theta_1} \tag{28a}$$

and

$$X = 2 \sum d_i \tan \Theta_i \quad (28b)$$

where,

$D(\Theta_1)$ = geometric divergence versus angle of incidence,
X = offset,
Θ_i = incidence angle in the ith layer, and
d_i = thickness of the ith layer.

Normal Moveout Errors

In order to measure the AVO on a sample by sample basis, it is necessary to perfectly correct the common depth point (CDP) gather for normal moveout (NMO) (Spratt, 1987; Swan, 1988). Conventional velocity analysis is generally not adequate to do the job and special care must be taken (Chiburis, 1984). Furthermore, as shown in Figure 13 (and Table 4), there is a fundamental ambiguity between stacking velocity and ODR (Spratt, 1987). NMO related problems are much less severe for methods which measure the energy within a window about the event of interest (Mazzotti, 1990) or for instantaneous AVO. Seismic modeling and waveform inversion can also be utilized to deal with the problem.

Overburden Effects

As transmission coefficients are angle dependent, the amplitude of the P-wave incident on the reflector of interest may also be angle dependent. This problem is most significant when the reflectivity above the target is very strong. Angle-dependent transmission losses are dependent on both the P-wave and S-wave velocity structures. Figure 14 and Table 5 show the effects of varying V_P and V_S over a particular target. Models A and B have the same V_P structure but different V_S structure. The reflection amplitudes are the same at near offset, but diverge significantly at far offsets. Similarly, models A and C have the same V_S structure but different V_P structures. The near-offset reflection amplitudes are very different, but the AVO is similar for both models. Figure 15 and Table 6 show that V_P/V_S variations in the very near surface can also significantly affect the AVO response. According to Gassaway (1984), transmission loss is the most signif-

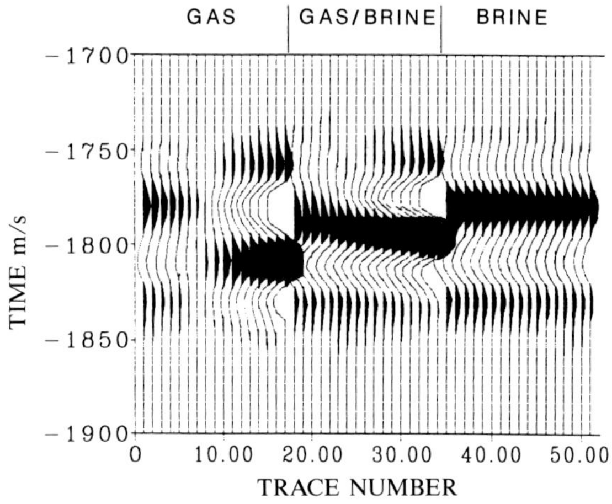

Fig. 13. Three NMO corrected synthetic CDP gathers for three models (A, B, and C) with identical P-wave velocity structure but differing S-wave velocity structures (see Table 4). V_P/V_S ratios are appropriate for gas sand (Model A), brine sand (Model C), and a sand with a gas-water contact (Model B).

Table 4. Model parameters for Figure 13.

Lithology	P-wave Velocity ft/s	P-wave Velocity m/s	S-wave Velocity ft/s	S-wave Velocity m/s	Density gm/cc	V_P/V_S	Thickness ft
Model A							
Shale	9000	2743	4575	1394	2.06	1.97	
Gas Sand	9300	2835	5782	1762	2.04	1.61	150
Shale	9000	2743	4575	1394	2.06	1.97	
Model B							
Shale	9000	2743	4575	1394	2.06	1.97	
Gas Sand	9300	2835	5782	1762	2.04	1.61	75
Brine Sand	9300	2835	4828	1472	2.08	1.93	75
Shale	9000	2743	4575	1394	2.06	1.97	
Model C							
Shale	9000	2743	4574	1394	2.06	1.97	
Brine Sand	9300	2835	4828	1472	2.08	1.93	150
Shale	9000	2743	4575	1394	2.06	1.97	

icant problem encountered in AVO analysis, and a layer stripping technique can be utilized to account for the overburden (see Figure 16).

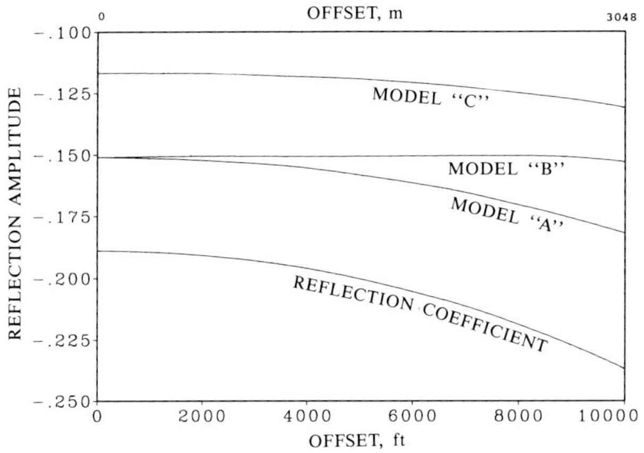

Fig. 14. Reflection amplitude variation with offset for three models (A, B, and C) with identical target reflection coefficient-versus-offset response but different overburden transmission losses (see Table 5). Model A has a shallow brine sand over the target gas sand. Model B shows the effect of increasing V_S in this layer, while Model C shows the effect of decreasing V_P.

Global transmission loss is very amenable to statistical correction. However, strong lateral change in overburden effects (as would result from shallow gas) is not easily corrected. Deterministic correction is problematical as the overburden cannot be perfectly characterized. Normalizing the measured AVO relative to a reference event is one way to tackle this problem (Chiburis, 1984 and 1987). The normalized AVO is given by

$$\frac{a(x)^{\text{tar}}/a_o^{\text{tar}}}{a(x)^{\text{ref}}/a_o^{\text{ref}}} \tag{29}$$

where $a(x)^{\text{tar}}$ and $a(x)^{\text{ref}}$ are the AVO for the target and reference events, respectively, and a_o^{tar} and a_o^{ref} are the normal incidence amplitudes. Troublesome variations in overburden can sometimes be detected by examining ties at line intersections (Huston and Backus, 1986).

Spatial variations in overburden velocities result in errors in the calculation of local angle of incidence which can be computed by ray tracing through a complex overburden. If the overburden can be represented by a linear velocity gradient

$$V_i = V_o + Kz \tag{30}$$

Table 5. Model parameters for Figure 14. * indicates the shallow layer for which the model parameters are varied.**

Depth ft	Depth m	Density gm/cc	P-wave Velocity ft/s	P-wave Velocity m/s	S-wave Velocity ft/s	S-wave Velocity m/s	V_P/V_S
Model A							
0	0	1.00	5000	1524	0	0	—
100	30	2.02	6000	1829	1326	404	4.52
1000	305	2.10	7000	2133	2188	667	3.20***
1200	366	2.18	8000	2438	3050	930	2.62
7000	2133	2.30	10000	3048	4774	1455	2.09
9000 (target)	2743	2.19	7161	2183	4774	1455	1.50
9100	2774	2.36	11000	3353	5636	1718	1.95
Model B							
0	0	1.00	5000	1524	0	0	—
100	30	2.02	6000	1829	1326	404	4.52
1000	305	2.10	7000	2133	4667	1422	1.50***
1200	366	2.18	8000	2438	3050	930	2.62
7000	2133	2.30	10000	3048	4774	1455	2.09
9000 (target)	2743	2.19	7161	2183	4774	1455	1.50
9100	2774	2.36	11000	3353	5636	1718	1.95
Model C							
0	0	1.00	5000	1524	0	0	—
100	30	2.02	6000	1829	1326	404	4.52
1000	305	2.10	3282	1000	2188	667	1.50***
1200	366	2.18	8000	2438	3050	930	2.62
7000	2133	2.30	10000	3048	4774	1455	2.09
9000 (target)	2743	2.19	7161	2183	4774	1455	1.50
9100	2774	2.36	11000	3353	5636	1718	1.95

where,

V_i = interval velocity,
z = depth,
V_o = velocity when $z = 0$, and
K is a constant, then (Ostrander, 1984)

$$\Theta_i = \tan^{-1}(zX + V_oX/K)/(z^2 + 2V_oz/K - X^2/4). \quad (31)$$

When velocities differ vertically or laterally from those used to compute angle of incidence, AVO errors will result (Xu and McDonald, 1988). When velocity increases with depth or decreases horizontally, an apparently bigger Poisson's ratio will be produced. The converse is also true.

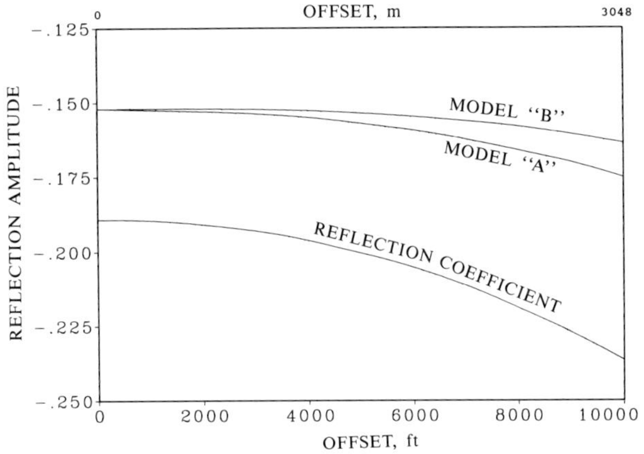

Fig. 15. Amplitude variation with offset for two models (A and B) with identical target reflection coefficient-versus-offset response but different overburden transmission losses (see Table 6). Model A has high near-surface V_P/V_S (wet sand) while model B has low near-surface V_P/V_S (dry sand).

Processing Considerations

Some conventional processing techniques are clearly only appropriate for relative AVO analysis: automatic gain control, whole trace equalization, non-amplitude preserving deconvolution or prestack migration, and all but the mildest velocity filtering. Muting may be too severe or not severe enough. Unrecorded processing gain adjustments can wreak havoc. Offset-dependent phase changes can result in incorrect stacking velocities and incorrect residual statics. NMO stretch increases the energy at far offsets and should be accounted for. Marine statics problems may obscure AVO responses (Treadgold et al, 1990b; Figure 17).

Duren (1990, 1991) applied the "range equation" to deterministically correct for "all AVO perturbing factors". The interpreter must keep in mind that such approaches can only be expected to apply a "ball-park" correction, and that significant residual problems must be expected.

Naive Application of Rock Physics Principles or Uncooperative Petrophysical Circumstances

According to Gassmann's equations, a gas sand with 1 percent gas saturation can have the same V_P/V_S as a commercial accumulation of gas. Thus, unless den-

Table 6. Model parameters for Figure 15.

Depth ft	Depth m	Density gm/cc	*P*-wave Velocity ft/s	*P*-wave Velocity m/s	*S*-wave Velocity ft/s	*S*-wave Velocity m/s	V_P/V_S
Model A							
0	0	2.00	5000	1524	550	168	9.09
100	30	2.02	6000	1829	1326	404	4.52
1000	305	2.10	7000	2133	2188	667	3.20
1200	366	2.18	8000	2438	3050	930	2.62
7000	2133	2.30	10000	3048	4774	1455	2.09
9000	2743	2.19	7161	2183	4774	1455	1.50
9100	2774	2.36	11000	3353	5636	1718	1.95
Model B							
0	0	2.00	5000	1524	3400	1036	1.47
100	30	2.02	6000	1829	1326	404	4.52
1000	305	2.10	7000	2133	2188	667	3.20
1200	366	2.18	8000	2438	3050	930	2.62
7000	2133	2.30	10000	3048	4774	1455	2.09
9000	2743	2.19	7161	2183	4774	1455	1.50
9100	2774	2.36	11000	3353	5636	1718	1.95

sity can be very accurately extracted utilizing far offset information, AVO cannot distinguish commercial and noncommercial gas accumulations. One wonders how many AVO failures were, in fact, technical successes where the "dry holes" actually contained small quantities of gas in the target zone.

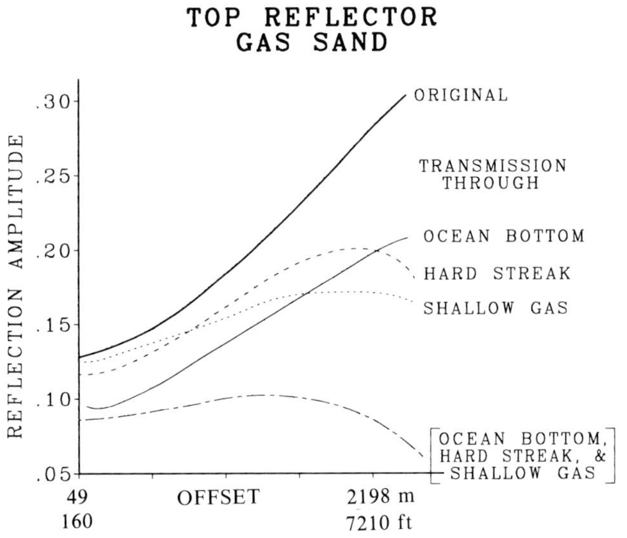

Fig. 16. Overburden transmission losses modify the target AVO. In this case, transmission through the ocean bottom, a hard streak, and a shallow gas sand suppresses the target gas sand amplitude increase with offset (from Gassaway, 1984).

Hilterman (1990) classified 5 different shale types with unique velocity-density trends. AVO response of a sand package depends on degree of shaliness and shale distribution within the package. Our petrophysical knowledge of shales remains inadequate.

According to Payne (1991): "Existing empirical relations do not provide S-wave velocity estimates with enough accuracy to yield quality model gathers." In light of the results of Greenberg and Castagna (1992), who showed S-wave velocity predictions arguably as good as direct S-wave sonic logs, one must ask if the direct "ground truth" is sufficiently adequate.

Hard sands and carbonates with strong frame moduli may show little gas effect. An unfortunate *fact of life* is the decreasing gas effect with increasing depth. However, gas brine contact reflections should be detectable in many instances.

Interpreters should also pay attention to expected Poisson's ratios for different lithologies. Abnormal AVO extracted Poisson's ratios (such as .49 for tight limestone) should be a warning flag that something is biasing the AVO analysis.

Inadequacies of the Forward Model

Wright (1984) concluded that anisotropy in overlying shales severely affects AVO (see Figure 18). The 20 percent P-wave anisotropy he assumed for shales seems somewhat extreme for the general in situ case,

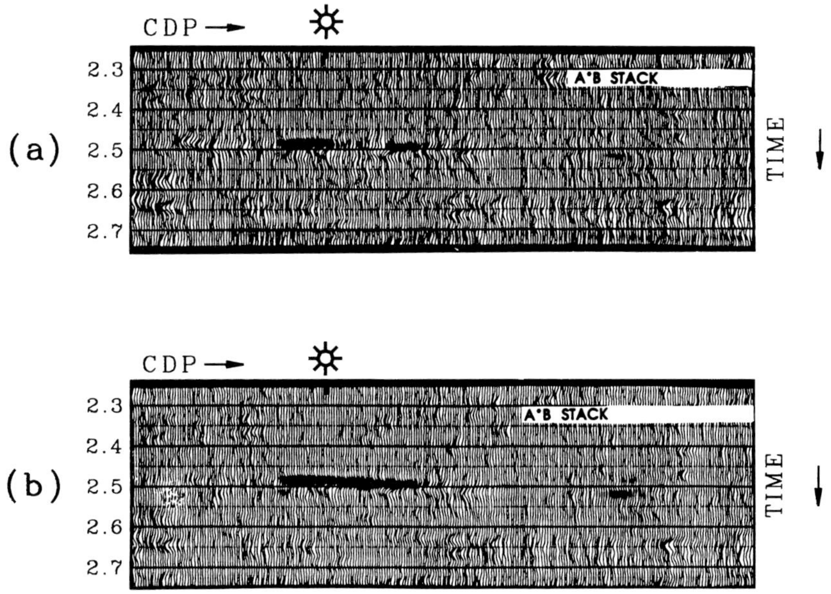

Fig. 17. A*B stack (normal incidence amplitude times AVO slope) before (a) and after (b) marine statics. A positive A*B stack indicates hydrocarbons in this case (from Treadgold et al., 1990a, b).

however, he points out that small amounts of gas in shales would increase anisotropy dramatically. This is a generally overlooked and perhaps very significant phenomenon. Anisotropy and attenuation can change the phase of the wavelet and can focus energy preferentially along the wavefront (Samec et al., 1990). Computing amplitudes within a small window about the event of interest, rather than on a sample by sample basis, will reduce this effect.

Inelastic attenuation and associated dispersion corrupt the AVO estimate (Luh, 1988). Overlying gas sands (which exhibit very high attenuation) can cause great difficulty. Q compensation and windowed amplitudes are recommended to combat this problem.

Theoretically, reflection amplitude at a porous interface is dependent on formation permeability (Stoll and Kan, 1981). However, Biot theory predicts a negligible permeability effect at seismic frequencies (Dutta and Ode, 1983).

For very shallow targets the plane-wave assumption may not be valid (Krail and Brysk, 1983; Rendleman and Levin, 1988; Wapenaar et al., 1990).

Interpretation Considerations

The interpreter must weigh all these complications while synthesizing additional dimensions of data. Consequently, the handling and display of the additional offset data becomes a major problem. Polarity remains an issue, as hard streak AVO can be mistaken for gas sand AVO when polarity is reversed. Gas leakage over the reservoir (Sengupta and Rendleman, 1989) can confound AVO analysis in a variety of ways: increased scattering, attenuation, and anisotropy and decreased P-wave velocity and V_P/V_S over the target. Finally, vertical and lateral resolution limits and variable S/N for various extracted quantities must be kept in mind.

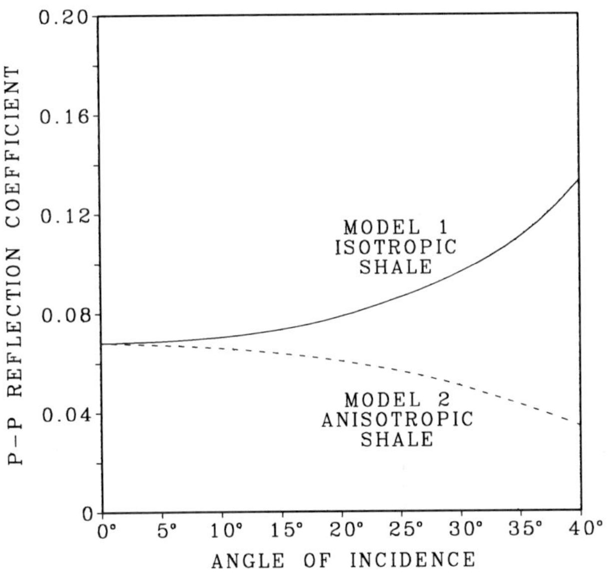

Fig. 18. Reflection coefficient versus offset for a shale/sand interface. Model 1: isotropic shale. Model 2: anisotropic shale (from Wright, 1984).

SEISMIC DATA PROCESSING FOR AVO ANALYSIS

When attempting to select an appropriate data processing scheme for AVO analysis, the processor must carefully balance two competing objectives: (1) Noise suppression and isolation of the reflectivity of the event of interest, versus (2) not biasing or otherwise corrupting the reflectivity variation with offset. This tradeoff usually leads to the selection of a basic but robust processing scheme (Ostrander, 1984; Chiburis, 1984; Long and Richgels, 1985; Yu, 1985; Dey-Sarkar and James, 1986; Fouquet, 1990). Keep in mind that the processor is deprived of his strongest weapon; CDP stacking. Table 7 gives some processing schemes presented in the literature.

In areas of significant dip or structural complexity, we are forced to utilize more sophisticated processing techniques. Resnick et al. (1987) and de Bruin et al. (1988, 1990) handle complex surfaces by prestack migration. Amplitude preserving prestack migration is recommended for all AVO projects. The use of Radon transform techniques for multiple suppression is also highly recommended (Kelamis et al., 1990). Resnick (1988) asks the important question: "When have we overprocessed the data?"

"Weighted-stacking" (Smith and Gidlow, 1987) is a means of reducing prestack information to AVO attribute traces (usually some form of Shuey's coefficients) versus time. This is accomplished by computing the local incident angle for each time sample and then performing regression analysis to solve for the first two or all three coefficients of an equation of the kind

$$R_{PP}(\Theta_1) \approx A + B \sin^2 \Theta_1 + C \sin^2 \Theta_1 \tan^2 \Theta_1 . \quad (32)$$

A is the "zero-offset" stack or "true" normal incidence reflectivity, B is commonly referred to as the AVO "slope" or "gradient". Following Smith and Gidlow (1987), we have from Wiggins (1983)

$$\Delta V_S / V_{Sa} = A - B \quad (33)$$

where $\Delta V_S = V_{S2} - V_{S1}$, and $V_{Sa} = (V_{S2} + V_{S1})/2$. Using Gardner's relation:

$$\Delta\rho/\rho_a \approx .25\ \Delta V_P/V_P \quad (34)$$

where $\Delta V_P = V_{P2} - V_{P1}$, and $V_{Pa} = (V_{P2} + V_{P1})/2$. We also have:

$$\Delta V_P/V_{Pa} \approx 8A/5. \quad (35)$$

Smith and Gidlow (1987) define the "fluid factor", ΔF, as the difference between observed $\Delta V_P/V_{Pa}$ and $\Delta V_P/V_{Pa}$ predicted from $\Delta V_S/V_{Sa}$. Using the "mudrock trend" (Castagna et al., 1985) they get:

$$\Delta F = \Delta V_P/V_{Pa} - 1.16(V_S/V_P)\Delta V_S/V_{Sa} \quad (36)$$

where V_S/V_P is the background S-wave to P-wave velocity ratio which can be predicted by application of the mudrock trend to interval velocities obtained from conventional velocity analysis. Of course, when necessary, a locally calibrated V_P versus V_S trend can be employed. An empirically derived function from the AVO data which usually does not match the petrophysical control is often used. This is necessitated by the biases in the seismic data resulting from the myriad of complicating factors discussed in the preceding section. Figure 19 shows $\Delta V_P/V_{Pa}$, $\Delta V_S/V_{Sa}$, $\Delta\sigma/\sigma_a$, and ΔF displays for a real data example (Smith and Gidlow, 1987). ΔF shows distinct anomalies which are not obvious on the other sections. Advantages and disadvantages of weighted stacking are listed in Table 8 (Lortzer et al., 1988).

INTERPRETATION OF AVO DATA

Gas Detection

By far, gas-sand detection is the most promising application of AVO analysis. It is hoped that the characteristically low V_P/V_S of gas sands should allow their differentiation from other low impedance layers such as coals and porous brine sands. Rutherford and Williams (1989) define three distinct classes of gas sand AVO anomalies (see Figure 20). Class 1 occurs when the normal incidence P-wave reflection coefficient is strongly positive and shows a strong amplitude decrease with offset and a possible phase change at far offset. Class 2, for small P-wave reflection coefficients, shows a very large percentage change in AVO. If the normal-incidence reflection coefficient is slightly positive, a phase change at near or moderate offsets will occur. Class 3 anomalies have a large negative normal-incidence reflection coefficient which becomes more negative as offset increases (these are classical bright spots). A simple rule of thumb which generally applies to shale over gas sand reflections: the reflection coefficient becomes more negative with increasing offset.

The literature is rich in gas sand detection examples from a variety of localities: the Sacramento Valley (Ostrander, 1984; Gassaway, 1984; Soroka et al., 1990;

Table 7. Processing schemes reported in the literature.

Ostrander, 1984:	1. spherical divergence correction
	2. exponential gain correction
	3. minimum-phase spiking deconvolution
	4. velocity analysis
	5. NMO correction
	6. trace equalization
	7. horizontal trace summing
Chiburis, 1984:	1. mild f-k multiple suppression
	2. spherical divergence and NMO correction
	3. whole-trace equalization
	4. flattening on a consistent reference event
	5. horizontal trace summing
	6. peak amplitudes picked interactively
	7. smoothed least-squares curve fitting
	8. despiking of outliers
	9. curve refitting
	10. results clipped and smoothed
Long and Richgels, 1985:	1. spherical divergence, attenuation, and emergence angle corrections determined by regression analysis
	2. variations in source strength, receiver sensitivity, overall offset amplitudes, consistently modelled and resolved
	3. deviations rejected as noisy
	4. inverse amplitude correction
	5. spectral surface array attenuation correction
	6. accurate NMO applied using very fine velocity sampling and geologically consistent velocity contouring
	7. surface consistent deconvolution for land data
	8. CDP residual statics carefully monitored for incorrect shifts due to AVO effects
	9. minimal muting
	10. common offset spatial filtering
	11. linear or non-linear fit for AVO coefficients
Yu, 1985, a and b:	1. apply exponential gain, suppress coherent noise, and remove gain
	2. offset compensation
	3. deconvolution,
	4. NMO correction
	5. surface consistent correction
	6. partial trace sum
	7. bandpass
	8. section dependent equalization
Todd, 1986:	1. spatial averaging
	2. gapped deconvolution
	3. geometrical divergence correction
	4. global velocity analysis and NMO correction
	5. partial stacking
	6. correction for average amplitude variation in time and offset

Gabay, 1990), the North Sea (Wrolstad, 1986; Snyder et al., 1989, Strudley, 1990), the Gulf Coast (Zilik, 1989; Hoopes and Aber, 1989; Burnett, 1989 and 1990; Rutherford and Williams, 1989), offshore Gulf of Mexico (Ostrander, 1984; Rutherford and Williams, 1989, Zimmerman and Fahmy, 1990), the Arkoma Basin (Rutherford and Williams, 1989), the Arabian Gulf (Chiburis, 1984; 1987), Alaska (Zimmerman and Fahmy, 1990), and Italy (Mazzotti, 1990). Notably, Soroka et al. (1990) claim a 90 percent success ratio with three dimensional (3-D) AVO in the Sacramento Valley. Similarly, in the Sacramento Valley, Gassaway (1984) was successful in predicting hydrocarbons 13 out of 15 times, and in predicting no hydrocarbons 8 out of 12 times. Table 9 shows the AVO track record in a particular Gulf Coast field (Burnett, 1990).

Successful hydrocarbon detection in carbonates has also been reported (Chiburis, 1987). Based on petrophysical arguments, we expect this to be possible only for restricted circumstances when the carbonate rock frame is sufficiently compressible for the fluid modulus to significantly influence the bulk modulus of the saturated rock. This would require significant amounts of low aspect ratio porosity. For porous high aspect ratio carbonates, it may be possible to exploit the effect of gas on bulk density.

Enhanced Oil Recovery Monitoring

As shown in Ito et al. (1979) and others, the effect of steam on rock V_P/V_S is very similar to the effect of

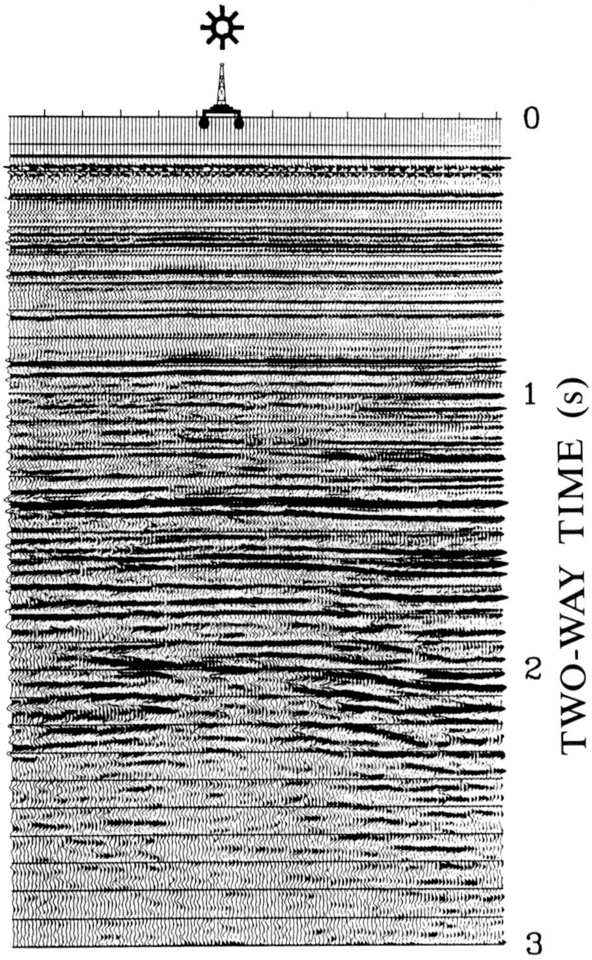

(a)

Fig. 19. Weighted stack outputs. (a) $\Delta V_P/V_{Pa}$, (b) $\Delta V_S/V_{Sa}$, (c) $\Delta\sigma/\sigma_a$, and (d) ΔF (from Smith and Gidlow, 1987).

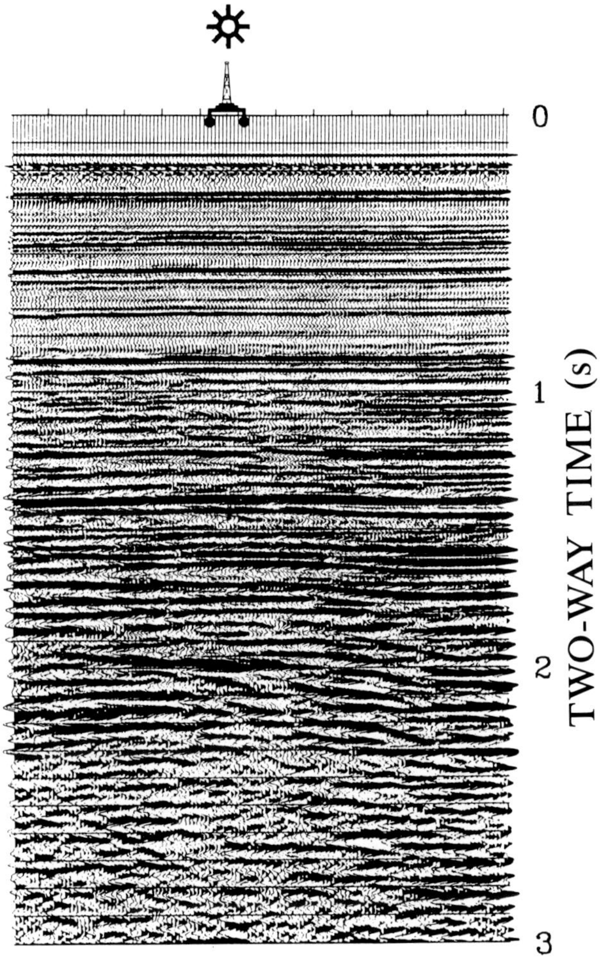

(b)

Fig. 19. continued

natural gas. Tsingas and Kanasewich (1991) showed that AVO can be very effective for monitoring steam flooding. As seismic is acquired at various stages of enhanced oil recovery (EOR), many of the perturbing factors we have discussed remain constant and cancel out when the AVO measurements are ratioed. This application utilizes reflection seismology's greatest strength; the ability to detect changes. Other EOR processes (such as water flooding), although more subtle in effect, may also be amenable to AVO analysis.

Oil Detection

Hwang and Lellis (1988) describe bright spots due to high gas-oil-ratio (GOR). A detailed review of factors affecting the acoustic properties of oils is presented in Castagna et al., this book pages 135–171 (1993). Under favorable circumstances, oil sands can exhibit V_P/V_S ratios about half-way between brine and gas sands. Gassaway (1984) and Chiburis (1987) report successful utilization of AVO to detect oil reservoirs.

Lithology Identification

The petrophysical ''signal'' for AVO lithology analysis is the lithology dependence of V_P/V_S. Inspection of V_P versus V_S trend curves for sandstones, shales, limestones, and dolomites, reveals that lithology discrimination is most robust at higher velocities where sandstones have low V_P/V_S (1.5–1.6) and the other lithologies have higher V_P/V_S (1.7–2.0). Some examples from the literature include Johnson (1988; sands versus carbonates), Fuller et al. (1989; sand versus

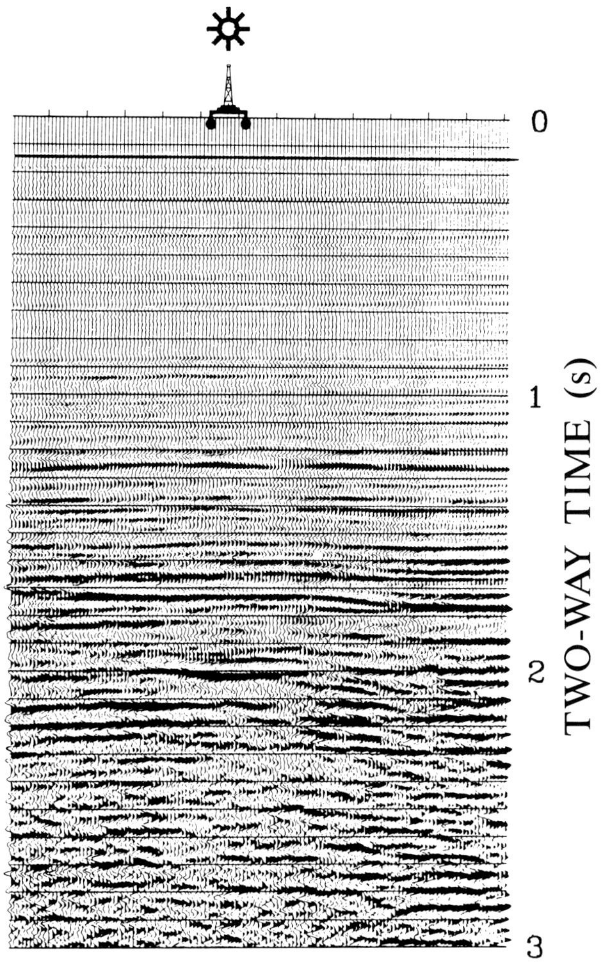

(c)

Fig. 19. continued

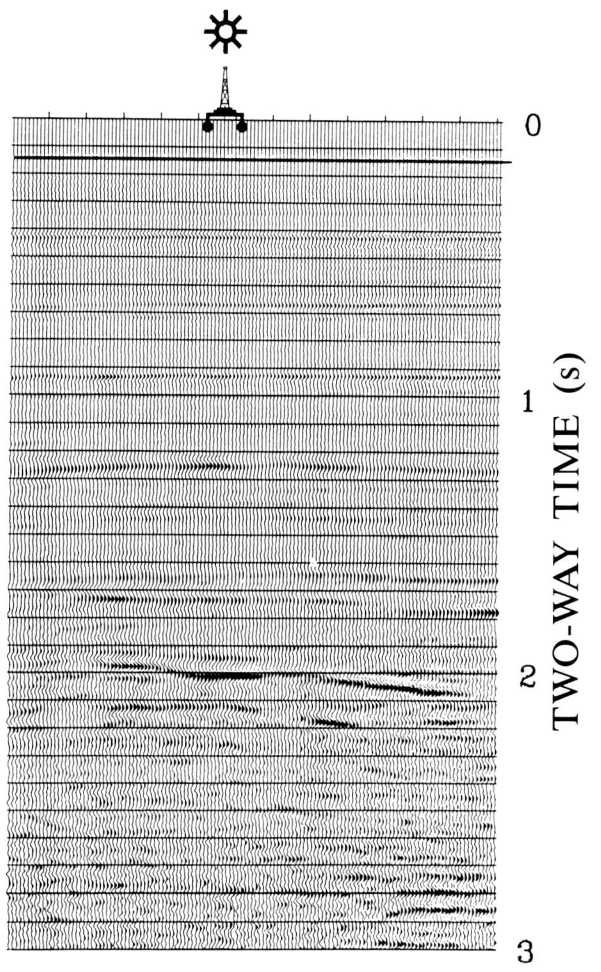

(d)

Fig. 19. continued

shale), Ramirez (1990; salt versus clastics), and Mazzotti (1984; high-velocity layers in general).

Porosity Identification

AVO can be used to detect porosity in carbonates (Chacko, 1989; Pigott et al.; 1990). Given the highly variable pore structure of carbonates, it is difficult to make generalizations regarding expected AVO response without local calibration. Usually V_P/V_S is insensitive to porosity and pore fluid content (with the sometimes important exception of unconsolidated carbonate sands or lithified carbonates with flat pores or microfractures). For vuggy porosity dominated by equant (spherical) pores, density may be the parameter which is most sensitive to porosity and pore fluid content; density becomes even more important when the rock has variable clay content. In these cases, Shuey's third coefficient (C) may be required to separate velocity and density contrasts.

Land Versus Marine AVO

One would expect, given the data quality concerns for land seismic profiles, that AVO would be primarily applied offshore. In fact, severe short period multiple problems may make marine data quality no better for AVO analysis than land data in many cases. Ostrander's original paper showed successful onshore examples. The Sacramento Valley, in particular, has yielded a plethora of successful AVO analyses (Ostrander, 1984; Gassaway, 1984; Gabay, 1990; Soroka et al., 1990; and various papers in this volume). The success ratio of AVO in the Yegua trend is a reported 80 percent (Allen, personal communication).

Interpretation Techniques

AVO analysis brings additional dimensions to the seismic interpretation problem. Uniqueness of the result is less important than the need to provide AVO information to the interpreter in a manageable form. Onstott et al. (1984) stated the problem very well: "The one-dimensional seismic inversion problem has not been solved perfectly and never will be. Nevertheless, seismic interpreters do a reasonable job of inferring subsurface structure and lithology from CDP

Table 8. Advantages and disadvantages of weighted stacking methods (after Lortzer et al., 1988).

Advantages:
- (1) fast
- (2) no local minima
- (3) wavelet not required
- (4) "zero offset stack" more accurate than conventional stack

Disadvantages:
- (1) outputs relative parameters only
- (2) wavelet is not removed
- (3) sensitive to NMO errors
- (4) output density and velocity contrasts are not independent
- (5) shear and V_P/V_S contrasts and fluid factor are noisy measurements

Table 9. Crosstabulation of well cumulative production versus AVO effect, Mestena Grande Field, Jim Hogg Co., Texas (from Burnett, 1989 and 1990). The AVO effect is defined as follows: Strong = high amplitude which increases with offset, Moderate = medium amplitude which varies little with offset, Weak = low amplitude which decreases with offset.

Cumulative Production		AVO Effect	
	Strong	Moderate	Weak
1–4 BCF	5	1	0
Less than 1 BCF	0	3	2
abandoned	0	1	3

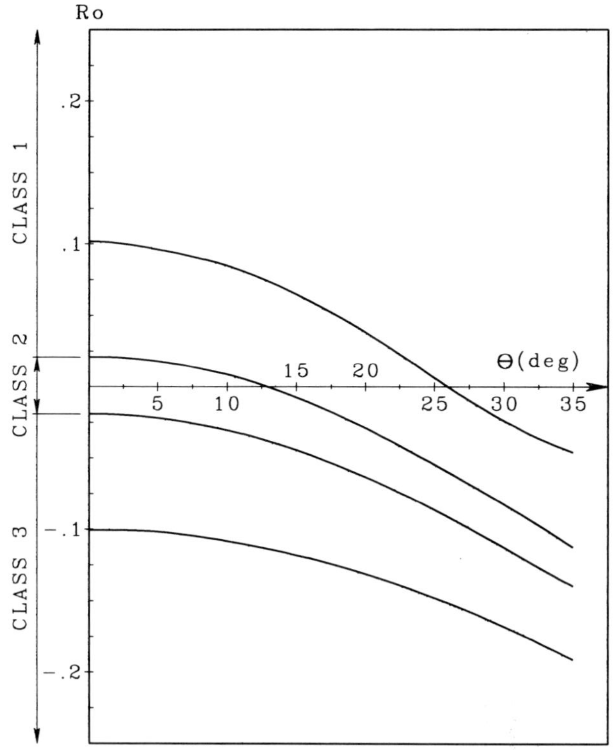

Fig. 20. Zoeppritz *P*-wave reflection coefficients for a shale over gas sand interface for a range of R_P values. The Poisson's ratio and density were assumed to be .38 and 2.4 gm/cc for shale and .15 and 2.0 gm/cc for gas sand (from Rutherford and Williams, 1989).

stack sections by applying knowledge and intuition. The same can be done, with more accurate results, from reflectivity-versus-offset sections, provided that an organized method of interpretation is developed." Toward this end, a variety of interpretive displays have been devised (see Table 10).

Full elastic-wave modeling techniques are described elsewhere in this volume. A simple and easy modeling technique is to calculate the AVO gradient B as a function of depth from well log data, and to convolve the B series with a wavelet. Johnson (1988) shows an excellent use of the B synthetic (see Figure 21).

Partial near, middle, and far offset range stacks simply and robustly show AVO variation in interpretable form (Onstott, 1984, see Figure 22; Todd and Backus, 1985). Figure 23 shows a partial stack display for a class II AVO anomaly (Rutherford and Williams, 1989). In certain circumstances, the partial stack may have better S/N than the full CDP stack (Todd, 1986). Plane-wave decomposition may be used to produce common ray parameter stacks. The colorstack (Onstott, 1984) superimposes near, middle, and far offset ranges; each range represented by a different color. Amplitudes are represented by color intensity and the partial stacks are superimposed. The result is a multicolor seismic section which reveals the amplitude variation with offset; the dominant color indicating the offset range with the greatest amplitude. A "white" result indicates constant amplitude with offset. The method also provides an excellent means of NMO quality control, as imperfectly aligned events show a distinct color fringing. The colorstack, with the help of the human eye, allows full exploitation of the key piece of AVO information which "... is the *difference* in AVO behavior along a seismic line when moving off a structure containing hydrocarbons" (Chiburis, 1987).

Table 10. Listing of AVO interpretation displays (after Wrolstad, 1988; and Nelson, 1989)

Raw CDP gathers
AVO processed and NMO corrected CDP gathers
Crossplot of event amplitude or energy versus offset
Common offset stack
Partial stack
Prestack difference gathers at line intersections
P-wave reflection coefficient (normal incidence; R_p or "A")
S-wave reflection coefficient (normal incidence; R_s)
Difference between P and S reflection coefficients ($R_P - R_S$)
Gradient stack (also called "B" stack or "AVO slope")
Product of *P*-wave reflection coefficient and AVO slope (A*B)
Colorstack
Change in Poisson's ratio section ($\Delta\sigma$)
Envelope of $\Delta\sigma$
Color enveloped $\Delta\sigma$
Fluid factor (ΔF)

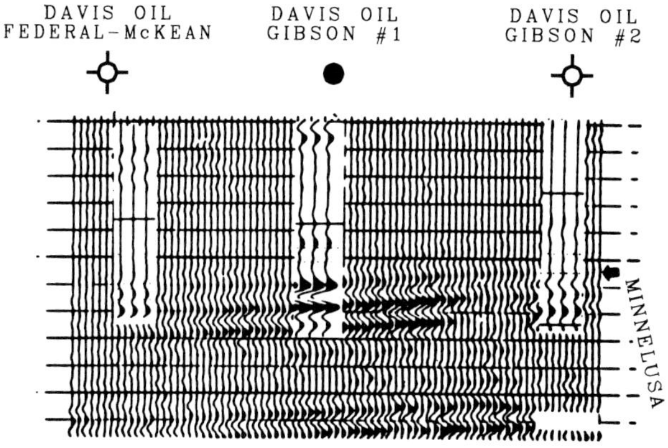

Fig. 21. AVO gradient (B) processing and modeling (from Johnson, 1988).

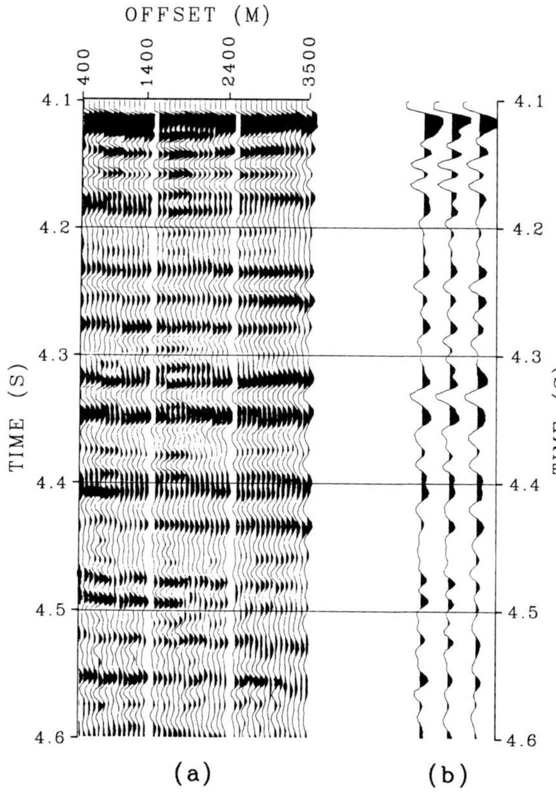

Fig. 22. (a) A portion of a normal moveout corrected common depth point gather divided into three offset ranges. (b) Partial stack traces formed by summing traces within the indicated offset ranges (from Onstott, 1984).

ELASTIC INVERSION

Seismic inversion is the construction of an earth model which is compatible with observed seismic data on the basis of an assumed functional relationship (forward model) between the earth model and "perfect" (noise free) data. It is interesting to note that conventional seismic interpretation is, by this description, a form of seismic inversion. In general, although the functional relationship is nonlinear, the inverse problem is linearized, and an iterative approach is used.

Given infinite computer time, one could model a very large number of combinations of V_P, V_S, ρ, and Q profiles. A combination of these parameters which results in acceptably small discrepancy between real and synthetic data would be taken as the inverted earth model. The objective of seismic inversion is to arrive at an adequate earth model with a limited number of iterations. This is often accomplished by starting with an initial guess and then converging to a solution by minimizing the error seismograms.

Elastic inversion of an entire CDP gather has several advantages over conventional AVO methods such as the weighted stack: (1) The wavelet is removed and interfering reflections are deciphered; tuning is no longer a fundamental limitation. (2) Differential NMO, mode conversions, and multiples become signal rather than noise, and (3) A priori information in the form of starting models and constraints of various kinds can be incorporated, thereby improving effective S/N and bandwidth.

Seismic inversion has the fundamental problem of nonuniqueness and the major disadvantage of high computer costs. It is likely that "automated" modeling, where inverse theory will guide the interpreter in refining his model, will emerge in the near future as a practical tool.

Mora (1984, 1986, 1987a, 1987b, 1987c, and 1988) illustrates elastic earth model reconstruction by full-waveform inversion. For land data, strong singly converted S-waves help to constrain the inversion. Approaches that fix the background velocity (Stolt and Weglein, 1985) will exhibit wild fluctuations if valid velocity-density relationships are not employed (Backus, 1987).

Inversion approaches that depend on amplitude only (Parsons, 1986; Gelfand et al., 1986; etc.) though computationally more acceptable at the present time, should be less reliable than full-waveform techniques which utilize time and moveout information.

Hampson and Russell (1990) combat the nonuniqueness problem by using a Monte Carlo approach, where

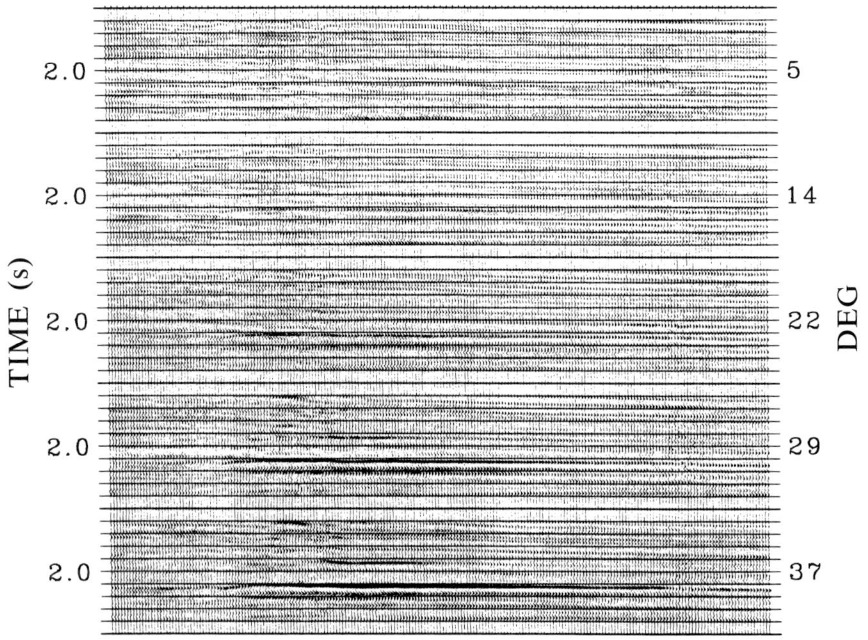

Fig. 23. Panel display of constant reflection angle sections. The angles posted on the right side of the figure refer to the centers of the reflection angle ranges in each panel. Each panel displays 1.6 to 2.4 s (from Rutherford and Williams, 1989).

a variety of starting models are randomly generated. Given enough starting models (and sufficient computer time) this approach can circumvent "local minima" to some extent. However, it is difficult to assess what constitutes "enough" starting models.

Carrazzone and Srnka (1989) and Assous and Richard (1989) have reported excellent headway in full-gather inversion. Tarantola et al. (1990) review the current state-of-the-art. Ultimately, inversion may yield four images of the subsurface: low frequency V_P and V_S, and high frequency P-wave and S-wave impedance. There is ambiguity between low-frequency P-wave velocity and high-frequency impedance. Further ambiguity results from attenuation, anisotropy, and wavelet uncertainty.

Increased inversion stability and robustness can be achieved if multicomponent data are available (Berkhout and Wapenaar, 1988; Lortzer and Berkhout, 1989; Tarantola et al., 1989; de Haas and Berkhout, 1989; Du and Spencer, 1990; Wang et al., 1990; among others).

Pan et al. (1990) studied the sensitivity of target-oriented full-waveform inversion. They found that pay thickness estimates for shallow gas sands were robust, even at thicknesses below 1/10th of a wavelength and in the presence of coherent noise (except at the tuning thickness; see Figure 24). The problem at tuning is attributed to the amplitude and time-thickness insensitivity which occurs on a Widess (1973) tuning plot at the tuning thickness. The pay thickness can be estimated (but not resolved) well below tuning from the inverted Poisson's ratio of an interval of which only a fraction is gas pay. Figure 25 shows a crossplot of inverted Poisson's ratios and P-wave impedances for a shallow gas prospect. A priori, we expect gas sands to have Poisson's ratios of about .1 and brine sands and shales to have Poisson's ratios above .4 in this locality. To the contrary, a continuum of Poisson's ratios are observed. This is attributed to intervals which are not entirely gas pay; these would exhibit a composite Poisson's ratio averaged in some way between gas sand and brine sand or shale. By assuming a time-average relationship for the layers for which the composite Poisson's ratio has been inverted and utilizing a priori V_P versus V_S trends, and Gassmann fluid substitution, the fractional gas sand thickness within each composite layer can be computed (the net thickness is determined—not the distribution). For the special case of a sand interval with S-wave velocities about equal above and below the gas/water contact and known average velocities for the gas and water-saturated intervals, the fractional gas sand thickness, X_{gas}, can be extracted from the inverted V_S/V_P by:

$$X_{gas} = \frac{1/V_P^{br} - (V_S/V_P)/V_S^{br}}{1/V_P^{br} - 1/V_P^{gas}} \qquad (37)$$

where,

V_P^{br} = brine-sand P-wave velocity,
V_P^{gas} = gas-sand P-wave velocity, and
V_S^{br} = brine-sand S-wave velocity.

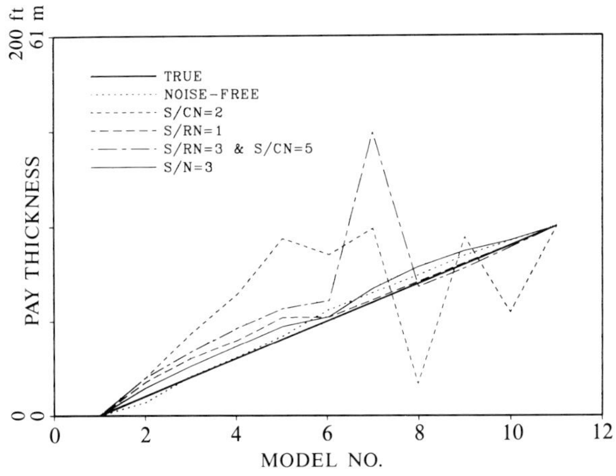

Fig. 24. Inverted gas sand pay thickness estimate for noise-free synthetic data, and data with added random and coherent noise (from Pan et al., 1990). CN = coherent noise. RN = random noise.

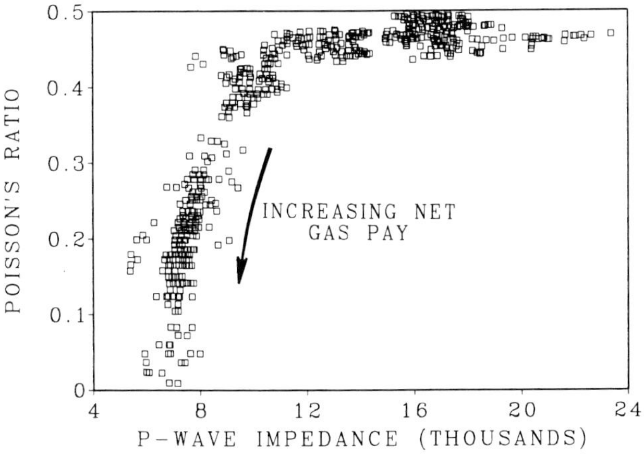

Fig. 25. Inverted Poisson's ratio versus inverted P-wave impedance in the vicinity of a shallow Gulf of Mexico bright spot. For this situation Poisson's ratios of about .1 are expected for gas sands and .4 or above are expected for brine sands and shales. Intermediate Poisson's ratios are caused by vertical averaging of thin layers.

As V_S/V_P is one of the more robust parameters obtainable from AVO data, the predicted X_{gas} is also robust, providing the above assumptions are valid. Figure 26 is a crossplot of observed versus predicted gas thicknesses for some shallow Miocene Gulf of Mexico sands. (Thanks to Al Garcia and Chi Young who performed the inversions). Please note that the two points with no sand present are localized "shale-outs" (verified by side-tracking) which are believed to be smaller than a Fresnel zone.

This example demonstrates the power of a priori constraints tied to borehole information. Such inversion schemes are probably most applicable in the vicinity of well control.

MULTICOMPONENT AVO

Multicomponent seismology and AVO analysis are intimately linked through the Zoeppritz equations. A review of multicomponent analysis techniques is beyond the scope of this book (the reader is referred to Danbom and Domenico, 1986).

Garotta and Granger (1987), and Miles and Gassaway (1989) among others have demonstrated the utility of combined AVO analysis of P, S_V, and S_H data. A few key points follow:

(1) Inversion of the Zoeppritz equations is unique for S_H-waves (Rosa, 1976).

(2) The slope of converted P to S_V-wave AVO is primarily dependent on the S-wave impedance contrast.

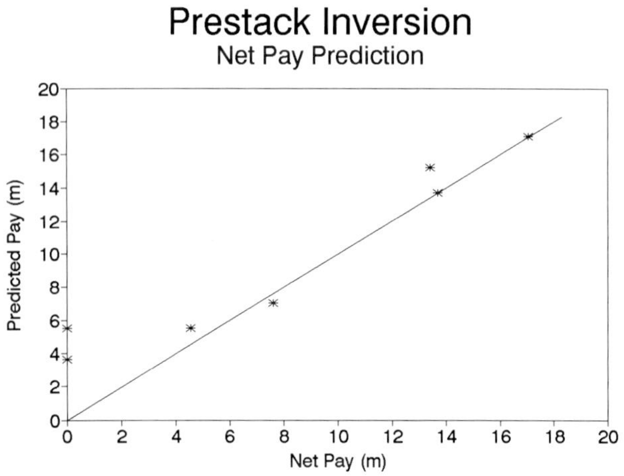

Fig. 26. Calculated net gas sand thickness versus observed net gas sand thickness. The erroneous sand thicknesses at zero observed net pay are localized shale outs smaller than a Fresnel zone (verified by further drilling).

(3) S-waves are less perturbed by transmission through overlying gas sands than are P-waves.

(4) The redundancy of multicomponent AVO measurements is advantageous. This is readily apparent from the following analysis (Corrigan, personal communication). Let us represent the converted P to S_V-wave AVO by:

$$R_{PS} \approx G \sin^2 \Theta_1 \tag{38}$$

where R_{PS} is the incident P-wave to reflected S_V-wave reflection coefficient. Remembering,

$$R_{PP} \approx A + B \sin^2 \Theta_1 \tag{39}$$

it can be shown (for $V_P/V_S \approx 2$) that

$$G \approx B - A \tag{40}$$

(5) Additional S-wave information such as reflection time ratios, normal incidence amplitude ratios, and S-wave moveout further constrains the analysis (Garotta, 1988).

OTHER APPLICATIONS

We have concentrated on the "seismic lithology" aspects of AVO analysis. In fact, comprehension of AVO effects leads to generally better data processing and interpretation and some unique applications in special circumstances.

Effects of AVO on Conventional Interpretation

When sands and shales have about the same impedance, sand-shale reflections may be weak at near offsets, and strong at far offsets. This variation will produce events on the stacked section which cannot be modeled with a simple normal incidence synthetic seismogram and is explained by the following argument: Consider a sand-shale interface. If impedance is constant across the interface, then velocity-density relationships tell us that the shale density should be greater than the sand density and the shale P-wave velocity should be less than the sand P-wave velocity. Using the mudrock V_P versus V_S line (or more precise sandstone and shale trends), it can be shown that the shale will have lower S-wave impedance and higher V_P/V_S than the sandstone. The V_P/V_S drop across the interface will result in the reflection coefficient becoming more negative with offset. As the reflection coefficient is near zero at normal incidence, it should be negative at far offsets. The stacked trace will show a relatively strong negative event.

Wrolstad (1986) showed a comparison of real North Sea seismic data with an elastic prestack synthetic gather computed with predicted S-wave velocities. A stack of the prestack synthetic shows a significantly better tie to the stacked field data than does the conventional synthetic seismogram.

In general, conventional synthetic seismograms should be augmented with synthetic prestack gathers, and the seismogram produced by stacking the synthetic prestack gather.

AVO Effects in Data Processing

It is well recognized that amplitude and phase variations versus offset may cause CDP stacking to degrade reflection data quality (Poley et al., 1989). Extraction of the "A" term by least squares regression produces a "zero offset" stack (Denham et al., 1985; Coruh and Demirbag, 1989, and Kolb et al., 1989). As elucidated in Denham et al. (1985), the zero-offset stack provides: (1) a better approximation to the normal incidence P-wave reflection coefficient, (2) broader bandwidth and higher frequencies, and (3) higher resolution. Noise can be combatted by restricting the rate-of-change of the AVO slope, thereby stabilizing the A term extraction. Seismic impedance traces inverted from zero-offset stack traces should be superior to conventional seismic impedance logs.

Varying reflection-coefficient-versus-offset also contradicts the assumptions of conventional velocity analysis (see Figure 14). In clastic sections exhibiting a well-defined V_P versus V_S relationship, it may be possible to improve velocity analysis using a model-based approach; this is a variation of full gather inversion, however, V_P is the only unknown parameter while V_S and ρ are specified as functions of V_P. Large residuals caused by gas sands would indicate a need to change the V_S and ρ functions.

Critical Angle Effects

The amplitude build up at critical angle for positive reflections can be used to identify polarity, as this amplitude increase does not occur for negative reflections, and to illuminate very weak reflectors at the critical angle (Anstey, 1977). Makris and Thiessen (1984) use critical reflections to image through high velocity evaporites. In some cases, the large amplitude build up can be used to determine the critical angle. Chiburis and Al-Faraj (1987) determined the critical angle in this way and then applied Snell's law to extract overburden velocity and reflector depth.

FUTURE DIRECTIONS

Successful AVO analysis requires a degree of integration beyond that which has been absolutely necessary for conventional seismic interpretation: well logging, rock physics, reservoir engineering, seismic data acquisition and processing, seismic modeling and inversion, conventional seismic interpretation, and a healthy dose of geological concepts and constraints must be incorporated into the analysis. A formula for failure is to couple these disciplines in a straight-line cascaded series of boxes: be they computer programs, departments, or experts. These disciplines must be intimately interwoven in the mind of the interpreter and in the form of truly integrated workstation software. Recent developments in computing hardware and software have brought us to the threshold of realizing this goal; however, as of this writing, no single software package can do all aspects of the job adequately. Sumner (1988) provides an excellent discussion in the area of AVO technology transfer.

The ongoing revolution in computing essentially guarantees significant improvement in AVO analysis. Workstation technologies allow the interpreter to effectively deal with more kinds and dimensions of data. Increased memory, computation speed, and storage capacity means there will be more channels, longer spread lengths, more 3-D data, better processing, more sophisticated modeling, and more ambitious inversion schemes. Expert systems will assist the interpreter in those disciplines in which he or she is not well versed, and will also provide localized expertise for particular areas or formations.

With the increasing computation speed and memory becoming available, AVO inversion techniques are becoming practically applicable. Appropriate selection of parameters, background velocity, wavelet estimation, application of a priori information, and methods for interpreter interaction are still important issues which remain to be resolved. We should not let nonuniqueness in AVO inversion blind us to the tremendous utility of AVO analysis in general. After all, let us not forget that generating structure maps from reflection times is an inherently nonunique process. Ultimately, geophysics is a science of anomalies, and anomalies are defined by our expectations. There is no rigorous solution and our expectations are inexact; nevertheless, AVO does have the ability to reduce risk and to "illuminate" new prospects which were previously overlooked.

ACKNOWLEDGMENTS

Special thanks to M. M. Backus whose many corrections, comments, and ideas made this review somewhat more acceptable. F. Hilterman's unpublished course notes were an invaluable reference. Thanks to ARCO Oil and Gas Co. for permission to publish this paper.

REFERENCES

Aki, K., and Richards, P. G., 1980, Quantitative seismology: Theory and methods: W. H. Freeman and Co.

Angerer, R. H., and Richgels, H. J., 1990, Reservoir states and elastic parameters: Factors which affect reflection AVO: 60th Ann. Internat. Mtg., Soc. Expl. Geophys., Expanded Abstracts, 1537–1538.

Anstey, N. A., 1977, Seismic interpretation: The physical aspects: International Human Resources Development Corporation.

Assous, F., and Richard, V., 1989, Full elastic inversion of multioffset marine data on a stratified medium: 59th Ann. Internat. Mtg., Soc. Expl. Geophys., Expanded Abstracts, 959–962.

Backus, M. M., 1987, Amplitude versus offset: A review: 57th Ann. Internat. Mtg., Soc. Expl. Geophys., Expanded Abstracts, 359–364.

Ball, V., 1988, Thin bed tuning analysis using AVO stratigraphy methods: 58th Ann. Internat. Mtg., Soc. Expl. Geophys., Expanded Abstracts, 1213–1216.

Balogh, D., and Snyder, G., 1986, Examples of a new approach to offset amplitude analysis: 56th Ann. Internat. Mtg., Expanded Abstracts, 350–351.

Berkhout, A. J., and Wapenaar, C. P. A., 1988, Delft philosophy on inversion of elastic data: 58th Ann. Internat. Mtg., Soc. Expl. Geophys., Expanded Abstracts, 831–833.

Bernitsas, N., 1990, Curvature effect of a reflecting surface for arbitrary offset and azimuth., 60th Ann. Internat. Mtg., Soc. Expl. Geophys., Expanded Abstracts, 1053–1055.

Bortfeld, R., 1961, Approximation to the reflection and transmission coefficients of plane longitudinal and transverse waves: Geophys. Prosp., **9**, 485–503.

Burnett, R. C., 1989, Seismic amplitude anomalies and AVO analysis at Mestena Grande field: 59th Ann. Internat. Mtg., Soc. Expl. Geophys., Expanded Abstracts, 690–694.

Burnett, R. C., 1990, Seismic amplitude anomalies and AVO analysis at Mestena Grande field: Geophysics, **55**, 1015–1025.

Carazzone, J. J., and Srnka, L. J., 1989, Elastic inversion of Gulf of Mexico data: 59th Ann. Internat. Mtg., Soc. Expl. Geophys., Expanded Abstracts, 956–958.

Castagna, J. P., Batzle, M. L., and Eastwood, R. L., 1985, Relationship between compressional and shear-wave velocities in clastic silicate rocks: Geophysics, **50**, 551–570.

Castagna, J. P., Batzle, M. L., and Kan, T. K., 1993, Rock physics: the link between rock properties and AVO response *in* Castagna, J. P., and Backus, M. M., Eds., Offset-dependent reflectivity—Theory and practice of AVO analysis, Soc. Expl. Geophys., 135–171.

Cerveny, V., and Ravindra, R., 1971, Theory of seismic head waves: Univ. of Toronto Press.

Chacko, S., 1989, Porosity identification using amplitude variations with offset: Examples from South Sumatra: Geophysics, **54**, 942–951.

Chiburis, E. F., 1984, Analysis of amplitude versus offset to detect gas-oil contacts in the Arabia Gulf: 54th Ann. Internat. Mtg., Soc. Expl. Geophys., Expanded Abstracts, 669–670.

Chiburis, E. F., 1987, Studies of amplitude versus offset in Saudi Arabia: 57th Ann. Internat. Mtg., Soc. Expl. Geophys., Expanded Abstracts, 614–616.

Chiburis, E. F., and Al-Faraj, 1987, Depthing shallow reflectors using postcritical P-wave reflection amplitudes: 57th Ann. Internat. Mtg., Soc. Expl. Geophys., Expanded Abstracts, 631–633.

Coruh, C., and Demirbag, E., 1989, Zoeppritz amplitude correction: Is it needed?: 59th Ann. Internat. Mtg., Soc. Expl. Geophys., Expanded Abstracts, 1160–1163.

Danbom, S. H., and Domenico, S. N., 1986, Shear-wave exploration: Geophysical Development Series, Volume 1, Soc. Expl. Geophys.

de Bruin, C. G. M., Wapenaar, C. P. A., and Berkhout, A. J., 1988, Angle-dependent reflectivity by shot record migration: 58th Ann. Internat. Mtg., Soc. Expl. Geophys., Expanded Abstracts, 1093–1096.

de Bruin, C. G. M., Wapenaar, C. P. A., and Berkhout, A. J., 1990, Imaging for angle-dependent reflectivity in the presence of dip: 60th Ann. Internat. Mtg., Soc. Expl. Geophys., Expanded Abstracts, 1503–1506.

de Haas, J. C., and Berkhout, A. J., 1988, On the information content of *P-P*, *P-SV*, *SV-SV*, and *SV-P* Reflections: 58th Ann. Internat. Mtg., Soc. Expl. Geophys., Expanded Abstracts, 1190–1194.

de Haas, J. C., and Berkhout, A. J., 1989, Practical approach to nonlinear inversion of amplitude versus offset information: 59th Ann. Internat. Mtg., Soc. Expl. Geophys., Expanded Abstracts, 839–842.

Denham, L. R., Palmeira, R. A. R., and Farrell, R. C., 1985, The zero-offset stack: 55th Ann. Internat. Mtg., Soc. Expl. Geophys., Expanded Abstracts, 624–625.

Dey-Sarkar, S. K., Dutta, N. C., Weglein, A. B., Backus, M. M., and Hilterman, F., 1988, Amplitude versus offset workshop: 58th Ann. Internat. Mtg., Soc. Expl. Geophys., Expanded Abstracts, 1353–1359.

Dey-Sarkar, S. K., and James, C. F., 1986, Prestack analysis: relevance of petrophysical properties: 56th Ann. Internat. Mtg., Soc. Expl. Geophys., Expanded Abstracts, 337–339.

Domenico, S. N., 1976, Effect of brine-gas mixture on velocity in an unconsolidated sand reservoir: Geophysics, **41**, 882–894.

Du, X., and Spencer, T. W., 1990, Quantitative analysis of amplitude versus offset: 60th Ann. Internat. Mtg., Soc. Expl. Geophys., Expanded Abstracts, 1459–1462.

Duren, R. E., 1990, Seismic range equation: 60th Ann. Internat. Mtg., Soc. Expl. Geophys., Expanded Abstracts, 1495–1498.

Duren, R. E., 1991, Seismic range equation: Geophysics, **56**, 1015–1026.

Duren, R. E., 1992, Range-equation weights for AVO: Geophysics, **57**, 1203–1208.

Dutta, N. C., and Ode, H., 1981, Biot loss mechanism and its importance in seismology: A review: 51st Ann. Internat. Mtg., Soc. Expl. Geophys., Abstracts, 103–104.

Dutta, N. C., and Ode, H., 1983, Seismic reflections from a gas-water contact: Geophysics, **48**, 148–162.

Fouquet, D. F., 1990, Principles of AVO processing: 60th Ann. Internat. Mtg., Soc. Expl. Geophys., Expanded Abstracts, 1486.

Frasier, C. W., 1988, AVO: usage/misusage: 58th Ann. Internat. Mtg., Soc. Expl. Geophys., Expanded Abstracts, 1356.

Fuller, B. N., Iverson, W. P., and Smithson, S. B., 1989, AVO for thin sand bed detection: 59th Ann. Internat. Mtg., Soc. Expl. Geophys., Expanded Abstracts, 826–828.

Gabay, S. H., 1990, Integrated lithologic interpretation of normal and nonnormal-incidence reflection character: 60th Ann. Internat. Mtg., Soc. Expl. Geophys., Expanded Abstracts, 1519–1522.

Garotta, R., 1988, Amplitude-versus-offset measurements involving converted waves: 58th Ann. Internat. Mtg., Soc. Expl. Geophys., Expanded Abstracts, 1357.

Garotta, R., and Granger, P. Y., 1987, Comparison of responses of compressional and converted waves on a gas sand: 57th Ann. Internat. Mtg., Soc. Expl. Geophys., Expanded Abstracts, 627–630.

Gassaway, G. S., 1984, Effects of shallow reflectors on amplitude versus offset (seismic lithology) analysis: 54th Ann. Internat. Mtg., Soc. Expl. Geophys., Expanded Abstracts, 665–669.

Gassmann, F., 1951, Elastic waves through a packing of spheres: Geophysics, **16**, 673–685.

Gelfand, V., Ng, P., Nguyen, H., and Larner, K., 1986, Seismic lithologic modeling of amplitude versus offset data: 56th Ann. Internat. Mtg., Soc. Expl. Geophys., Expanded Abstracts, 334–337.

Greenberg, M. L., and Castagna, J. P., 1992, Shear-wave velocity estimation in porous rocks: Theoretical formulation, preliminary verification and applications: Geophys. Prosp., **40**, 195–209.

Gregory, A. R., 1976, Fluid saturation effects on dynamic elastic properties of sedimentary rocks: Geophysics, **41**, 895–921.

Hales, A. L., and Roberts, J. L., 1974, The Zoeppritz amplitude equations: more errors: Bull., Seis. Soc. Am., **64**, 285

Hampson, D., and Russell, B., 1990, AVO inversion: theory and practice: 60th Ann. Internat. Mtg., Soc. Expl. Geophys., Expanded Abstracts, 1456–1458.

Herman, G. C., and Bionk, B., 1990, Influence of small-scale interface roughness in the estimation of lithologic parameters from reflection coefficients: 60th Ann. Internat. Mtg., Soc. Expl. Geophys., Expanded Abstracts, 1475–1478.

Hilterman, F., 1975, Amplitudes of seismic waves—a quick look., Geophysics, **40**, 745–762.

Hilterman, F., 1989, Is AVO the seismic signature of rock properties?: 59th Ann. Internat. Mtg., Soc. Expl. Geophys., Expanded Abstracts, 559.

Hilterman, F., 1990, Is AVO the seismic signature of lithology? A case history of Ship Shoal—South Addition, Leading Edge, 9, no. 6, 15–22.

Hindlet, F. J. F., and McDonald, J. A., 1986, Thin layer analysis using offset/amplitude data: 56th Ann. Internat. Mtg., Soc. Expl. Geophys., Expanded Abstracts, 346–349.

Hoopes, R., and Aber, W. M., 1989, Integrated multianalysis procedure that qualifies DHIs: A case history of the Louisiana Hackberry Trend: 59th Ann. Internat. Mtg., Soc. Expl. Geophys., Expanded Abstracts, 843–847.

Huston, D. C., and Backus, M. M., 1986, Interpretation of seismic signal and noise through line intersection analysis: 56th Ann. Internat. Mtg., Soc. Expl. Geophys., Expanded Abstracts, 501–503.

Hwang, L. F., and Lellis, P. J., 1988, Bright spots related to high GOR oil reservoir in Green Canyon: 58th Ann. Internat. Mtg., Soc. Expl. Geophys., Expanded Abstracts, 761–763.

Ikelle, L. T., Diet, J. P., and Tarantola, A., 1986, Linearized inversion of multioffset seismic reflection data in ω-k domain with depth-dependent reference medium: 56th Ann. Internat. Mtg., Soc. Expl. Geophys., Expanded Abstracts, 530–533.

Ito, H., DeVilbiss, J., and Nur, A., 1979, Compressional and shear waves in saturated rock during water-steam transition: J. Geophys. Res., **84**, 4731–4735.

Johnson, W. E., 1988, Amplitude versus offset for sandstones in shales and carbonates., 58th Ann. Internat. Mtg., Soc. Expl. Geophys., Expanded Abstracts, 593–595.

Kelamis, P. G., Chiburis, E. F., and Shabryar, S., 1990, Radon multiple elimination, a practical methodology for land data: 60th Ann. Internat. Mtg., Soc. Expl. Geophys., Expanded Abstracts, 1611–1613.

Knott, C. G., 1899, Reflexion and refraction of elastic waves with seismological applications: Phil. Mag., **48**, 64–97.

Koefoed, O., 1955, On the effect of Poisson's ratios of rock strata on the reflection coefficients of plane waves: Geophys. Prosp., **3**, 381–387.

Kolb, O., Chapel, F., and Picart, I., 1989, Lithologic inversion: a reflectivity versus angle (RVA) approach: 59th Ann. Internat. Mtg., Soc. Expl. Geophys., Expanded Abstracts, 695–699.

Krail, P. M., and Brysk, H., 1983, Reflection of spherical seismic waves in elastic layered media: Geophysics, **48**, 655–664.

Long, A., and Richgels, J., 1985, A practical application of amplitude versus offset effects: 55th Ann. Internat. Mtg., Soc. Expl. Geophys., Expanded Abstracts, 594–598.

Lortzer, G. J. M., and Berkhout, A. J., 1989, Linear AVO inversion of multicomponent seismic data: 59th Ann. Internat. Mtg., Soc. Expl. Geophys., Expanded Abstracts, 967–972.

Lortzer, G. J. M., de Haas, J. C., and Berkhout, A. J., 1988, Evaluation of weighted stacking techniques for AVO inversion: 58th Ann. Internat. Mtg., Soc. Expl. Geophys., Expanded Abstracts, 1204–1208.

Luh, P. C., 1988, Wavelet dispersion and bright spot detection: 58th Ann. Internat. Mtg., Soc. Expl. Geophys., Expanded Abstracts, 1217–1220.

MacLeod, M. K., and Martin, H. L., 1988, Amplitude changes due to bed curvature: 58th Ann. Internat. Mtg., Soc. Expl. Geophys., Expanded Abstracts, 1209–1212.

Makris, J., and Thiessin, J., 1984, Wide-angle reflections: a tool to penetrate horizons with high acoustic impedance contrasts: 54th Ann. Internat. Mtg., Soc. Expl. Geophys., Expanded Abstracts, 672–674.

Mazzotti, A., 1984, Prestack analysis of seismic amplitude anomalies: 54th Ann. Internat. Mtg., Soc. Expl. Geophys., Expanded Abstracts, 659–661.

Mazzotti, A., 1990, Prestack amplitude analysis methodology and application to seismic bright spots in the Po Valley, Italy: Geophysics, **55**, 157–166.

Miles, D. R., and Gassaway, G. S., 1989, Three-component AVO analysis: 59th Ann. Internat. Mtg., Soc. Expl. Geophys., Expanded Abstracts, 706–708.

Mora, P., 1984, Inversion of CMP gathers for P- and S-velocity: 54th Ann. Internat. Mtg., Soc. Expl. Geophys., Expanded Abstracts, 730–733.

Mora, P., 1986, Nonlinear elastic inversion of multioffset seismic data: 56th Ann. Internat. Mtg., Soc. Expl. Geophys., Expanded Abstracts, 533–537.

Mora, P., 1987a, Elastic inversion of multi-offset seismic amplitude data for *P*-wave velocity, *S*-wave velocity, and density *in* Danbom, S. H., and Domenico, S. N., Editors, Shear-wave Exploration, Geophysical Developments No. 1, 199–213, Soc. Expl. Geophys.

Mora, P., 1987b, Nonlinear two-dimensional elastic inversion of multioffset seismic data: Geophysics, **52**, 1211–1228.

Mora, P., 1987c, Nonlinear elastic inversion of real data: 57th Ann. Internat. Mtg., Soc. Expl. Geophys., Expanded Abstracts, 430–432.

Mora, P., 1988, Elastic wave-field inversion of reflection and transmission data: Geophysics, **53**, 750–759.

Morley, L., 1984, Invertibility of elastic layered earth parameters from precritical *P*-wave reflection amplitudes: 54th Ann. Internat. Mtg., Soc. Expl. Geophys., Expanded Abstracts, 641–643.

Murphy, W. F., 1982, Effects of microstructure and pore fluids on acoustic properties of granular sedimentary materials: Ph.D. dissertation, Stanford Univ.

Muskat, M., and Meres, M. W., 1940, Reflection and transmission coefficients for plane waves in elastic media: Geophysics, **5**, 115–148.

Nelson, V., 1989, Display techniques for AVO analysis: 59th Ann. Internat. Mtg., Soc. Expl. Geophys., Expanded Abstracts, 447–448.

Newman, P., 1973, Divergence effects in a layered earth: Geophysics, **38**, 481–488.

Onstott, G. E., 1984, Processing and display of offset dependent reflectivity in reflection seismograms: M.Sc. thesis, Univ. of Texas at Austin.

Onstott, G. E., Backus, M. M., Wilson, C. R., and Phillips, J. D., 1984, Color display of offset dependent reflectivity in seismic data: 54th Ann. Internat. Mtg., Soc. Expl. Geophys. Expanded Abstracts, 674–675.

Ostrander, W. J., 1982, Method for interpretation of seismic records to yield indication of gaseous hydrocarbons: United States Patent No. 4,316,268.

Ostrander, W. J., 1984, Plane-wave reflection coefficients for gas sands at nonnormal angles of incidence: Geophysics, **49**, 1637–1648.

Pan, G. S., Young, C. Y., and Castagna, J. P., 1990, Sensitivity and resolution of an integrated target-oriented prestack elastic inversion: 60th Annual Internat. Mtg., Soc. Expl. Geophys., Expanded Abstracts, 1173–1176.

Parson, R., 1986, Estimating reservoir mechanical properties using constant offset images of reflection coefficients and incident angles: 56th Ann. Internat. Mtg., Expanded Abstracts, 617–620.

Payne, M. A., 1991, Shear-wave logging to enhance seismic modeling: Geophysics, **56**, 2129–2138.

Pigott, J. D., Shrestha, R. K., and Warwick, R. A., 1990, Direct determination of carbonate reservoir porosity and pressure from AVO inversion: 60th Ann. Internat. Mtg., Soc. Expl. Geophys., Expanded Abstracts, 1533–1536.

Poley, D. F., Lawton, D. C., and Blasco, S. M., 1989, Amplitude-offset relationships over shallow velocity inversions: Geophysics, **54**, 1114–1122.

Rafipour, B., 1987, Seismic response for reservoir fluid evaluation: 57th Ann. Internat. Mtg., Soc. Expl. Geophys., Expanded Abstracts, 377–380.

Ramirez, L. C., 1990, AVO statistical analysis: a useful tool for interbedded salt body detection: 60th Ann. Internat. Mtg., Soc. Expl. Geophys., Expanded Abstracts, 1483–1485.

Rendleman, C. A., and Levin, F. K., 1988, Reflection maxima for reflections from single interfaces: Geophysics, **53**, 271–275.

Resnick, J. R., Ng, P., and Larner, K., 1987, Amplitude versus offset analysis in the presence of dip: 57th Ann. Internat. Mtg., Soc. Expl. Geophys., Expanded Abstracts, 617–620.

Resnick, J. R., 1988, Seismic data processing for amplitude-versus-offset analysis: 58th Ann. Internat. Mtg., Soc. Expl. Geophys., Expanded Abstracts, 1354.

Richards, P. G., and Frasier, C. W., 1976, Scattering of elastic waves from depth-dependent inhomogeneities: Geophysics, **41**, 441–458.

Richards, T. C., 1961, Motion of the ground on arrival of reflected longitudinal and transverse waves at wide-angle reflection distances: Geophysics, **26**, 277–297.

Rosa, A. L. R., 1976, Extraction of elastic parameters using seismic reflection amplitude with offset variation: M.A. Thesis, Univ. of Houston.

Rutherford, S. R., and Williams, R. H., 1989, Amplitude-versus-offset variations in gas sands: Geophysics, **54**, 680–688.

Samec, P., Blangy, J. P., and Nur, A., 1990, Effect of viscoelasticity and anisotropy on amplitude-versus-offset interpretation: 60th Ann. Internat. Mtg., Soc. Expl. Geophys., Expanded Abstracts, 1479–1482.

Sengupta, M. K., and Rendleman, C. A., 1989, Case study: the importance of gas leakage in interpreting amplitude-versus-offset (AVO) analysis: 59th Ann. Internat. Mtg., Soc. Expl. Geophys., Expanded Abstracts, 848–850, and Geophys., **56**, 1886–1895.

Sherwood, J., Kan, T. K., and Crampin, S., 1983, Synthetic seismograms in a solid layered earth: 53rd Ann. Internat. Mtg., Soc. Expl. Geophys., Expanded Abstracts, 636

Shuey, R. T., Banik, N. C., and Lerche, I., 1984, Amplitude from curved reflectors: 54th Ann. Internat. Mtg., Soc. Expl. Geophys., Expanded Abstracts, 664–665.

Shuey, R. T., 1985, A simplification of the Zoeppritz equations: Geophysics, **50**, 609–614.

Smith, G. C., and Gidlow, P. M., 1987, Weighted stacking for rock property estimation and detection of gas: Geophys. Prosp., **35**, 993–1014.

Snyder, A. G., Kelsey, D. J., and Wrolstad, K. H., 1989, Direct detection using AVO, Central Graben, North Sea: 59th Ann. Internat. Mtg., Soc. Expl. Geophys., Expanded Abstracts, 700–701.

Soroka, W. L., Fitch, T. J., Van Sickle, K. H., and North, P. D., 1990, Three-dimensional AVO analysis: 60th Ann. Internat. Mtg., Soc. Expl. Geophys., Expanded Abstracts, 1530–1532.

Spratt, S., 1987, Effect of normal moveout errors on amplitude versus offset-derived shear reflectivity: 57th Ann. Internat. Mtg., Soc. Expl. Geophys., Expanded Abstracts, 634–637.

Spratt, R. S., Goins, N. R., and Fitch, T. J., 1993, Pseudo-shear: The analysis of AVO *in* Castagna, J. P., and Backus, M. M., Eds., Offset-dependent reflectivity, Soc. Expl. Geophys. 37–56.

Stoll, R. D., and Kan, T. K., 1981, Reflection of acoustic waves at a water-sediment interface: J. Acoust. Soc. Am., **70**, 149–156.

Stolt, R. H., and Weglein, A. B., 1985, Migration and inversion of seismic data: Geophysics, **50**, 2458, 2472.

Strudley, A., 1990, Amplitude versus offset: Methodology and application to pore fill prediction in the Danish Central Graben: 60th Ann. Internat. Mtg., Soc. Expl. Geophys., Expanded Abstracts, 1523–1525.

Sumner, J. R., 1988, Amplitude versus offset: The transfer of a complex technology: 58th Ann. Internat. Mtg., Soc. Expl. Geophys., Expanded Abstracts, 1358–1359.

Swan, H. W., 1988, Amplitude versus offset analysis in a finely layered media: 58th Ann. Internat. Mtg., Soc. Expl. Geophys., Expanded Abstracts, 1195–1198.

Tarantola, A., Crase, E., Jervis, M., Konen, Z., Lindgren, J., Mosegaard, K., and Noble, M., 1990, Nonlinear inversion of seismograms: State of the art: 60th Ann. Internat. Mtg., Soc. Expl. Geophys., Expanded Abstracts, 1193–1198.

Tarantola, A., Beydoun, W., Cao, D., Crase, E., Koren, Z., Mendes, M., Mora, P., Mosegaard, K., Noble, M., Singh, S., and Trezeguet, D., 1989, Present know-how and difficulties in elastic nonlinear inversion of seismograms: 59th Ann. Internat. Mtg., Soc. Expl. Geophys., Expanded Abstracts, 980–982.

Todd, C. P., and Backus, M. M., 1985, Offset-dependent-reflectivity in a structural context: 55th Ann. Internat. Mtg., Soc. Expl. Geophys., Expanded Abstracts, 586–589.

Todd, C. P., 1986, Isolation, display and interpretation of offset dependent phenomena in seismic reflection data using offset to depth (ODR) range partial stacking., M.Sc. thesis, Univ. Texas Austin.

Treadgold, G. E., Dey-Sarkar, S. K., Smith, S. W., and Swan, H. W., 1990a, Amplitude versus offset and thin beds: 60th Ann. Internat. Mtg., Soc. Expl. Geophys., Expanded Abstracts, 1463–1466.

Treadgold, G. E., Ritchie, K., and Dey-Sarkar, S. K., 1990b, AVO: An example of processing pitfalls: 60th Ann. Internat. Mtg., Soc. Expl. Geophys., Expanded Abstracts, 1487–1490.

Tsingas, C., and Kanasewich, E. R., 1991, Seismic reflection amplitude versus angle variations over a thermally enhanced oil recovery site: Geophysics, **56**, 292–301.

Wang, D., Beylkin, G., Burridge, R., and Chang, H. W., 1990, GRT inversion for elastic parameters using *PP*, *PS*, *SP*, and *SS* seismic data: 60th Ann. Internat. Mtg., Soc. Expl. Geophys., Expanded Abstracts, 1199–1202.

Wapenaar, C. P. A., Verschuur, D. J., and Herrmann, P., 1990, Amplitude preprocessing of 2-D seismic data: 60th Ann. Internat. Mtg., Soc. Expl. Geophys., Expanded Abstracts, 1491–1494.

Waters, K. H., 1981, Reflection Seismology: A tool for energy resource exploration: John Wiley & Sons, Inc.

Widess, M. B., 1973, How thin is a thin bed?: Geophysics, **38,** 1176–1180.

Wiggins, R., Kenny, G. S., and McClure, C. D., 1983, A method for determining and displaying the shear-velocity reflectivities of a geologic formation: European Patent Application 0113944.

Wright, J., 1984, The effects of anisotropy on reflectivity-offset: 54th Ann. Internat. Mtg., Soc. Expl. Geophys., Expanded Abstracts, 670–672.

Wrolstad, K. H., 1986, Offset-dependent amplitude analysis by visco-elastic lithological modeling: 56th Ann. Internat. Mtg., Soc. Expl. Geophys., Expanded Abstracts, 341–346.

Wrolstad, K. H., 1988, AVO representations: 58th Ann. Internat. Mtg., Soc. Expl. Geophys., Expanded Abstracts, 1355.

Xu, Y., and McDonald, J. A., 1988, Effects of velocity variation on AVO and AVA calculations: 58th Ann. Internat. Mtg., Soc. Expl. Geophys., Expanded Abstracts, 1221–1223.

Young, G. B., and Braile, L. W., 1976, A computer program for the application of Zoeppritz' amplitude equations: Seis. Soc. Am. Bull., **52**, 923–956.

Yu, G., 1985(a), Offset amplitude variation and controlled amplitude processing: 55th Ann. Internat. Mtg., Soc. Expl. Geophys., Expanded Abstracts, 591–594.

Zilik, R. P., 1989, Seismic bright spots: detecting Yegua channel sands using velocity inversion: 59th Ann. Internat. Mtg., Soc. Expl. Geophys., Expanded Abstracts, 829–831.

Zimmerman, C. J., and Fahmy, B. A., 1990, An effective new method of amplitude-versus-offset analysis for marine data: 60th Ann. Internat. Mtg., Soc. Expl. Geophys., Expanded Abstracts, 1499–1502.

Zoeppritz, K., 1919, Erdbebenwellen VIIIB, On the reflection and propagation of seismic waves: Gottinger Nachrichten, I, 66–84.

REFERENCES FOR GENERAL READING

Backus, M. M., and Chen, R. L., 1975, Flat-spot exploration: Geophys. Props., **23**, 533–577.

Backus, M. M., Garcia, A. I., and Huston, D. C., 1987, On deconvolution and inversion in a one-dimensional earth: Proceedings SEG-EAEG workshop on deconvolution and inversion, Rome.

Backus, M. M., and Goins, N. R., 1983, Research workshop 2—the change in reflectivity with offset: 53rd Ann. Internat. Mtg., Soc. Expl. Geophys., Expanded Abstracts, 636.

Backus, M. M., Nepomuceno, F., and Cao, J., 1982, The reflection seismogram in a solid layered earth: 52nd Ann. Internat. Mtg., Soc. Expl. Geophys., Expanded Abstracts, 218–220.

Backus, M. M., Nepomuceno, F., and Cao, J., 1983, Reflection seismogram in a solid layered earth: Abstract, Bull. Am. Assn. Petr. Geol., 416–417.

Ball, V., 1987, Depth inversion of impedance and Poisson's Ratio using the Insight interactive modeling system: 57th Ann. Internat. Mtg., Soc. Expl. Geophys., Expanded Abstracts, 624–626.

Carrion, P. M., and Foster, D. J., 1985, Inversion of seismic data using precritical reflection and refraction data: Geophysics, **50**, 759–765.

Clark, V. A., 1992, The effect of oil under in-situ conditions on the seismic properties of rocks: Geophysics, **57**, 894–901.

Cunha, C. A., Jr., 1990, Snell-beam transform: retrieving the angle-dependent reflectivity: 60th Ann. Internat. Mtg., Soc. Expl. Geophys., Expanded Abstracts, 1507–1510.

Daggett, P. H., Young, C. Y., and Kan, T. K., 1983, Comparison of prestack acoustic and elastic synthetic seismograms: 53rd Ann. Internat. Mtg., Soc. Expl. Geophys., Expanded Abstracts, 457–458.

de Beukelaar, P., and Richard, V., 1990, Source estimation and lateral sensitivity of nonlinear inversion for multioffset seismic surface data: 60th Ann. Internat. Mtg., Soc. Expl. Geophys., Expanded Abstracts, 1161–1164.

de Haas, J. C., and Berkhout, A. J., 1990, Local inversion of *PP*, *PS*, *SP*, and *SS* reflections: 60th Ann. Internat. Mtg., Soc. Expl. Geophys., Expanded Abstracts, 1189–1192.

Demirbag, E., and Coruh, C., 1988, Inversion of Zoeppritz equations and their approximations: 58th Ann. Internat. Mtg., Soc. Expl. Geophys., Expanded Abstracts, 1199–1203.

Demirbag, E., and Coruh, C., 1989, Inversion of multilayer amplitude versus offset data: 59th Ann. Internat. Mtg., Soc. Expl. Geophys., Expanded Abstracts, 709–712.

Dey-Sarkar, S. K., 1986, Prestack analysis: Data processing: 56th Ann. Internat. Mtg., Soc. Expl. Geophys., Expanded Abstracts, 339–341.

Domenico, S. N., 1974, Effect of water saturation on seismic reflectivity of sand reservoirs encased in shale: Geophysics, **39**, 759–769.

Foster, D. J., and Mosher, C. C., 1992, Suppression of multiple reflections using the Radon transform: Geophysics, **57**, 386–395.

Fulton, T. K., and Darr, K. M., 1984, Offset panel: Geophysics, **49**, 1140–1152.

Garcia, A., and Backus, M. M., 1985, Thin layer resonse catalog, 55th Ann. Internat. Mtg., Soc. Expl. Geophys., Expanded Abstracts, 588–590.

Gardner, G. H. F., Gardner, L. W., and Gregory, A. R., 1974, Formation velocity and density—the diagnostic basis for stratigraphic traps: Geophysics, **39**, 770–780.

Gassaway, G. S., 1983, Comparison and application of synthetic plane wave front CDP seismograms and observed CDP gathers from the Sacramento Valley and Denver-Julesburg Basin: 53rd Ann. Internat. Mtg., Soc. Expl. Geophys., Expanded Abstracts, 638.

Gassaway, G. S., Brown, R. A., and Bennett, L. E., 1986, Pitfalls in seismic amplitude-versus-offset analysis: case histories: 56th Ann. Internat. Mtg., Soc. Expl. Geophys., Expanded Abstracts, 332–334.

Gassaway, G. S., and Richgels, H. J., 1983, SAMPLE: Seismic amplitude measurement from primary lithology estimation: 53rd Ann. Internat. Mtg., Soc. Expl. Geophys., Expanded Abstracts, 610–613.

Gassaway, G. S., and Richgels, H. J., 1984, Seismic amplitude measurement for primary lithology estimation (SAM-

PLE): Case histories from Tertiary western basins: Offshore Tech. Conf. 4784.
Han, D. H., Nur, A., and Morgan, D., 1986, Effects of porosity and clay content on wave velocities in sandstones: Geophysics, **51**, 2093–2107.
Hindlet, F., Chapel, F., and Picart, I., 1990, Increasing the sensitivity of AVO data for reservoir characterization: 60th Ann. Internat. Mtg., Soc. Expl. Geophys., Expanded Abstracts, 1457–1470.
Huston, D. C., and Backus, M. M., 1989, Offset-dependent mis-tie analysis at seismic line intersections: Geophysics, **54**, 962–972.
Jervis, M., 1990, Simultaneous recovery of the source wavelet and seismic impedances by nonlinear elastic inversion: 60th Ann. Internat. Mtg., Soc. Expl. Geophys., Expanded Abstracts, 1158–1160.
Johnson, W. E., 1990, Amplitude-versus-offset analysis for sandstones encased in carbonates and shales., 60th Ann. Internat. Mtg., Soc. Expl. Geophys., Expanded Abstracts, 1526–1529.
Kanasewich, E. R., and Tsingas, C., 1990, Amplitude and ratio for partial angle (ARPA): a new method for imaging amplitude-versus-offset seismic reflection data: 60th Ann. Internat. Mtg., Soc. Expl. Geophys., Expanded Abstracts, 1516–1518.
Kind, R., 1976, Computation of reflection coefficients for layered media, J. of Geophysics, **42**, 191–200.
Koefoed, O., 1962, Reflection and transmission coefficients for plane longitudinal and incident waves: Geophys. Prosp., **10**, 304–351.
Kolb, P., and Canadas, G., 1986, Least-squares inversion of prestack data: simultaneous identification of density and velocity of stratified media: 56th Ann. Internat. Mtg., Soc. Expl. Geophys., Expanded Abstracts, 604–607.
Kolb, P., Collino, F., and Lailly, P., 1986, Prestack inversion of 1-D medium: Proc. Inst. Electr. Electron. Eng., Special Issue, Seismic Inverse Problems.
Levin, F. K., 1986, When reflection coefficients are zero: Geophysics, **51**, 736–741.
Lin, T. L., 1989, AVO modeling by domain classification: 59th Ann. Internat. Mtg., Soc. Expl. Geophys., Expanded Abstracts, 702–705.
Lortzer, G. J. M., and Berkhout, A. J., 1990, Influence of lithology and data quality on lithologic inversion results: 60th Ann. Internat. Mtg., Soc. Expl. Geophys., Expanded Abstracts, 1185–1188.
Luh, P. C., 1992, True-amplitude signal recovery: The Leading Edge, **11**, no. 2, 13–24.
Martinez, R. D., and McMechan, G. A., 1988, Multiparameter inversion for viscoelastic media: 58th Ann. Internat. Mtg., Soc. Expl. Geophys., Expanded Abstracts, 853–855.
Masuda, R. M., 1990, Patterm recognition of AVO and amplitude anomalies in exploration seismology: 60th Ann. Internat. Mtg., Soc. Expl. Geophys., Expanded Abstracts, 1539–1540.
Mazzotti, A., 1988, Methodology of prestack amplitude analysis and its application to seismic bright spots in the Po Valley, Italy: 58th Ann. Internat. Mtg., Soc. Expl. Geophys., Expanded Abstracts, 756–760.
McCarthy, C. J., 1984, Seismic prediction of pore fluid and gas thickness: 54th Ann. Internat. Mtg., Soc. Expl. Geophys., Expanded Abstracts, 326–328.
McCauley, A. D., 1984, Importance of shear in plane layer point source modeling for prestack inversion: 54th Ann. Internat. Mtg., Soc. Expl. Geophys., Expanded Abstracts, 541–544.
McCauley, A. D., 1985, Prestack inversion with plane-layer point source modeling, Geophys., **50**, 77–89.
McCauley, A. D., 1986, Plane-layer point-source prestack inversion of marine data with unknown initial profile: 56th Ann. Internat. Mtg., Soc. Expl. Geophys., Expanded Abstracts, 537–539.
Miles, D. R., Gassaway, G. S., Bennett, L. E., and Brown, R. A., 1988, Detecting hydrocarbons in reefs using AVO analysis: A case history from Alberta, Canada: 58th Ann. Internat. Mtg., Soc. Expl. Geophys., Expanded Abstracts, 753–755.
Muskat, M., and Meres, M. W., 1940, The seismic wave energy reflected from various types of stratified horizons: Geophysics, **5**, 149–155.
Neidell, N. S., 1985, Geophysical determination of lithology, using shear waves and amplitudes with offsets—a progress update: Continuing Education Course Note Series #29, Am. Assn. Petr. Geol.
Neidell, N. S., 1985, A progress report on use of shear waves and amplitude with offset in seismic exploration: Leading Edge, **4**, 10, 92–93.
Neidell, N. S., 1986, Amplitude variation with offset: Leading Edge, **5**, 1, 47–51.
Neidell, N. S., 1986, Research elements in amplitude variation with offset: Leading Edge, **5**, 4, 42–44.
Normington, E. J., 1990, Seismic model studies of the overburden bedrock reflection: *P*-wave and *S*-wave: 60th Ann. Internat. Mtg., Soc. Expl. Geophys., Expanded Abstracts, 376–379.
Ostrander, W. J., 1982b, Plane-wave reflection coefficients for gas sands at nonnormal angles of incidence: 52nd Ann. Internat. Mtg., Soc. Expl. Geophys., Expanded Abstracts, 216–220.
Ostrander, W. J., and Gassaway, G., 1983, The use of offset dependent reflectivity in exploration: 53rd Ann. Internat. Mtg., Soc. Expl. Geophys., Expanded Abstracts, 637.
Pan, G. S., and Phinney, R. A., 1986, Full waveform inversion of p-τ seismogram: 56th Ann. Internat. Mtg., Soc. Expl. Geophys., Expanded Abstracts, 614–617.
Pan, N. D., and Gardner, G. H. F., 1986, The p,q algorithm for seismic data analysis: 56th Ann. Internat. Mtg., Soc. Expl. Geophys., Expanded Abstracts, 413–417.
Pigott, J. D., 1989, Young's modulus from AVO inversion: 59th Ann. Internat. Mtg., Soc. Expl. Geophys., Expanded Abstracts, 832–835.
Pullan, S. E., and Hunter, J. A., 1985, Seismic model studies of the overburden bedrock reflection: Geophysics, **50**, 1684–1688.
Rafipour, B., and Herrin, E., 1986, Phase offset indicator (POI): A study of phase shift versus offset and fluid content: Geophysics, **51**, 679–688.
Regueiro, J., 1989, Gas sand reservoirs: low versus high porosity in their seismic perspective: 59th Ann. Internat. Mtg., Soc. Expl. Geophys., Expanded Abstracts, 822–825.
Richter, C. F., 1958, Elementary seismology: San Francisco, Freeman and Co., 670–672.
Ross, C. P., 1992, Incomplete AVO near salt structures: Geophysics, **57**, 543–553.
Samec, P., and Blangy, J. P., 1992, Viscoelastic attenuation, anisotropy, and AVO: Geophysics, **57**, 441–450.
Schneider, W. A., and Goins, N., 1983, General implications of offset-dependent reflectivity and converted waves: 53rd Ann. Internat. Mtg., Soc. Expl. Geophys., Expanded Abstracts, 638.
Sengupta, M. K., 1987, Sensitivity analysis of amplitude versus offset (AVO) method: 57th Ann. Internat. Mtg., Soc. Expl. Geophys., Expanded Abstracts, 621–623.
Shang, Z., McDonald, J. A., and Gardner, G. H. F., 1989, Extraction of amplitude versus incident angle information in the presence of structure: 59th Ann. Internat. Mtg., Soc. Expl. Geophys., Expanded Abstracts, 713–715.
Shang, Z., McDonald, J. A., and Gardner, G. H. F., 1990,

Automated extraction of amplitude versus incident angle information in the presence of structure: 59th Ann. Internat. Mtg., Soc. Expl. Geophys., Expanded Abstracts, 713–715.
Sheriff, R. E., 1975, Factors affecting seismic amplitudes: Geophys. Prosp., **23**, 159–186.
Sherwood, J., 1983, Synthetic seismograms with offset for a layered elastic medium: Offshore Technology Conference, Paper Offshore Tech. Conf., 4508.
Sherwood, J., Hilterman, F. J., Neale, R., and Chen, K. C., 1983, Synthetic seismograms with offset for a layered elastic medium: 53rd Ann. Internat. Mtg., Soc. Expl. Geophys., Expanded Abstracts, 444–447.
Silva, R., and Ahmed, H., 1989, Application of AVO technique in production geophysics., 59th Ann. Internat. Mtg., Soc. Expl. Geophys., Expanded Abstracts, 836–838.
Singh, S. C., West, G. F., and Bregman, N. D., 1989, Full waveform inversion of reflection data: J. Geophys. Res., **94**, 1777–1794.
Snyder, A. G., and Wrolstad, K. H., 1992, Direct detection using AVO, Central Graben, North Sea: Geophysics, **57**, 313–323.
Swan, H. W., 1990, Noise sensitivity of linear seismic inversion: 60th Ann. Internat. Mtg., Soc. Expl. Geophys., Expanded Abstracts, 1177–1180.
Tarantola, A., 1986, A strategy for nonlinear elastic inversion of seismic reflection data: 56th Ann. Internat. Mtg., Soc. Expl. Geophys., Expanded Abstracts, 527–530.
Tatham, R. H., and Stoffa, P. L., 1976, V_p/V_s—A potential hydrocarbon indicator: Geophysics, **41**, 837–849.
Thomsen, L., and Tatham, R., 1983, The change in reflection coefficient with angle of incidence: 53rd Ann. Internat. Mtg., Soc. Expl. Geophys., Expanded Abstracts, 636.
Tooley, R. D., Spencer, T. W., and Sagoci, H. F., 1965, Reflection and transmission of plane compressional waves: Geophysics, **30**, 552–570.
Tura, M. A. C., 1990, Multiparameter elastic inversion: A stable method: 60th Ann. Internat. Mtg., Soc. Expl. Geophys., Expanded Abstracts, 1170–1172.
Ursin, B., 1989, Offset-dependent geometrical spreading in a layered medium: 59th Ann. Internat. Mtg., Soc. Expl. Geophys., Expanded Abstracts, 1156–1159.
Ursin, B., 1990, Least-squares estimation of reflectivity polynomials: 60th Ann. Internat. Mtg., Soc. Expl. Geophys., Expanded Abstracts, 1069–1071.
Vail, P. J., Strauss, P. J., Levitt, P. R., Smith, G. C., and Gidlow, P. M., 1990, Extraction of P- and S-wave velocities from a 3-D reflection data set and its application to direct hydrocarbon detection: 60th Ann. Internat. Mtg., Soc. Expl. Geophys., Expanded Abstracts, 1181–1184.
Wright, J., 1987, The effects of transverse isotropy on reflection amplitude versus offset: Geophysics, **52**, 564–567.
Young, C. Y., Jarrard, R. D., and Kan, T. K., 1983, Prestack synthetic seismogram of elastic-layered earth: 53rd Ann. Internat. Mtg., Soc. Expl. Geophys., Expanded Abstracts, 455–457.
Yu, G., 1985(b), Offset-amplitude variation and controlled amplitude processing: Geophysics, **50**, 2697–2708.

APPENDIX A—MATRIX REPRESENTATION OF THE KNOTT-ZOEPPRITZ EQUATIONS

Aki and Richards (1980) give the Knott-Zoeppritz equations in convenient matrix form. For completeness they are repeated here.

For an interface between two infinite elastic half-spaces, there are sixteen reflection and transmission coefficients (see Figure A-1). In the Aki and Richards notation, the coefficients are represented by two letters (e.g. $\grave{P}\acute{S}$). The first letter indicates the type of incident wave and the second letter represents the type of derived wave. The acute accent (´) indicates an upgoing wave while a downgoing wave has a grave accent (`). Thus, $\grave{P}\acute{S}$ is the downgoing P-wave to upgoing S-wave coefficient. With this notation, the scattering matrix is

$$\mathbf{Q} = \begin{pmatrix} \grave{P}\acute{P} & \grave{S}\acute{P} & \acute{P}\acute{P} & \acute{S}\acute{P} \\ \grave{P}\acute{S} & \grave{S}\acute{S} & \acute{P}\acute{S} & \acute{S}\acute{S} \\ \grave{P}\grave{P} & \grave{S}\grave{P} & \acute{P}\grave{P} & \acute{S}\grave{P} \\ \grave{P}\grave{S} & \grave{S}\grave{S} & \acute{P}\grave{S} & \acute{S}\grave{S} \end{pmatrix} = \mathbf{P}^{-1}\mathbf{R} \qquad \text{(A-1)}$$

where $\mathbf{P}$ is the matrix

$$\begin{pmatrix} -\sin\Theta_1 & -\cos\Phi_1 & \sin\Theta_2 & \cos\Phi_2 \\ \cos\Theta_1 & -\sin\Phi_1 & \cos\Theta_2 & -\sin\Phi_2 \\ 2\rho_1 V_{S1}\sin\Phi_1\cos\Theta_1 & \rho_1 V_{S1}(1-2\sin^2\Phi_1) & 2\rho_2 V_{S2}\sin\Phi_2\cos\Theta_2 & \rho_2 V_{S2}(1-2\sin^2\Phi_2) \\ -\rho_1 V_{P1}(1-2\sin^2\Phi_1) & \rho_1 V_{S1}\sin 2\Phi_1 & \rho_2 V_{P2}(1-2\sin^2\Phi_2) & -\rho_2 V_{S2}\sin 2\Phi_2 \end{pmatrix}$$

and $\mathbf{R}$ is the matrix

$$\begin{pmatrix} \sin\Theta_1 & \cos\Phi_1 & -\sin\Theta_2 & -\cos\Phi_2 \\ \cos\Theta_1 & -\sin\Phi_1 & \cos\Theta_2 & -\sin\Phi_2 \\ 2\rho_1 V_{S1}\sin\Phi_1\cos\Theta_1 & \rho_1 V_{S1}(1-2\sin^2\Phi_1) & 2\rho_2 V_{S2}\sin\Phi_2\cos\Theta_2 & \rho_2 V_{S2}(1-2\sin^2\Phi_2) \\ \rho_1 V_{P1}(1-2\sin^2\Phi_1) & -\rho_1 V_{S1}\sin 2\Phi_1 & -\rho_2 V_{P2}(1-2\sin^2\Phi_2) & \rho_2 V_{S2}\sin 2\Phi_2 \end{pmatrix}$$

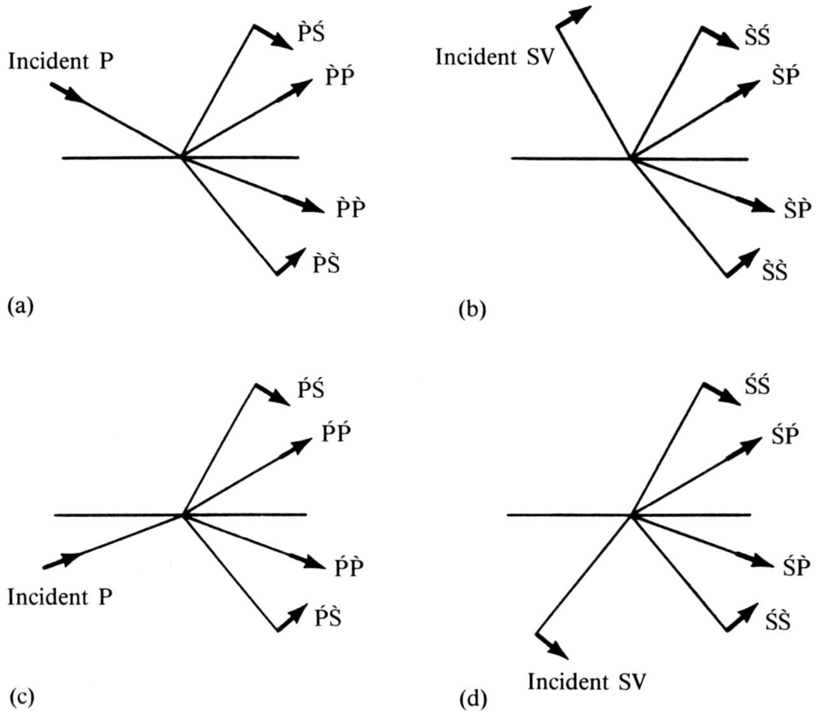

Fig. A-1. Notation for the sixteen possible reflection/transmission coefficients for *P-SV* waves at an interface between two different solid half-spaces. Short arrows show the directions of particle motion (from Aki and Richards, 1980).

Pseudo-Shear—The Analysis of AVO

R. S. Spratt, N. R. Goins‡, and T. J. Fitch§*

INTRODUCTION

In the last 10–15 years, there has been increased interest in the use of shear (S)-wave information to constrain the properties of the subsurface. A number of authors have reported on successful S-wave applications (Backus and Goins, 1984; Domenico, 1984; Yale, 1985). However, in many situations (particularly in hard rocks) both acoustic log and laboratory rock data indicate that V_p and V_s are highly correlated as illustrated in Figure 1. In such cases, only a part of the S-wave data is independent information, and to extract that part requires precise measurement and interpretation of relatively small changes in V_p/V_s.

One situation where V_s behaves quite differently from V_p is at a bright event resulting from hydrocarbon accumulations where the pore fluid in weakly consolidated rocks changes from brine to gas. Figure 2a shows the contrast in P-wave reflectivity between gas and brine saturation for a suite of reservoir sandstones. (Induration is a measure of rock "strength" and is defined as the ratio of the measured frame modulus to the maximum frame modulus allowed by the Gassmann-Biot model.) In the weak rocks on the left, the difference in reflectivity between the brine- and gas-saturated rock can exceed 20 percent. In contrast, changing from brine to gas causes only a small change in V_s, resulting in negligible changes in S-wave reflectivity in weak as well as strong rocks (Figure 2b). Thus, while gas or gas-charged oil accumulations in weak rocks will often stand out on the P-wave section as an amplitude anomaly (a bright event), there should be little if any anomaly on the corresponding S-wave section.

This feature should be useful in distinguishing between amplitude anomalies that correspond to hydrocarbon accumulations and those that result from anomalous lithologies such as salt or coal, because the latter will typically have both anomalous P- and S-wave velocities and so appear as similar amplitude anomalies on both sections. A good seismic demonstration of this is shown in Figure 3. Two bright events are visible on the P-wave section in the boxes at about 0.5 s; drilling results show that the right event corresponds to a gas sand while that on the left is from a coal seam. The corresponding S-wave seismic data at about 1.4 s clearly show this distinction (Ensley, 1985).

Thus, a prime application for S-wave data should be in distinguishing between amplitude anomalies that are indicative of hydrocarbons and those that are not. Unfortunately, many bright spots occur offshore, particularly in the relatively unconsolidated sediments of the Gulf of Mexico, where direct S-wave acquisition is not possible. One solution to this problem is to use amplitude variation with offset (AVO) on conventional P-wave data; a variety of efforts (Hilterman, 1983, personal communication; Ostrander, 1984; Shuey, 1985; Gassaway, 1984; Smith and Gidlow, 1987) have shown that these changes contain S-wave velocity information.

However, other authors have pointed out that extracting this information is difficult. The AVO that carries the shear information is small and difficult to extract from random noise. Worse, many other factors cause AVO and thus act as sources of systematic noise. This paper presents a method to extract shear information in the presence of these noise sources, estimate the expected accuracy, and test the process on both synthetic and field data. Our goal is to formulate a method that extracts the shear information quantitatively and to obtain a shear reflectivity seismic section that is a (noisy) estimate of what would have been recorded using shear vibrators and geophones in the field. We term this result pseudo-shear to distinguish it from shear seismic recorded in the field. If successful, this approach would provide a major exploration benefit at little cost since the information is already contained in standard P-wave common depth point (CDP) gathers.

*Diasonics Inc., Milpitas, CA.
‡Mobil Oil Corporation, Fairfax, VA.
§Mobil Exploration and Production, Dallas, TX.

The first section briefly reviews the analysis for an isolated reflector. The following section deals primarily with other factors that influence AVO. An analysis of how to improve the pseudo-shear results by using statistical methods and other considerations follows. The paper concludes with an example from tests on field data.

ANALYSIS FOR AN ISOLATED REFLECTOR

When a plane seismic P-wave encounters a reflector at normal incidence, the amplitude of the reflection is given by the standard expression

$$R(\theta = 0) = \frac{\rho_2 V_{p2} - \rho_1 V_{p1}}{\rho_2 V_{p2} + \rho_1 V_{p1}} \quad (1)$$

where ρ_1 and V_{p1} are the density and P-wave velocity above the interface and ρ_2 and V_{p2} are the values

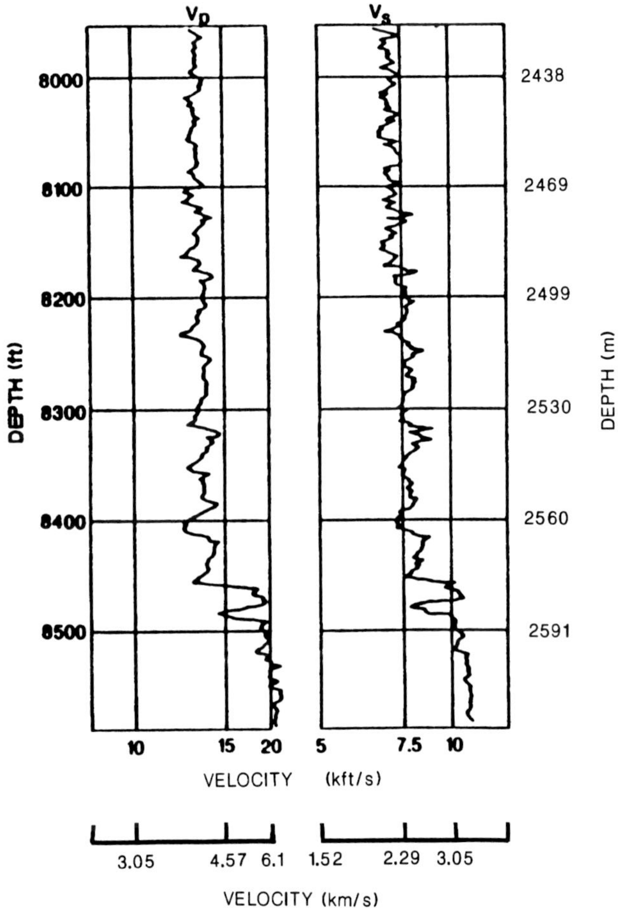

Fig. 1. An example of closely correlated west Texas P- and S-wave log velocities.

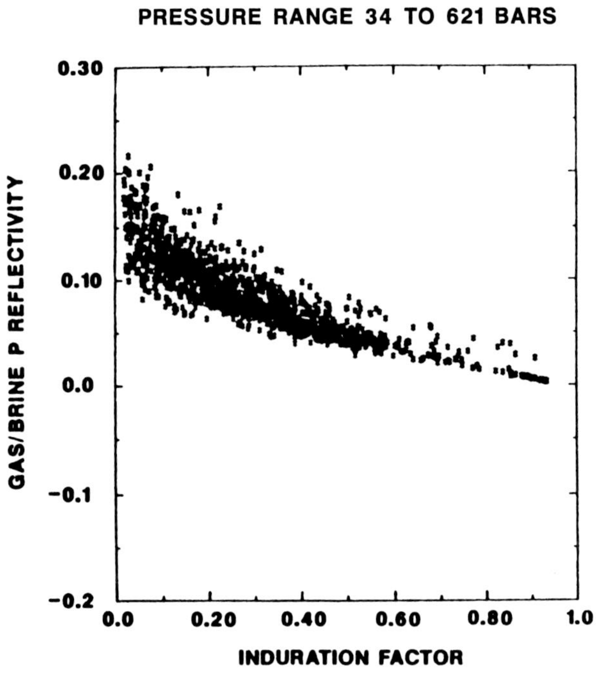

Fig. 2a. Laboratory measurements of P-wave reflectivity at a gas-brine interface in sandstones.

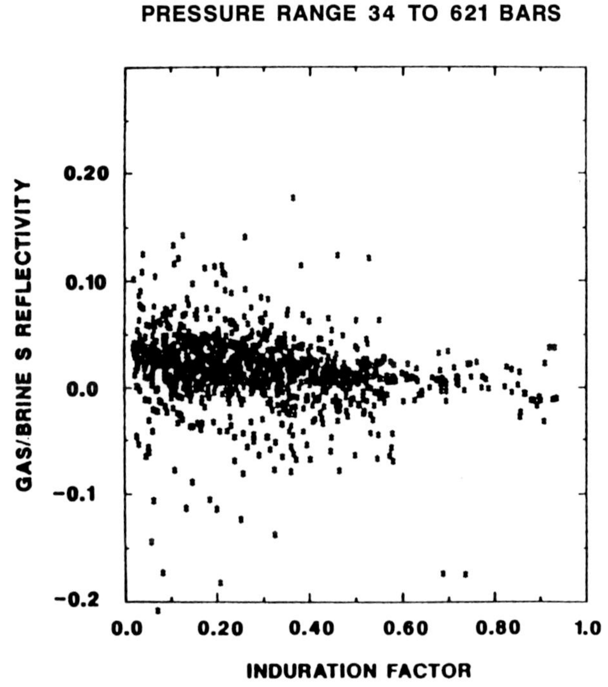

Fig. 2b. Laboratory measurements of S-wave reflectivity at a gas-brine interface in sandstones.

below. However, when the P-wave arrives at an angle to the reflector, as in the case of nonzero source-receiver offset, a shearing force is applied to the boundary and part of the P-wave energy is converted to shear energy as illustrated in Figure 4. This mode conversion removes energy from the P-wave reflection amplitude; the amount removed depends in part on the S-wave velocities on either side of the reflector and the incident angle. Thus, the P-wave reflection amplitude at nonzero offset depends additionally on the angle of incidence and the S-wave velocities V_{s1} and V_{s2}.

The principle of pseudo-shear is to use this relationship to extract shear information from the AVO. Amplitude variation with angle of incidence (AVA) θ and with material ρ_1, ρ_2, V_{p1}, V_{p2}, V_{s1}, and V_{s2} is given by the Zoeppritz equations. Figure 5 shows three theoretical curves from these expressions. Each of the curves has the same P velocity and density contrast, so that they all have the same normal-incidence ($\theta = 0$) reflection amplitude of -0.1. At large angles of incidence (but less than grazing incidence), the amplitude is controlled only by the P-wave velocity contrast and so again the three curves converge. Only the S-wave velocity contrasts differ among the curves, and this causes the separation at intermediate angles. It is this separation that allows us to use amplitude-offset data from P-wave common depth point (CDP) gathers to obtain S-wave information.

The Zoeppritz equations are highly nonlinear with respect to the velocities and densities. More importantly, the presence of noise and other amplitude effects makes determining which if any of these parameters can be uniquely found from a measured AVO difficult. To simplify the inversion problem and reduce the number of parameters, we used an expansion of Zoeppritz equations in even powers of sin θ and retained only terms that included first order changes in

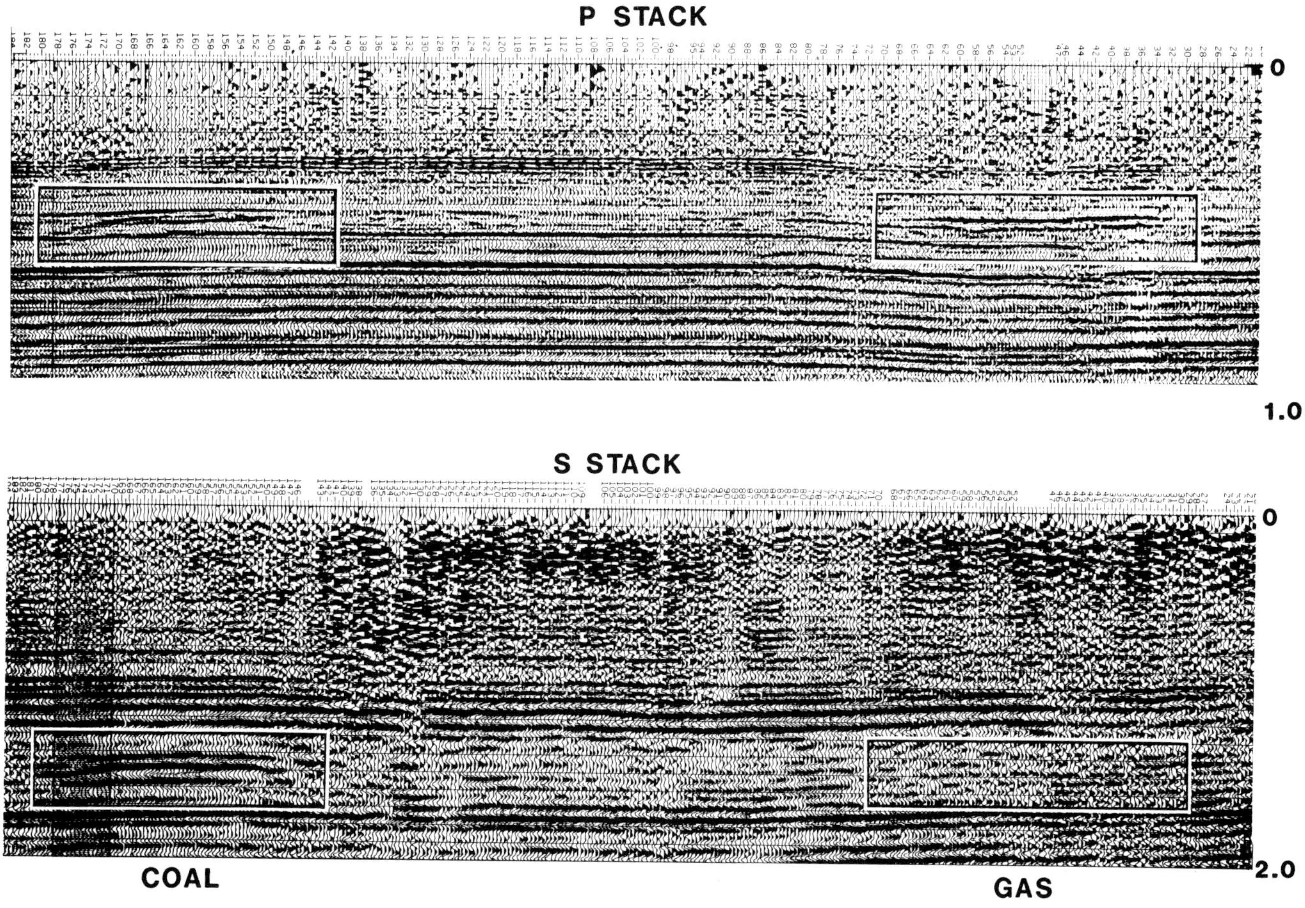

Fig. 3. The Alberta Mynam Line as an illustration of how contrasting P- and S-wave reflectivity can discriminate between a gas reservoir and an anomalous lithologic unit.

the velocities and densities (for details see Appendix A). The simplified expression for the case of a P-to-P reflection can be found in Aki and Richards (1980) as

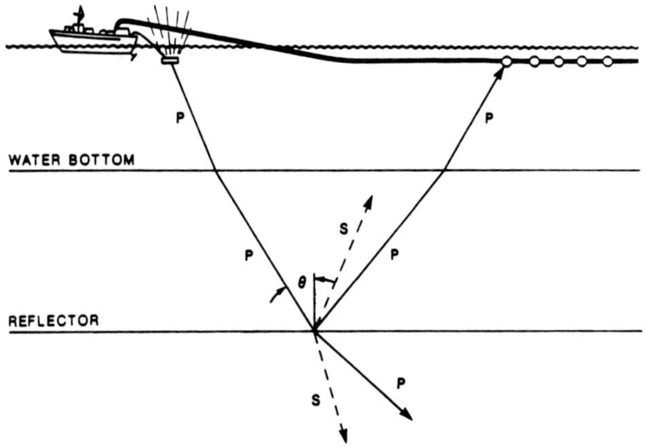

Fig. 4. Marine acquisition and P-wave scattering at a reflecting boundary.

$$R(\theta) \approx \frac{1}{2}\left[1 - 4\left(\frac{V_s}{V_p}\right)^2 \sin^2\theta\right]\frac{\Delta\rho}{\rho} + \frac{1}{2\cos^2\theta} \times \frac{\Delta V_p}{V_p} - 4\left(\frac{V_s}{V_p}\right)^2 \frac{\Delta V_s}{V_s}\sin^2\theta \qquad (2)$$

where $R(\theta)$ is the reflection amplitude, $\Delta\rho/\rho$, $\Delta V_p/V_p$, and $\Delta V_s/V_s$ are the fractional changes in density and velocity across the reflector, and V_s/V_p is the local average ratio. Earlier work (Bortfeld, 1961) contains a similar first order approximation. If we additionally assume that the incident angle θ is relatively small and the V_p/V_s approximation is equal to 2, the expression further simplifies to:

$$R(\theta) \approx R_p + (R_p - 2R_s)\sin^2\theta \qquad (3)$$

where, to first order in the Δ's, R_p is the normal-incident P-wave reflectivity given by equation (1) and R_s is the corresponding normal-incidence shear reflectivity. R_s is the pseudo-shear result referred to in the

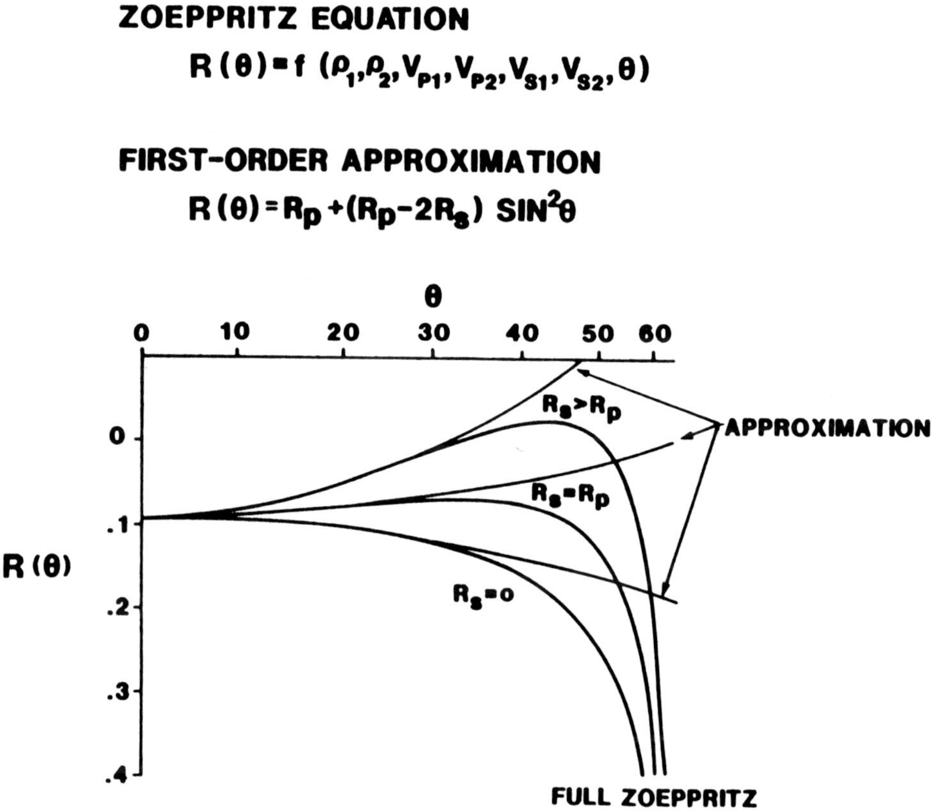

Fig. 5. An example of the range of P-wave reflectivity predicted by the Zoeppritz equation, $R(\theta) = f(\rho_1, \rho_2, V_{p1}, V_{p2}, V_{s1}, V_{s2}, \theta)$. The curve for an interface with no change in S-wave reflectivity (S-wave reflection coefficient is zero) corresponds to an interface between a brine and gas saturated sandstone.

introduction and like R_p can be displayed as a seismic section in which case R_s is referred to as the pseudo-shear section. If V_p/V_s is not two, then there is a small additional term involving $\Delta\rho/\rho$. (In the pseudo-shear algorithm the smoothed average V_p/V_s ratio is a time-depth dependent input variable and the density term is handled by using a statistical relationship between density and velocity described in Appendix A.)

Equation (3) shows that the reflection amplitude on the near-offset traces ($\theta = 0$) primarily represents the normal incidence P-wave reflectivity R_p. The AVO is controlled by a linear combination of R_p and the normal-incidence shear reflectivity R_s. Thus, the new information in the offset behavior is the shear reflectivity. The same information is what we attempt to measure by recording field S-wave data with horizontal vibrators and geophones, except that R_s in equation (3) refers to S-waves polarized in the plane of incidence, S_v, while typically field shear records S-waves polarized transverse to that plane, S_h. In the absence of anisotropy, these measurements are identical. As a matter of interest, in Appendix A we also include a table showing the angular dependence of all reflection coefficients which shows that only the P-to-P reflection amplitude contains both R_p and R_s information.

Figure 5 shows that the simplified theory closely reproduces the results from the full Zoeppritz equations at angles of incidence less than 20 to 30 degrees depending on the size of the velocity/density changes. At worst the percentage error in R_s, due to the small reflectivity approximation, is roughly equal to the magnitude of R_s. For example, an R_s of .25 should be in error by about 25 percent or .06. (A higher order expansion indicates that the actual error is much smaller when R_p and R_s are about equal—see the last section of Appendix D.) A complete error analysis described in Appendix B also shows that including higher-order angular terms actually produces poorer pseudo-shear estimates than the two term expression in equation (3). Furthermore, optimal results are obtained by using incident angles out to about 20–25 degrees. Including traces from larger angles increases the systematic error due to failure of the small angle assumptions, whereas muting more closely increases the random noise in the pseudo-shear estimates by decreasing the amount of data used. Fortunately, angles of 20–25 degrees correspond typically to offsets equal to about the target depth, and are recorded on standard seismic data. Thus, no special recording effort is required to obtain P-wave data suitable for pseudo-shear processing. However, the results in Appendix B clearly show that pseudo-shear, in common with most prestack processes, requires good signal-to-noise ratios (S/N) on the individual prestack traces. In general, the S/N on a pseudo-shear trace estimated from 50–100 fold P-wave data will be about equal to the average S/N of a single P-wave trace. Thus, a quick estimate of the expected pseudo-shear section quality can be obtained by looking at a single-fold P-wave section

APPLICATION TO SEISMIC DATA

Real seismic data does not result from P-waves imaging an isolated reflector separating two constant velocity media. A more realistic model must be developed. The basic model for our pseudo-shear process consists of a smoothly varying background velocity/density trend, upon which are superimposed small fractional changes in density and velocity which cause reflections. In equation (3), the smoothly varying trend is represented by the local average V_p/V_s, while R_p and R_s are the fractional change terms. We assume a P-wave stacking velocity function is available. A generalization of the Dix method, described in Appendix C, is used to convert this function to an interval velocity profile. Ray tracing can then be applied to calculate angle of incidence as a function of time/depth and offset, and mute curves designed to eliminate data beyond 20–25 degrees as discussed previously. Dip effects are presupposed to be negligible.

However, many factors beside shear reflectivity and the reflection coefficient effect described by equation (3) cause reflection AVO. In Appendix D we derive the effects of surface boundary conditions, spherical divergence, array length, source and receiver directivity, and inelastic attenuation. A correction for the omitted higher order angular terms is also derived using results from the next section. In all cases we assume that angular dependence at a given depth can be represented in the form $1 + g_{i*} \sin^2 \theta$ so that terms of order $\sin^4 \theta$ and above are ignored. Then the reflection AVA is given by

$$R(\theta) = [R_p + (R_p - 2R_s) \sin^2 \theta] \times \prod_i (1 + g_i \sin^2 \theta) \tag{4}$$

or

$$R(\theta) \approx R_p + (GR_p - 2R_s) \sin^2 \theta \tag{5}$$

where $G = 1 + \sum_i g_i$.

Note that all effects are normalized to one at zero offset because any scaling that is applied to all traces equally, even if time dependent, has no effect on the pseudoshear analysis.

Thus, the net result of all these other effects is to add the factor G in equation (5). This factor can be quite significant. For example, consider just directivity. If the data are recorded with deep explosive sources and omnidirectional hydrophones, G is 1. For a typical bright event, such as that in Figure 3, R_p is large and positive while R_s is near zero. For $G = 1$, the amplitude variation with angle is thus given by

$$R(\theta) = R_p(1 + \sin^2 \theta) \tag{6}$$

so that the amplitude increases with offset. However, if the data are recorded on land with a vertical vibrator and geophones, then both source and receiver have g_i of $-1/2$ and $G = 0$. In this case the amplitude-offset behavior would be given by $R(\theta) = R_p$, so that the very same bright event would show no AVO. Thus, these effects must be considered if valid S-wave information is to be extracted, and indeed if amplitude-offset information is to be interpreted reliably.

Once the function G is calculated by use of Appendix D, a preliminary shear-reflectivity trace can be extracted from the P-wave data. For each time point of a muted, normal moveout (NMO) corrected, CDP gather with offset converted to incident angle, the amplitude variation across the gather along a time line should vary:

$$R(\theta) = P + Q \sin^2 \theta \tag{7}$$

where $P = R_p$ and $Q = GR_p - 2R_s$.

P and Q can be determined by fitting equation (7) to the amplitude variation using a least-squares criterion. Then the normal-incidence shear reflectivity is simply

$$R_s = -\frac{1}{2}(Q - GP) \tag{8}$$

and the normal-incidence P reflectivity (uncontaminated by AVO effects that are present in the stacked section) is $R_p = P$.

Thus, we can produce two sections. The R_s section (the pseudo-shear section) is an estimate of what would have been recorded in a S-wave field experiment, except that this section will have events at the same time as those on the P-wave section. Consequently, the R_P and R_s sections will be automatically aligned in time. This alignment avoids the need for correlating P and S events which is a major obstacle in exploiting field S-wave data. Additionally, the pseudo-shear process will extract a section showing R_p. The extracted section will differ from the standard stacked section because the amplitudes will be uncontaminated by variations with offset. Assuming both sections receive true amplitude processing, the R_p section should be a better estimate of coincident source-receiver P-wave reflectivity than the conventional stacked section. This approach to the analysis of P-wave AVO is disclosed in Wiggins et al. (1985).

OTHER CONSIDERATIONS

Experience has shown that several potential difficulties in the straightforward estimation procedure of equation (8) must be addressed. Probably the most serious problem is that the estimate of G can easily be incorrect. Some effects are difficult to compute as discussed in Appendix E and therefore are not included in G, and undoubtedly there are others that we have overlooked entirely. Fortunately, there is another way to estimate G and evaluate the applicability of pseudo-shear to a particular dataset. Specifically, equation (7) shows that values of P and Q are directly related to R_p and R_s by G. By taking averages over time (depth) and assuming G is constant over the averaging interval we can obtain the relationship

$$\frac{\langle PQ \rangle}{\langle P^2 \rangle} = G - 2\frac{\langle R_p R_s \rangle}{\langle R_p^2 \rangle} \tag{9}$$

where $\langle\ \rangle$ represents averaging. Thus, G can be estimated using the P and Q values determined from the CDP gathers if the statistical value $\langle R_p R_s \rangle / \langle R_p^2 \rangle$ is known. In practice care must be taken in calculating $\langle PQ \rangle$. Since P and Q are determined from the same dataset, the simple average of their product is not an unbiased estimate of $\langle PQ \rangle$ in the presence of random noise. The details of this problem and the solution coded in the pseudo-shear program are given in Appendix F. This analysis also leads to a method for including a g_i factor that partially corrects for the omission of higher order angular terms, as detailed in Appendix D.

In principle, the quantity $\langle R_p R_s \rangle / \langle R_p^2 \rangle$ could be determined from regional well logs if both P- and S-wave logs are available. For the purposes of this paper, we used laboratory measurements on a worldwide suite of reservoir sandstones. Computer experiments calculating reflection coefficients from pairs of combinations of rocks in this database consistently gave values of $\langle R_p R_s \rangle / \langle R_p^2 \rangle$ near unity. This result agrees within the error range with the few available log results.

Thus, there are two philosophically different approaches to estimating the effects of other factors on the AVO, and two possible pseudo-shear estimates. In the deterministic method, we assume that all factors can be identified and their effects calculated. For the statistical approach, we assume that $\langle R_p R_s \rangle / \langle R_p^2 \rangle$ is known. In the pseudo-shear method, we do both. We first calculate a deterministic G factor G_{det} from Appendix D. A δG term is then calculated from the statistical information using

$$\delta G = \frac{\langle PQ \rangle}{\langle P^2 \rangle} + 2\,\frac{\langle R_p R_s \rangle}{\langle R_p^2 \rangle} - G_{\mathrm{det}}. \qquad (10)$$

We assume that δG is small and varies smoothly with depth. Either a constant or linearly varying in time (depth) value is fit to the right hand side of equation (10) to get an estimate of δG for a CDP gather. If not expected to vary along the line, δG can be spatially averaged.

The value δG is a measure of the discrepancy between the deterministic calculation and the comparison between the reflectivity statistics $\langle R_p R_s \rangle / \langle R_p^2 \rangle$ and the measured statistics $\langle PQ \rangle / \langle P^2 \rangle$. A value near zero indicates that the two methods agree closely. In initial tests, we found δG values ranging from -1 to -5. (To put these δG values in perspective, an amplitude decay of unknown origin that has the form $\cos^2(\theta)$ would yield a δG of -1 provided this decay is pervasive in the time windows that include the reflections of interest.) It is inconceivable that the large negative values of δG we computed are caused by an incorrect value for the reflectivity statistic $\langle R_p R_s \rangle / \langle R_p^2 \rangle$; if so, the true P and S reflectivities would have to range from uncorrelated to highly anticorrelated. Indeed, most physical processes tend to attenuate amplitudes with offset, so we could expect δG in general to be somewhat negative indicating some ignored effects. Values as large as -5 indicate that some mechanism is destroying the amplitude of the far offsets or greatly exaggerating the amplitude of the near offsets. In either case the pseudo-shear approach is not applicable, but there is little S-wave information available in such data anyway. Values around -1 are reasonable and suggest that the pseudo-shear theory is appropriate for the data set. Ongoing work on a variety of seismic data has shown that usually δG is acceptably small, suggesting that the theory developed here is adequate.

Another major problem is the sensitivity of the pseudo-shear estimate to small errors in NMO velocity (Spratt, 1987). The pseudo-shear estimate is derived by measuring amplitude behavior of the P-wave data across timing lines in an NMO-corrected CDP gather. If the data are not moved out precisely, an error is introduced that has the same form as the derivative of the estimated P-wave trace, and an amplitude that is proportional to the error in moveout velocity. The size of this error can be estimated by fitting the pseudo-shear wave trace to the P-wave trace derivative. When the velocity error is small, the errors can be subtracted directly from the pseudo-shear trace. A better approach is to use the coefficient of the fit to determine a residual normal moveout correction (RNMO) and move out the CDP data again which can be done iteratively as illustrated in Figure 6. In practice we have found that five iterations are sufficient to remove up to 20 ms of residual moveout in typical marine data sampled every 4 ms. (Of course, although developed to aid the pseudo-shear analysis, this process is also a high-resolution method for RNMO correction.) A more subtle result discussed in Spratt (1987) is that any part of the true shear reflectivity sequence that has the same form as the derivative of the P reflectivity cannot be distinguished from residual NMO and, therefore, cannot be extracted using the pseudo-shear method. In practice this situation probably occurs infrequently and is not a significant limitation.

Finally, we consider the combination of differential tuning and NMO. Differential tuning results because reflections tend to pinch together with increasing offset. The NMO process as a correction stretches the traces. The net effect on pseudo-shear is to add another g_i factor and to distort the pseudo-shear traces at frequencies on the edges of the P-wave wavelet bandwidth. This distortion can be eliminated by filtering the pseudo-shear section so that the effective wavelet has a somewhat narrower bandwidth than that on the CDP gathers. In practice, residual static errors will cause a greater loss in high-frequency content than this effect.

ACQUISITION AND PREPROCESSING RECOMMENDATIONS

When recording P-wave data that will be processed for pseudo-shear information, high fold, long offsets, and short arrays are desirable. In practice, we are limited by the number of available channels. A good compromise is to fix the maximum offset so that the angle of incidence is 20–25 degrees for the deepest events of interest. This angle typically corresponds to a maximum offset equal to the depth, and guarantees that all events of interest will have the full range of offsets used in the pseudo-shear calculation. The minimum offset should be the nearest point that is not overwhelmed by shot noise. The number of channels then determines the trace spacing. Long receiver arrays will attenuate the dipping arrivals of shallow events at far offsets, and so the shortest practical arrays should be used.

Pseudo-shear is obviously sensitive to amplitude changes introduced by processing. The only amplitude scaling that should be applied is a divergence and/or anelastic attenuation correction that is the same for all traces. In addition, surface consistent amplitude balancing may be necessary to remove amplitude variations caused by changes in shot strength, receiver

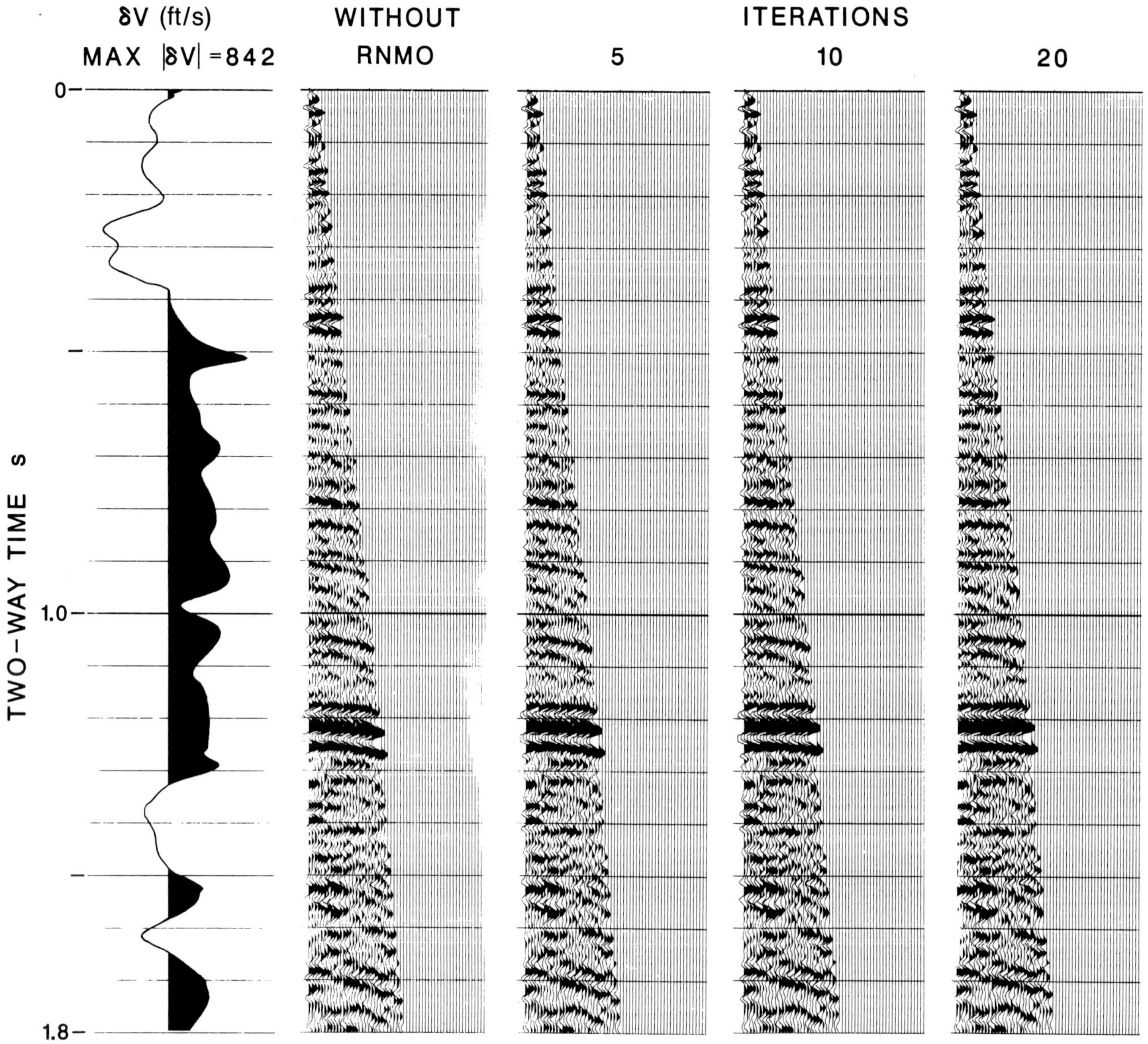

Fig. 6. The effect of residual normal moveout (RNMO) on a marine gather with the cumulative adjustments to the stacking velocities. The mute corresponds to an angle of incidence of 24 degrees. In this particular example, we have applied RNMO to a poorly moved out gather to illustrate the stability of the process through 20 iterations.

sensitivity, ground coupling, and near-surface attenuation (Taner and Koehler, 1981) and should be applied to land data, as well as to any data where shot energy varies along the line.

We recommend other processing steps as follows. Since pseudo-shear does not enjoy the noise rejection capability of stacking, removal of all noisy traces is vital. Editing should be ruthless. Spike deconvolution with minimum bandpass filtering should be applied before NMO to ensure that maximum possible pseudo-shear bandwidth is obtained. To better preserve amplitude information, the deconvolution operator should be designed for an entire shot or receiver gather as in the case of surface consistent operators. Additional filtering can be applied directly to the pseudo-shear section as necessary. Aggressive steps should be taken to remove multiples because they can bias or even obscure the AVO from a primary reflection. Pseudo-shear is also very sensitive to trace misalignment caused by statics. All appropriate statics algorithms should be applied, including residual statics. Additionally, since the structure can be mapped independently from standard stacked sections, nonsurface-consistent or even time-varying statics algorithms can be applied if the events cannot be aligned with more conventional techniques. If *f-k* filters are used to supplement short ground arrays, the reject band should be as narrow as possible so as to preserve the reflection amplitude at far offsets.

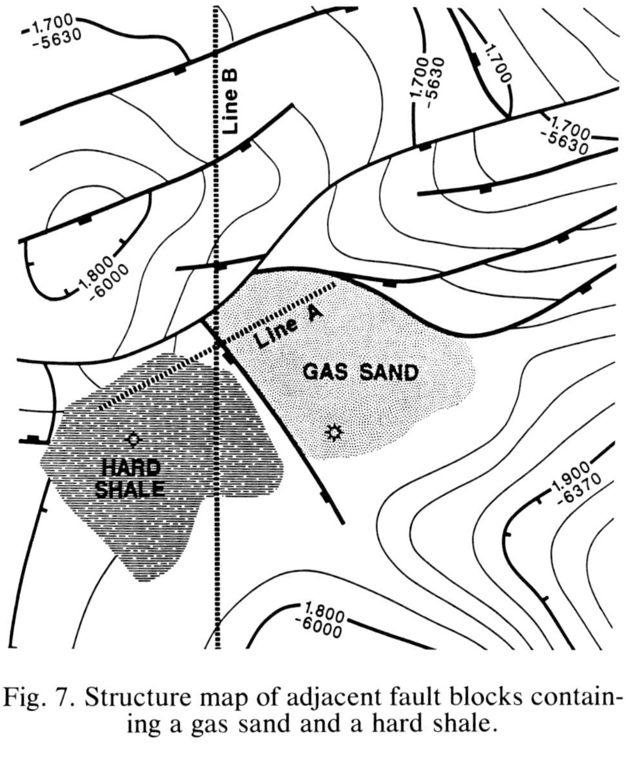

Fig. 7. Structure map of adjacent fault blocks containing a gas sand and a hard shale.

APPLICATION TO REAL DATA

As a test case we chose adjacent bright events located in the Texas offshore; one event is a true hydrocarbon indicator and the other is a false indicator. Figure 7 shows a map view of two seismic lines crossing these anomalous bodies, which are at about 1.8 s (gas sand) and 1.9–2.0 s (shale). Well logs (not shown) indicate a −15 percent reflection coefficient for the gas sand and +9 percent value for the shale. The insets below the events of interest in Figures 8 and 9 show that both events are relatively bright when the data are processed to preserve relative amplitudes.

The pseudo-shear results are shown on the right sides of Figures 8 and 9. We have displayed them in the form $R_p(R_p - R_s)$, that is by multiplying the vertical-incidence section R_p by the difference between R_p and R_s. The purpose is to emphasize events that are both bright (large R_p) and with S-wave reflectivity much less than P-reflectivity (large $R_p - R_s$). This display will attenuate both weak events (small R_p) and normal lithologic events that have similar R_p and R_s. The results clearly show that the hydrocarbon event is a dominant event while the bright event from the shale is dimmed to a background level. In this case, pseudo-shear analysis would have helped make the correct drilling decision.

Note that the difference $(R_p - R_s)$ is a useful measure of the contrast between R_p and R_s because the reflectivities are perfectly aligned which would not be the case if the P- and S-wave information came from separate experiments.

CONCLUSIONS

S-wave information is encoded onto standard P-wave reflection data in the form of AVO. Pseudo-shear is a technique that tries to extract this information by processing. The result is a noisy estimate of the S-wave reflection section. The pseudo-shear section approximates the shear section that would be recorded by reshooting the P-wave line with S-wave vibrators and geophones, except that it is produced at the P-wave time scale with a band-limited version of the P wavelet. Thus, the correlation problem associated with field shear data is avoided. The necessary low-frequency information is input explicitly in the form of stacking velocities and a smoothed V_p/V_s. The output is in common seismic section form and illuminates the

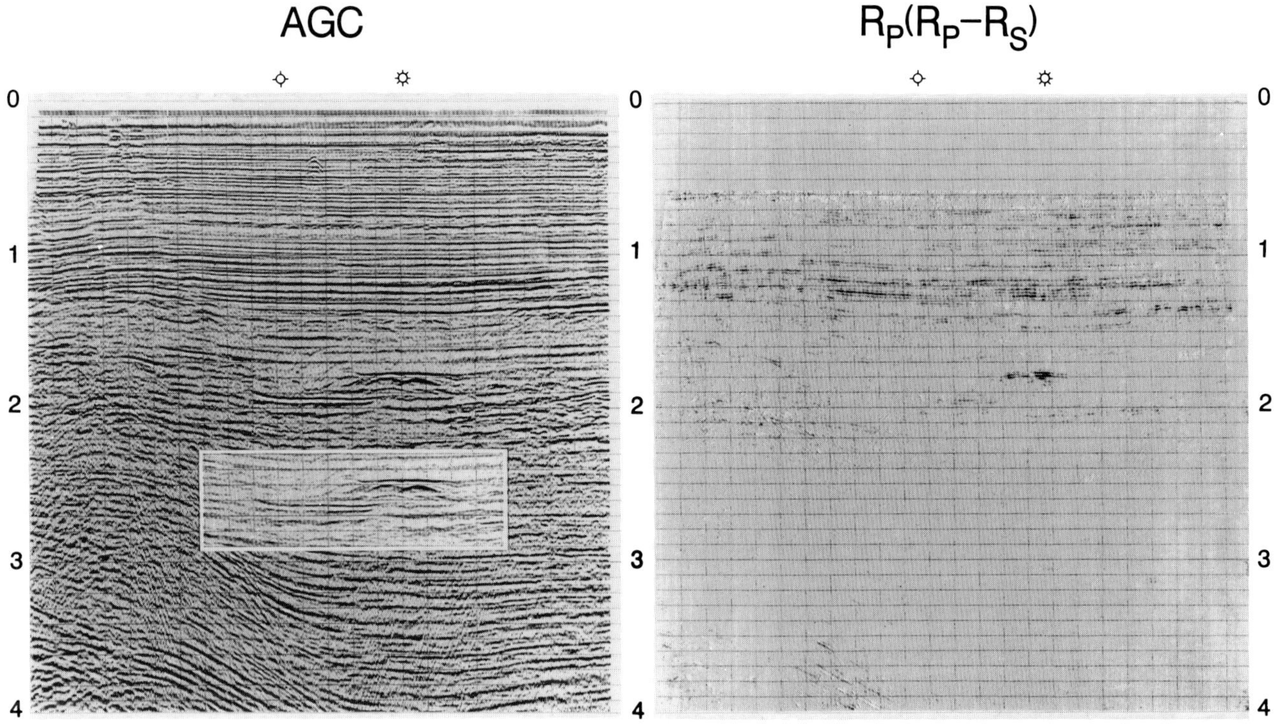

Fig. 8. Line A: Stacked seismic section with inset containing "true amplitude" stack and pseudo-shear display section.

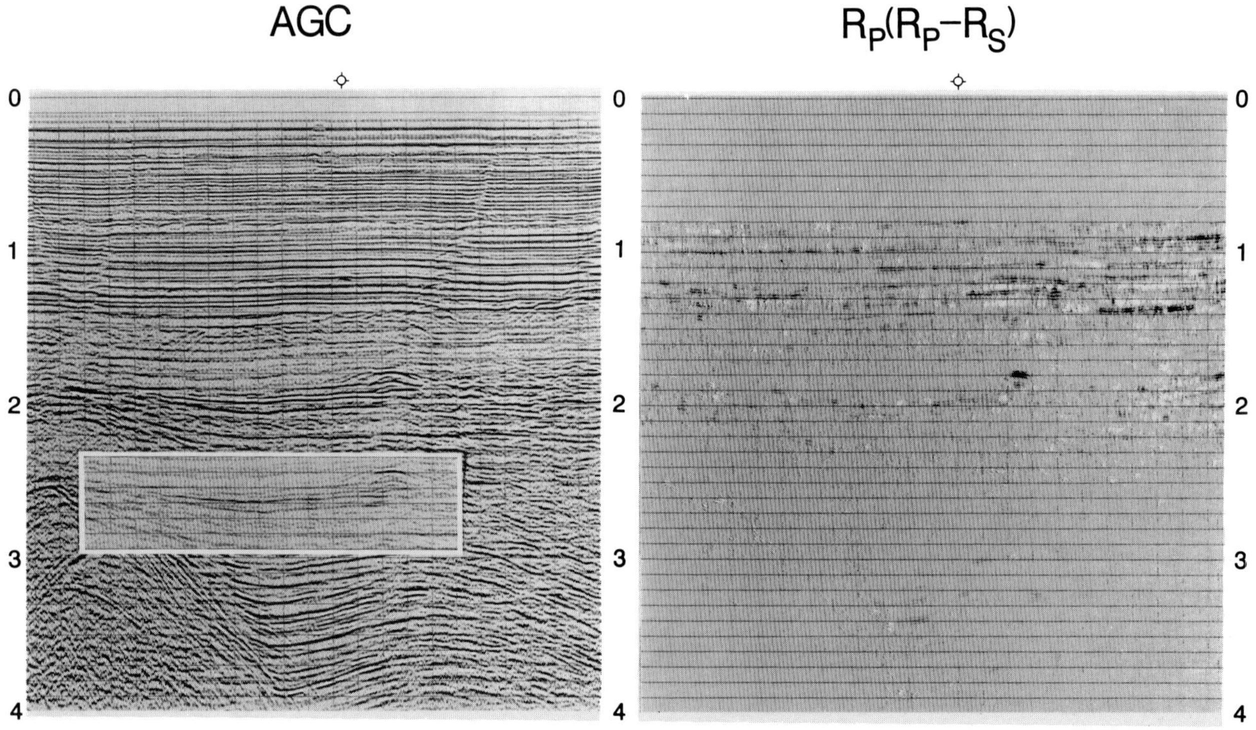

Fig. 9. Line B: Stacked seismic section with inset containing "true amplitude" stack and pseudo-shear display section.

true noisy and band-limited nature of the available information.

Pseudo-shear information is admittedly difficult to extract. To achieve a unique solution the Zoeppritz equations have been reduced to a form that will fit AVOs that display no more complicated pattern than an intercept and a trend. There are many spurious factors that cause AVO. The need to estimate move-out velocities from the same data makes sacrifice of some of the S-wave information necessary (Spratt, 1987). A number of different effects can bias the results in the presence of noise.

However, nature has not thrown a shut-out, particularly in the case of discriminating between true (hydrocarbon) and false (lithologic) bright events. These events generally occur in poorly consolidated regions that, except for them, have relatively small reflection contrasts. This suggests that the small reflectivity assumption will hold and that transmission losses, which we have ignored, will not be large. The dramatic difference between the shear response of a true bright event (not much) and a false one (same as the P response) should be detectable over the approximation errors and noise in the pseudo-shear information. Furthermore, the information can be obtained relatively quickly at little cost.

ACKNOWLEDGMENTS

The analysis of AVO reported here is a further development of work completed under the direction of R. W. Wiggins in the early 1980s at Mobil's Research Lab in Dallas.

APPENDIX A—FIRST ORDER APPROXIMATIONS TO THE ZOEPPRITZ EQUATIONS

In Appendix A we derive the first order reflectivity expressions from the Zoeppritz equations. The Zoeppritz equation governing the reflection amplitude of plane P-waves at the interface between two thick homogeneous, elastic, isotropic layers depends on two densities, two P-wave velocities, two S-wave velocities and one angle (scale invariance reduces the number of independent parameters by two). A much simpler expression arises if we assume that the fractional changes in material parameters are small across the interface. How small is small in general depends on the purpose of the analysis and whether or not the analysis is linear or nonlinear. Pseudo-shear is a linear process designed to qualitatively evaluate a "bright" event for the presence or absence of gas. Thus, larger errors in the approximations can be ignored than would be the case if a quantitative result were desired. Linearity means that approximation errors are not compounded as would be the case if an iterative nonlinear solution were sought.

Specifically with regard to "bright" events P-wave reflectivities well away from critical points are expected to be on the order of 10 percent. At normal incidence R_p or R_s is represented by the series

$$\frac{\Delta V}{2V}\left(1 - \frac{\Delta V}{2V} + \frac{\Delta V^2}{(2V)^2}\right)$$

assuming density contrasts can be ignored. If for example the fractional change in compressional velocity, $\Delta V_p/V_p$, across a gas sand/shale interface were 10 percent, then the error in the small change approximation to R_p would be 1 percent which is acceptably small for the purposes of pseudo-shear.

The following small change expansion for P-wave reflectivity is given in Aki and Richards (1980)

$$R(\theta) = \frac{1}{2}\left\{1 - 4\left(\frac{V_s}{V_p}\right)^2 \sin^2\theta\right\}\frac{\Delta\rho}{\rho} + \frac{1}{2\cos^2\theta}\frac{\Delta V_p}{V_p} - 4\left(\frac{V_s}{V_p}\right)^2\frac{\Delta V_s}{V_s}\sin^2\theta$$

where Δ represents the change in density ρ, compressional velocity V_p, and shear velocity V_s, and θ is the incident angle.

If we further assume that the incident angle is small and retain only terms to first order in $\sin^2\theta$ the equation becomes

$$R(\theta) = \frac{1}{2}\left[1 - 4\left(\frac{V_s}{V_p}\right)^2 \sin^2\theta\right]\frac{\Delta\rho}{\rho} + \frac{1}{2}(1 + \sin^2\theta)\frac{\Delta V_p}{V_p} - 4\left(\frac{V_s}{V_p}\right)^2\frac{\Delta V_s}{V_s}\sin^2\theta.$$

Rearranging the terms gives

$$R(\theta) = \frac{1}{2}\left(\frac{\Delta\rho}{\rho} + \frac{\Delta V_p}{V_p}\right) + \left[\frac{1}{2}\left(\frac{\Delta\rho}{\rho} + \frac{\Delta V_p}{V_p}\right) - 8\left(\frac{V_s}{V_p}\right)^2\frac{1}{2}\left(\frac{\Delta\rho}{\rho} + \frac{\Delta V_s}{V_s}\right)\right]\sin^2\theta + \left[2\left(\frac{V_s}{V_p}\right)^2 - \frac{1}{2}\right]\frac{\Delta\rho}{\rho}\sin^2\theta$$

or

$$R(\theta) = R_p + (R_p - 2^*R_s)\sin^2\theta + 0^*\frac{\Delta\rho}{\rho}\sin^2\theta$$

where

$$R_p = \frac{1}{2}\left(\frac{\Delta\rho}{\rho} + \frac{\Delta V_p}{V_p}\right),\ R_s = \frac{1}{2}\left(\frac{\Delta\rho}{\rho} + \frac{\Delta V_s}{V_s}\right),$$

$$2^* = 8\left(\frac{V_s}{V_p}\right)^2,\ 0^* = 2\left(\frac{V_s}{V_p}\right)^2 - \frac{1}{2}.$$

In the preceeding expression R_p and R_s are just the P and shear reflectivities, respectively, (correct to first order in the Δ's). Notice that when $V_p/V_s = 2$ then $2^* = 2$ and $0^* = 0$, and the expression reduces to the simple formula (equation 3) given in the text.

The pseudo-shear analysis allows a time (depth) dependent V_p/V_s to be input and uses the full expression for 2^*. The extra density term can be handled in the following manner. Rock property measurements show that the density and P-wave velocity of rocks are highly correlated. To a good approximation they are related by the following empirical formula

$\rho = \gamma V_p^{.25}$ where γ is a scale factor that depends on the units of ρ and V_p.

Differentiating and dividing gives

$$\frac{\Delta\rho}{\rho} = .25\frac{\Delta V_p}{V_p}.$$

Therefore, this effect has the same generic form gR_p as those described in Appendix D, where

$$g = .25\left[2\left(\frac{V_s}{V_p}\right)^2 - \frac{1}{2}\right].$$

For V_p/V_s's ranging from 1.5 to 3, the value of g ranges from 0.1 to −0.07 which is much smaller than most known effects.

It is also of interest to apply the small change and small angle approximation to the Zoeppritz equations and determine the approximate behavior of other reflection coefficients. Again assuming $V_p/V_s = 2$, we find the reflection coefficient approximations given in Table A-1.

In Table A-1 S_v and S_h indicate source or receiver orientation along and perpendicular to the source-receiver line. Note that to first order P-S_v converted waves contain the same shear reflectivity information R_s as standard recording using S_h vibrators and receivers (assuming isotropy).

APPENDIX B—ERROR ANALYSIS

Appendix B estimates the random and systematic errors in the pseudo-shear estimate. The amplitude of a recorded reflection coefficient can be represented by

$$R_{\text{rec}}(\theta) = R_A + R_B\sin^2\theta + R_C\sin^4\theta + \cdots.$$

We wish to get an estimate of the error in the least-squares determination of the coefficients in the above expansion. The estimate of each coefficient is a weighted sum of the measurements. For example, R'_B (the estimate of R_B) is calculated by

$$R'_B = \sum W_\theta R_{\text{rec}}(\theta)$$

Table A-1. Small change and small angle reflections. Coefficient approximation for P, S_v, and S_h waves.

	Receiver		
Source	P	S_v	S_h
P	$R_p + (R_p - 2R_s)\sin^2\theta$	$-2R_s\sin\theta$	—
S_v	$-2R_s\sin\theta$	$-R_s + \left[7R_s + \frac{1}{2}\left(\frac{\Delta\rho}{\rho}\right)\right]\sin^2\theta$	—
S_h	—		$-R_s + \frac{1}{2}\left(\frac{\Delta V_s}{V_s}\right)\sin^2\theta$

where the weights W_θ are determined by the least-squares criterion. The rms error in R'_B due to random errors in the data is given simply by

$$\sigma_{\text{stack}} = \sigma_{\text{data}}\left(\sum W_\theta^2\right)^{1/2}$$

where σ_{data} is the rms level of the uncorrelated, stationary noise in the data. These estimates for R_A, R_B, and R_C in 24 fold data are shown in Figure B-1 and will be discussed. Figure B-2 shows estimates of the systematic errors due to fitting a finite number of terms to the angular dependence of the reflection coefficient. These errors were determined by applying the weights calculated for a given order model to data containing the next higher term in the expansion.

Figure B-1 shows the random error in the fit of the dc term R_A versus the angular aperture for 24-fold data taken at evenly spaced offsets. All of the random noise results scale as the square-root with increasing fold.

For the 2-term fit, the noise is about one-third of the noise in the single trace for an improvement equal to 11 fold, almost one-half of the full 24 fold used in this model. For the 3-term fit the random error is about 0.37 for an effective fold of about 8. Figure B-1b shows the same curves for R_B. Notice the difference in vertical scale compared to Figure B-1a. Both curves decrease with increasing angle but are quite large for all angles. The minimum error in the 2-term fit is

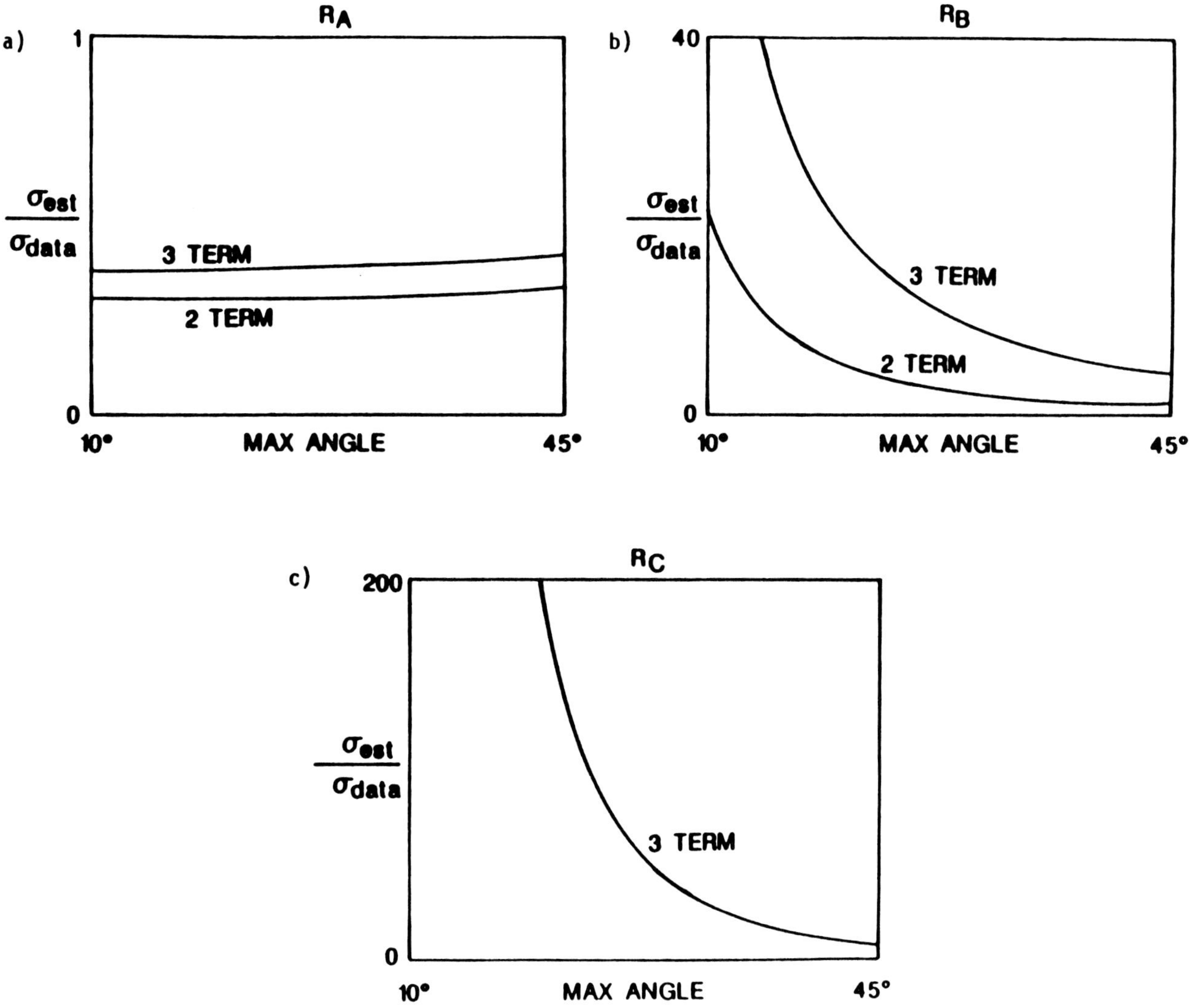

Fig. B-1. The effect of random errors on low order fits to the Zoeppritz equation for P-wave reflectivity.

slightly more than 1 while the result of a 3-term fit will be 4.5 times as noisy for (24-fold data) as the individual input traces. Figure B-1 shows the 3-term result for R_C. This coefficient is very poorly determined and is more than nine times as noisy as single fold data at 45 degrees maximum angle.

Including the systematic errors makes the situation worse. Figure B-2 shows the systematic errors for the 2- and 3-term fits for the coefficient R_A. The systematic errors are independent of fold and input noise and are, therefore, normalized to the particular coefficient they are associated with. The errors increase with increasing angle but are generally quite small for this term and remain below 0.03 at 45 degrees for the 2-term fit and 0.003 for the 3-term fit. Figure B-2 shows the same curves for R_B. Notice again the increase in vertical scale over that of the dc term. The systematic error gets large very fast with increasing angle for the 2-term fit, reaching 0.2 by 24 degrees. The systematic error of the 3-term fit increases less dramatically and reaches 0.13 at 45 degrees. The systematic error in the 3-term fit for R_C showing in Figure B-2 is quite large at all angles reaching 0.73 at 45 degrees.

The possibility of getting accurate estimates of R_A, R_B, and R_C is quite different for each coefficient. The dc term R_A shows good improvement over single fold data and small systematic error, and should be easy to estimate accurately. There is little use in trying to extract the $\sin^4 \theta$ term R_C since it will be buried by both random and systematic errors. The critical ques-

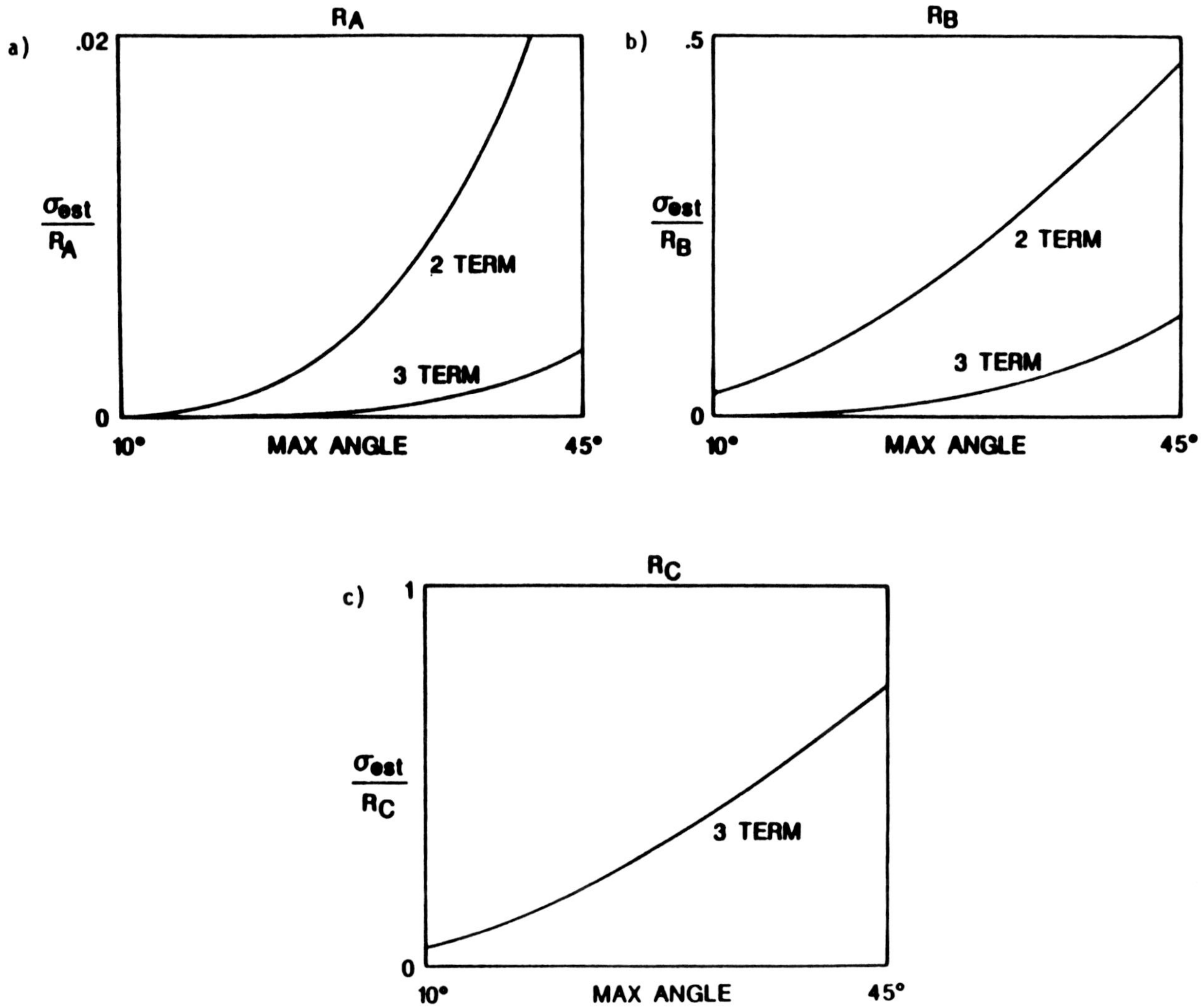

Fig. B-2. Effect of systematic data errors on low order fits to the Zoeppritz equation for *P*-wave reflectivity.

tion then is how well the $\sin^2 \theta$ coefficient R_B can be determined, since this represents the quality of the S-wave information that we can extract from the data. Figure B-3 shows the minimum total error (random plus systematic) in the R_B estimate, and the aperture angle where it occurs, as a function of the rms random error on the input traces. This figure is for 192-fold data and represents an estimate of the best we could hope to do with "state of the art" seismic systems (any other fold can be examined by scaling the horizontal axis by the square-root of the ratio of the fold divided by 192). Since $R_S \sim \frac{1}{2}(R_A - R_B)$, the error in the estimate of R_S will be about one-half the error in the estimate of R_B. Therefore, for an accuracy of 0.1 in R_S, we can tolerate a total error of 0.2 in R_B. For

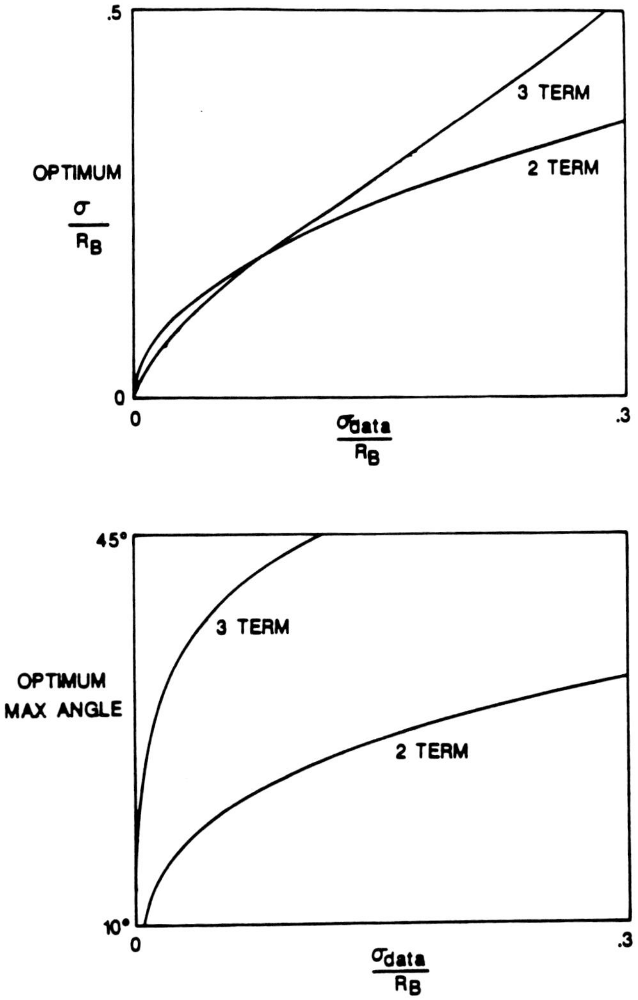

Fig. B-3. The combined effect of random and systematic errors on low order fits to the Zoeppritz equation for P-wave reflectivity.

both the 2- and 3-term fits, this occurs when the random error in the input data is about 0.125. For the 2-term fit, the optimal angle is 22 degrees while for the 3-term fit, it is 45 degrees.

The result of this analysis is that a high quality pseudo-shear section can only be obtained from high-fold data with a very high S/N on the individual traces. There seems to be little advantage to including a $\sin^4 \theta$ term in the fit since the associated coefficient is very poorly determined, and the reduction in systematic error in R_B is overcome by the loss of effective fold for all but very high S/N ratios.

Note that this analysis assumes that the data are initially unweighted. In fact, the earth applies a variety of weights some of which are removed by pseudo-shear. In addition, random and coherent noise present different problems.

APPENDIX C—ESTIMATION OF INTERVAL VELOCITIES AND ANGLES OF INCIDENCE

In Appendix C we derive an improved Dix method for estimating interval velocities and critical angles. The standard Dix formulation assumes a model of plane constant velocity layers. Such a discontinuous interval velocity is inconsistent with the pseudo-shear assumption of a smooth background velocity plus first order reflections. Even after smoothing, the Dix velocities can vary rapidly in time (depth). A better model allows the velocity in each layer to increase linearly with time. If U_i is the velocity at the top of layer i which begins at time t_i, then the velocity V_i within the layer is

$$V_i = U_i + \alpha_i(t - t_i)$$

where α has the units of acceleration and is an input parameter.

The values U_i can be solved for in much the same way as in the standard Dix solution. Beginning with the definition of the mean-squared velocity

$$V_i^{(2)} = \frac{1}{t_i} \int_0^{t_i} V^2 \, dt$$

$$t_{i+1} V_{i+1}^{(2)} - t_i V_i^{(2)} = \int_{t_i}^{t_{i+1}} V^2 \, dt.$$

Using our assumption of linearly increasing velocities gives

$$t_{i+1} V_{i+1}^{(2)} - t_i V_i^{(2)} = \int_{t_i}^{t_{i+1}} [U_i + \alpha_i(t - t_i)]^2 \, dt$$

which results in the following quadratic expression for U_i

$$(t_{i+1} - t_i)U_i^2 + \alpha_i(t_{i+1} - t_i)^2 U_i + \frac{\alpha_i^2}{3}(t_{i+1} - t_i)^3 - t_{i+1}V_{i+1}^{(2)} + t_i V_i^{(2)} = 0.$$

After the U_i are calculated, a smoothed interval velocity function is obtained by averaging the two velocity estimates at each interface and interpolating the results in time with a cubic spline. In principle the α_i values should be area dependent; however, 100 ft/s^2 for water and 700 ft/s^2 for rocks seems to give good results for a variety of areas.

Incident angles can then be calculated from the offset and moments of the smoothed interval velocity function. We use the following notation:

ρ = density,
V_p = compressional-wave velocity,
V_s = shear-wave velocity,
θ = incident angle (subscript i will denote ith reflector, o will denote the surface),
$p = \sin\theta_i/V_{pi} = \sin\theta_0/V_{po}$,
$V^{(n)} = \frac{1}{t}\int_0^t V_p^n(t')\,dt'$ (t' is the one way traveltime),
V_p, V_s, θ_i, and p are functions of depth, $V^{(n)}$ are functions of traveltime,
Z_i = thickness of the ith layer,
X_i = lateral distance traversed by ray in the ith layer, and
X = 1/2 the total offset.

Then for any layer

$$X_j = Z_j \tan\theta_j = Z_j[pV_{pj} + 1/2p^3V_{pj}^3 \cdots].$$

Integrating yields

$$X = \int_0^z pV_p\,dz + 1/2\int_0^z p^3V_p^3\,dz + \cdots.$$

Substitute $dz = V_p\,dt$, and

$$X = pV^{(2)}t + 1/2p^3V^{(4)}t + \cdots$$

Inverting and substituting $p = \sin\theta/V_p$ gives

$$\sin\theta = \frac{V_pX}{V^{(2)}t} - \frac{V_p}{2}\frac{V^{(4)}}{V^{(2)}}\left(\frac{X}{V^{(2)}t}\right)^3 + \cdots.$$

While the first term is sufficient for first order consistency, the program actually uses

$$\sin^2\theta = \frac{V_p^2}{V^{(2)}}\left[\frac{(2X)^2}{(2X)^2 + 4V^{(2)}t^2}\right]$$

which is correct to better than first order when $V^{(4)} \approx [V^{(2)}]^2$.

APPENDIX D—DETERMINISTIC CORRECTIONS TO *P*-WAVE REFLECTED AMPLITUDES

In Appendix D we derive the effects of several factors which contribute to the observed AVO. The notation is the same as that in Appendix B.

Surface Boundary Conditions

On land, only the vertical component of the motion is detected (see Figure D-1).

$$Z = (1 - \mathrm{PP})\cos\theta_0 + \mathrm{PS}\sin\theta_0$$

From Aki and Richards (1980) we obtain

$$\mathrm{PP} = \frac{-\left(\frac{1}{V_s^2} - 2p^2\right)^2 + 4p^2\frac{\cos\theta_0}{V_p}\frac{\cos\theta_0'}{V_s}}{\left(\frac{1}{V_s^2} - 2p^2\right)^2 + 4p^2\frac{\cos\theta_0}{V_p}\frac{\cos\theta_0'}{V_s}} \approx -1 + 8\left(\frac{V_s}{V_p}\right)^3\sin^2\theta_0$$

and

$$\mathrm{PS} = \frac{4\left(\frac{V_p}{V_s}\right)p\frac{\cos\theta_0}{V_p}\left(\frac{1}{V_s^2} - 2p^2\right)}{\left(\frac{1}{V_s^2} - 2p^2\right)^2 + 4p^2\frac{\cos\theta_0}{V_p}\frac{\cos\theta_0'}{V_s}}$$

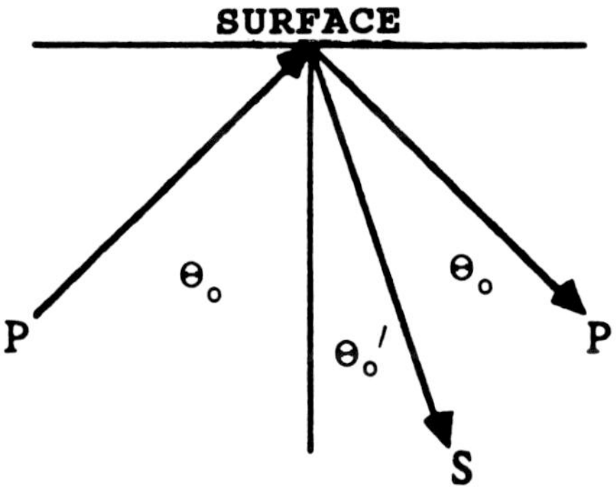

Fig. D-1. Surface boundary conditions. PP = *P*-wave reflection coefficient; PS = *P* to *S* conversion coefficient. $\sin\theta_0/V_p = \sin\theta_0'/V_s$

$$\approx 4\left(\frac{V_s}{V_p}\right)\sin\theta_0$$

so

$$Z = 2\left[1 - \left(4\frac{V_s^3}{V_p^3} - 2\frac{V_s^2}{V_p^2} + \frac{1}{2}\right)\sin^2\theta_0\right]$$

and for

$$\frac{V_s}{V_p} \approx \frac{1}{2} \text{ or } \frac{V_s}{V_p} \ll 1$$

$$Z \approx 2\left(1 - \frac{1}{2}\sin^2\theta_0\right) = 2\left(1 - \frac{1}{2}\frac{V_{po}^2}{V_p^2}\sin^2\theta\right).$$

Spherical Divergence

The spherical divergence correction for non-normal incidence is given in Newman (1973)

$$\frac{\tan\theta_0}{\left[(2X)^2 + 4X\sum_{i=1}^{n} z_i\tan\theta_i^3\right]^{1/2}}$$

Using $z_i = V_p t_i$, expanding in $\sin\theta$ and substituting $p = \sin\theta_i/V_{pi}$ where convenient then becomes

$$\frac{\sin\theta_0}{2X}\left[1 + \frac{1}{2}\sin^2\theta_0 - \frac{p^3}{x}tV^{(4)}\right]$$

where t is one way vertical time. Using $X \approx tpV^{(2)}$ (from ray tracing) and rewriting in terms of $\sin\theta$ gives

$$\frac{V_{po}}{V^{(2)}(2t)}\left[1 + \frac{1}{2}\frac{V_{po}^2}{V^{(2)}}\sin^2\theta - \frac{V^{(4)}}{V^{(2)}V_p^2}\sin^{(2)}\theta\right].$$

Often a simple spherical divergence correction

$$\frac{2tV^{(2)}}{V_{po}}$$

is applied to the data before move-out. Using

$$tV^{(2)} = \sum t_i V^2$$

and

$$t \sim t_0\left(1 + \frac{X^2}{V^{(2)}t_0^2}\right)$$

the correction becomes simply

$$\frac{2t_0V^{(2)}}{V_{po}}\left(1 + \frac{1}{2}\sin^2\theta\right).$$

If the simple divergence correction is applied, the residual correction necessary is

$$\left[1 + \left(\frac{1}{2} + \frac{1}{2}\frac{V_{po}^2}{V^{(2)}} - \frac{V^{(4)}}{V^{(2)}V_p^2}\right)\sin^2\theta\right].$$

Directivity and Arrays

Any source or receiver directivity will directly affect the offset behavior of the recorded reflections. Linear arrays can have a large, frequency-dependent effect.

The response of a linear array of length L is given by

$$R(\theta) = \frac{\sin\left(\frac{\pi L}{\lambda}\sin\theta\right)}{\left(\frac{\pi L}{\lambda}\sin\theta\right)}$$

where λ = wavelength.

Expanding to first order in $\sin^2\theta$ gives

$$R(\theta) = 1 - \frac{1}{6}\left(\frac{\pi L}{\lambda}\right)^2\sin^2\theta$$

or

$$R(\theta) = 1 - \frac{1}{6}\left(\frac{\pi f L}{V_p}\right)^2\sin^2\theta$$

where f is the frequency. In the program, a representative frequency is input and a frequency-independent correction is made.

Anelastic Attenuation

Attenuation will also affect the AVO behavior of recorded reflections. For a constant Q, the amplitude A will decay according to

$$A \sim \exp\left(-\frac{2\pi}{Q}\sum\frac{R_j}{\lambda_j}\right)$$

where R_j is the length of the one-way travel path. Using

$$\frac{R_j}{\lambda_j} = \frac{z_j f}{V_p}\left(1 + \frac{1}{2}\sin^2\theta_j \cdots\right)$$

and integrating yields

$$\sum\frac{R_j}{\lambda_j} \approx ft + \frac{1}{2}ftV^{(2)}p^2$$

or finally

$$A \sim \exp\left(-\frac{z\pi f t}{Q}\right)\left[\exp\left(-\frac{\pi f t}{Q} V^{(2)} p^2\right)\right].$$

Expanding the second exponential gives the angle dependence

$$A \sim A_0 \left\{1 - \left(\frac{\pi f t}{Q}\right)\frac{V^{(2)}}{V_p^2}\sin^2\theta\right\}.$$

As with the array response, only a frequency-independent correction is made.

Higher-Order Correction

A correction can be made for some of the $\sin^4\theta$ terms in an expanded theory. Maintaining these higher order terms in the reflectivity and the angle dependent effects gives the following expression

$$R(\theta) \approx [R_p + (R_p - 2{*}R_s)\sin^2\theta + R_p\sin^4\theta]\,[1 + g\sin^2\theta + h_4\sin^4\theta]$$

where g and h are the expansion coefficients for the other effect. Gathering terms of various orders gives

$$P = R_p$$

$$Q = [(1+g)R_p - 2{*}R_s]\sin^2\theta + [R_p + g(R_p - 2{*}R_s) + h_4 R_p]\sin^4\theta.$$

Part of the $\sin^4\theta$ term will be interpreted by the least-squares fit as a $\sin^2\theta$ behavior. A simple estimate of the magnitude of this effect is

$$\frac{d}{d\varepsilon}\langle(\sin^4\theta - \varepsilon\sin^2\theta)^2\rangle = 0$$

$$\varepsilon = \frac{\langle\sin^6\theta\rangle}{\langle\sin^4\theta\rangle} \sim \frac{5}{7}\sin^2\theta_{\max}.$$

The $\langle\rangle$ represent an average over time. Including the effect of a constant term in the fit gives $\varepsilon \approx .85\sin^2\theta_{\max}$.

As shown in the text in equation 10,

$$\frac{\langle PQ\rangle}{\langle P^2\rangle} = 1 + g - 2{*}\frac{\langle R_p R_s\rangle}{\langle R_{p2}\rangle}, \text{ where } 1 + g = G.$$

Substituting $\varepsilon\sin^2\theta$ for $\sin^4\theta$ gives the new expression

$$\frac{\langle PQ\rangle}{\langle P^2\rangle} = (1+g)(1+\varepsilon) - 2{*}\frac{\langle R_p R_s\rangle}{\langle R_{p2}\rangle}(1+\varepsilon g) + \varepsilon h_4.$$

The pseudo-shear program uses this last expression with $\varepsilon h = 0$. The only place where the results are affected significantly is very shallow in the data if the mute is allowed to be at a greater offset than $\theta = 20$–25 degrees in order to preserve fold.

APPENDIX E—OTHER AMPLITUDE EFFECTS

In Appendix E we derive the effects of several factors that are not handled by the current pseudo-shear method. The notation is the same as in Appendices B and D.

Transmission Effects

The ith reflected signal on a seismogram undergoes a transmission, a reflection, and another transmission, in addition to all of the transmissions and reflections that the $(i-1)$th signal underwent. Therefore, if the transmission coefficient of the previous layer is angle dependent, it might affect the AVO behavior of the deeper reflectors. To first order the PP transmission coefficient is given by

$$\mathrm{PP} = 1 - \frac{1}{2}\left(\frac{\Delta\rho}{\rho} + \frac{\Delta V_p}{V_p}\right) + \frac{1}{2}\frac{\Delta V_p}{V_p}\sin^2\theta.$$

Fortunately each reflection is transmitted through each shallower interface twice, once on the way down and once on the way up. On the return trip each of the first-order changes $\Delta\rho$, ΔV_p, and ΔV_s reverse sign and hence there are no first-order transmission effects. Second-order effects will cause a gradual, smooth, angle-dependent loss of amplitude. The angle dependence will be of the form

$$A_0 - \left(\sum_{i,j} A_{ij}\frac{\Delta_i}{i}\frac{\Delta_j}{j}\right) V_p^2 p^2$$

where i and j represent ρ, V_p, and V_s. The coefficients A_{ij} can be derived (with great difficulty) from the Zoeppritz equations. If we assume $\Delta_i\Delta_j$ are random variables which are stationary with depth and that the reflection density does not vary with depth, then the smoothed effect is given by

$$1 - 2\sum_{i,j} A_{ij}\left\langle\frac{\Delta_i}{i}\frac{\Delta_j}{j}\right\rangle\int_0^z V_p^2 p^2\,dz$$

$$\approx A_0 - 2(A_T V^{(3)} t)p^2$$

where $A_T = \sum_{i,j} A_{ij} \langle \Delta i/i \; \Delta j/j \rangle$ and $\langle \; \rangle$ = average depth so that A_T has the units of length^{-1}. At present the program does not estimate transmission effects.

Anisotropy

Backus (1962) shows that the variations of velocity with angle for a weakly anisotropic material are given by

$$V^2 = A + C \cos\theta + D \sin 2\theta + E \cos 4\theta + F \sin 4\theta.$$

From the symmetry of transverse isotropy this equation can be rewritten as

$$V^2 = (A + C + E) - (2C + 8E) \sin^2\theta$$

or

$$V^2 = V_0^2 \left(1 - 2 \frac{\Delta V^2}{V_0^2} \sin^2\theta \right).$$

Where $\rho V^2 = A$ and $\rho \Delta V^2 = A - (F + 2L)$ where A, F, and L (and C and N) are the standard elastic moduli. Note that V^2 is independent of C which is proportional to the square of the P-wave velocity at 90°. The effect of this anisotropy on the P-wave reflection coefficient is calculated assuming that we can substitute $V(\theta)$ for one V in the isotropic formula.

$$PP_{\text{Anisotropic}} = PP_{\text{Isotropic}} + \frac{\Delta V^2}{V^2} \sin^2\theta.$$

This term sits in the same place as $-2R_s$ in the equation and will have the same effect as a shear velocity contrast.

An important type of anisotropy occurs when the seismic wavelength is longer than the distance between reflectors in an interval of plane layered isotropic material. The magnitude of the effect was just shown to be

$$\rho V^2 = A - (F + 2L).$$

From Backus (1962) this equation becomes, for a layered medium $\langle \rangle$ denotes averaging over relative abundance),

$$\begin{aligned} A - (F + 2L) = {} & 4\langle \rho V_s^2 \rangle - 4\langle \rho V_s^4 V_p^{-2} \rangle \\ & - 2\langle 1/\rho V_p^2 \rangle^{-1} \langle V_s^2/V_p^2 \rangle + 4\langle 1/\rho V_p^2 \rangle^{-1} \\ & \times \langle V_s^2/V_p^2 \rangle^2 - 2\langle 1/\rho V_s^2 \rangle^{-1}. \end{aligned}$$

If we set the material parameters in this equation equal to $\rho + \Delta\rho$ in each interval, where ρ is the average density (similarity for V_p and V_s), we can expand the expression in powers of $\Delta\rho/\rho$ etc. The zero order terms are the averages and all cancel. Since $\langle \Delta\rho \rangle = 0$ (etc.) by definition, all of the first order terms are zero. What remains is

$$A - (F + 2L) = <\rho V_s^2 \sum A_{ij} \frac{\Delta_i}{i} \frac{\Delta_j}{j} + \text{higher order}$$

where i and j = 1, 2, 3 represent ρ, V_p, and V_s, respectively. The term $\langle \Delta_i \Delta_j \rangle$ is second order and does not depend on reflection density, nor does the magnitude of the effect increase in time as the transmission effect does. This effect is therefore ignored.

APPENDIX F—NONDETERMINISTIC AMPLITUDE CORRECTION

The calculation of δG involves the product of P with itself and the product of P and Q. Here we derive unbiased estimates of these products. P and Q are the coefficients in the linear model of amplitude variation with offset which in terms of the incidence angle θ_i is

$$A(\theta_i) = P + Q \sin^2\theta_i. \tag{F-1}$$

The index i ranges from the near offset trace number to the farthest offset trace number consistent with the mute designed to retain only data for which the θ_i is less than 25°–30°.

P and Q are not only linear transformations on the data, the $A(\theta_i)$'s, but because of the least-squares estimation scheme, they are simply weighted sums of the $A(\theta_i)$'s. The weighting factors for P and Q are

$$W_P(i) = \frac{1}{\det} \left(\sum_{k=1}^{N} \sin^4\theta_k - \sin^2\theta_i \sum_{k=1}^{N} \sin^2\theta_k \right)$$

and (F-2)

$$W_Q(i) = \frac{1}{\det} \left(\sum_{k=1}^{N} \sin^2\theta_k - N \sin^2\theta_i \right),$$

respectively, where N is the total number of traces retained after the mute is applied. det is the determinant in the least-squares estimation of P and Q from equation (F-1). Although least squares estimation ensures that the expected values of P and Q are zero, the expected values of the products PP and PQ are not zero but are proportional to the variance in the amplitudes. The proportionality factor is the product of the weighting factors given in equation (F-2). The resulting unbiased estimates of PP and PQ are

$$[W_P(i) A(\theta_i)]^2 - \sum \frac{W_P(i)^2}{N - 2} \times [A(\theta_i) - P - Q \sin^2\theta_i]^2$$

and

$$W_P(i)W_Q(i)A^2(\theta_i) - \sum \frac{W_P(i)W_Q(i)}{N-2}$$

$$\times [A(\theta_i) - P - Q \sin^2 \theta_i]^2,$$

respectively.

REFERENCES

Aki, K. and Richards, P. G., 1980, Quantitative seismology: Freeman & Co.

Backus, G. E., 1962, Long-wave elastic anisotropy produced by horizontal layering: J. Geophys. Res., **67**, 4427–4440.

Backus, M. M. and Goins, N. R., 1984, Summary of change in reflectivity with offset research workshop: Geophysics, **49**, 838–839.

Bortfeld, R., 1961, Approximation to the reflection and transmission coefficients of plane longitudinal and transverse waves, Geophys. Prosp., **9**, 485–502.

Domenico, S. N., 1984, Rock lithology and porosity determination from shear and compressional wave velocity: Geophysics, **49**, 1188–1195.

Ensley, R., 1985, Evaluation of direct hydrocarbon indicators through comparison of compressional- and shear-wave seismic data: A case study of the Myrnam gas field, Alberta: Geophysics, **50**, 37–48.

Gassaway, G. S., 1984, Effects of shallow reflectors on amplitude versus offset (seismic lithology) analysis: 54th Ann. Internat. Mtg., Soc. Expl. Geophys., Expanded Abstracts, 665–669.

McCormack, M. D., Dunbar, T. A. and Sharp, W. W., 1984, A case study of stratigraphic interpretation using shear and compressional seismic data: Geophysics, **49**, 509–520.

Newman, P., 1973, Divergence effects in a layered earth: Geophysics, **38**, 481–488.

Ostrander, W. J., 1984, Plane-wave reflection coefficients for gas sands at non-normal angles of incidence: Geophysics, **49**, 1637–1648.

Robertson, J. D. and Pritchett, W. C., 1985, Direct hydrocarbon detection using comparative P-wave and S-wave seismic sections: Geophysics, **50**, 383–393.

Shuey, R. T., 1985, A simplification of the Zoeppritz equations: Geophysics, **50**, 609–614.

Spratt, S., 1987, Effect of normal moveout errors on amplitude versus offset-derived shear reflectivity: 57th Ann. Internat. Mtg., Soc. Expl. Geophys., Expanded Abstracts, 634–637.

Smith, G. C. and P. M. Gidlow, 1987, Weighted stacking for rock property estimation and detection of gas: Geophys. Prosp., **35**, 993–1014.

Taner and Koehler, 1981, Surface consistent corrections: Geophysics, **46**, 17–21.

Prestack Analysis—An Integrated Approach for Seismic Interpretation in Clastic Basins

*Samir K. Dey-Sarkar and Steve V. Svatek**

INTRODUCTION

Direct determination of lithology and pore-fluid content from seismic data has been a goal of geophysicists for a long time. One specific phenomenon that has been extensively utilized is the occurrence of a "bright spot" or amplitude anomaly in a true-amplitude processed seismic section. For some time the bright-spot phenomenon has been recognized as being often associated with the occurrence of gas in a sand/shale sequence. Introduction of gas to a porous sand unit decreases the unit's velocity and density relative to the surrounding lithologic units. This decrease produces a high-impedance contrast at the shale/gas sand boundary resulting in a high-reflectivity zone or bright spot. A lucid description of this technique and its historical development is presented in Sengbush (1983). This technique is entirely based on stacked data or equivalent normally incident (zero angle of incidence) seismograms.

Initially, the exploration industry had tremendous success in finding sizeable reserves based on the bright-spot technique. However, soon pitfalls began to emerge in this simple technique. It was realized that often bright spots were not associated with gas sands, but with other lithologic changes such as coal seams, overpressured shales, high-porosity sands, etc.

Examples of two such false bright spots are presented in Figures 1 and 2. In Figure 1, the bright spot is entirely due to a shale/shale interface. In Figure 2, the bright spot (2.8 s) on the downthrown side of the fault is due to gas, whereas the bright spots on the upthrown side are due to porous wet sands. Thus, interpretation of bright spots from stacked data could often lead to wrong results.

Recently, various authors (Backus, 1982; Gassaway, 1983; Ostrander, 1984) have attempted a different approach for the interpretation of these bright spots. Conventional bright-spot interpretation is based on stacked data. The stacked data are produced by summing a group of traces that belong to a common-depth-point gather. The new approach bases the interpretation on the relative amplitude and phase variation among the individual traces of the gather. The analysis is now performed in the prestack domain rather than in the stack domain.

The rationale behind this approach is based on a rather simple concept of differential behavior of compressional-wave and shear-wave velocity in gas-saturated sands. Several authors in the past [Gardner and Harris, 1968; Gregory, 1976; Domenico, 1976, 1977] have demonstrated that introduction of gas to water-saturated sand reduces the sands P-wave velocity (α or V_p) and density (ρ), but does not change the shear-wave velocity (β or V_s). Shear-wave velocity in a formation depends on the formation's shear modulus and density. Replacement of brine by gas should hardly change the shear modulus of the formation, and changes the density only slightly, thus keeping the S-wave velocity intact. In most other cases of lithologic changes, if there is a change in P-wave velocity and density, there will be a corresponding change in S-wave velocity.

This interesting behavior of S-wave velocity cannot be observed from a stacked data set. Ideally, stacked data essentially represents the response along the raypath that would be obtained if source and receiver were coincident. The reflection coefficient or amplitude is given by the equation in Figure 3a. This equation does not have any S-wave term, therefore, as a result the stack data does not have any information about S-wave velocity behavior.

On the other hand, if reflection coefficients are evaluated for the individual rays corresponding to different source-receiver offsets, then the formulas are given in Figure 3b (Dey-Sarkar and Chapman, 1978). These reflection coefficients now contain the S-wave term. Thus, prestack amplitudes contain S-wave velocity information, which may be exploited to differentiate between true bright spots (hydrocarbons) and false bright spots (lithologic changes). Figure 4 represents reflection coefficients evaluated for several litho-

*ARCO Exploration and Production Technology, Plano, Texas

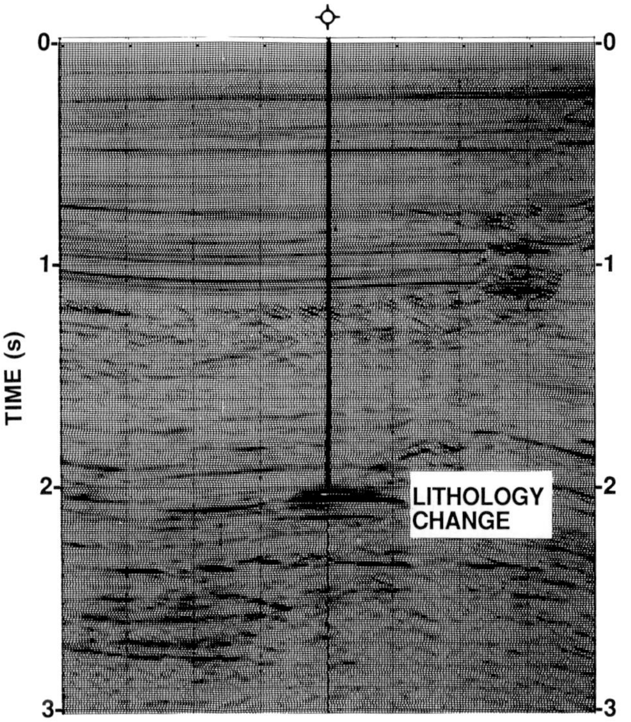

Fig. 1. Bright spot due to lithology change. Peak equals negative reflection zone.

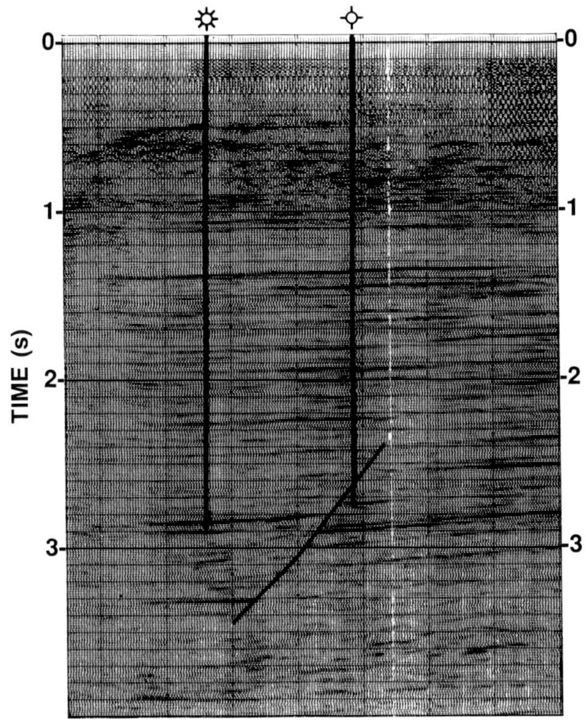

Fig. 2. Bright spots in a gas and brine saturated zone. Peak equals negative reflection zone.

$$R(\theta_1) = \frac{\rho_2 V_{P2} - \rho_1 V_{P1}}{\rho_2 V_{P2} + \rho_1 V_{P1}} \qquad \theta_1 \cong 0$$

(a)

$$R(\theta_1) = -\frac{[\rho_1 \eta_{\alpha 2} - \rho_2 \eta_{\alpha 1}][\rho_2 \eta_{\beta 1} + \rho_1 \eta_{\beta 2}] + P^2(\rho_1 + \rho_2)^2}{[\rho_2 \eta_{\alpha 1} + \rho_1 \eta_{\alpha 2}][\rho_2 \eta_{\beta 1} + \rho_1 \eta_{\beta 2}] + P^2(\rho_1 + \rho_2)^2}$$

$$\eta_{\alpha 1} = \left[\frac{1}{V^2_{P1}} - P^2\right]^{1/2} \qquad \eta_{\alpha 2} = \left[\frac{1}{V^2_{P2}} - P^2\right]^{1/2}$$

$$\eta_{\beta 1} = \left[\frac{1}{V^2_{S1}} - P^2\right]^{1/2} \qquad \eta_{\beta 2} = \left[\frac{1}{V^2_{S2}} - P^2\right]^{1/2}$$

$$P = \frac{\sin \theta_1}{V_{P1}}$$

(b)

Fig. 3. Reflection coefficient formula for (a) stack and (b) prestack data. ρ equals density, V_p equals compressional velocity, V_s equals shear wave velocity, 1 equals incident medium, 2 equals reflecting medium, and θ_1 equals angle of incidence.

Table 1. Velocity/density data for various lithologies utilized in computing reflection coefficient diagrams of Figure 4.

Lithology	*P*-Wave velocity (m/s)	*S*-Wave velocity (m/s)	Density (gms/cc)
Shales	3811	2263	2.40
Limestone	5335	2957	2.65
Sand	3811	2302	2.25
Salt	4573	2729	2.05
Coal	2652	1280	1.27
Gas Sand	3453	2302	2.10
Limestone (Gas)	5043	2957	2.49

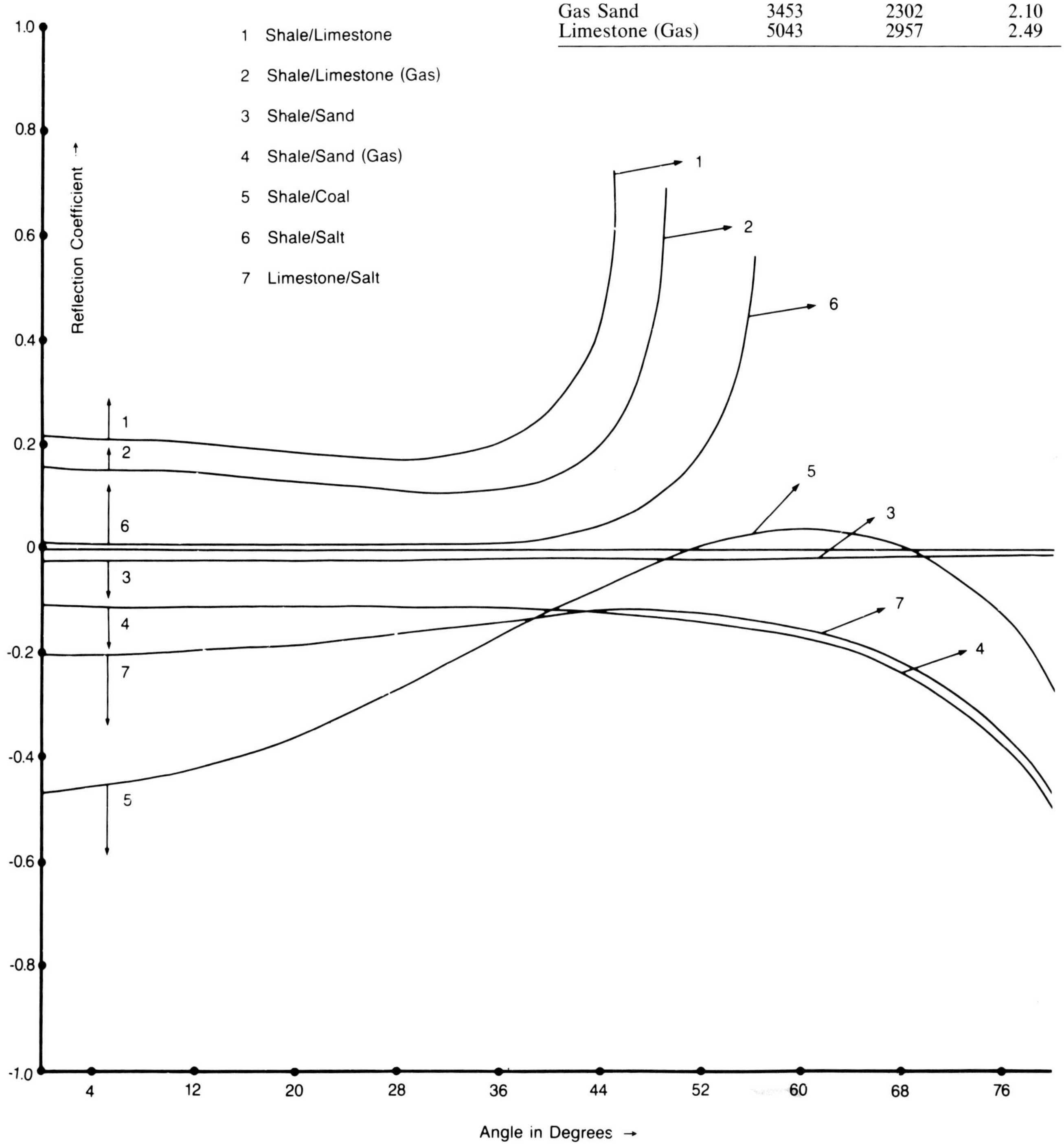

Fig. 4. Reflection coefficient versus angle diagram for various combinations of lithologies.

logic scenarios for a typical stratigraphic sequence in a basin and demonstrates clearly the utility of prestack data. Table 1 presents the relevant velocity and density data appropriate for Figure 4.

There are three major issues in the prestack analysis scheme. These are:

(1) Data processing for preservation of amplitudes and phase in the prestack domain;
(2) Analysis of petrophysical parameters and integration with processed seismic data; and
(3) Representation of prestack data in a comprehensive manner for routine interpretation.

In this paper, we discuss only the first two issues. A data processing strategy will be discussed which satisfies the internal consistency of the data. Analysis of petrophysical properties will lend itself to some simple interpretational concepts.

DATA PROCESSING

The basic objective of the data processing is to preserve relative amplitude information for all offsets at all times for any depth point gather. Additionally, relative amplitudes among all depth points need to be preserved. In the case of stacked data, the process of CDP stacking cancels many types of noise. In the prestack domain, various types of noise that distort the true amplitudes of the seismic data have to be removed by different noise suppression processes.

There are three basic types of noise that distort the amplitudes in the prestack domain. We define these as Type I, Type II, and Type III noise.

Type I Noise

Type I noise phenomena can be removed without any prior knowledge of subsurface velocities and densities. These phenomena are:

(a) source generated noise,
(b) multiples,
(c) surface-consistent source-receiver effects, and
(d) source signature variation with offset.

Current marine data acquisition procedures in conjunction with digital processing techniques can handle most of these noises. Usual techniques to remove these effects are (1) two-dimensional Fourier techniques; and (2) surface-consistent computation.

One such example is presented in Figure 5. In Figure 5a, a CDP gather is shown before the multiple removal and, in Figure 5b, the same CDP gather is shown after the multiple removal by frequency-wavenumber filtering. The corresponding velocity analyses are shown in Figures 6a and 6b, respectively. This type of multiple removal technique attenuates the multiples in the far traces adequately, but performs poorly in the near traces (Figure 5). In this example there are two sets of multiples. One set is generated around 1.0 s and the other set around 2.25 s (Figure 6). The rejection process has performed quite well for 1.0 s multiple (lower apparent velocity) but rather poorly for the 2.25 s multiple (higher apparent velocity). Some of these problems can be better resolved with the Radon transform techniques (Hampson, 1986) currently used in the industry. Additionally, far traces in Figure 5 seem to be contaminated by aliased energy which results because of the presence of high-frequency noise that was not filtered out. In actual cases

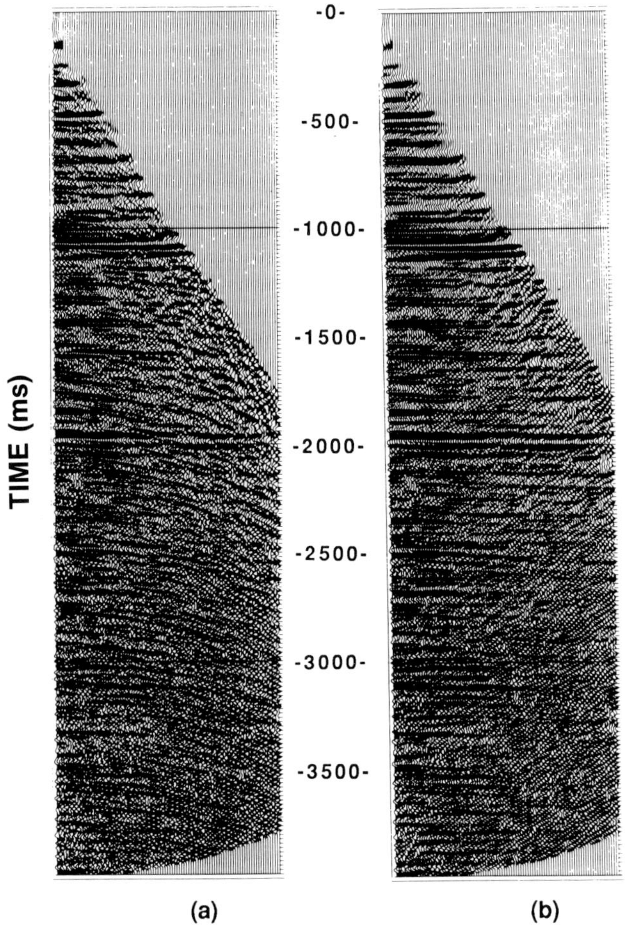

Fig. 5. Examples of multiple removal by the two-dimensional Fourier technique (a) with no multiple attenuation, and (b) with *f-k* filtered.

this type of noise can be monitored and data are filtered prior to any multiple rejection.

Another example of Type I noise is presented in Figures 7a, 7b, and 7c. This particular dataset was acquired in very shallow water, requiring the use of a drag cable. This data acquisition technique had created a considerable amount of amplitude variation throughout the length of the line. In Figure 7a, rms amplitude values are plotted for a particular time window in CDP-offset space. In this type of representation, common-source variations will follow linear trends from left to right and common-receiver trends will follow linear trends from right to left. Common-offset trends will be parallel to the CDP axis. Figure 7a shows one example of each type of trend.

Figure 7b shows the trends that were removed by the surface consistent technique. Note that the common-receiver trend indicated in Figure 7a has been effectively removed. Additionally, it is important to note that all the trends removed are either common-source or -receiver trends. The process cannot handle common-offset trends. The common-offset trends in Figure 7a do not seem to persist for all the CDPs. The reason they don't persist is not known.

The source-generated coherent noise can also be attenuated by the two-dimensional (2-D) Fourier tech-

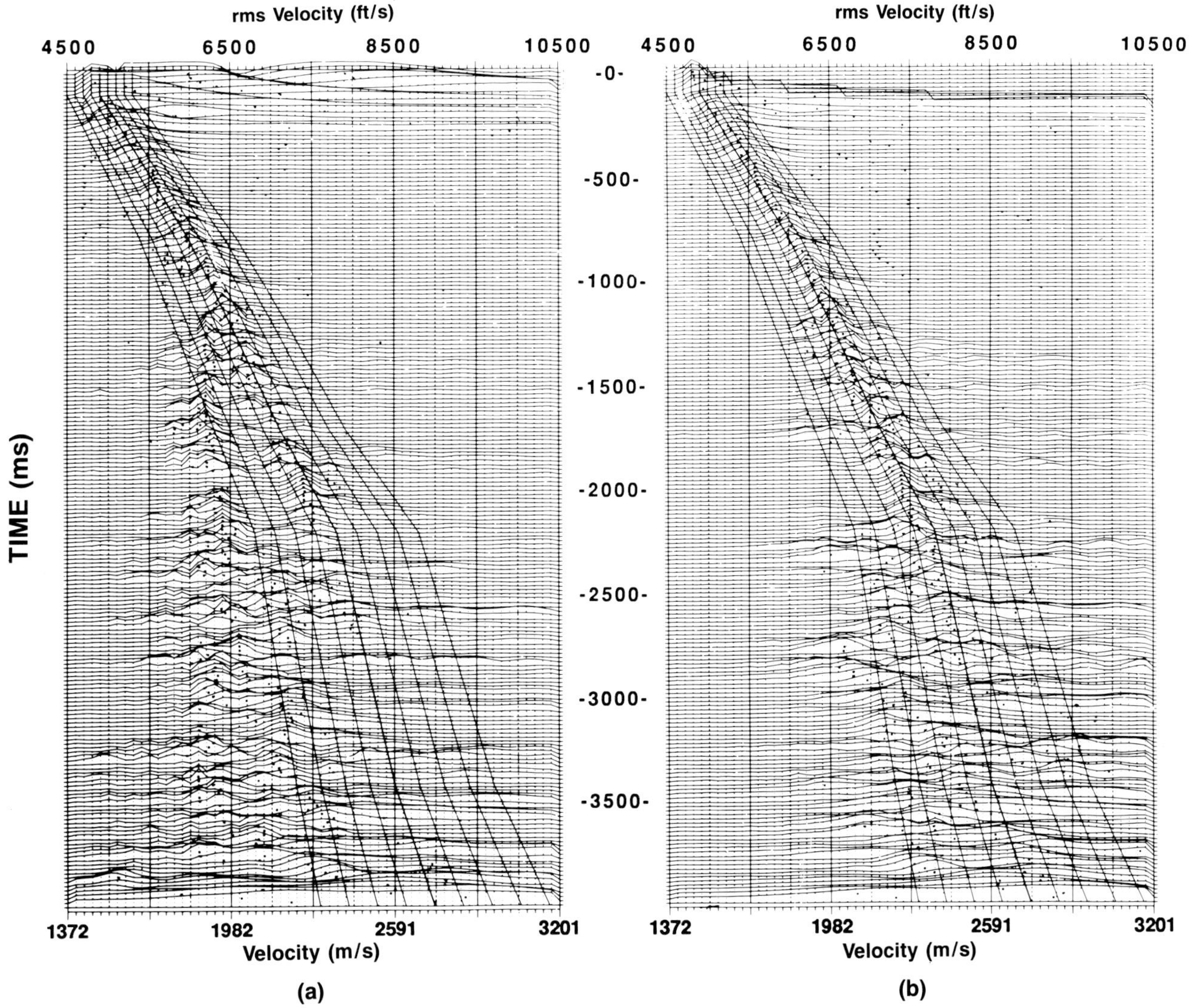

Fig. 6. Velocity analysis used to discriminate multiples in Figure 5, (a) with no multiple attenuation, and (b) *f-k* filtered.

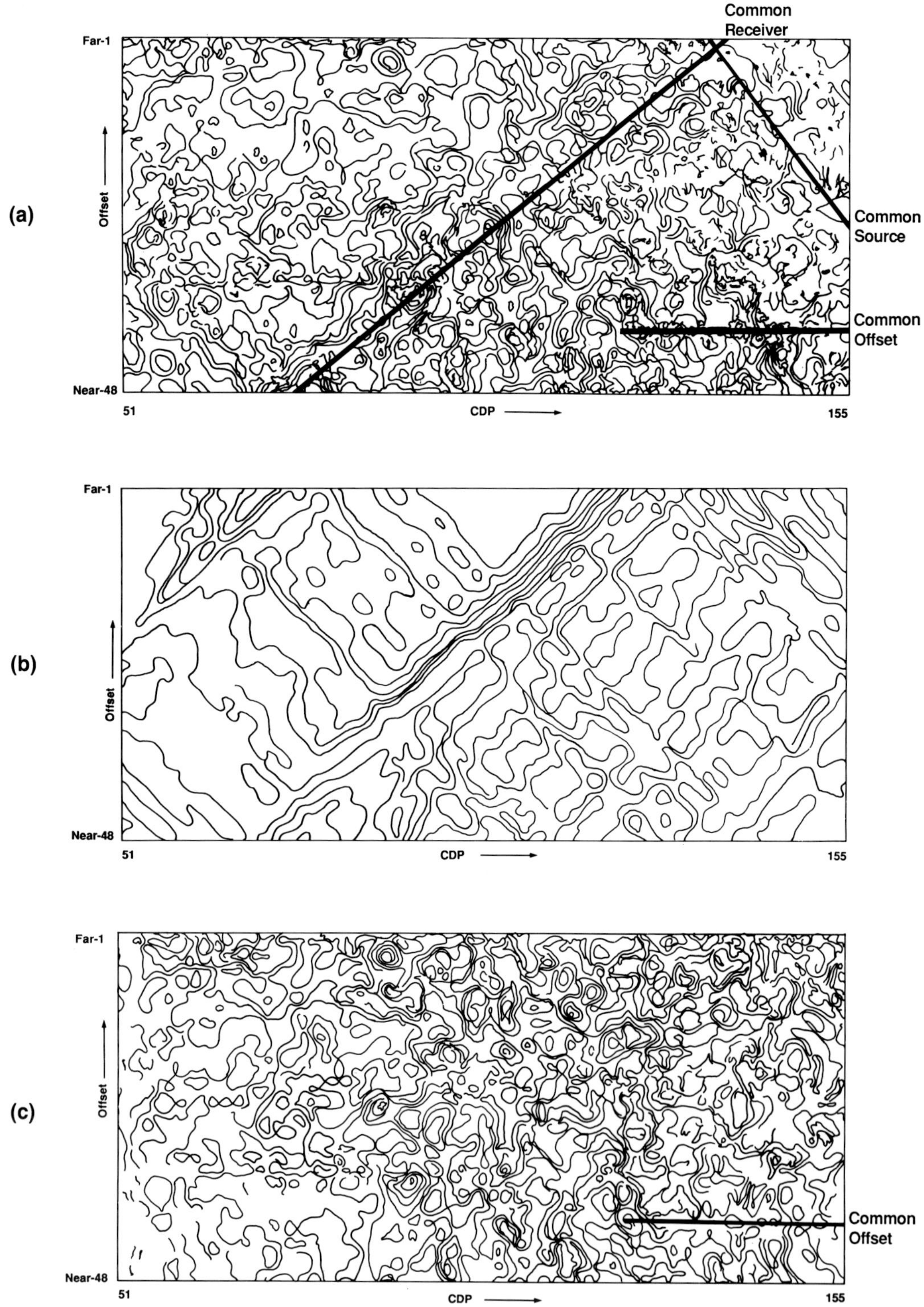

Fig. 7(a). The rms amplitude plot in CDP-offset space before surface consistent correction. (b) for the derived surface consistent corrections. (c) after the surface consistent corrections.

niques and source signature variation can be handled using the surface-consistent deconvolution technique.

Type II Noise

Type II noise is generally associated with instrumentation or cultural noise during data acquisition. Some examples are

(a) High-frequency noise,
(b) Low-frequency noise,
(c) Drilling noise, and
(d) Channel imbalance problems.

High-frequency noise is due to electronics problems and can be edited out in some instances.

Low-frequency noise is due to flexural waves generated at the bulkhead in the streamers. In many instances the noise has very high amplitudes saturating the channels, thus, making attenuation very difficult.

Drilling noise, as the result of seismic data acquired near actively drilling wells, frequently will have a very high-frequency hyperbolic event associated with the drilling operations. Some of the high frequencies can be filtered but low frequencies will remain.

Channel imbalance problems (Figure 8) are very often due to either faulty electronics or to sections of cable being bad which result in certain channels that have anomalous amplitudes. The result is basically an offset-consistent problem rather than a surface-consistent problem and, therefore, is difficult to handle. The offset-consistent problem is quite common and could be channel related, cable related, or have something to do with odd pressure variations on the surface. Offsets having anomalous amplitudes have been observed to shift in some unpredictable manner (for all CDPs the same channels are not anomalous). This type of amplitude variation is not source/receiver consistent, and, therefore, not related to shallow subsurface conditions.

Type III Noise (wave propagation effects)

Type III noises are entirely due to wave propagation effects in a visco-elastic medium. They are

(a) two-dimensional spherical divergence,
(b) thin bed attenuation,
(c) array attenuation,
(d) inelastic attenuation,
(e) curvature effects, and
(f) overburden transmission effects.

In a complicated geological situation, evaluating the effect of all these phenomena is difficult. However, for a vertically inhomogeneous medium, analytical techniques exist to compute the effect of these phenomena. The computation requires knowledge of velocities, densities, and attenuation parameters. Figures 9 through 11 present examples of such effects for the first three categories described. These computations

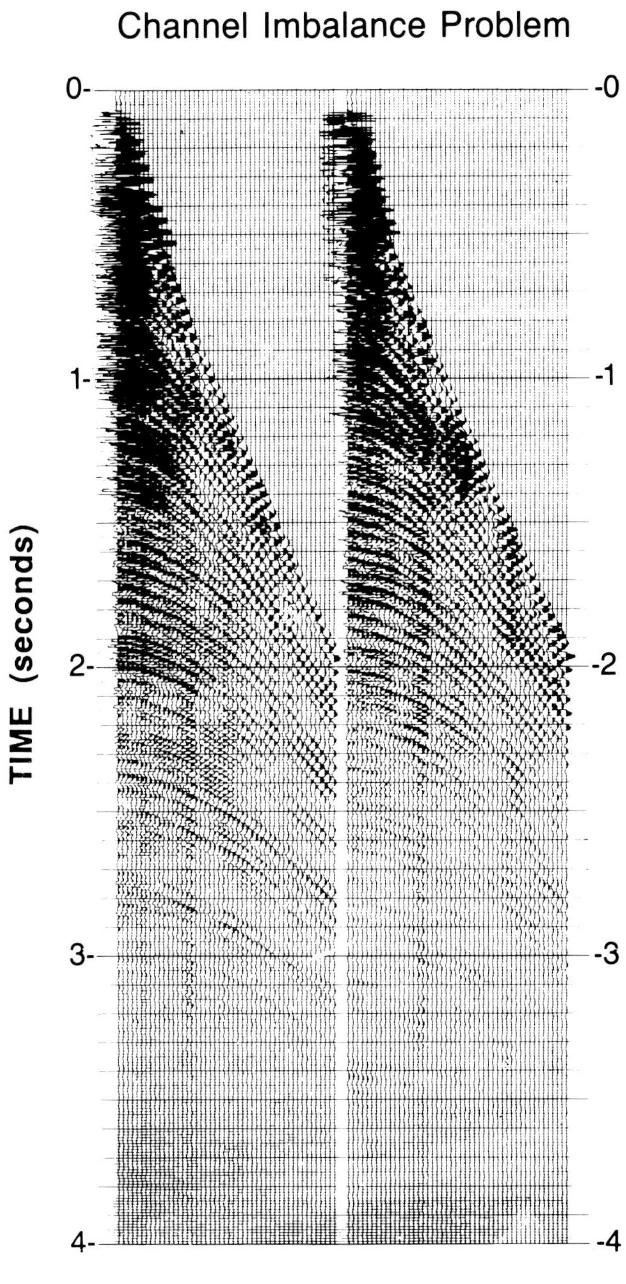

Fig. 8. Amplitude imbalance among various offsets in shot records.

were performed based on well data. Figures 9 and 10 present equal-amplitude contour plots for residual spherical divergence and residual thin-bed effect, respectively. In each of these figures, the 0 time axis corresponds to zero-offset time. The term residual means the corrections are normalized to zero offset. Thus, these figures demonstrate the decay of amplitudes with respect to zero offset for every time. Since these figures are for hyperbolic events, actual computation will require superposition of these plots on hyperbolic moveout functions.

Figure 11 represents amplitude attenuation at three particular frequencies superimposed on the hyperbolic moveout functions as determined from a well velocity function. The attenuation functions are derived for a particular source/receiver combination commonly used in the Gulf. Figure 11 illustrates how various frequencies will be attenuated with offsets. For example, for an event at 0.8 s, the 60 Hz component of the signal will be attenuated by 6 dB around an offset of 3520 ft (1073 m). Whereas for an event at 1.8 s, the 60 Hz component will be attenuated by 6 dB around 8800 ft (2682 m) offset. Due to the natural filtering effect of the subsurface, shallower events have higher frequency content relative to the deeper events. So, as a result of the array attenuation, the shallower events will be more effected than the deeper events.

Another type of phenomenon that needs investigation is the effect of transmission loss from the overly-

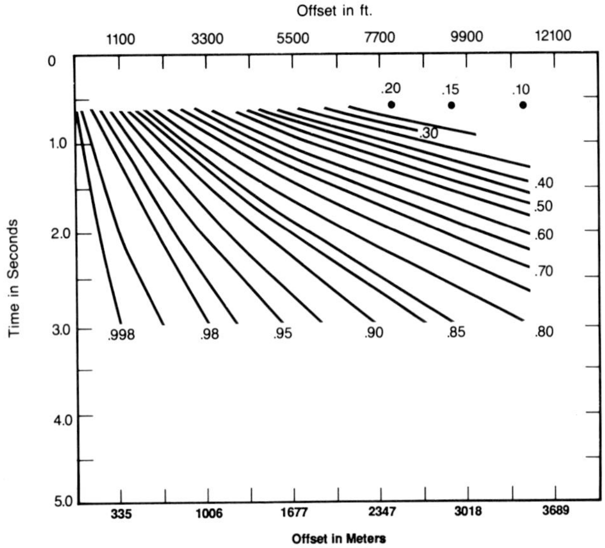

Fig. 9. Contour diagram of residual spherical divergence in time-offset space.

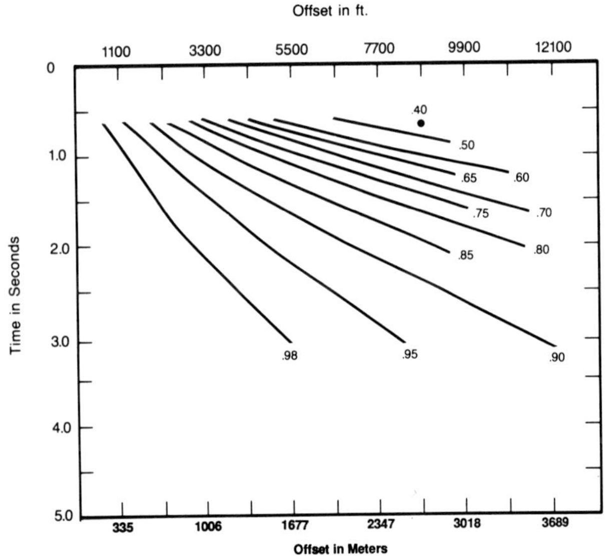

Fig. 10. Contour diagram of residual thin bed effect in time-offset space.

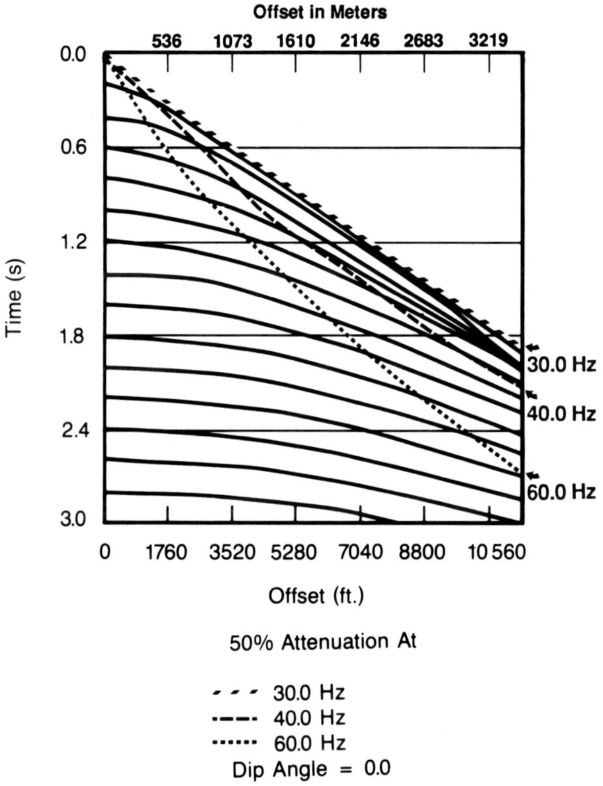

Fig. 11. Source/receiver array attenuation for various frequencies.

ing media. To evaluate this loss, a velocity model (Figure 12) was constructed for a typical well in the Gulf. Three different variations of this model were considered. The first version did not include a water layer or low-velocity layer shown in Figure 12. The second version included the low-velocity layer only. The third version included the low-velocity layer and a water/hard sedimentary interface. Initially reflection coefficients were computed for a shale/sand interface around 9300 ft (2835 m) depth, with velocities and densities as shown in the inset in Figure 13. This computation was performed as if two half spaces were welded at the interface and is shown as curve 1 in Figure 13. This response is for infinite frequency and represents the ideal amplitude-variation-with-offset (AVO) for the particular interface in question. Curves 2, 3, and 4 represent the modified reflection coefficients for three variations of the model as discussed earlier. In each case a ray-tracing algorithm was applied to trace rays to the appropriate interface. As a ray with particular angle of emergence propagated to the shale/sand interface, the reflection coefficient (amplitude) at the interface was modulated according to the product law of transmission through all the overlying interfaces. These modified amplitudes were then plotted for various emergent angles. Note that this analysis is strictly valid only at infinite frequency.

(a) In general for every case there is a reduction in absolute amplitudes at all offsets and some change in relative amplitudes among various offsets.
(b) In the absence of any strong velocity contrasts in the overlying media, the shape of the reflection-coefficient function (Curve 2, Figure 13) is not strongly distorted relative to the theoretical reflection-coefficient function (Curve 1, Figure 13). In a geologic basin like the Gulf of Mexico, which is characterized by weak reflectivity, the implication is strong that the overall effect due to transmission loss is not very important.
(c) In the presence of large velocity contrasts, the reflection coefficient is strongly distorted (Curves 3, 4, Figure 13). The presence of shallow gas sands or hard water bottom conditions implies that there will be severe distortion in the reflection coefficient function.

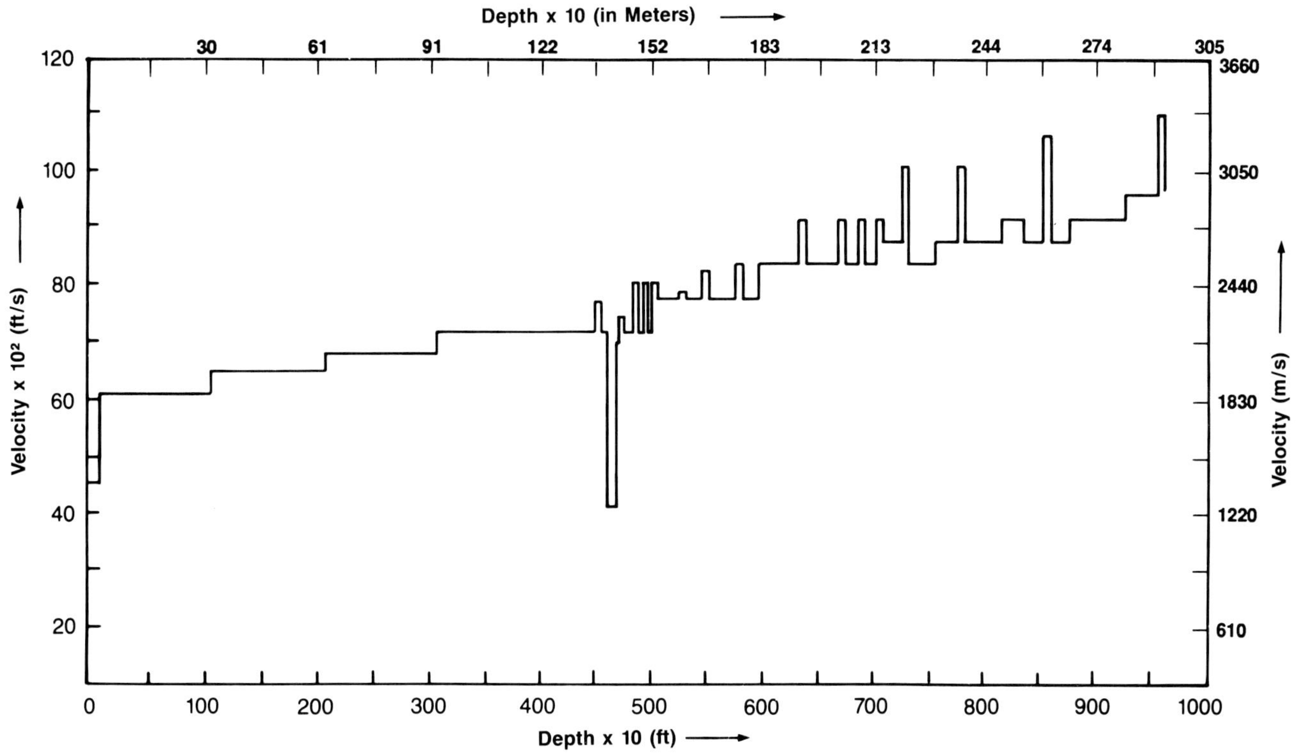

Fig. 12. Velocity-depth model used to calculate the transmission loss in Figure 13.

The effect of inelastic attenuation of Q (dissipation factor) can be evaluated provided a reliable estimate of Q can be made from the seismic data. In a frequency independent attenuation, for a given Q in the overburden, the amount of attenuation will be directly proportional to the length of raypath traversed through the medium. In general, there will again be a loss of amplitude and some phase distortion in the far offsets relative to the near offsets.

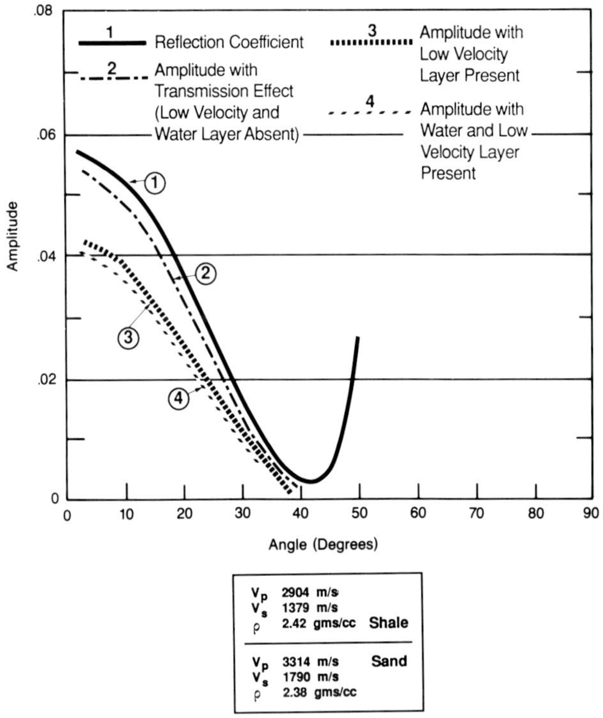

Fig. 13. Amplitude attenuation due to transmission effect (2440–3659 m) for various cases.

The effect of reflector curvature is more complicated. Shuey and Banik (1984) have computed the curvature effect from geometrical optics principle. They have demonstrated that amplitude loss or gain is dependent on the sign of the curvature (positive or negative). It is a well-known phenomenon that for a normal incidence (zero angle) reflection, amplitudes could be defocussed (anticline) or focussed (syncline). This focussing and defocussing effect could produce either apparent gain or loss of amplitude in the far offsets relative to the near offsets.

Thus, in general, wave propagation phenomena have a tendency to reduce the amplitude in the far offsets relative to near offsets. The trouble with this type of deterministic correction is two-fold:

(1) The knowledge of subsurface parameters necessary to compute the amplitude corrections is often inadequate and sometimes nonexistent.
(2) The two-dimensional inhomogeneity causes rapid variation of these parameters, thus making the deterministic corrections unreliable outside the point of control.

An alternate approach is to calibrate the amplitudes using some statistical procedure. After the Type I and Type II noises are removed from the prestack data, there are two components of the amplitude. The first component is associated with the amplitude variation due to Type III noise. The second component is associated with the reflection coefficient variation with offset. We would like to extract the second component and remove the first component. This cannot be achieved by analyzing an event of interest which is contaminated by the reflection coefficient variation. We estimate the first component (Type III noise) from a window of events spanning about 250–500 ms above the target event.

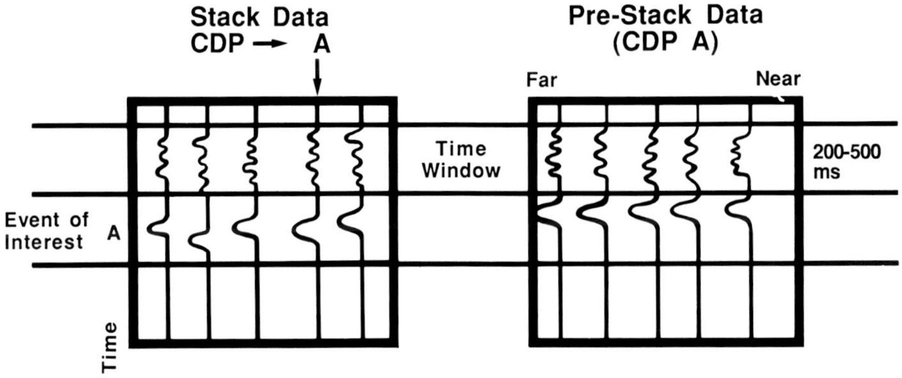

Fig. 14. Schematic representation of amplitude recovery process in the prestack domain.

Figure 14 illustrates the methodology for this estimate. Event (A) in Figure 14 is the event of interest for analysis. A time window of 250–500 ms is selected above the event (A). The rms amplitudes are computed for each offset and an exponential decay function is fitted through the data points. The coefficient of this decay function is the amplitude correction factor for the event (A). The coefficient is spatially averaged to obtain a smoothly varying function. The advantages of this technique are, (1) the robustness, (2) no subsurface parameters are assumed, and (3) no distortion is produced in the data because of the slowly varying function. The disadvantage is that the technique is not time varying and only valid temporarily in the vicinity of the event of interest. Should the normalization window contain some component of offset-dependent reflectivity, the results will still be valid in a relative sense.

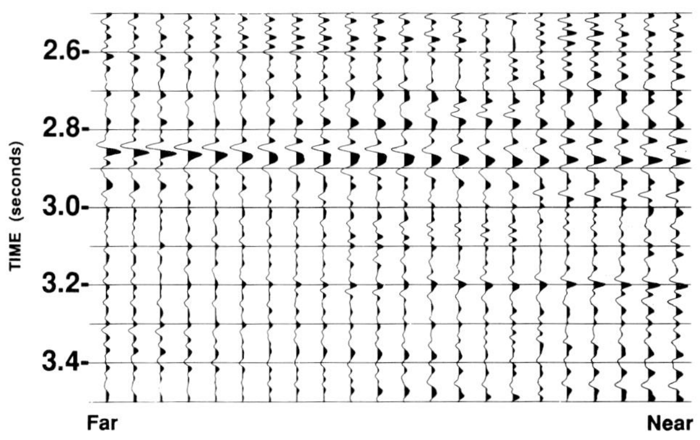

Fig. 15. CDP gather before amplitude calibration. Peak equals positive reflection zone.

Figures 15 and 16 illustrate two data sets before the amplitude calibration. Figures 17 and 18 represent the same two data sets after the amplitude calibration. In both data sets the calibration window was about 500 ms wide and above the target horizon.

This section discussed various types of noise phenomena that complicate data processing in the prestack domain. The strategy for data processing is to:

(1) remove the Type I and II noise, and
(2) apply an offset-dependent regional correction to the amplitude which is computed based on a statistical procedure.

If the data are heavily contaminated by noise, processing alternate data sets may be better.

PETROPHYSICAL ANALYSIS

The amplitude response in the prestack domain is dictated by three key petrophysical parameters. These are *P*-wave velocity (α), *S*-wave velocity (β), and density (ρ). The amplitude response is dependent on

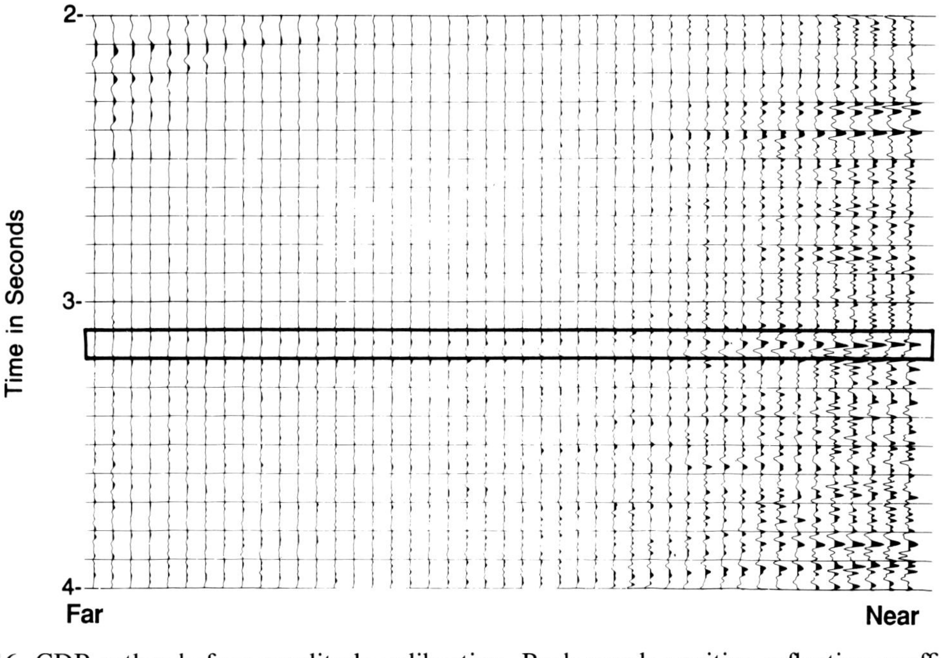

Fig. 16. CDP gather before amplitude calibration. Peak equals positive reflection coefficient.

the initial values of these physical parameters in the rocks, their spatial location with respect to other rocks, and the percentage change of these properties when hydrocarbons are introduced.

In the Gulf of Mexico, the entire Tertiary section is composed of clastic sediments. Productive sands are encountered at all depths. They belong to a large age group ranging from upper Pleistocene to lower Miocene. The sands and shales have a very wide range of physical and acoustical properties. Under these variable conditions, it is necessary to obtain a proper understanding of (1) the velocity-density relationship between shales and sands for various depths; (2) the relationship among velocity, porosity, and clay content for sands; and (3) behavior of S-wave velocity in sands and shales.

Once these properties are understood, it is possible to perform sensitivity analyses and obtain an understanding of the discriminating properties in the prestack domain.

P-Wave Velocities and Densities in Sands and Shales

Figure 19 is a typical example of velocity-density-depth relationships for shales and sands in the Gulf of Mexico. A key feature of this plot is the crossover of sand/shale velocity functions. This crossover plays an important role both in stack and prestack domains. In general, our experience has shown that:

(1) For both sands and normally pressured shales, density increases with depth.
(2) Densities of porous sands are most often lower than densities of shales at the corresponding depth.
(3) The difference in density between shales and sands gradually decreases with depth.
(4) Velocities for both sands and shales are an increasing function of depth (except for overpressure zones). Sands could be *slower*, *equal to*, or *faster* than the shales at the equivalent depth. This velocity relationship between shales and sands is significant for discriminating characteristics in the prestack domain.

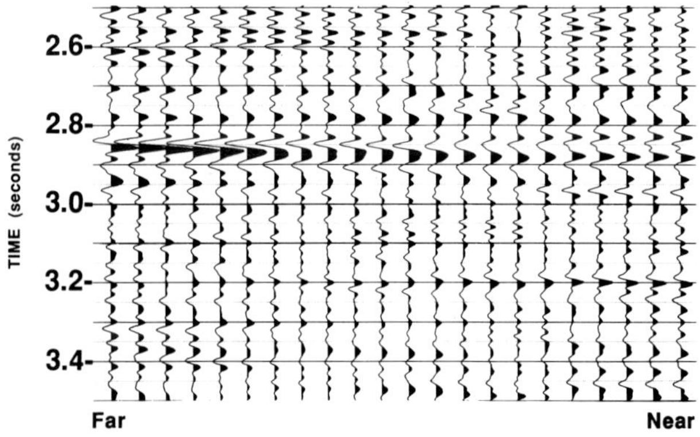

Fig. 17. CDP gather (same as in Figure 15) after amplitude calibration. Peak equals positive reflection coefficient.

Velocity-Porosity-Shaliness Relationship

To determine the velocity-porosity-shaliness relationship, a standard technique from log analysis was used. Neutron porosity and density porosity were determined for all sands and shales for every well in a particular area. Figure 20 shows an example cross-plot of these two porosities. In Figure 20, if a sand has zero clay content, it will fall on the 45 degree line. Once all the points for sands and shales are plotted, a clay point is chosen corresponding to 100 percent clay and zero porosity. This clay point should be lower than any density porosity and higher than any neutron porosity for all shales determined from the log. Given the coordinates of the clay point, true porosity and clay content for all the sands can be determined by the following formulas:

$$\Phi_T = \frac{\Phi_D - K\Phi_N}{1 - K}, \tag{1}$$

$$C = \frac{\Phi_D - \Phi_N}{\Phi_{\mathrm{DSC}} - \Phi_{\mathrm{NSC}}}, \tag{2}$$

and

$$K = \frac{\Phi_{\mathrm{DSC}}}{\Phi_{\mathrm{NSC}}}, \tag{3}$$

where

Φ_T = true porosity,
C = clay content,
Φ_D = density porosity,
Φ_N = neutron porosity,
Φ_{DSC} = density porosity of the clay point, and
Φ_{NSC} = neutron porosity of the clay point.

Once the true porosity and clay content are determined, a contour diagram could be obtained among velocity, porosity, and clay content. Figure 21 shows that velocity, clay content, and porosity are linearly

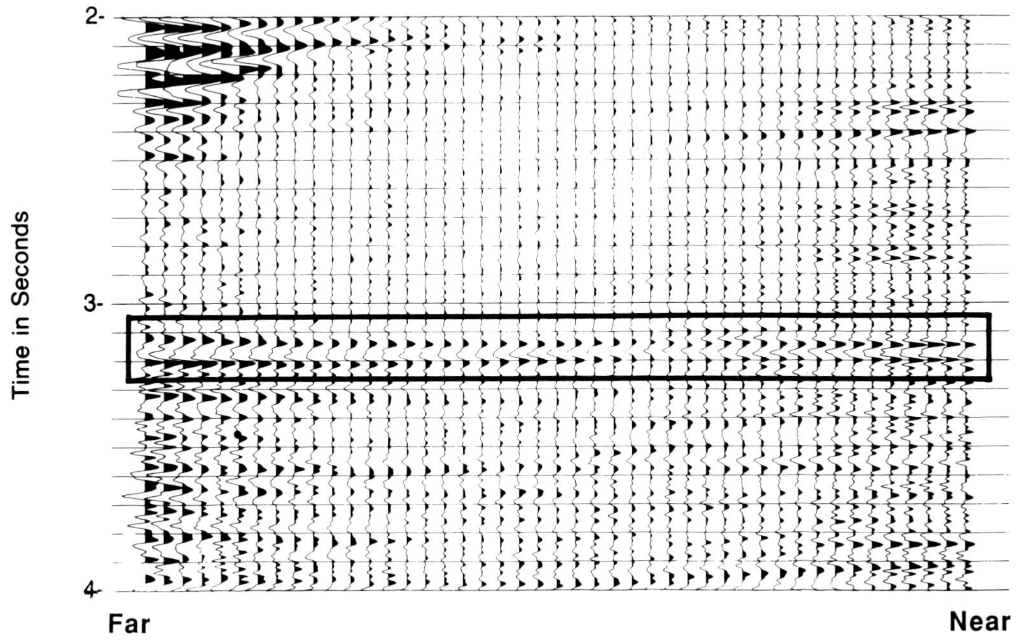

Fig. 18. CDP gather (same as in Figure 16) after amplitude calibration. Peak equals positive reflection coefficient.

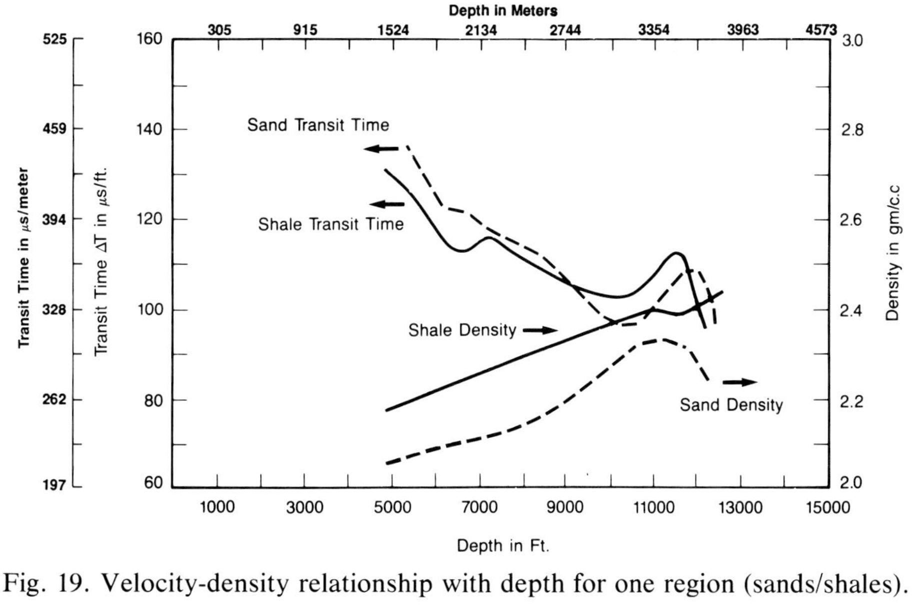

Fig. 19. Velocity-density relationship with depth for one region (sands/shales).

related. For a fixed porosity (inclined lines), increasing the clay content decreases the velocity linearly. On the other hand, for a fixed clay content, as the porosity is increased, velocity also decreases linearly. This implies that from a *P*-wave velocity alone, it will be difficult to distinguish porosity from shaliness.

These types of linear relationships have been independently observed by a variety of workers (e.g., Han et al., 1986).

S-Wave Velocities

As pointed out previously, *S*-wave velocity plays a key role in governing amplitude response in the prestack domain. Reliable measurements of S-wave velocity in unconsolidated sediments (seismic or well log frequencies) are hard to come by. Recently there has been substantial technological development in obtaining some measurements of S-wave velocity in a borehole. Castagna et al. (1985) found P-wave and S-wave velocities for mud rocks to be linearly related with a slope of 1.16 and intercept of 4462 ft/s (1.36 km/s), corresponding to a zero S-wave velocity in fluid. A "time average" line was also found for clean competent sandstones with an effective fluid transit-time of 304 μs/ft for S-waves.

Thus, as a first order approximation, we may assume that (a) P-wave velocity and S-wave velocity in sands and shales follow a linear relationship, and (b) S-wave velocity in sands could be *slower*, *equal to*, or *faster* than the shales of the same depth and will follow the corresponding P-wave velocity relationship.

Sensitivity Analysis

Results from the petrophysical analysis can be summarized as follows:

1. With respect to P-wave velocity, sands could be *slower*, *equal to*, or *faster than* the shales at the equivalent depth. This defines three classes of sands for prestack analysis.
2. Since S-wave velocity is approximately linearly related to P-wave velocity, it will have a similar kind of relationship with respect to shales.
3. In general, sands will have a lower density than the shales at the same depth.
4. P-wave velocity in sands bears a linear relationship with porosity and clay content.
5. For S-waves, Castagna et al. (1985), Han et al. (1986), and others have demonstrated a linear relationship of velocity with porosity and clay content.
6. Introduction of gas to the sands will lower the P-wave velocity and density, keeping the S-wave velocity relatively intact. On the other hand, *lithologic changes will alter all three parameters in the same direction simultaneously.*

The sensitivity analysis is performed with respect to the least known parameter, i.e., S-wave velocity. Since there are three classes of sands, it should be performed separately for each class of sand.

Figures 22, 23 and 24 represent the sensitivity analysis for the three different classes of sands. In Figure 25, cases of high-velocity/low-velocity shale and gas/water contact are presented. In each case, both brine and gas-sand scenarios are presented. In each case, the change in direction of reflection coefficient function is illustrated with respect to changes in S-wave velocity of the reflecting media (equivalent to

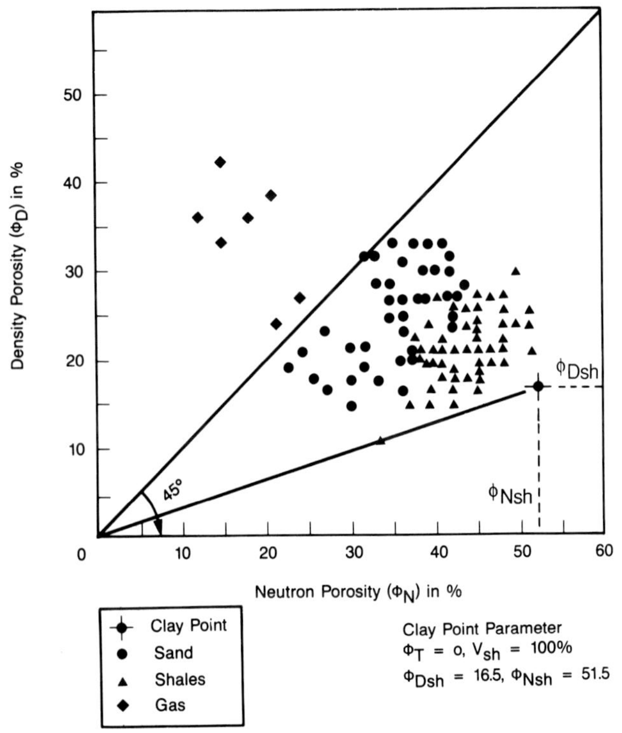

Fig. 20. Neutron-density crossplot for one region.

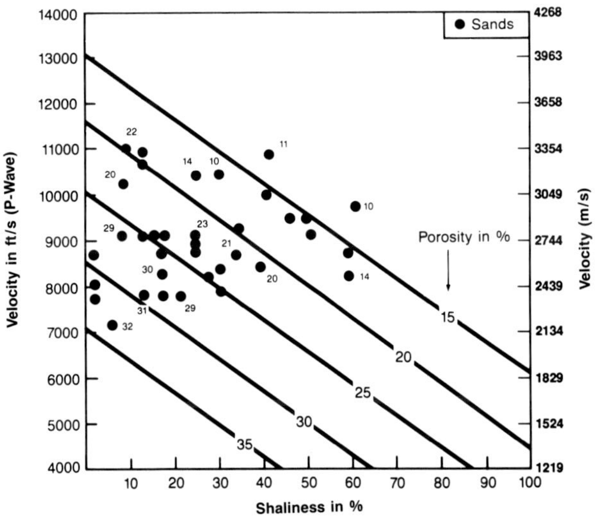

Fig. 21. Velocity-shaliness-porosity plot for the region corresponding to Figure 20.

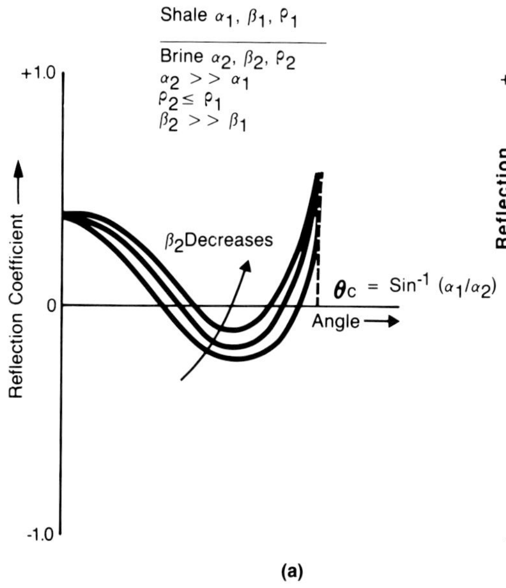

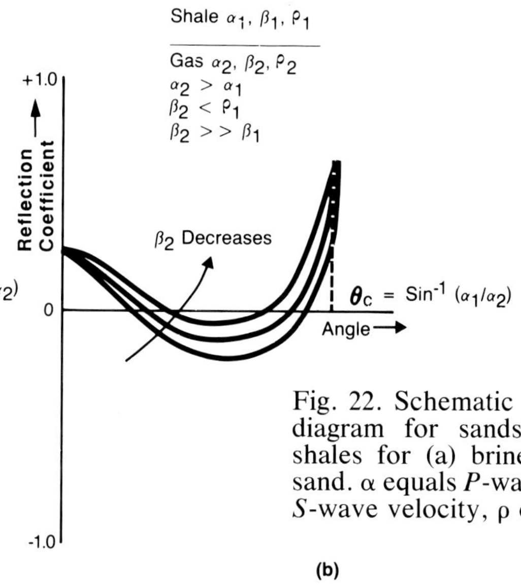

Fig. 22. Schematic reflection coefficient diagram for sands much faster than shales for (a) brine sand, and (b) gas sand. α equals *P*-wave velocity, β equals *S*-wave velocity, ρ equals density.

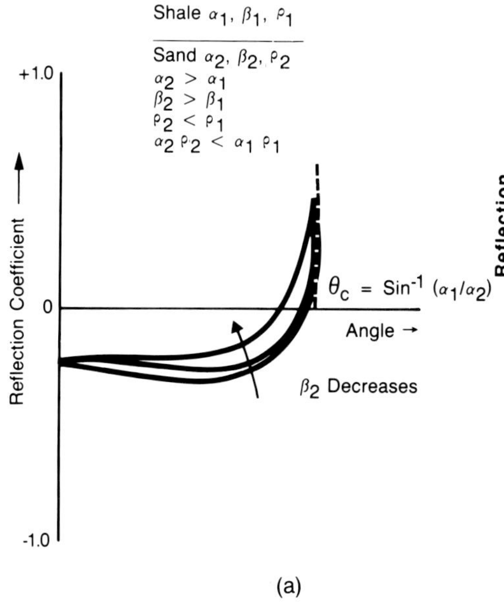

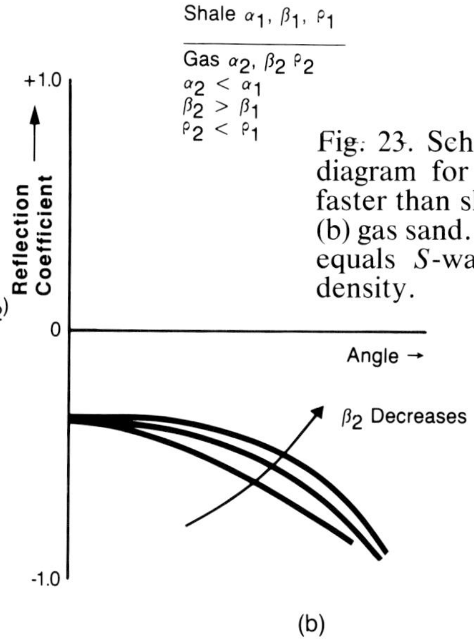

Fig. 23. Schematic reflection coefficient diagram for sands equal to or slightly faster than shales for (a) brine sand, and (b) gas sand. α equals *P*-wave velocity, β equals *S*-wave velocity, and ρ equals density.

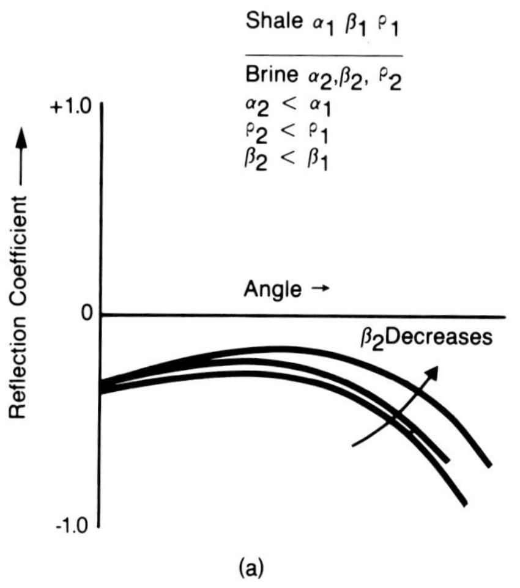

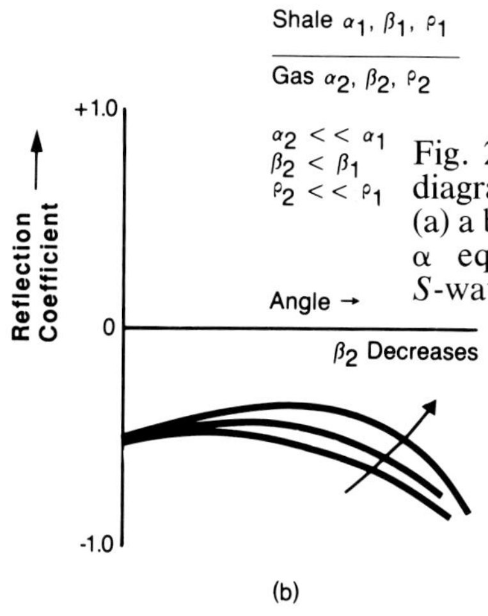

Fig. 24. Schematic reflection coefficient diagram for sands slower than shales for (a) a brine sand, and (b) a brine/gas sand. α equals *P*-wave velocity, β equals *S*-wave velocity, ρ equals density.

the *S*-wave velocity contrast across the reflecting interface). The results can be summarized as follows.

1. When the brine sands are slightly faster than the shales at the equivalent depth, the gas sands can be clearly recognized from the corresponding brine sands.
2. In all other cases, the discriminating properties are subtle. The discriminating properties are the zero offset (zero incident angle) reflection coefficient and rate of change of the reflection coefficient with angle or offset.
3. The most difficult case occurs when the sands are much faster than the shales, such that a positive reflection coefficient is produced at zero offset. This is the case of *dim spots*.
4. In all cases, the magnitude and polarity of the zero offset data are of extreme importance.
5. Bright spots due to lithologic changes like high-velocity/low-velocity shale (Figure 25) *can be easily distinguished from bright spots due to gas sands (Figure 24)*. Bright spots due to lithologic changes will have a rapid decrease of amplitude with offset. *Thus, it is more difficult to distinguish gas sands from brine sands than it is to distinguish gas sands from lithologic changes.*
6. To distinguish gas sands from brine sands, it is necessary (a) to know the velocity regime of sand/shale, and (b) to use both zero-offset reflection coefficient and rate of change of reflection coefficient with offset.

EXAMPLES

This section presents examples of discrimination based on the models developed in the previous section. All data sets were processed in exactly the same manner.

Figures 2 and 26 show a case where sands are slightly faster than shales. Figure 2 shows the stacked seismic section with well locations on both the upthrown and downthrown sides of the fault. The sands in the downdip direction (2.85 s two-way time) are filled with gas and oil, and, on the upthrown side, they are wet. Figures 26(a) and 26(b) present the prestack data for CDPs corresponding to gas and brine, respectively. (All stack data are presented with a negative reflection coefficient corresponding to a peak, whereas the prestack data are presented with a negative reflection coefficient corresponding to a trough.) Notice the clear increase of amplitude with offset in the gas sand [Figure 26(a)] and the absence of such an increase in the wet sand case [Figure 26(b)].

Figure 27 shows the case for a sand that is considerably slower than the shales. Figure 27(a) shows the stacked data set and 27(b) and 27(c) show the gas and brine sands, respectively. The gas sand has higher zero offset (near trace) amplitude and a much slower rate of amplitude variation with offset with respect to these parameters in the brine sand.

Figures 28 and 29 show two cases of false bright spots due to lithologic changes. In Figure 28(a), the event around 1.9 s (deviated well) is productive,

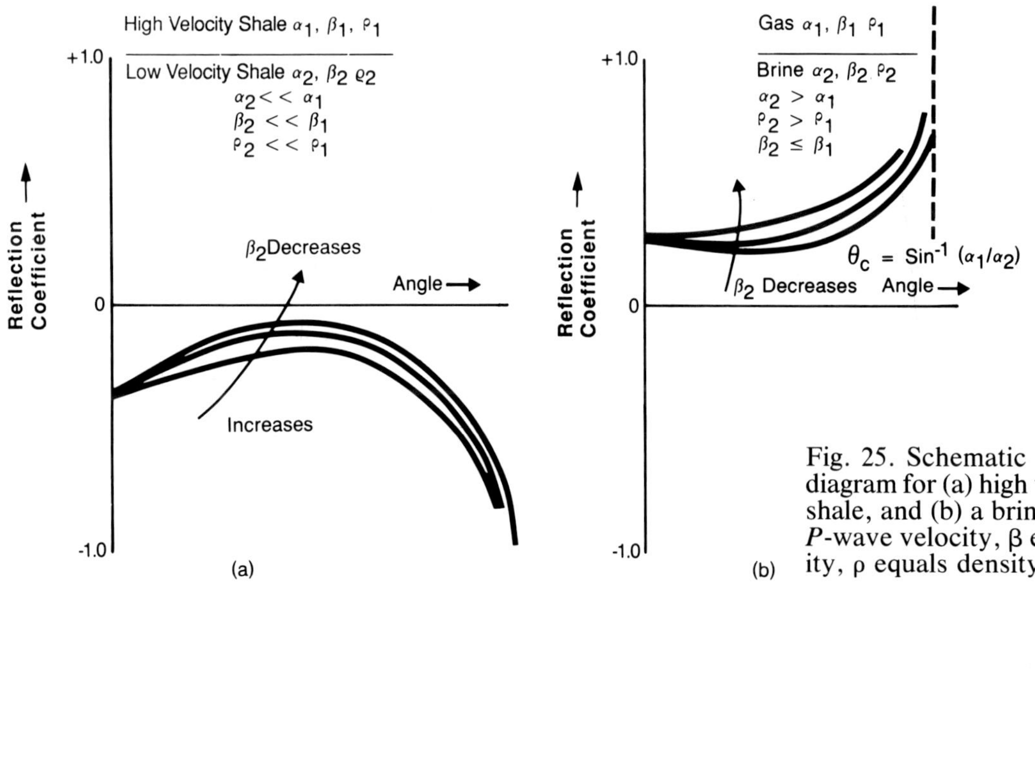

Fig. 25. Schematic reflection coefficient diagram for (a) high velocity/low velocity shale, and (b) a brine interface. α equals *P*-wave velocity, β equals *S*-wave velocity, ρ equals density.

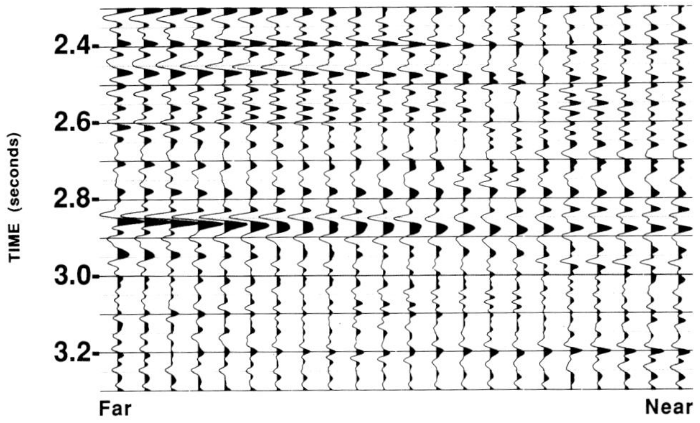

Fig. 26(a). CDP gather at the production well location. Peak equals positive reflection coefficient.

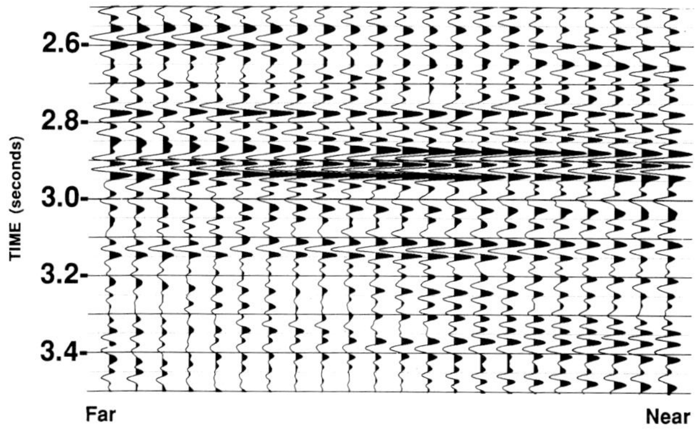

Fig. 26(b). CDP gather at the dry well location. Peak equals positive reflection coefficient.

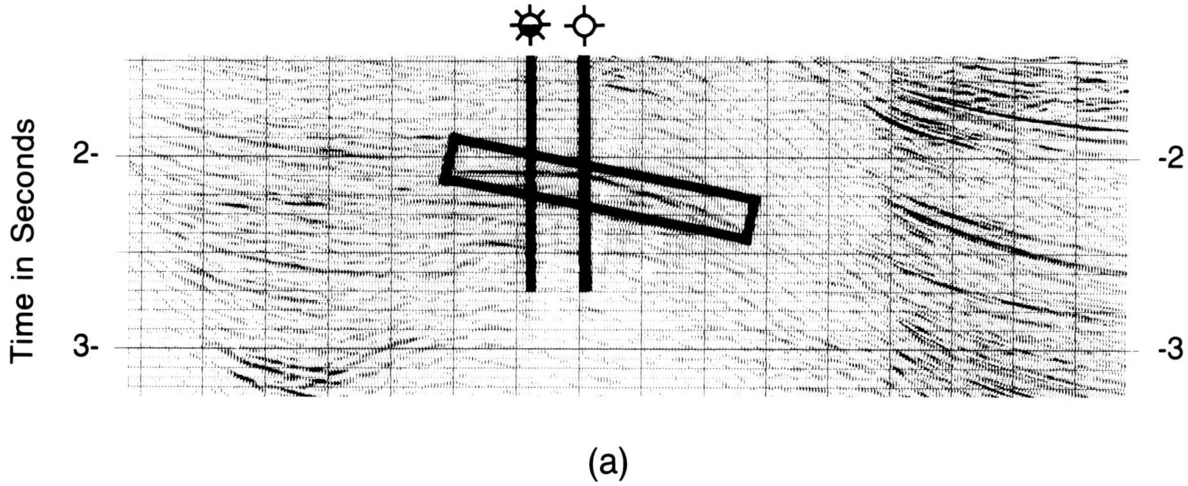

Fig. 27(a). Stack data showing locations of wells. Peak equals negative reflection coefficient.

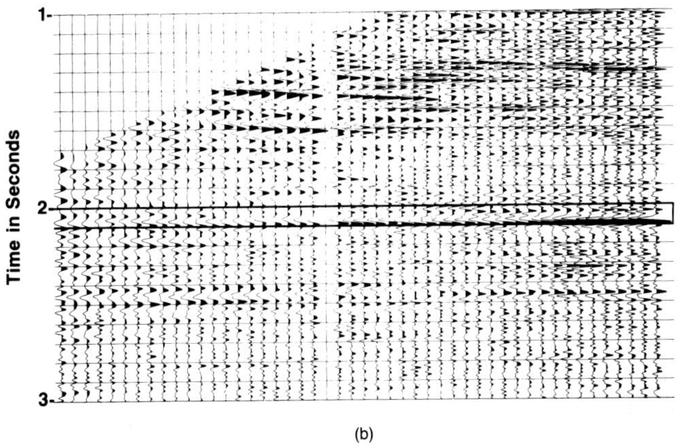

Fig. 27(b). CDP gather at the gas well location. Peak equals positive reflection coefficient.

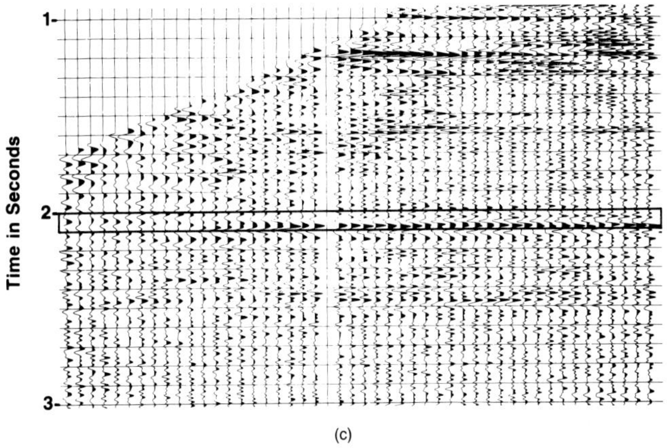

Fig. 27(c). CDP gather in a downdip location, where the well is dry. Peak equals positive reflection coefficient.

whereas the event around 2.0 s is due to lithologic change from a hard streak to shale.

Figure 28(b) represents the prestack data corresponding to the productive horizon and Figure 28(c) represents the prestack response of the false bright spot. In this case, the amplitude decreases rapidly with offset as predicted by the petrophysical model in Figure 25(a).

Figure 1 shows the stacked data for a high-velocity to low-velocity shale interface. Figure 29(a) shows the prestack response at the well location. The rapid decrease of amplitude variation with offset is very distinctive.

Figure 30(a) shows the stacked data (positive reflection coefficient is represented as a peak) response for a sequence of shale/limestone horizon in a sedimentary basin not similar to the Gulf. Figure 30(b) shows the theoretical reflection coefficient diagram for the prestack case and Figure 30(c) shows the prestack data. The similarity between the theoretical and the actual response is evident.

Figure 31 shows the prestack response of a complicated sequence of shale/limestone units. Various parts of the response are quite remarkable in the sense that as a result of stacking it is possible to obtain almost any kind of response ranging from bright spot to dim spot, to no response (zero amplitude).

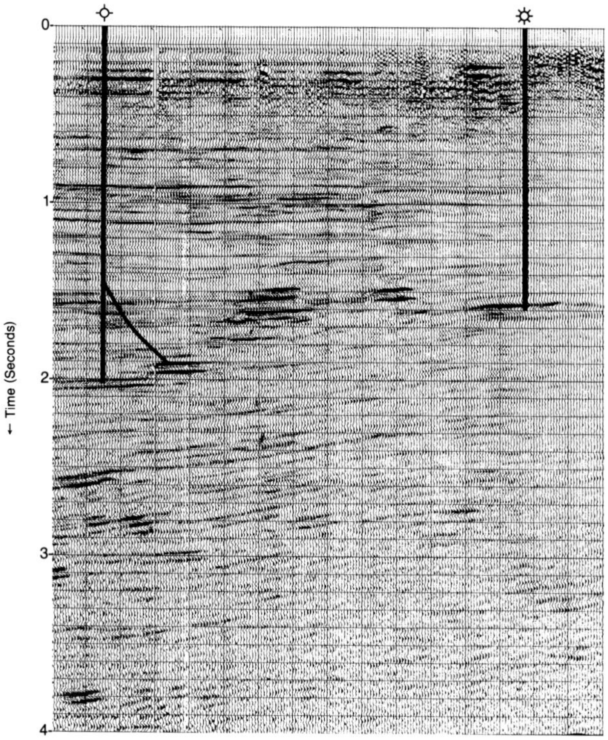

(a)

Fig. 28(a). Stack data showing the location of bright spots (1.9 s) due to gas (deviated well) and bright spots due to lithologic changes (straight well). Peak equals negative reflection coefficient.

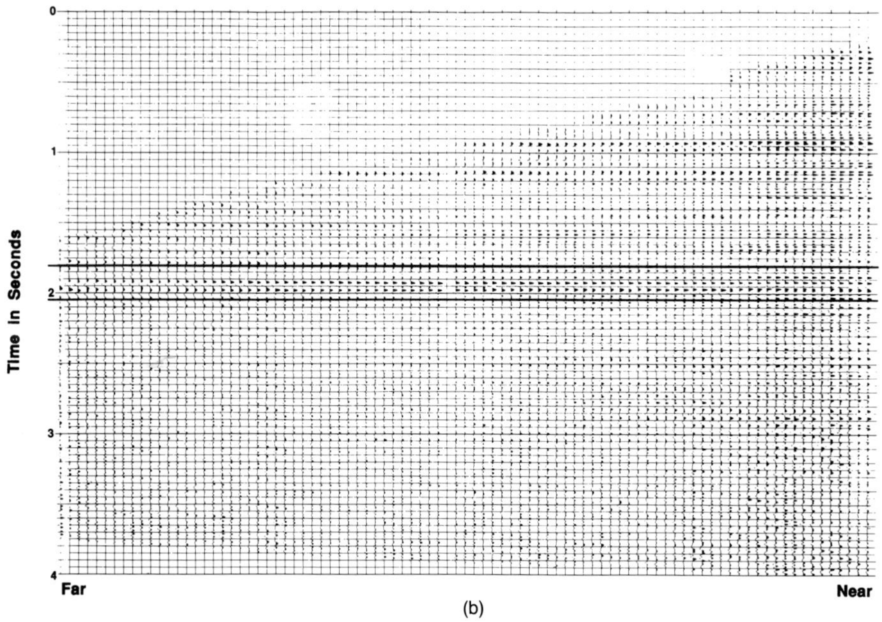

(b)

Fig. 28(b). CDP gather at the location of gas. Peak equals positive reflection coefficient.

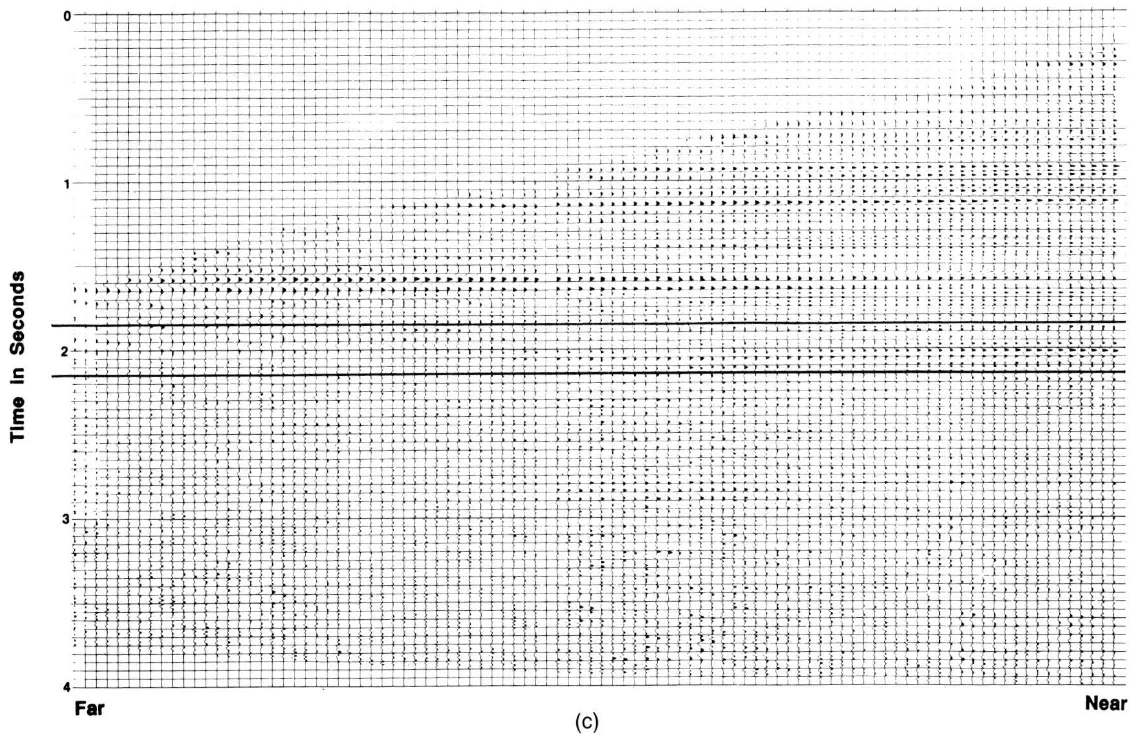

Fig. 28(c). CDP gather at the location of a false bright spot (Lithology change). Peak equals positive reflection coefficient.

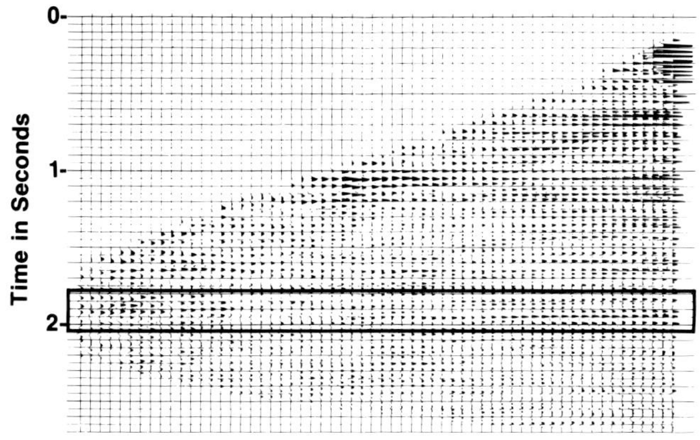

Fig. 29. CDP gather (high velocity shale/low velocity shale) at the location of well in Figure 1. Peak equals positive reflection coefficient.

CONCLUSIONS

In this article, we have discussed

(a) A data processing approach to retrieve true seismic amplitude in the prestack domain;
(b) Analysis of petrophysical properties to demonstrate the significance of optimum velocity (sand/shale) regime for prestack analysis; and
(c) Analysis of real data to integrate the data processing technique and petrophysics demonstrating the feasibility of prestack analysis.

We conclude that prestack analysis can be a powerful exploration tool when the data are processed in a robust manner and the variable effects of rock properties are properly considered.

ACKNOWLEDGMENTS

The authors thank the management of Atlantic Richfield Company for giving permission to publish this paper. The authors are indebted to many colleagues in the Exploration and Production Research organization, without whose help the project would not have been completed in a timely manner. A sincere thank you to Omri C. Crew and Collie F. James II for their contributions to this work. Patty Bogg and Suzan

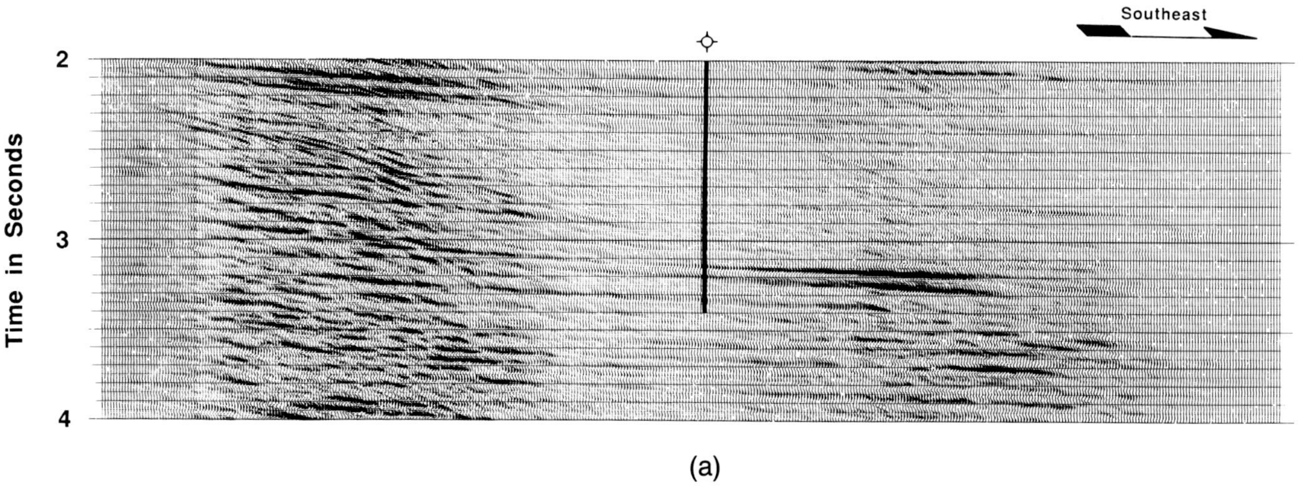

Fig. 30(a). Stack data showing a shale/limestone sequence (3.1 s) at the well location. Peak equals positive reflection coefficient.

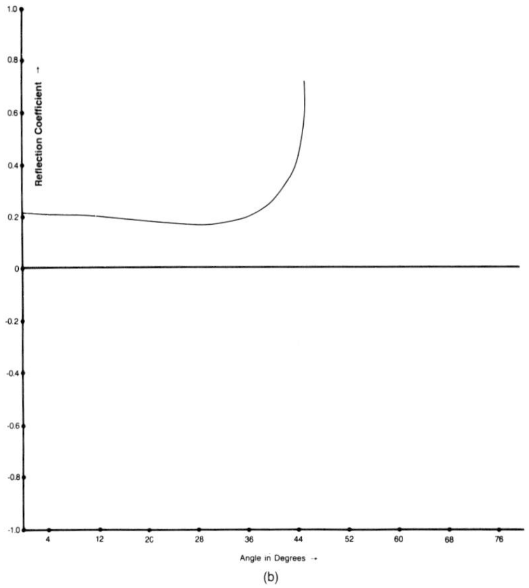

Fig. 30(b). Reflection coefficient diagram for the shale/limestone interface.

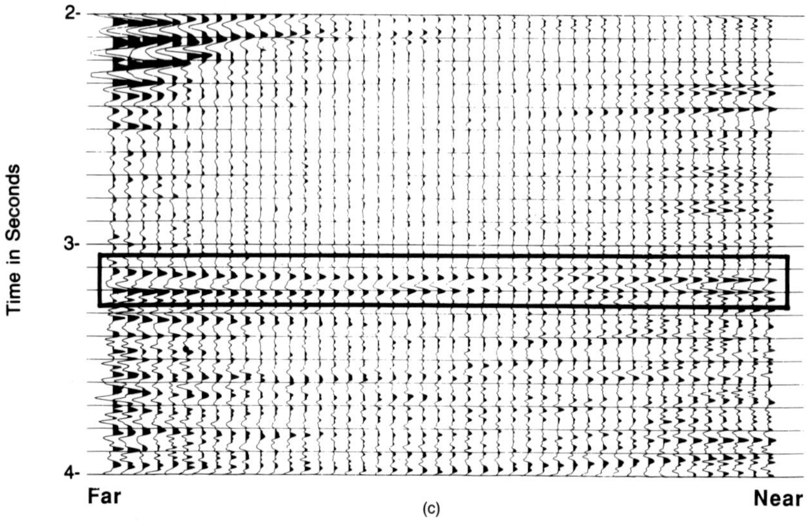

Fig. 30(c). CDP gather for the shale/limestone sequence. Peak equals positive reflection coefficient.

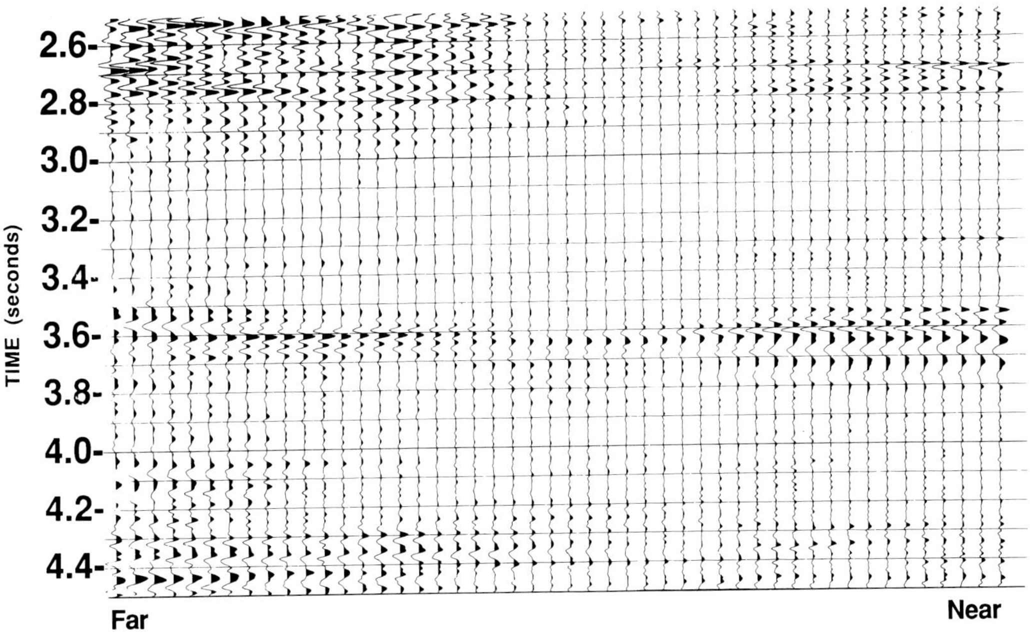

Fig. 31. CDP gather for a sequence of shale/limestone units. Peak equals positive reflection coefficient.

Fleischman deserve special thanks for drafting the manuscript and figures.

REFERENCES

Backus, M. M., 1982, The reflection seismogram in a solid layered earth: 52nd Ann. Internat. Mtg., Soc. Expl. Geophys., Expanded Abstracts, 218–220.

Castagna, J. P., Batzle, M. L., Eastwood, R. L., 1985, Relationships between compressional- and shear-wave velocities in clastic silicate rocks: Geophysics, **50**, 571–581.

Dey-Sarkar, S. K., Chapman, C. H., 1978, A simple method for the computation of body wave seismograms: Bull. Seis. Soc. of America, **68**, 1577–1593.

Domenico, S. N., 1976, Effect of brine-gas mixture on velocity in an unconsolidated sand reservoir: Geophysics, **41**, 882–894.

Domenico, S. N., 1977, Elastic properties of unconsolidated porous sand reservoirs: Geophysics, **42**, 1339–1368.

Gardner, G. H. F., and Harris, M. H., 1968, Velocity and attenuation of elastic waves in sand: Soc. Prof. Well Log Analysts, 9th Ann. Logging Symposium, M1–M19.

Gassaway, G. S., 1983, SAMPLE: Seismic amplitude measurements for primary lithology estimation: 53rd Ann. Internat. Mtg., Soc. Expl. Geophys., Expanded Abstracts, 610–613.

Gregory, A. R., 1976, Fluid saturation effects on dynamic elastic properties of sedimentary rocks: Geophysics, **41**, 895–921.

Han, D., Nur, A., and Morgan, D., 1986, Effects of porosity and clay content on wave velocities in sandstones: Geophysics, **51**, 2093–2107.

Hampson, D., 1986, Inverse velocity stacking for multiple elimination: 56th Ann. Internat. Mtg., Soc. Expl. Geophys., Expanded Abstracts, 422–424.

Ostrander, W. J., 1984, Plane wave reflection coefficients for gas sands at non-normal angles of incidence: Geophysics, **49**, 1637–1648.

Sengbush, R. L., 1983, Seismic exploration methods: Internat. Human Res. Dev. Corp. Publication, Boston, 187–230.

Shuey, R. T., and Banik, N. C., 1984, Amplitude from curved reflections: 54th Ann. Internat. Mtg., Soc. Expl. Geophys., Expanded Abstracts, 664–665.

Properties of Direct AVO Hydrocarbon Indicators

*Herbert W. Swan**

INTRODUCTION

An important clue in the detection of hydrocarbon deposits is the way in which seismic waves vary in amplitude as a function of incidence angle when they are reflected from a buried rock interface. Careful measurements of these small amplitude variations may be combined to form a variety of *hydrocarbon indicators* (HCIs), which ideally yield a strong positive response where hydrocarbons are present, and a weak or negative response elsewhere. Figure 1 shows one way in which HCIs may be categorized. *Linear* indicators are comprised of additive contributions from each reflector. These indicators can be based on known geologic trends, and thus have a strong petrophysical basis. However, they can be tedious to interpret in large volume, since it is impossible to tell from a single linear indicator itself whether the reflection *magnitude* increases or decreases with angle. Two or more indicators must be compared side-by-side to get this information. They are also highly sensitive to velocity errors in their generation. Nonlinear *product* indicators get around this problem, at the expense of introducing thin-bed interference effects, which also complicate the interpretation in a different way. Other nonlinear algorithms, similar to that of Biondi and Kostov (1989) which exploit the eigenstructure of the prestack data, are also beginning to appear.

Linear HCIs may be further subdivided into constrained and unconstrained types. Unconstrained types are formed by performing a least-squares parametric fit of the prestack data. As long as there is no systematic (deterministic) error in the data, only zero-mean random noise, these indicators will be completely unbiased. They will average out to the correct answer after a large number of statistical trials. Unfortunately, they can also be very ill-conditioned. Hard constraints improve conditioning by eliminating some of the unknowns, but lead to unbiased results only if the constraints are indeed correct. Soft constraints provide a compromise between the unconstrained and the tightly constrained solutions by simultaneously minimizing both the model-data distance and the model-constraint distance. However, they must be used with caution since they do bias the answers in the direction of the constraints. Any data could thus be made to support any conclusion, if constraints are applied too forcefully.

The choice of an "optimal" HCI is data and area dependent, but all indicators are subject to deterministic and statistical errors to some degree. This paper discusses the characteristics of some of the more common indicators, and quantifies the uncertainties they are likely to have.

PROBLEM FORMULATION

A simple interface has a reflectivity for plane waves that may be approximated by

$$r(\phi) = A + B \sin^2 \phi + C \sin^2 \phi \tan^2 \phi, \qquad (1)$$

where ϕ is the incidence angle. One may refer to the coefficients (A, B, C) as the *angular* reflection coefficients, or the angular RCs, for short. This sequence is useful because: (1) Its members may be expressed as linear combinations of the medium property changes at each interface, provided these changes are small compared with their background values, and (2) Prestack seismic data, which have been corrected for normal moveout, may be least-squares fitted to equation (1) in order to obtain empirical estimates of its members. These data fits are equivalent to weighted stacks, designed to enhance A, B, and C as functions of time.

Zoeppritz Approximations

Aki and Richards (1980) demonstrated point 1 by approximating the angular RCs in the following form:

$$A = \frac{1}{2}\left(\frac{\Delta V_p}{V_p} + \frac{\Delta\rho}{\rho}\right); \qquad (2)$$

*ARCO Exploration and Production Technology, Division of Atlantic Richfield Company, 2300 W. Plano Pkwy, Plano, TX 75075-8427.

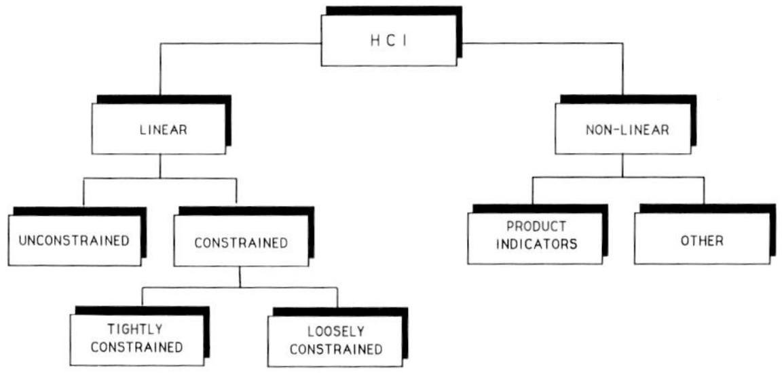

Fig. 1. A classification of hydrocarbon indicators.

$$B = \frac{1}{2}\frac{\Delta V_p}{V_p} - 2\left(\frac{V_s}{V_p}\right)^2\left(2\frac{\Delta V_s}{V_s} + \frac{\Delta \rho}{\rho}\right); \quad (3)$$

$$C = \frac{1}{2}\frac{\Delta V_p}{V_p}, \quad (4)$$

where V_p, V_s, and ρ are the average values across the boundary for the compressional velocity, shear velocity, and density, respectively. Shuey (1985), while retaining the same expressions for A and C, derived an alternate expression for the "slope" term B, in terms of Poisson's ratio, σ_p:

$$B = -\left(\frac{3 - 7\sigma_p}{2 - 2\sigma_p}\right)\frac{\Delta V_p}{V_p} - \left(\frac{1 - 2\sigma_p}{1 - \sigma_p}\right)\frac{\Delta \rho}{\rho} + \frac{1}{(1 - \sigma_p)^2}\Delta\sigma_p. \quad (5)$$

Poisson's ratio may be obtained from the V_p/V_s ratio as follows:

$$\sigma_p = \frac{1}{2}\left[\frac{(V_p/V_s)^2 - 2}{(V_p/V_s)^2 - 1}\right]. \quad (6)$$

By solving these equations, any HCI which is expressed in terms of rock properties can also be expressed as a linear combination of A, B, and C.

Least-Squares Solution

Point (2) above, may be shown by writing (A, B, C) as the vector $\mathbf{c}^3$. Then at each instant of time, the moved-out common midpoint gather may be modeled as:

$$\mathbf{d} = \underset{\sim}{\mathbf{S}}_3\mathbf{c}^3 + \mathbf{n}. \quad (7)$$

Here $\mathbf{d}$ and $\mathbf{n}$ are the $M \times 1$ data and noise vectors, respectively, and $\underset{\sim}{\mathbf{S}}_3$ is the "system matrix:"

$$\underset{\sim}{\mathbf{S}}_3 = \begin{bmatrix} 1 & \alpha_1^2 & \alpha_1^4/(1 - \alpha_1^2) \\ 1 & \alpha_2^2 & \alpha_2^4/(1 - \alpha_2^2) \\ \vdots & \vdots & \vdots \\ 1 & \alpha_M^2 & \alpha_M^4/(1 - \alpha_M^2) \end{bmatrix} \quad (8)$$

Each α_i represents the sine of the ith incidence angle in the midpoint gather.

Sometimes C is completely neglected in the analysis. In this case the solution vector becomes $\mathbf{c}^2 = (A, B)$, and the system matrix becomes $\underset{\sim}{\mathbf{S}}_2$, which consists of the first two columns of $\underset{\sim}{\mathbf{S}}_3$. In the future, I use subscripts on $\mathbf{c}$ and $\underset{\sim}{\mathbf{S}}$ only when I wish to distinguish between the two and three-parameter cases. Without the scripts, they refer equally well to both cases.

The classic solution for $\mathbf{c}$ which minimizes

$$E = \|\underset{\sim}{\mathbf{S}}\mathbf{c} - \mathbf{d}\|^2 + \gamma M\|\mathbf{c}\|^2, \quad (9)$$

is

$$\hat{\mathbf{c}}_c = (\underset{\sim}{\mathbf{S}}^T\underset{\sim}{\mathbf{S}} + \gamma M\underset{\sim}{\mathbf{I}})^{-1}\underset{\sim}{\mathbf{S}}^T\mathbf{d}, \quad (10)$$

where the subscript "c" denotes "softly constrained." This procedure has become the standard method of amplitude variation with offset (AVO) analysis.

The γ term of equation (9) limits the norm of the solution vector. It usually is added to provide stability of the solution where it would otherwise be lacking, e.g., in low-fold or low-angle areas. γ should be chosen as small as possible in the zones of interest; otherwise, the solution will be biased in ways we shall explore next. As γ approaches zero, $\hat{\mathbf{c}}$ approaches the unconstrained least-squares solution,

$$\hat{\mathbf{c}} = (\underset{\sim}{\mathbf{S}}^T\underset{\sim}{\mathbf{S}})^{-1}\underset{\sim}{\mathbf{S}}^T\mathbf{d}. \quad (11)$$

Soft Constraints

Soft constraints increase the stability of the results, but also bias them. In other words, the expected value of the estimates no longer equals the true answer. To see this, let the constrained solution be expressed in terms of the unconstrained one:

$$\hat{\mathbf{c}}_c = (\underset{\sim}{\mathbf{S}}^T\underset{\sim}{\mathbf{S}} + \gamma M\underset{\sim}{\mathbf{I}})^{-1}(\underset{\sim}{\mathbf{S}}^T\underset{\sim}{\mathbf{S}})\hat{\mathbf{c}} \quad (12)$$

$$= [\mathbf{I} + \gamma M(\underset{\sim}{\mathbf{S}}^T\underset{\sim}{\mathbf{S}})^{-1}]^{-1}\hat{\mathbf{c}} = \underset{\sim}{\mathbf{R}}\hat{\mathbf{c}}. \quad (13)$$

$\underset{\sim}{\mathbf{R}}$ is called the *model resolution matrix*, after Menke (1989).

For a 2-parameter inversion ($N = 2$), assuming that $C = 0$ and that the noise has zero mean, the expected values of the softly constrained estimates may be written as

$$E\{\hat{A}_c\} = r_{11}A + r_{12}B, \tag{14}$$

$$E\{\hat{B}_c\} = r_{21}A + r_{22}B, \tag{15}$$

where $\{r_{ij}\}$ are the elements of $\underset{\sim}{\mathbf{R}}$, and A and B are the true angular RCs of equation (1). For a uniformly spaced receiver cable which fixes the α_i's of equation (8) to be

$$\alpha_i = \frac{i \tan \theta_M}{\sqrt{M^2 + i^2 \tan^2 \theta_M}}, \qquad i = 1, \ldots, M, \tag{16}$$

all the elements of $\underset{\sim}{\mathbf{R}}$ are positive for all positive γs. It is thus meaningful to discuss the mean percentages of the constrained slope and zero-offset response (P_B and P_A, respectively) which are actually determined by their underlying values:

$$P_B(\gamma) = r_{22}/(r_{21} + r_{22}); \tag{17}$$

$$P_A(\gamma) = r_{11}/(r_{11} + r_{12}). \tag{18}$$

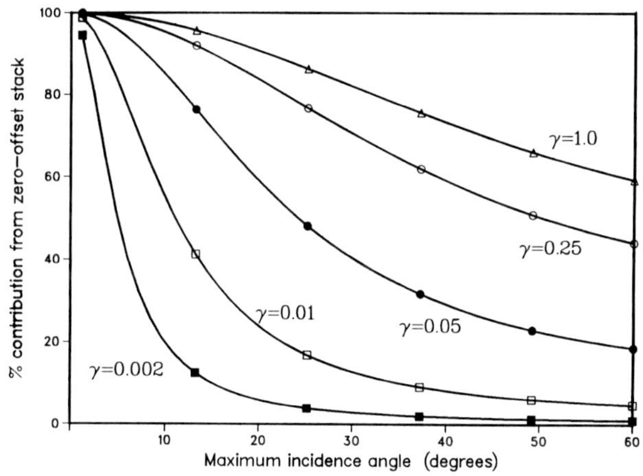

Fig. 2. The percentage of constrained two-parameter slope estimate which is actually contributed by the zero-offset response, under five degrees (γ) of soft constraints.

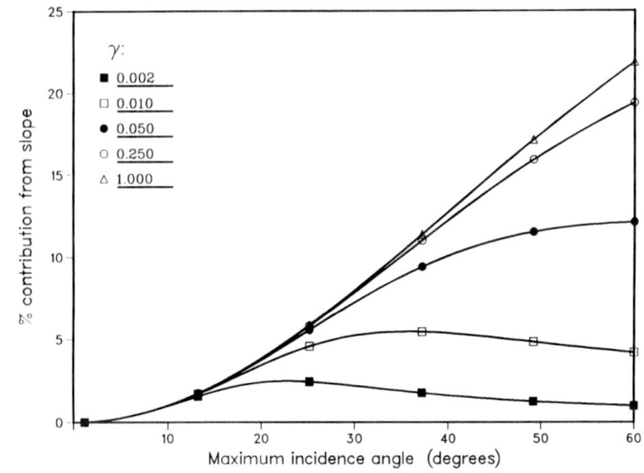

Fig. 3. The percentage of constrained two-parameter zero-offset estimate which is actually contributed by the AVO slope, under five degrees (γ) of soft constraints.

These percentages are plotted in Figures (2) and (3), respectively, as a function of the maximum incidence angle. The fold M, is assumed to be >20, large enough to make the Riemann-like sums of the matrix multiplication in equation (10) approach integrals.

It is easy to see how a soft constraint on the solution magnitude stabilizes the slope solution. At low to moderate angles, it decreases the contribution from the true slope (the unstable part), while it mixes in a portion of the zero offset response, which is much more stable. Figure 2 shows that for $\gamma = 0.05$, and a maximum incidence angle of 30 degrees only 60 percent of the mean slope estimate actually comes from the underlying slope. The other 40 percent is derived from A. It thus becomes difficult to *resolve* the effect of slope from that of the zero-offset response.

Transformations and Hard Constraints

Very often the data are best modeled not as equation (1), but as the sum of P linear combinations of (1, $\sin^2 \phi$, $\sin^2 \phi \tan^2 \phi$). In this case, there is a transformation matrix $\underset{\sim}{\mathbf{T}}$ which expresses the original angular RCs in terms of the new model parameters, ($p_1 \cdots p_P$); i.e., $\mathbf{c} = \underset{\sim}{\mathbf{T}}\mathbf{p}$. The data model equation (7) can then be revised to utilize this new parameterization:

$$\mathbf{d} = \underset{\sim}{\mathbf{S}}\underset{\sim}{\mathbf{T}}\mathbf{p} + \mathbf{n}. \tag{19}$$

For example, suppose it were known that $\Delta\rho = 0$. Then $A = C$, and the model for the reflectivity function equation (1) would become

$$r(\phi) = p_1(1 + \sin^2 \phi \tan^2 \phi) + p_2 \sin^2 \phi. \tag{20}$$

The transformation matrix would be found from,

$$\mathbf{c} = \begin{bmatrix} A \\ B \\ C \end{bmatrix} = \underbrace{\begin{bmatrix} 1 & 0 \\ 0 & 1 \\ 1 & 0 \end{bmatrix}}_{T} \begin{bmatrix} p_1 \\ p_2 \end{bmatrix}. \tag{21}$$

Substituting equation $\mathbf{c} = \underset{\sim}{\mathbf{T}}\mathbf{p}$ into equation (9), and assuming that γ is small enough to be neglected in the zone of interest, the error function now being minimized is

$$E = \|\underset{\sim}{\mathbf{S}}\underset{\sim}{\mathbf{T}}\mathbf{p} - \mathbf{y}\|^2. \tag{22}$$

Differentiating this expression with respect to $\mathbf{p}$ leads to the optimal parameter vector,

$$\hat{\mathbf{p}} = \underset{\sim}{\mathbf{V}}\hat{\mathbf{c}}, \tag{23}$$

where the $\underset{\sim}{\mathbf{V}}$ matrix projects and transforms the optimal angular RCs into coordinates of the new parameter space:

$$\underset{\sim}{\mathbf{V}} = (\underset{\sim}{\mathbf{T}}^T\underset{\sim}{\mathbf{S}}^T\underset{\sim}{\mathbf{S}}\underset{\sim}{\mathbf{T}})^{-1}\underset{\sim}{\mathbf{T}}^T\underset{\sim}{\mathbf{S}}^T\underset{\sim}{\mathbf{S}}. \tag{24}$$

If the number of parameters P, is less than the number of angular RCs, N (either 2 or 3), then the $\underset{\sim}{\mathbf{T}}$ matrix may be said to be a *hard* constraint upon the data, since it reduces the order of the inversion. Hard constraints are often necessary to produce useful HCIs in the presence of noise, as we shall see. However, transforming the unconstrained RC estimates via $\underset{\sim}{\mathbf{V}}$ is thus equivalent to solving for the constrained parameters directly using equation (19). If $\underset{\sim}{\mathbf{T}}$ is of full rank, the inversion remains unconstrained, and the $\underset{\sim}{\mathbf{V}}$ matrix becomes $\underset{\sim}{\mathbf{T}}^{-1}$.

LINEAR HYDROCARBON INDICATORS

Gardner's Relationship

This section introduces specific constraints and transformations applicable to hydrocarbon detection. One commonly used constraint is derived from Gardner's relationship as reported in Gardner et al. (1974):

$$\rho \propto V_p^{1/4}. \tag{25}$$

Although this formula is reasonably accurate for water-filled sand/shale sequences, it is not very accurate for gas sands. However, HCIs are designed to eliminate water-filled bright spots and tight-streaks from consideration as exploration targets. It is, therefore, desirable if gas sands do not conform to Gardner, because it makes them less likely to be screened out by the HCI.

Let us now differentiate equation (25) and divide it by itself to obtain:

$$\frac{\Delta\rho}{\rho} \approx \frac{1}{4}\frac{\Delta V_p}{V_p}. \tag{26}$$

There are considerable perils in doing this, since small deviations about the Gardner trend are amplified by its differentiation. On the average, though, equation (26) should be valid.

Fractional Velocity Changes

Gardner's differentiated relation may be used to obtain robust estimates for the fractional changes in P and S-wave interval velocities. Substituting equation (26) into equations (2) and (3), it is seen that the hard constraint matrix $\underset{\sim}{\mathbf{T}}$, and the new parameter vector $\mathbf{p}$ may be expressed as

$$\underset{\sim}{\mathbf{T}} = \begin{bmatrix} 5/8 & 0 \\ \frac{1}{2}(1 - v_r^2) & -4v_r^2 \\ 1/2 & 0 \end{bmatrix}; \quad \mathbf{p} = \begin{bmatrix} \Delta V_p/V_p \\ \Delta V_s/V_s \end{bmatrix}, \tag{27}$$

where v_r is the V_s/V_p ratio. If $\hat{A}$ and $\hat{B}$ have already been determined by another method, then

$$\frac{\Delta V_p}{V_p} = \frac{8}{5}\hat{A}; \tag{28}$$

$$\frac{\Delta V_s}{V_s} = \frac{1}{5}\left[\left(\frac{V_p}{V_s}\right)^2 - 1\right]\hat{A} - \frac{1}{4}\left(\frac{V_p}{V_s}\right)^2\hat{B}. \tag{29}$$

This formulation is useful since a sudden drop in V_p/V_s is often associated with the presence of gas in sands.

The Fluid Factor

Smith and Gidlow (1987) introduced an indicator they called the "fluid factor," ΔF, which was designed to emphasize deviations from the $\Delta V_p = k\Delta V_s$ mudrock trend of Castagna et al. (1985):

$$\Delta F = \frac{\Delta V_p}{V_p} - k\left(\frac{V_s}{V_p}\right)\frac{\Delta V_s}{V_s}. \tag{30}$$

Substituting equations (28) and (29) into this expression, results in an equivalent form of the fluid factor in terms of $\hat{A}$ and $\hat{B}$:

$$\Delta F = \frac{1}{5}\left[8 - k\left(\frac{V_p}{V_s}\right) + k\left(\frac{V_s}{V_p}\right)\right]\hat{A} + \frac{1}{4}k\left(\frac{V_p}{V_s}\right)\hat{B}. \tag{31}$$

For example, if $k = 1.16$ and $V_p/V_s = 2$ (typical values), then

$$\Delta F = 1.252\hat{A} + 0.58\hat{B}, \tag{32}$$

which is zero if the Castagna trend holds. If the fluid factor is to be solved in conjunction with P-impedance changes, ΔZ_p, the constraint matrix and parameter vector for equation (19) are

$$\underline{\mathbf{T}} = \begin{bmatrix} 1 & 0 \\ 0.8(1 - 8v_r/k - v_r^2) & 4v_r/k \\ 0.8 & 0 \end{bmatrix}; \tag{33}$$

$$\mathbf{p} = \begin{bmatrix} \Delta Z_p/Z_p \\ \Delta F \end{bmatrix}.$$

P- and *S*-Wave Reflection Coefficients

Another candidate HCI combination is the P- and S-wave reflection coefficients:

$$R_{pp} = \frac{1}{2}\left(\frac{\Delta V_p}{V_p} + \frac{\Delta \rho}{\rho}\right); \tag{34}$$

$$R_{ss} = \frac{1}{2}\left(\frac{\Delta V_s}{V_s} + \frac{\Delta \rho}{\rho}\right). \tag{35}$$

Using the Gardner constraint, I obtain after Wiggins et al. (1984),

$$R_{pp} = \hat{A}; \tag{36}$$

$$R_{ss} = \frac{1}{10}\left[\left(\frac{V_p}{V_s}\right)^2 + 1\right]\hat{A} - \frac{1}{8}\left(\frac{V_p}{V_s}\right)^2 \hat{B}. \tag{37}$$

If $V_p/V_s = 2$, then

$$R_{ss} = \frac{1}{2}(\hat{A} - \hat{B}). \tag{38}$$

If hard constraints are used to estimate R_{pp} and R_{ss}, then

$$\underline{\mathbf{T}} = \begin{bmatrix} 1 & 0 \\ \frac{4}{5}(1 + v_r^2) & -8v_r^2 \\ 4/5 & 0 \end{bmatrix}; \quad \mathbf{p} = \begin{bmatrix} R_{pp} \\ R_{ss} \end{bmatrix}. \tag{39}$$

The shear reflectivity is useful because it may be expressed in terms of the shear modulus μ_s and density ρ:

$$R_{ss} = \frac{1}{4}\left(\frac{\Delta \rho}{\rho} + \frac{\Delta \mu_s}{\mu_s}\right). \tag{40}$$

Gassman (1951) showed that the shear modulus within a sandstone is relatively independent of fluid content, if other factors remain constant. Thus $\Delta \mu_s$ from shale to sand remains relatively independent of gas content, and R_{ss} variations become dominated by density variations within the sand. Gas-filled sands are characterized by abnormally low densities.

Inversion Strategies

There are several strategies to implement these indicators, each with its own advantages and disadvantages:

1. Perform an unconstrained or softly constrained three-parameter inversion for $\mathbf{c}_3$ using equation (10), then use a 3×3 transformation matrix, obtainable from equations (2) thru (5) to yield the HCIs.
2. Perform a hard-constrained estimate using equation (23), using one of the constraint matrices: equations (27), (33), or (39).
3. Perform an unconstrained or softly constrained 3-parameter inversion for $\mathbf{c}_3$, throw away the estimate of C and apply a 2×2 transformation on $(\hat{A}, \hat{B})$, such as equation pairs (28) and (29), (36) and (31), or (36) and (37).
4. Perform an unconstrained or softly constrained 2-parameter inversion for $\mathbf{c}_2$, and use one of the 2×2 transformations listed under item 3 to obtain the indicators.

This article does not advocate one strategy over another. However, some general statements about their respective strengths and weaknesses can be made. These are tabulated in Table (1). The performance criteria used were:

Noise immunity and bias—The degree of sensitivity of the estimates to noise or bias, depending on the soft constraint parameter γ.

Table 1. Relative performance of competing HCI computation strategies

Performance category	Strategy 1	Strategy 2	Strategy 3	Strategy 4
Noise immunity/bias	poor	good	fair	best
$\text{Sin}^4\,\phi$ immunity	poor	fair	good	poor
Assumptions	good	fair	fair	fair

Sin^4 ϕ immunity—The susceptibility of the method to spurious data components which vary proportionally to the fourth or higher powers of sin ϕ.

Assumptions—The number of apriori assumptions (e.g., constraints, trend curves) that were needed to justify the method.

Method 1 requires the fewest assumptions, but with the same degree of input noise, yields either the most noise-ridden or the most biased estimates, depending on the soft constraints. This method is typically useful for estimating only P-impedance changes, or as we will see, Poisson ratio changes. Method 2 achieves relatively good noise immunity, but is susceptible to higher-order data components through the third column of the $\underset{\sim}{\mathbf{V}}$ matrix equation (24). $\hat{C}$ is the most ill-conditioned angular RC, and is, therefore, the most easily contaminated by these deterministic components. Method 3 is the most immune to sin^4 ϕ errors, while method 4 is the most statistically robust. Since the latter two methods do not involve the estimate of C, they may generically be expressed as:

$$\text{Indicator} = \alpha \hat{A} + \beta \hat{B}, \tag{41}$$

with the coefficients α and β plotted in Figure 4, as a function of the background V_p/V_s ratio.

PRODUCT INDICATORS

Linear HCIs exhibit the usual properties of any linear data transformation, namely scalability and superposition. Their resolution limits are, therefore, well understood. (See Widess, 1973.) These indicators, however, have four other properties, which make them somewhat difficult to interpret:

1. They are easily contaminated by out-of-phase error components, caused (as we will see) by small velocity errors.
2. They are wavelet phase dependent. If the wavelet phase is rotated by an arbitrary angle, the linear HCI is rotated by the same angle. Since phase is difficult to determine without well control, it becomes unclear what character of event to look for: a peak, a trough, or a doublet.
3. To determine whether the magnitude of a reflection coefficient increases or decreases with offset is impossible from a linear HCI alone. Meticulous comparison with the conventional or zero-offset stack is essential, which is tedious to do over wide areas.
4. If the elastic parameters of a thin bed are largely different from its encasement (typical for gas beds), then a linear HCI would display opposing signs at the top and the bottom, thereby changing the phase of the response as a function of bed thickness.

To circumvent these problems, a product indicator may be used:

$$P(t) = \mathcal{R}e[A_c(t)B_c^*(t)]. \tag{42}$$

Here A_c and B_c represent the complex *analytical* zero-offset and slope RC estimates:

$$A_c(t) = \hat{A}(t) = i\mathcal{H}i[\hat{A}(t)]; \tag{43}$$

$$B_c(t) = \hat{B}(t) + i\mathcal{H}i[\hat{B}(t)], \tag{44}$$

where $\hat{A}$ and $\hat{B}$ are the estimated (real) angular RCs, $\mathcal{H}i$ represents the Hilbert transform, and $i = \surd(-1)$. The product indicator measures the component of slope which is in phase with the zero-offset response. Because of the conjugation of B_c, it is independent of the wavelet phase, and its sign makes it readily apparent whether the magnitude of the reflection coefficient increases or decreases with offset. Furthermore, for thin beds with opposing reflectivity functions, the direct product terms at the top and the bottom of the bed reinforce each other, thus preserving its sign. A final advantage of the product indication is that it is less sensitive to small stacking velocity errors than the AVO slope itself.

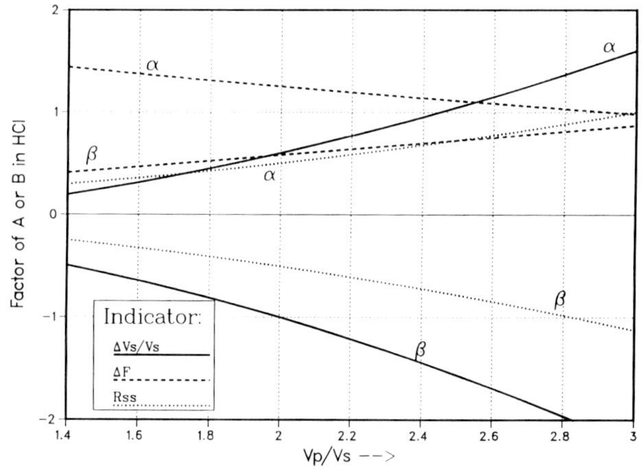

Fig. 4. Contributions to several linear hydrocarbon indicators from unconstrained angular reflection coefficients. α reflects the contribution from the zero-offset response, while β reflects that from the AVO slope.

Spratt (1987) showed that in the presence of a small error in stacking velocity, the estimated slope is given approximately by

$$\hat{B}(t) \approx B(t) - \Delta t A'(t), \tag{45}$$

where $B(t)$ and $A(t)$ are the true slope and zero-offset responses, and Δt is the maximum NMO time misalignment of the event. If the seismic wavelet is relatively narrowband, the second term of equation (45) is in phase quadrature with the first. Hence if A_t and B_t represent the analytic zero-offset and slope responses which would have resulted with perfect stacking velocities, then

$$B_c(t) \approx B_t(t) + KA_t(t), \tag{46}$$

where K is imaginary. Under this approximation, the product HCI [equation (42)] is *independent* of Δt even though the slope itself is strongly dependent:

$$P(t) \approx \mathcal{R}e[A_t(t)B_t^*(t)]. \tag{47}$$

To illustrate this effect, refer to Figure 5. A single seismic event caused by a small negative V_p velocity shift was processed with a series of moveout velocities ranging from 2985 m/s to 3012 m/s in steps of 3 m/s. The correct velocity was 3000 m/s. The first group of traces represents the product indicator, which aside from a 6 ms shift varies very little as a function of velocity. The second set representing the slope, on the other hand, varies dramatically over this range of velocities. Notice that the errors introduced by velocity are in phase quadrature with the zero-offset response.

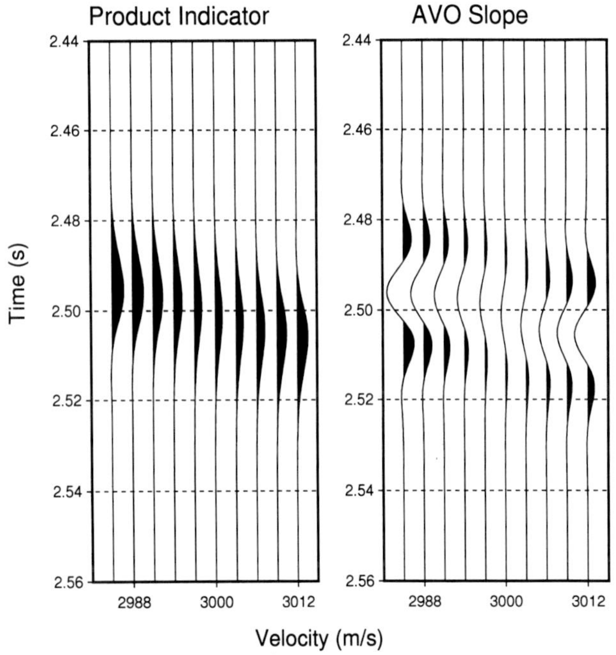

Fig. 5. A comparison of the sensitivity to stacking velocity errors between the product indicator and the AVO slope.

NOISE IMMUNITY OF INDICATORS

Previous sections have shown how all constrained linear hydrocarbon indicators may be expressed as linear combinations of the unconstrained angular reflection coefficients equation (23). The variance of each element of $\mathbf{p}$ is given by the corresponding diagonal element of its covariance matrix,

$$\underline{\Phi}_p = E\{\mathbf{p}, \mathbf{p}^H\} = \underline{V}E\{\mathbf{c}, \mathbf{c}^H\}\underline{V}^T = \underline{V}\underline{\Phi}_c\underline{V}^T. \tag{48}$$

Thus, the problem of estimating the noise characteristics of any linear HCI is reduced to estimating the covariance, $\underline{\Phi}_c$, of the unconstrained angular RCs.

Relative Signal-to-Noise Ratio of Angular RCs

Using equation (10), $\underline{\Phi}_c$ may be obtained from

$$\underline{\Phi}_c = (\underline{S}^T\underline{S})^{-1}\underline{S}^T\underline{\Phi}_d\underline{S}(\underline{S}^T\underline{S})^{-1}, \tag{49}$$

where $\underline{\Phi}_d$ is the $M \times M$ covariance matrix of the noise in the original seismic data. If the noise is mutually uncorrelated, with a uniform rms value ν, then equation (49) simplifies to

$$\underline{\Phi}_c = (\underline{S}\underline{S}^T)^{-1}\nu^2. \tag{50}$$

A numerical evaluation of the diagonal terms of this matrix is plotted in Figures (6) and (7), for three- and two-parameter inversions, respectively. The quantities plotted are *relative* signal-to-noise ratios (S/N), defined as

$$\text{Relative S/N} = \frac{\text{Estimate S/N}}{\text{Input data S/N}}. \tag{51}$$

The input data S/N is the ratio of the magnitude of underlying signal being measured, e.g., $|A|$, divided by ν. The estimate S/N is the same signal magnitude divided by the corresponding diagonal element of $\underline{\Phi}_c$. For the case of linear estimates, the "signal" is the same in both numerator and denominator; hence equation (51) reduces to the ratio of the standard deviations of the estimate and data noise; i.e., $\nu/\underline{\Phi}_c(i, i)$, for the ith parameter. It is, therefore, independent of the S/N of the input data. The off-diagonal terms of equation

(50), which are also nonzero, show how the noise in the coefficient estimates are correlated with each other.

From Figure 6, the relative S/N of an unconstrained 50-fold zero-offset estimate A, is about 3.7 times that of the input data, and is almost independent of the maximum incidence angle θ_M. By comparison, the normal stack has an S/N improvement factor of $\sqrt{50}$. These results contrast sharply with those for the slope trace B, whose S/N is only 17.8 percent of the input data, for $\theta_M = 30$ degrees. It is also highly dependent upon this angle, plummeting to 1 percent at 7 degrees.

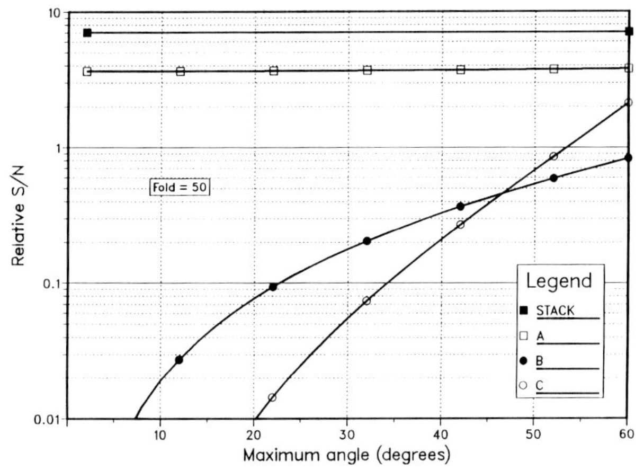

Fig. 6. The S/N of a conventional CDP stack, and of the angular reflection coefficients for a 3-parameter inversion normalized to the S/N of the prestack data. The CDP fold is 50.

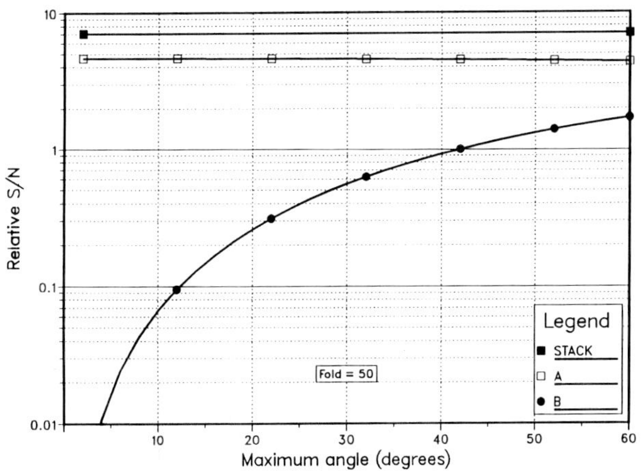

Fig. 7. The S/N of a conventional CDP stack, and of the angular reflection coefficients for a 2-parameter inversion, normalized to the S/N of the prestack data. The CDP fold is 50.

The quartic slope term C is even worse. At 30 degrees C can be estimated with a relative S/N of only 0.055. A minimum of 35 degrees of coverage is needed in order to make the relative S/N of C climb above 10 percent. In the case of a two-parameter inversion, the relative S/Ns for A and B are 4.6 and 0.56, respectively, for 30 degree coverage: both higher than for the three-parameter case.

S/Ns of Unconstrained HCIs

Totally unconstrained HCIs, computed by method (1), have the advantage of being completely unbiased, and determined solely from the data. However, they tend to be so dominated by noise as to be unusable. The transformation matrix $\underset{\sim}{\mathbf{U}}_s$, which converts the angular RC vector into the vector of Shuey's (1985) parameters,

$$\mathbf{p} = [\Delta V_p/V_p \quad \Delta\rho/\rho \quad \Delta\sigma_p]^T, \qquad (52)$$

is found by inverting equations (2), (5), and (4):

$$\underset{\sim}{\mathbf{U}}_s = \begin{bmatrix} 0 & 0 & 2 \\ 2 & 0 & -2 \\ -2(1-\sigma_p)(1-2\sigma_p) & (1-\sigma_p)^2 & (1-\sigma_p)(1-3\sigma_p) \end{bmatrix}. \qquad (53)$$

Figure 8 plots the relative S/Ns of $\mathbf{p}$, once $\underset{\sim}{\mathbf{U}}_s$ is substituted for $\underset{\sim}{\mathbf{V}}$ in equation (48) and definition (51) is used. Here the fold is 50, and the background V_p/V_s ratio is 2.0. If we arbitrarily declare a relative S/N which is 10 percent of the original data to be the minimum which can be used in practice, then only

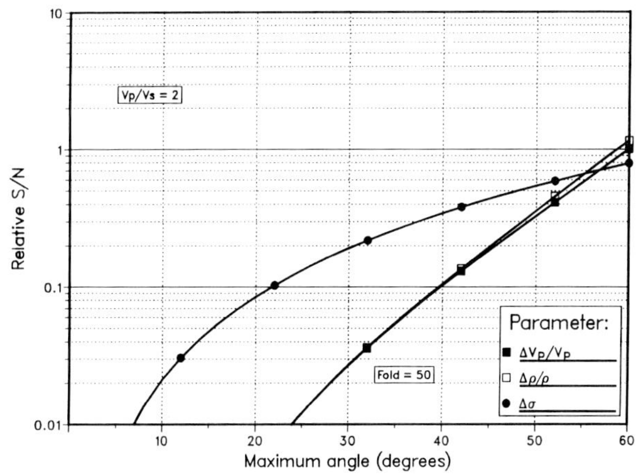

Fig. 8. S/Ns of unconstrained linear hydrocarbon indicators, normalized to the S/N of the prestack data.

acoustic impedance (A) and Poisson ratio estimates exceed this threshold for incidence angles less than 40 degrees. This is because all the other indicators heavily depend upon the quartic slope C, whereas $\Delta\sigma_p$ does not, since the (3, 3) element of $\underset{\sim}{\mathbf{T}}_s$ vanishes when $\sigma_p = 1/3$.

For this reason it might appear that the high quality of the Poisson estimates are due to the particular choice of the background σ_p. However, this is not true. Figure 9 displays the relative S/N of $\Delta\sigma_p$ for a fixed angle coverage (30 degrees), as a function of the background Poisson's ratio. The relative S/N of Poisson's ratio actually *increases* as σ_p drops below 1/3, until it reaches its peak at around $\sigma_p = 0.28$. In other words, its increased dependence on the relatively noisy quartic C slope *improves* the reliability of $\Delta\sigma_p$. This is because the Poisson estimate also depends on B, and the noise in C is correlated with that in B, according to the (2, 3) element of Equation (49). The transformation matrix from RCs to Shuey parameters causes some of the noise in B to cancel the noise in C.

S/N of Constrained Linear Indicators

The use of soft constraints reduces the variance of the estimated indicators, while it increases their bias. Figure 10 shows the effect the γ parameter has on the relative S/Ns, by evaluating the constrained covariance matrix,

$$\underset{\sim}{\mathbf{\Phi}}_c = (\underset{\sim}{\mathbf{S}}^T\underset{\sim}{\mathbf{S}} + \gamma M\underset{\sim}{\mathbf{I}})^{-1}\underset{\sim}{\mathbf{S}}^T\underset{\sim}{\mathbf{S}}(\underset{\sim}{\mathbf{S}}^T\underset{\sim}{\mathbf{S}} + \gamma M\underset{\sim}{\mathbf{I}})^{-1}, \tag{54}$$

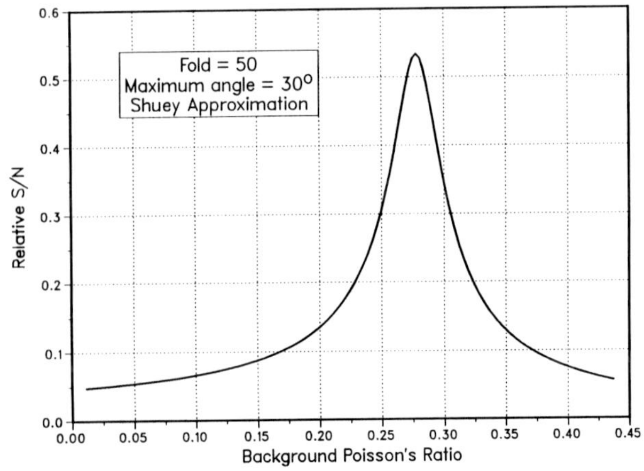

Fig. 9. Noise sensitivity of unconstrained estimates of changes in Poisson's ratio, as a function of the background Poisson's ratio.

with the fold $M = 50$ and the maximum angle is 30 degrees. Curiously, the performance of the two- and three-parameter A and B estimates track each other very closely for $\gamma > 10^{-2}$, but sharply split at values below this. Of the three angular RCs, $\hat{C}$ is the most sharply affected by the choice of γ.

The robustness of indicators is also improved by hard constraints, such as Gardner's relation in equation (26). Using equation (52), and the expression for the HCIs previously derived, it is easy to obtain relative S/N for tightly constrained estimates of $\Delta V_s/V_s$, R_{ss}, and the fluid factor. These ratios are shown in Figure 11. For all of these ratios, there is about a 20-fold improvement at 30 degrees over the unconstrained method. Method (3) was used in this analysis,

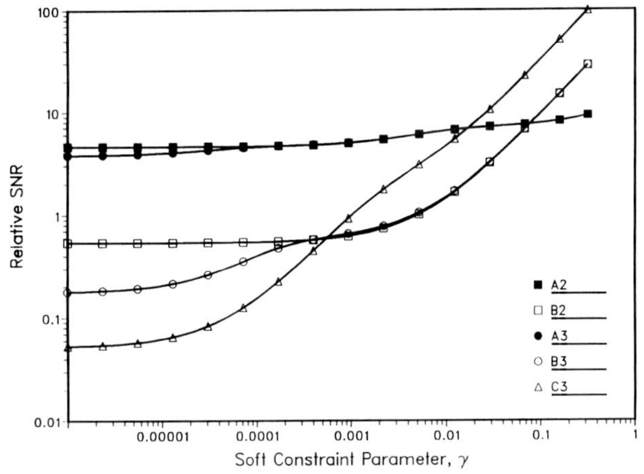

Fig. 10. Normalized S/Ns of softly constrained angular RC, as a function of the constraint parameter, γ.

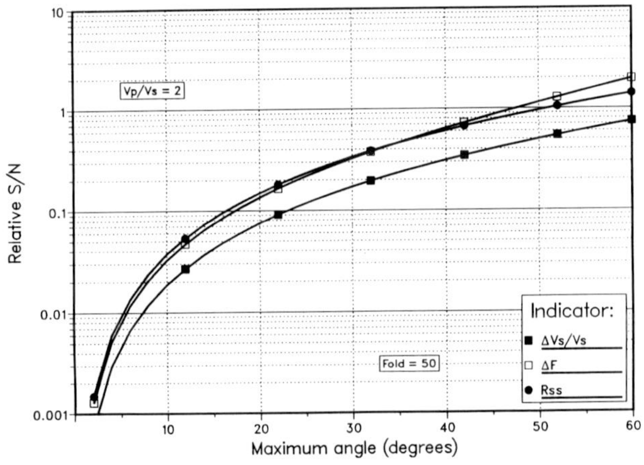

Fig. 11. Normalized S/Ns of selected constrained linear hydrocarbon indicators.

but similar results would be obtained with methods (2) and (4) as well.

S/N of Product Indicators

This section derives expressions for the noise uncertainty of the complex product trace, AB^*. Since the principle of superposition is no longer in effect, this indicator's noise properties will depend upon the underlying signal. They will also depend upon the higher-order statistics of the data noise.

Even the definition of the signal is subject to some debate. Although we are trying to estimate AB^*, it would be a mistake to define this quantity to be the signal. For example, if A were large but B were zero, should this imply that the signal is also zero? Certainly not, since this would yield a zero S/N no matter how small the noise in the estimate was! A better definition of signal is η:

$$\eta = \sqrt{|A|^2 + |B|^2}. \quad (55)$$

Adopting this definition, then without loss of generality, A and B may be written in the form,

$$A = \eta \cos \phi e^{i\lambda}; \quad (56)$$

$$B = \eta \sin \phi [e^{i\lambda - \zeta}] \quad (57)$$

where ϕ is the angle made by $|A|$ and $|B|$ in the $A - B$ plane, λ is the wavelet phase (cancelled out in product traces) and ζ is the phase difference between A and B. Combining equations (56) and (57), I obtain the following expression for the HCI:

$$\mathcal{R}e\{AB^*\} = \frac{1}{2}\eta^2 \sin 2\phi \cos \zeta. \quad (58)$$

Let us introduce the variable μ as the expected complex product of the estimated noises of $\hat{A}$ and $\hat{B}$:

$$\mu = E(n_a n_b^*). \quad (59)$$

As before, let σ_a and σ_b be the standard deviations of these estimates, respectively. Assuming Gaussian noise, the joint second moment of the product may be obtained from Price's theorem (1958), described in Papoulis (1965),

$$\text{Var}\{\hat{A}\hat{B}^*\} = \underbrace{2\mu^2 + \sigma_a^2\sigma_b^2}_{\nu^4 N_{ab}} \underbrace{+ |A|^2\sigma_b^2 + |B|^2\sigma_a^2 + 2\mu\mathcal{R}e\{AB^*\}}_{\nu^2\eta^2\Omega(\phi)} \quad (60)$$

The first collection of terms, $\nu^4 N_{ab}$ is independent of the signal, whereas the second, $\nu^2\eta^2\Omega$, depends on both the signal and the noise. The unitless quantities, N_{ab} and $\Omega(\phi)$, have been normalized to depend only on the system matrix [equation (8)], and not on the noise level ν, or the signal level η. From Equation (58), $\Omega(\phi)$ may be cast into the following form:

$$\nu^2\Omega(\phi) = \sigma_b^2 \cos^2 \phi + \sigma_a^2 \sin^2 \phi + \mu \sin 2\phi \cos \zeta. \quad (61)$$

As before, I wish to express noise sensitivity in terms of a relative S/N, as in Equation (51). In this case, the original S/N, which is denoted as γ_p, is simply,

$$\text{Original S/N} = \gamma_p = \eta/\nu. \quad (62)$$

However, as far as the estimate S/N is concerned, remember that $\hat{A}\hat{B}^*$ has units of signal *squared*, and consequently so does its standard deviation, σ_{ab}. In order to be consistent with the above equation, I must define the estimate S/N to be:

$$\text{Estimate S/N} = \eta/\sigma_{ab}^{1/2} = \eta/\text{Var}^{1/4}\{\hat{A}\hat{B}^*\}. \quad (63)$$

Using Equation (51), I conclude that the relative S/N of the AB^* estimate is

$$R_{ab} = \nu/\sigma_{ab}^{1/2} = [N_{ab} + \gamma_p^2\Omega(\phi)]^{-1/4}, \quad (64)$$

where

$$N_{ab} = \frac{2\mu^2 + \sigma_a^2\sigma_b^2}{\nu^4}. \quad (65)$$

Unfortunately, expression (64) still depends upon the input S/N, γ_p. Thus, a value for S/N, γ_p will have to be assumed, when R_{ab} is evaluated numerically. Although the estimate S/N increases as γ_p increases, it does not do so as fast. Another complication is that R_{ab} also depends on ϕ, i.e., the angle made by A and B in the $A - B$ plane. Clearly, certain angles will yield higher relative S/Ns than others.

R_{ab} is at a minimum when $\Omega(\phi)$ is at a maximum, and vice versa. Therefore, let us try to evaluate the minimum and maximum values that Ω can assume, and the corresponding angles. To do this, write Ω in the following form:

$$\Omega(\phi) = a \sin^2 \phi + b \cos^2 \phi + c \sin 2\phi, \quad (66)$$

where

$$a = (\sigma_a/\nu)^2; \quad (67)$$

$$b = (\sigma_b/\nu)^2; \quad (68)$$

$$c = (\mu/\nu^2)\cos\zeta. \tag{69}$$

Differentiating with respect to ϕ yields,

$$\frac{d\Omega}{d\phi} = (a - b)\sin 2\phi + 2c\cos 2\phi \tag{70}$$

$$= \sqrt{(a-b)^2 + 4c^2}\sin(2\phi + \psi), \tag{71}$$

where

$$\tan\psi = \frac{2c}{a-b}. \tag{72}$$

This expression [equation (72)] vanishes when

$$2\phi + \psi = n\pi, \qquad \text{for } n = 0, 1, \ldots. \tag{73}$$

Ω is at a minimum (implying a high S/N) when $\phi = \frac{1}{2}(\pi - \psi)$, and is at a maximum (implying a low S/N) when $\phi = -\frac{1}{2}\psi$.

The circle of Figure 12 represents constant signal strength. The portions of the $A - B$ plane so marked are the regions where the S/N of the complex product traces are either relatively high or low. ψ is the angle by which these regions are rotated clockwise from what would be symmetric about the axes. It is intuitively correct that the region close to the A axis would have a lower S/N than a region close to the B axis because there is more noise in the B trace than in the A trace. If the signal were distributed more toward A than B, then the noise in B would be multiplied by a larger signal in A, and hence appear to be more amplified in the AB^* product. On the other hand, if the signal were distributed more toward B than to A, then the noise in A would be amplified by the strength of B. But this noise is typically very small. Hence the overall S/N of AB^* would be comparatively high.

Figure 13 plots the best and worst relative S/N, R_{ab}, for a two-parameter inversion, depending on the underlying answer. These curves were generated by assuming $\gamma_p = 10$, $\zeta = 0$, and $M = 50$. Comparing Figure 13 to figure 7, we see that the best case product indicator is more robust than the slope itself at 30 degrees, but the worst case indicator is less robust. However, all three indicators are usable for high data quality at angle ranges above about 18 degrees. Surprisingly, there is not much change in this relative S/N for product terms for incidence angles of beyond about 30 degrees.

TEMPORAL RESOLUTION OF INDICATORS

As mentioned earlier, the AVO product indicator contains cross terms which invalidate the principle of superposition. This section will explore how these cross terms affect the temporal resolution of the product indicator and will lead to an answer to the question, "How thin can a gas sand bed be and still be detected?"

Suppose that two seismic reflectors separated in time by 2Δ are present. Then the *total* zero-offset reflectivity and slope responses can be described by

$$A_t = A_1 w(t - \Delta) + A_2 w(t + \Delta); \tag{74}$$

$$B_t = B_1 w(t - \Delta) + B_2 w(t + \Delta), \tag{75}$$

where the time origin has been shifted for convenience to the midpoint of the bed. Here w represents the analytic wavelet. The AVO product indicator may then be written as

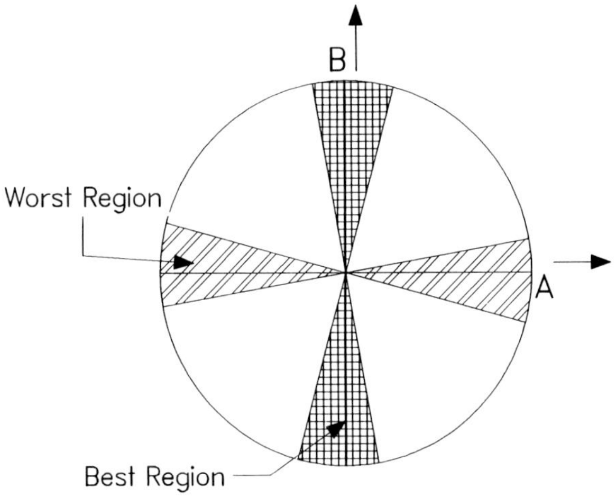

Fig. 12. Best and worst locations for A and B from the perspective of the normalized S/N ratio of AB^*.

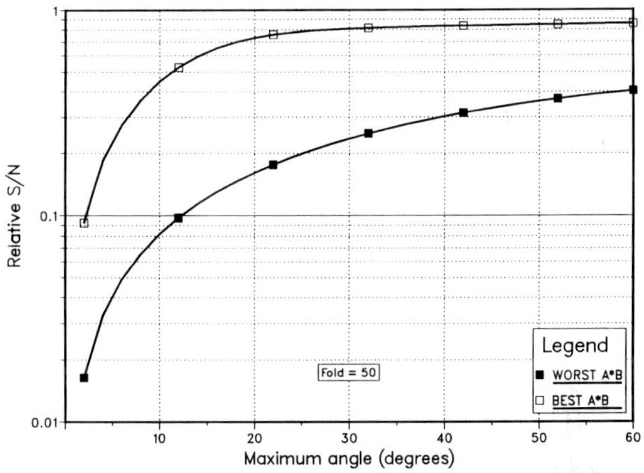

Fig. 13. Best and worst S/Ns of complex product indicators, depending on the location in the $A - B$ plane of the underlying answer.

$$P_t = A_t B_t^* = A_1 B_1^* |w(t - \Delta)|^2 + A_2 B_2^* |w(t + \Delta)|^2 + A_1 B_2^* w(t - \Delta) w^*(t + \Delta) + A_2 B_1^* w(t + \Delta) w^*(t - \Delta). \quad (76)$$

The first two terms of equation (76) represent superposition of the product terms of the individual events, while the last two terms represent the cross terms. If the analytic wavelet is Hermitian, i.e.,

$$w(-t) = w^*(t), \quad (77)$$

then the cross terms will both be symmetric in time about the bed center, and the real part of their sum will have its maximum magnitude there.

To offer a comparison between the resolution of a linear HCI such as the fluid factor, and that of the product indicator, three models representative of those faced by the explorationist were constructed. These models, shown in Table 2, were first used in Treadgold et al. (1990). The first model is a shale-embedded sand formation, which is 100 percent gas saturated. The second model is also a sand formation embedded in shale, but it is 100 percent water saturated. The last model is a sand formation which is partially gas saturated and partially water saturated, and is sealed from above by shale. The zone of water below the gas saturation is assumed to be thick in comparison to that of the gas. The elastic properties of each of these lithologies are shown in Table 3.

Model 1 is the classic case 3 AVO regime, as described in Rutherford and Williams (1989). It is typical of relatively shallow Pliocene and Pleistocene gas formations in the Gulf of Mexico and in South Texas. In this situation, the gas sand has a lower acoustic impedance than its encasing shale, and also a significantly lower Poisson's ratio. Its zero-offset reflection coefficient and AVO slope are both negative at the top of the bed, and both positive at the bottom. Consequently, one would expect a strongly positive product indicator at both the top and bottom of the bed. Furthermore, since $A_1 B_2$ and $A_2 B_1$ are both negative, the two cross terms of equation (76) serve to *enhance* temporal resolution, as we shall see.

Note that not all gas sand formations comply with this scenario. Older sands which are more tightly compacted can have either a positive impedance contrast with their shale, or no contrast at all (cases 1 and 2 from Rutherford and Williams, 1989). In both of these cases, the product indicator will fail to distinguish between wet sand and gas sand. For example, case 1 wet and brine sands will both exhibit $A > 0$ and $B < 0$ at the top of the bed, and both yield a negative product indicator. The fluid factor, on the other hand, still has a chance to succeed, since its response is based upon A and B deviating from a regional trend.

For each model, approximate angular RCs were computed from equations (2) and (3). These coefficients were then used as weights for zero-offset and slope spike-grams. The time difference between spikes was set to be the two-way traveltimes through a sequence of gas or water sand beds whose thickness varied from 0 to 30 m in steps of 3 m. Synthetic seismograms for zero-offset, $A_m(t)$, and slope, $B_m(t)$, were then obtained by convolving these spike-grams with a 30 Hz Ricker wavelet. This wavelet was arbitrarily phase shifted by −45 degrees to make the problem more representative of real data, whose true wavelet phase is often unknown. Next, the analytic A and B traces were computed from equations (43) and (44), and the fluid factor was obtained from equation (32). The results are plotted as a function of bed thickness, for all three models, in the left-hand sides of Figures 14, 15, and 16, respectively. The product indicator was computed from equation (42), and similarly plotted in the right-hand side of these figures. A scale factor of 25 was applied to the product indicators, to bring them to comparable numerical values as the fluid factors.

Let us first focus on models 1 and 2, shown in Figures 14 and 15. Both of these models consist of a sand formation sealed from above and below by the same shale. This means that their seismic response must vanish at zero bed thickness, as it indeed does. Both hydrocarbon indicators seem to be effective in differentiating the gas sand from the brine, but they do

Table 2. Three thin-bed lithology models.

	Model 1	Model 2	Model 3
Layer 1	Shale	Shale	Shale
Layer 2	Gas sand	Brine sand	Gas sand
Layer 3	Shale	Shale	Brine sand

Table 3. Elastic parameters corresponding to the lithologies of Table 2.

	V_p(m/s)	V_s(m/s)	ρ(g/cc)	σ_p
Shale	2,377	941	2.27	0.4069
Brine sand	2,664	1,253	2.23	0.3580
Gas sand	2,249	1,301	2.06	0.2484

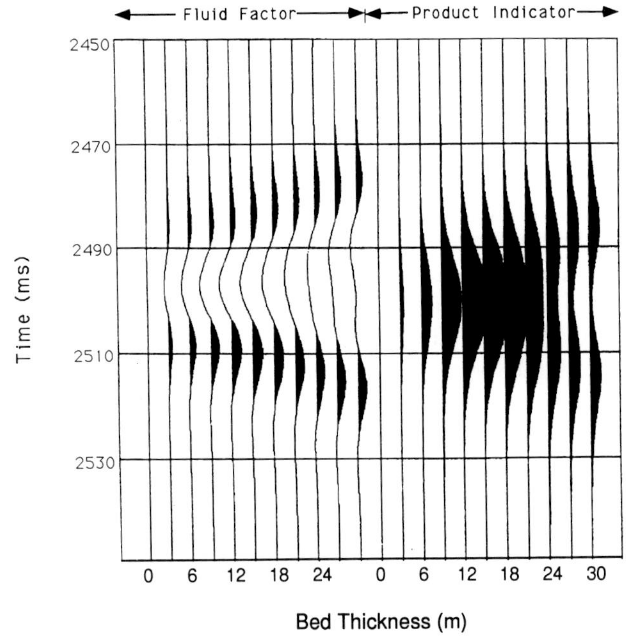

Fig. 14. A comparison of the fluid factor and product indicator for shale-gas-shale beds of varying thicknesses.

it in different ways. The product indicator does it by showing a strongly positive response for gas and a slightly negative response for brine. The fluid factor does it by showing a gas response which is about five times the strength of the brine response. (Note the difference in scale factors between these two figures.) The shape and sign of the fluid response, however, are virtually identical in both cases, the only difference being the time separation of the events caused by the disparity in bed velocities. Thus the fluid factor is *scale dependent*. It must be utilized by looking for relative bright spots in the section, ignoring sign. Both indicators detect their respective gas responses at all bed thicknesses. Although there is no fundamental limitation on the thinnest bed which is detectable, there is a practical limitation governed by noise. At first glance, the product indicator of the gas model, Figure 14, appears to closely resemble the complex envelope of the fluid factor. However, simply taking its envelope would lose the sign information of the product indicator.

Turning now to model 3 of Figure 16, we see that in the gas/brine case, both indicators also produce their respective positive responses for beds greater than about 5 m. However, for beds thinner than about 2 m, the product indicator is actually negative, being overwhelmed by the negative response from a shale/brine

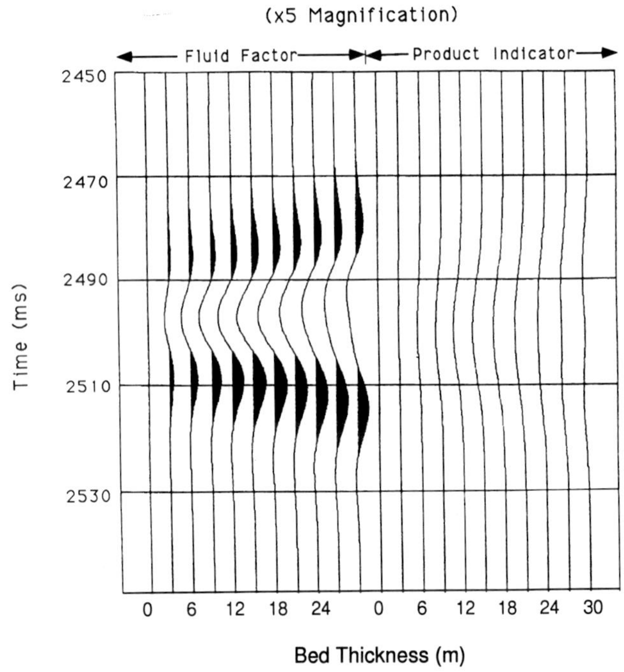

Fig. 15. A comparison of the fluid factor and product indicator for shale-brine-shale beds of varying thicknesses.

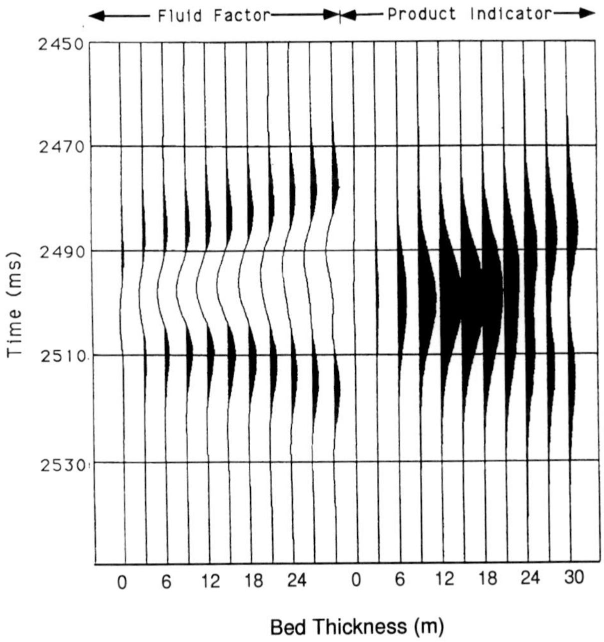

Fig. 16. A comparison of the fluid factor and product indicator for shale-gas-brine beds of varying thicknesses.

interface. This effect is made clearer by examining Figure 17. Here the maximum magnitude over time of the product indicator is plotted as a function of bed thickness. All of the models peak at their respective tuning thicknesses (about 15 m), but model 3 changes sign from negative to positive at about 3 m.

Before leaving the subject of resolution, note that a number of thorny issues, discussed in Swan (1991), remain:

Transmission coefficient loss and intrabed multiples;
Energy dissipation through the overburden;
Geometric spreading and directivity patterns of the source and receiver array;
NMO stretch of the seismic wavelet when AVO is performed after NMO;
Thin-bed tuning effects when AVO is performed before NMO.

These and other issues ensure that AVO analysis methods will continue to be studied for some time to come.

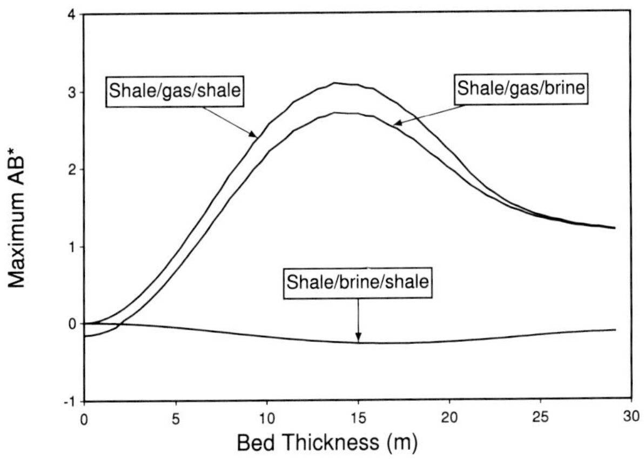

Fig. 17. The maximum signed magnitude the AVO product indicator can take over time, plotted as function of bed thickness.

CONCLUSIONS

Direct AVO hydrocarbon indicators may be divided into linear and nonlinear types. Linear types may be expressed as additive combinations of the angular reflectivity estimates, which in turn may be obtained from a least-squares fit of the moved out prestack seismic data. A covariance matrix of the angular RC estimates is easy to evaluate, and shows that even uncorrelated data noise leads to correlated noise between the estimated coefficients. This covariance matrix may be used to evaluate similar matrices for any other combination of linear indicators.

Unconstrained parameter inversions are unbiased and completely determined by the data. However, of all the elastic parameters considered by this chapter, only acoustic impedance and possibly Poisson ratio changes can be robustly estimated in the presence of realistic amounts of noise. All the other parameters, namely changes in V_p, V_s, density, shear-impedance, and the so-called "fluid factor" require a priori constraints to be measured in practice. These constraints may either be hard or soft, with their corresponding tradeoffs between resolution and robustness.

The only nonlinear indicator discussed by this chapter is the AVO product indicator, AB^*. This HCI is shown to possess superior immunity to residual velocity errors relative to linear techniques, and to be independent of wavelet phase. Furthermore, its sign is scale independent. Being nonlinear, it does possess thin bed interference effects, but these do not necessarily degrade temporal resolution. The noise characteristics of the product indicator depend upon the underlying signal being estimated, but the resulting S/N is generally similar to an unconstrained slope estimate, but with less angle dependence beyond 30 degrees. A disadvantage of product indicators is that they fail to detect gas in older, more consolidated formations where the sand impedance is greater than, or nearly equal to, that of the encasing shales.

REFERENCES

Aki, K., and Richards, P. G., 1980, Quantitative seismology theory and methods: W. H. Freeman and Co.

Biondi, B. L., and Kostov, C., 1989, High-resolution velocity spectra using eigenstructure methods: Geophysics, **54**, 832–842.

Castagna, J. P., Batzle, M. L., and Eastwood, R. L., 1985, Relationships between compressional-wave and shear-wave velocities in clastic silicate rocks: Geophysics, **50**, 571–581.

Gardner, G. H. F., Gardner, L. W., and Gregory, A. R., 1974, Formation velocity and density—the diagnostic basics for stratigraphic traps: Geophysics, **39**, 770–780.

Gassmann, F., 1951, Elastic waves through a packing of spheres: Geophysics, **16**, 673–685.

Menke, W., 1989, Geophysical data analysis: Discrete inverse theory: International Geophysics Series, **45**, Academic Press.

Papoulis, A., 1965, Probability, random variables and stochastic processes: McGraw-Hill.

Rutherford, S. R., and Williams, R. H., 1989, Amplitude-versus-offset variations in gas sands: Geophysics, **54**, 680–688.

Smith, G. C., and Gidlow, P. M., 1987, Weighted stacking for rock property estimation and detection of gas: Geophys. Prosp., **35**, 993–1014.

Shuey, R. T., 1985, A simplification of the Zoeppritz equations: Geophysics, **50**, 609–614.

Spratt, S., 1987, Effect of normal moveout errors on amplitude versus offset-derived shear reflectivity: 57th Ann. Internat. Mtg., Soc. Expl. Geophys., Expanded Abstracts, 634–637.
Swan, H. W., 1991, Amplitude-versus-offset measurement errors in a finely layered medium: Geophysics, **56**, 41–49.
Treadgold, G. E., Dey-Sarkar, S. K., Smith, S. W., and Swan, H. W., 1990, Amplitude versus offset and thin beds: 60th Ann. Internat. Mtg., Soc. Expl. Geophys., Expanded Abstracts, 1463–1466.
Widess, M. B., 1973, How thin is a thin bed?: Geophysics, **38**, 1176–1180.
Wiggins, R., Kenny, G. S., McClure, C. D., 1984, A method for determining and displaying the shear wave reflectivity of a geologic formation: European patent application #83300227.2, Bull. 84-30, July 15.

Prestack Synthetic Seismogram of Finely Layered Elastic Earth

T. K. Kan and C. Y. Young**

INTRODUCTION

In recent years amplitude variation with offset (AVO) has drawn substantial attention within the exploration industry (Ostrander, 1982, 1984; Backus et al., 1982). The ultimate goal of AVO is to detect the in-situ elastic property contrasts of the subsurface. As with many geophysical techniques, AVO can not be applied without making assumptions. One of the most common is that AVO responses are due to the reflectivity of a single elastic interface. The reality is that targets are normally overlain by thousands of layers of sedimentary rocks, and we have to deal with this complex reflection layering. Transmission loss of the signal through these layers is angle dependent, affecting the AVO response at the target. The interference effects of multiples and converted waves further obscure the reflectivities of the target interval. For these reasons it is important to model the prestack seismic amplitude and the waveforms using detailed stratigraphic layering. These prestack synthetic seismograms can help determine the consequences of certain processing and interpretational approximations.

Sonic logs reveal that the subsurface is characterized by rapid variations in P-wave velocities, on a scale smaller than the sampling interval (e.g., .305 m) whereas a seismic wavelength may be tens of meters. Attempts to accurately model the subsurface by synthetic seismograms must account for the effects of such fine stratigraphic layering. Comparison of one dimensional (1-D) conventional synthetics, based on the blocked models with those based on the raw logs, confirmed (Young et al., 1983) that a 500 layer block model contains almost all the relevant information of the 10 000 layer raw logs sampled at the .305 m interval. In the prestack domain, the synthetic seismogram is generated by using this kind of finely layered model with appropriate S-wave information. The S-wave velocity profiles are generated from published V_p/V_s relationships (Castagna et al., 1993) or from full waveform sonic measurements.

By assuming a plane-layered earth, the modeling technique can be used to address many AVO issues: simple interface versus complex layered target; abrupt versus transitional reflection; attenuation in the overburden; transmission loss in the overburden; and interference of multiple and converted waves with the target reflections.

COMPUTATION METHODS

We assume a model consisting of n plane, homogeneous, and isotropic layers over a half space. The ith layer is characterized by the P-wave velocity V_{pi}, the S-wave velocity V_{si}, the density ρ_i, and the thickness D_i (Figure 1). For the model input we use blocked conventional sonic and density logs, and derived S-wave logs from lithology and sonic logs. An explosive source S is located at $X = 0$ and $Z = 0$. The problem is to find the horizontal displacement U and the vertical displacement W for an arbitrary X position at the surface where $Z = 0$. In cylindrical coordinates, U and W can be expressed in terms of two elastic potentials, the scaler potential Φ, and vector potential Ψ (Ewing et al., 1957):

$$U = \frac{\partial \Phi}{\partial r} - \frac{\partial \Psi}{\partial Z}, \qquad (1)$$

$$W = \frac{\partial \Phi}{\partial Z} + \frac{\partial \Psi}{\partial r} + \frac{\Psi}{r}.$$

The differential equations for Φ and Ψ which can be derived by inserting equation (1) into the equations of motion are:

$$\nabla^2 \Phi = \frac{1}{V_p^2} \frac{\partial^2 \Phi}{\partial t^2}, \qquad (2)$$

and

*ARCO Exploration and Production Technology, Plano, Texas

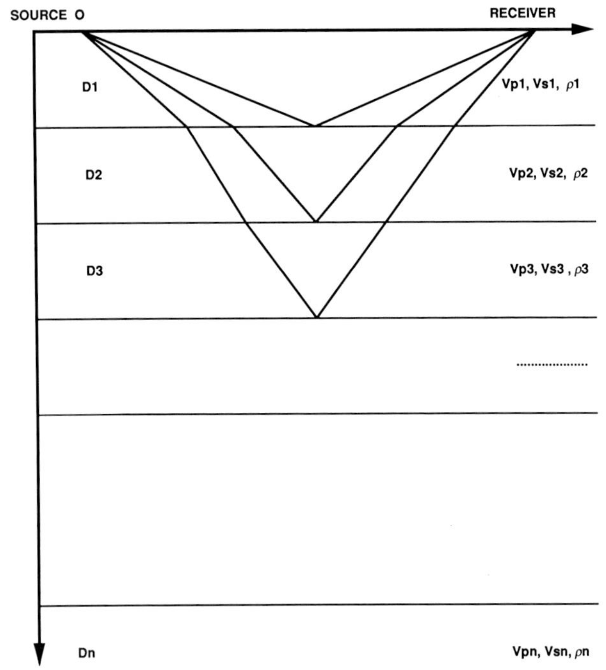

Fig. 1. Geometry of an elastic plane layer earth model for a reflection seismic survey.

$$\nabla^2\Psi - \frac{\Psi}{r^2} = \frac{1}{V_s^2}\frac{\partial^2\Psi}{\partial t^2}.$$

The explosive point-source S radiates a spherically symmetric P-wave with the compressional potential

$$S(t) = \frac{1}{r}F\left(t - \frac{r}{V}\right), \tag{3}$$

where r is the radial distance of a point in layer 1 from the source S, and $F(t)$ is the source time function.

The solutions of Φ and Ψ for equation (2) are:

$$\Phi(\omega) = F(\omega)\int_0^\infty \frac{K}{jv_1} H_0(Kr)R_{pp} \exp(jv_1 Z)\, dK, \tag{4}$$

and

$$\Psi(\omega) = F(\omega)\int_0^\infty \frac{K}{jv_1'} H_1(Kr)R_{ps} \exp(jv_1' Z)\, dK,$$

where

$$v_1 = \sqrt{\left(\frac{\omega}{V_{p1}}\right)^2 - K^2}, \quad \left(\frac{\omega}{V_{p1}}\right)^2 \ge K^2,$$

$$v_1 = -j\sqrt{-\left(\frac{\omega}{V_{p1}}\right)^2 + K^2}, \quad \left(\frac{\omega}{V_{p1}}\right)^2 \le K^2,$$

and

$$v_1' = \sqrt{\left(\frac{\omega}{V_{s1}}\right)^2 - K^2}, \quad \left(\frac{\omega}{V_{s1}}\right)^2 \ge K^2,$$

$$v_1' = -j\sqrt{-\left(\frac{\omega}{V_{s1}}\right)^2 + K^2}, \quad \left(\frac{\omega}{V_{s1}}\right)^2 \le K^2.$$

H_0 and H_1 are Hankel functions of the zero and first kind, respectively. R_{pp} and R_{ps} are the phase shifted complex reflectivities in $\omega - K$ (frequency-wave number) domain for the plane layered model (Kennett, 1974). Substituting equation (4) into equation (1) we obtain the horizontal and the vertical displacement U and W at $Z = 0$:

$$U(r, 0, \omega) = F(\omega)\int_0^\infty \frac{K}{jv_1} \times (KR_{pp} - jv_1' R_{ps})H_1(Kr)\, dK, \tag{5}$$

and

$$W(r, 0, \omega) = F(\omega)\int_0^\infty \frac{K}{jv_1'} \times (-jv_1' R_{pp} + KR_{ps})H_0(Kr)\, dK.$$

We rewrite equation (5) in discrete form (equation 6). For a two dimensional isotropic model, we replace horizontal distance r to X. We solve for the plane-layer response of R_{pp} and R_{ps} and then numerically integrate equation (6) to obtain U and W for every frequency within the seismic band. Finally, the displacements U and W as functions of space and time can be obtained by transforming equation (6) from frequency to the time domain using Fast Fourier Transform (FFT) algorithm. The final synthetic seismogram is expressed in terms of vertical and horizontal displacement as a function of traveltime and offset distance. The computation effort is proportional to the product of N (number of layers of the model), M (number of discrete frequency points), and L (discrete horizontal wave number);

$$U(X, t) = \sum_{\ell=0}^{L}\sum_{m=0}^{M} \frac{\ell\Delta K}{jv_1}(\ell\Delta K R_{pp} - jv_1' R_{ps}) \times H_1(\ell\Delta K X) \exp(-jm\Delta\omega t)\Delta K\Delta\omega, \tag{6}$$

and

$$W(X, t) = \sum_{\ell=0}^{L} \sum_{m=0}^{M} \frac{\ell \Delta K}{j\upsilon'_1} (j\upsilon'_1 R_{pp} + \ell \Delta K R_{ps}) \times H_0(\ell \Delta K X) \exp(-jm\Delta\omega t) \Delta K \Delta\omega.$$

ACOUSTIC VERSUS ELASTIC MODELING

A prestack synthetic for an acoustic model computed using this algorithm is shown in Figure 2c. The input impedance (product of sonic velocity and density) log from well A used for this synthetic is displayed as Figure 2b. Figure 2a is the elastic prestack synthetic seismogram for the vertical component, generated using the same acoustic impedance, and assuming the V_p/V_s ratio to be two. The elastic synthetic seismograms differ from the acoustic synthetics by the addition of converted-wave events and the offset dependent amplitude effects for each interface. The converted-wave generated at the shallow interfaces interferes with the primary P-wave events at later times. We perform velocity analyses for acoustic and elastic synthetic seismograms. The semblances and rms velocity picks are displayed in Figures 3 and 4 for which both P- and converted P- to SV-wave events have been identified in the velocity spectrum.

LAYERED MODEL SAMPLING

As stated, the objective of this modeling study is to synthesize prestack seismograms using realistic earth models. The trade-off between the scientific accuracy and economic efficiency (computation time) should be balanced. To avoid spatial and temporal aliasing, we are unable to reduce the number of horizontal wavenumbers (related to spatial sampling in X-direction, Wenzel et al., 1982) or the number of frequencies. We want to learn the optimum size for the realistic earth model by conducting a simple numerical experiment. We chose a well, and generated three earth models. The first model is represented by the raw sonic and density logs, with 0.305 meter sampling, having a total of 10 000 layers. For the second model, sonic and density logs were blocked (Young et al., 1983), having 500 layers with variable depth sampling rate. The third model used sonic and density logs blocked to 50 layers. Figure 5 presents three impedance logs for the three earth models. In Figure 6, we show three 1-D synthetic seismograms from the three earth models. The synthetic seismogram of the 10 000 layer model is almost identical to the synthetic of the 500 layer model. It is fair to say that within the normal frequency bandwidth of reflection seismology, the second model contains almost all the stratigraphic detail

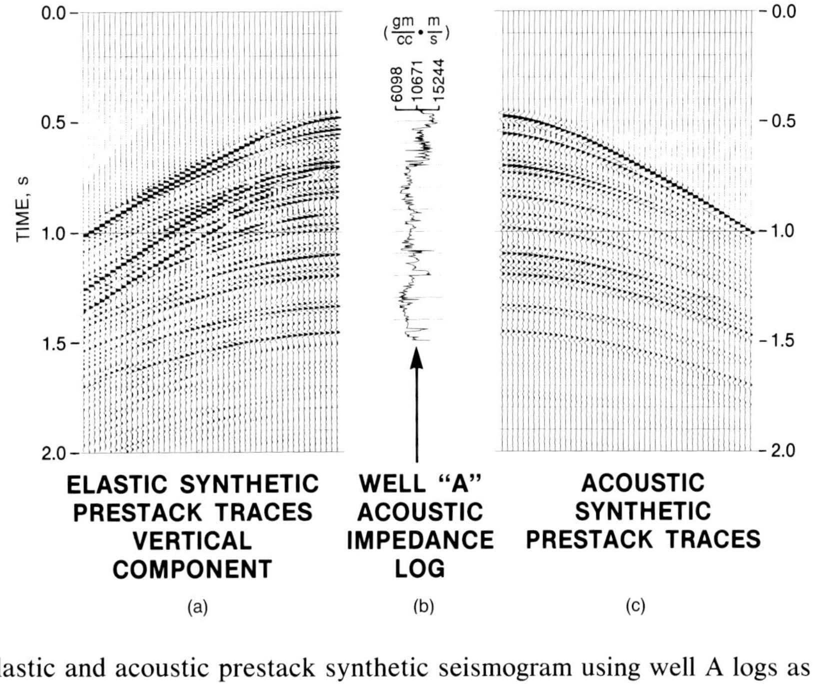

Fig. 2. Elastic and acoustic prestack synthetic seismogram using well A logs as a model.

of the 10 000 layer model. The 50 layer model does not have enough sampling points to represent details of the original log, and the resulting synthetic is not adequate.

The vertical component prestack synthetic seismograms of the 500 layer model are shown in Figure 7a and those for the 50 layer model as Figure 7b. For both models, a V_p/V_s ratio of two has been assumed. The detailed reflection pattern can not be represented by the 50 layer model in prestack seismic modeling. The AVO responses of the event at 1.5 s are different for the two synthetics. The converted SV arrivals for the 50 layer model are much stronger than those of the 500 layer model. Boundaries with gradual changing impedance in the 500 layer model are replaced with abruptly changing impedance boundaries, generating stronger converted waves. Strong converted wave events in real prestack seismic records are seldom encountered except in a few cases (Tatham and Stoffa, 1976). In real clastic sedimentary columns, abrupt impedance changes occur less frequently than would be implied by the model constructed by coarsely blocked logs.

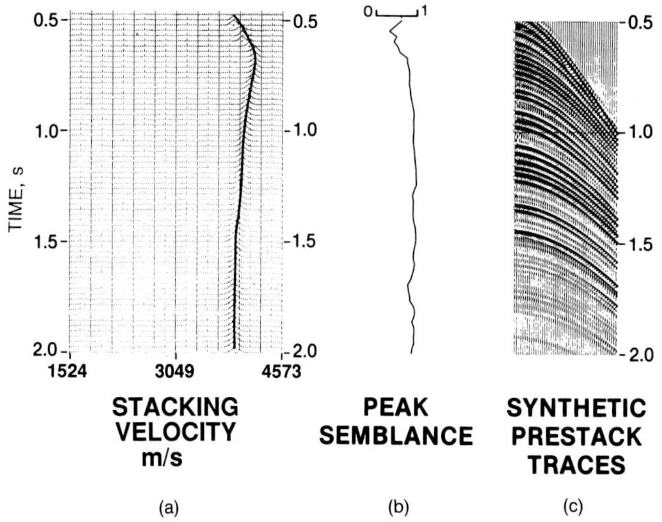

Fig. 3. Velocity analysis of the acoustic prestack synthetic of well A.

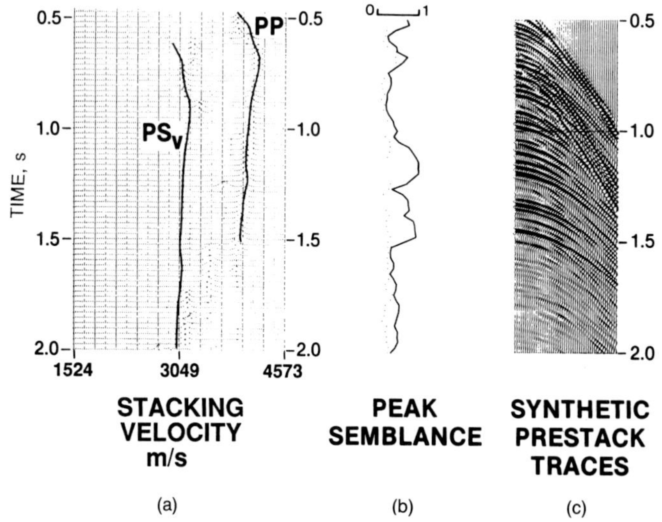

Fig. 4. Velocity analysis of the elastic prestack synthetic of well A.

OVERBURDEN TRANSMISSION EFFECTS

AVO technique attempts to extract the contrast of the elastic parameters for the target interval buried beneath thousands of meters of sedimentary rocks. Before analyzing the AVO response, the overburden effects should be compensated for, which will alter the amplitude and waveform of the signal.

The overburden effects fall into four categories: (1) intrinsic attenuation of the seismic energy; (2) transmission losses of the elastic wave propagation; (3) interference of the events such as converted *S*-waves and multiples; (4) scattering due to overburden heterogeneities. The fourth item involves 2-D or 3-D seismic modeling and is beyond the scope of this paper. In this section, we address the first three issues by analyzing AVO effects from three different synthetics. We use two acoustic model synthetic seismograms, one with attenuation and another without, to illustrate item one. We then compare the elastic and acoustic model synthetics to address the elastic transmission effects.

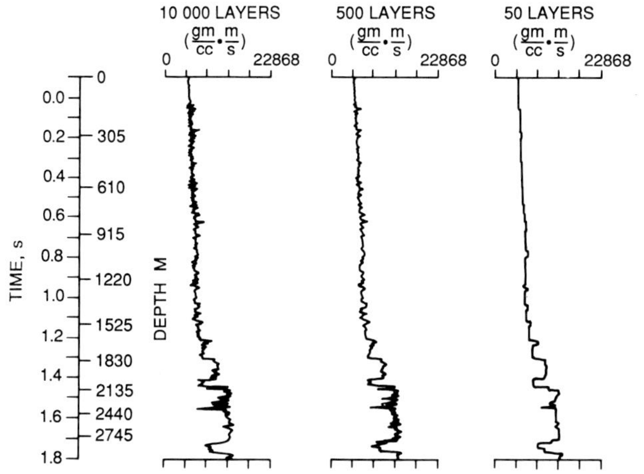

Fig. 5. Acoustic impedance log from well M. 10 000 layer is the original log sampled at one foot interval. 500 layers and 50 layers are the two blocked logs derived from the original log.

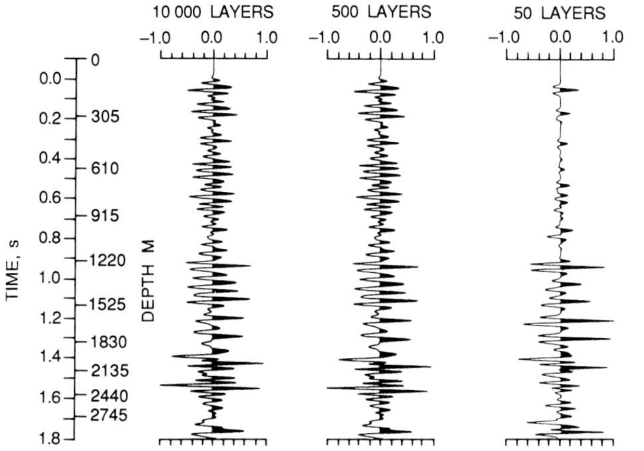

Fig. 6. Three 1-D zero offset synthetic seismograms generated from three impedance logs shown in Figure 5.

Figure 8(a) shows a common depth point (CDP) gather from offshore the Gulf of Mexico area. Plotted in the center is the time domain acoustic impedance profile of well B, located at this common-depth-point (CDP) location. In well B, the lithology is predominately shale and no hydrocarbon was encountered. A prestack synthetic seismogram was generated using the blocked sonic and density logs, and V_p/V_s ratio was assumed to be 1.8. The prestack synthetic seismogram (model B) is shown in Figure 8(b). The velocity model for the interval between sea level and top of the logs was assumed to account for the correct time shift. The free surface boundary condition was not imposed in the prestack synthetic generation.

We want to investigate the elastic transmission, attenuation, and interference of the converted *S*-waves and interbedded multiples. Figures 9 and 10 show comparison of the AVO responses of real data, two acoustic and one elastic prestack synthetic seismograms. The elastic synthetic is shown in Figure 8(b). We also generated two acoustic synthetic seismograms. The first acoustic synthetic, Figure 9 model (b), was generated from the blocked sonic and density logs from well B. The second model (c) was generated from the blocked sonic and density logs with attenuation effect (a constant Q value of 75). In Figure 9, we show AVO response (a) from the event at 1.8 s of the real data at CDP 388; and AVO responses of acoustic synthetic with no attenuation (b), and with attenuation (c). Note the AVO response of the real data exhibits undulating behavior, and amplitudes decay rapidly as a function of offset. The AVO response of acoustic synthetic (b) also shows undulating variation. Since there is no converted wave interference in the acoustic case, this undulating behavior is due to the interference of primaries and multiples. The average AVO response for this case is fairly flat. The AVO response of the acoustic model with Q included (c) has more decay at far offset than (b) because of the attenuation effect. Different from the real case, the AVO response of (c) has less undulating behavior.

We now compare the AVO response for the same event with elastic synthetics (Figure 10). In the elastic synthetic (b) the data show a steeper decay of AVO response than the acoustic synthetic (c). This is be-

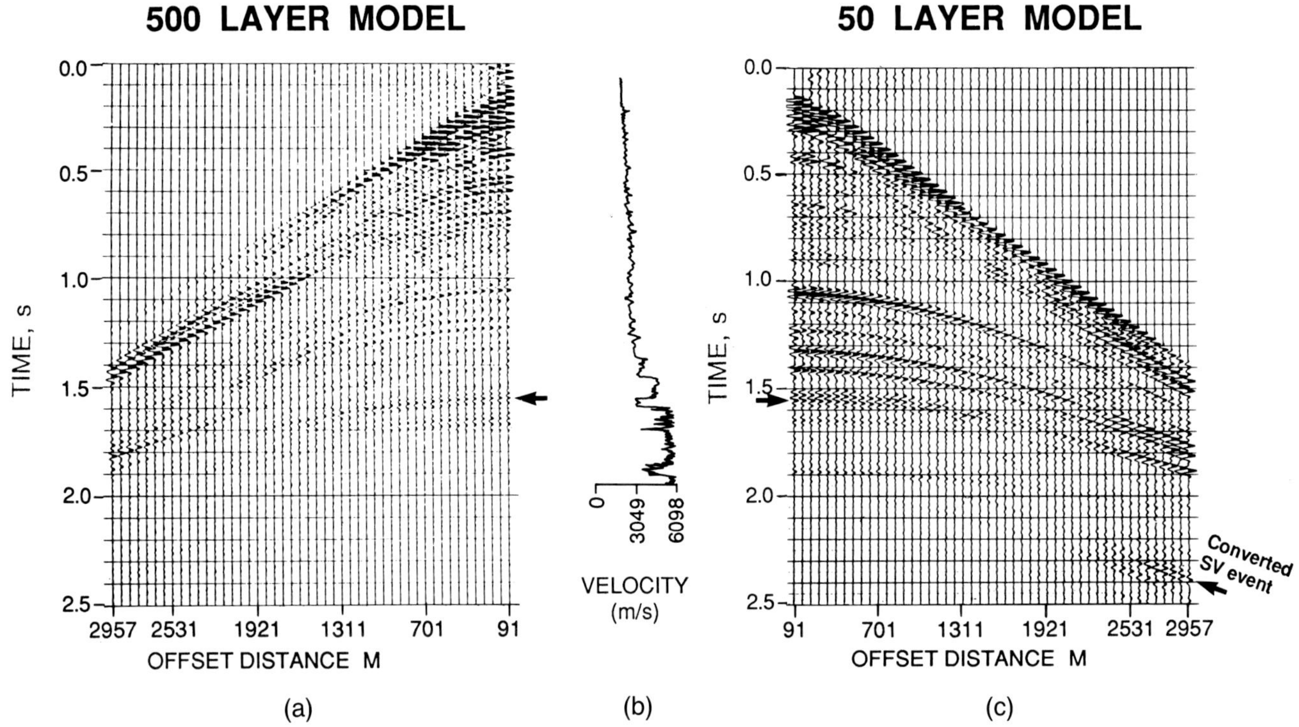

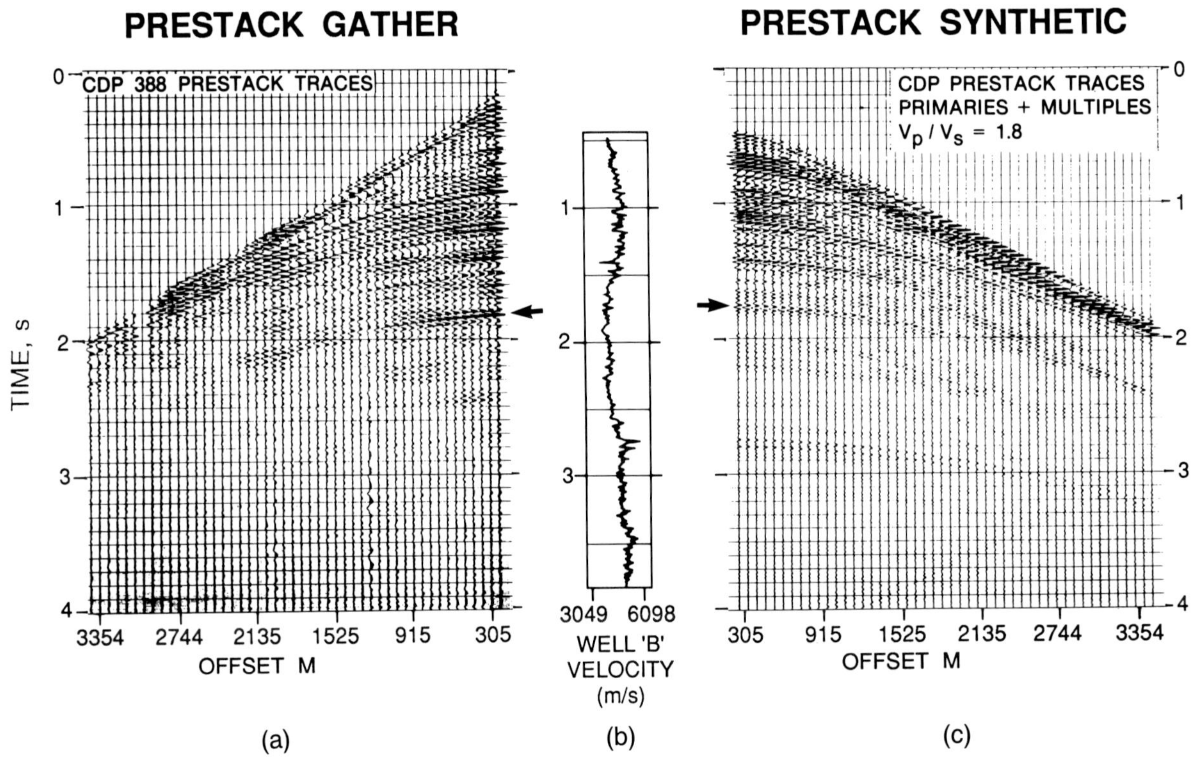

Fig. 8. Comparison of prestack seismic gather and an elastic prestack synthetic seismic gather from well B.

cause part of the energy of the initial *P*-wave was converted into other modes of propagation at each layer boundary. The transmission loss in elastic layered medium is related to continuous partition of energy from *P*- to *S*-waves through mode conversion from the surface to the target and back to the surface. The interference from the converted waves is more severe than acoustic short path multiples. The attenuation or the elastic transmission alone cannot account for the total overburden effect. The combined overburden effects of attenuation and elastic transmission are significant. Both of the overburden effects must be corrected to recover the true AVO response of the target events.

AVO TARGET RESPONSES

After the overburden effects have been accounted for, the AVO responses can be interpreted in terms of the contrast in the elastic parameters for a specific target layer. Exploration targets are complicated geologic features. For example, a single gas reservoir may be composed of several individual sand lenses stacked on top of each other and separated by shale streaks. The degree of complexity depends upon the depositional environment. The elastic properties can vary within the sand unit, due to different fluid content, porosity, grain size distribution of the sand body, and the degree of cementation of the sand. The bedding layer geometry further complicates the AVO response. To relate the AVO response directly to lithologic or rock/fluid factors of a simple half space may not be realistic.

In this section six models (Figure 11) are used to generate prestack synthetic seismograms. All models

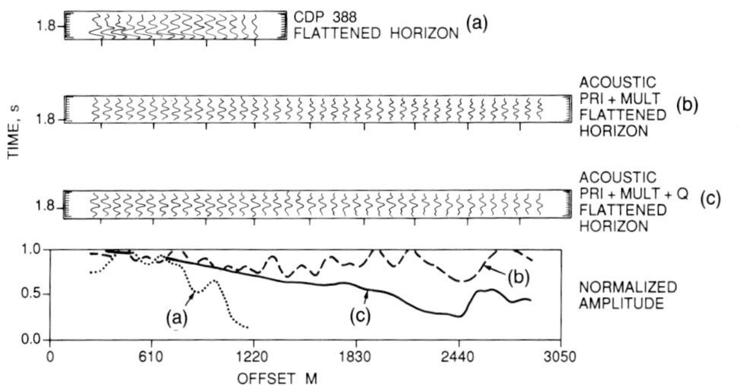

Fig. 9. Windowed seismic traces from flatten seismic gather and acoustic model (b) and (c). The AVO function of (a) real seismic data; (b) and (c).

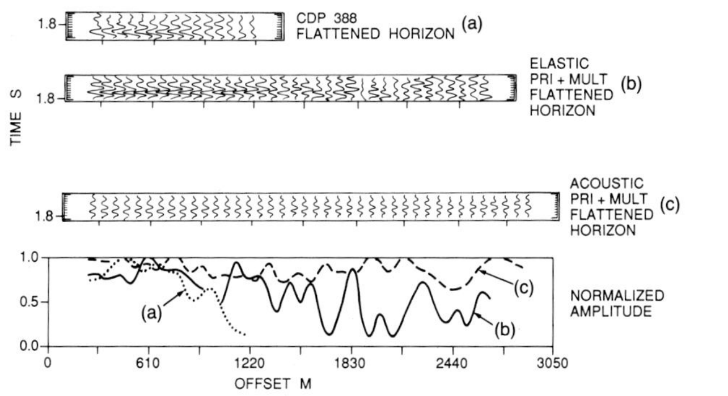

Fig. 10. Windowed seismic traces from flatten seismic gather and acoustic model (c) and elastic model (b). The AVO function of (a) real seismic data; (c) and (b).

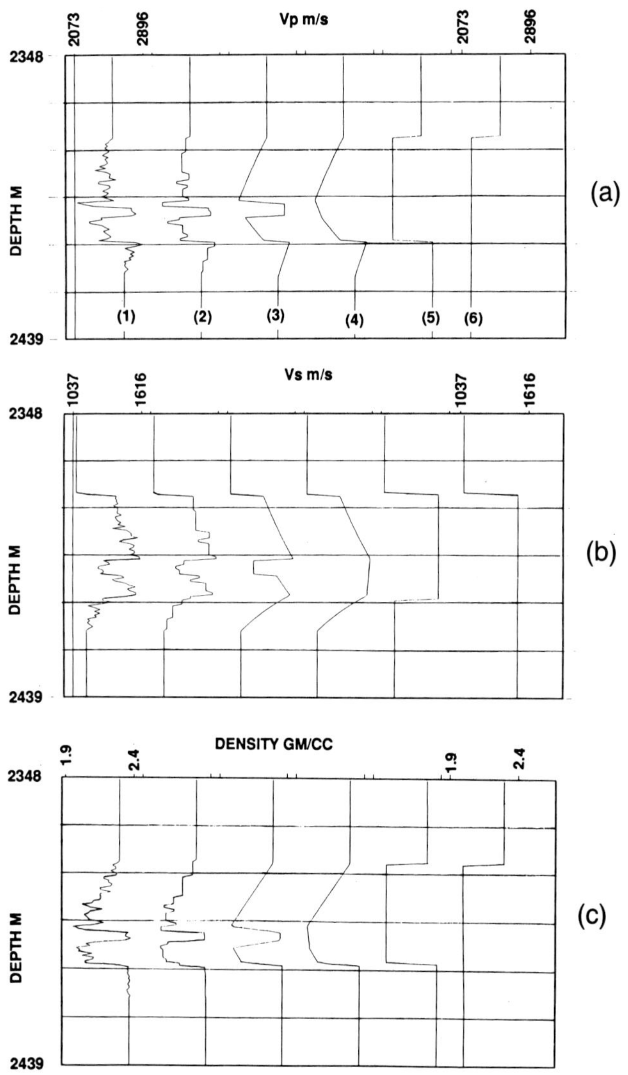

Fig. 11. V_p, V_s, and density profiles of six models. Model No. (1) is the original log.

were put under the same simplified overburden model, not shown in the figure. The overburden is designed such that no events are generated which interfere with the target reflection.

The original model (Model 1) is a deep water turbidite sand in a channel-levee system from offshore Gulf of Mexico. There is gas in two sand lenses, separated by a shale streak. The upper sand package coarsens downward, and the lower sand package coarsens upward. The total sand thickness is about 55 m thick, and is sampled every 2 ft (.61 m). The *S*-wave sonic (Figure 11b) is generated using lithology, *P*-wave sonic log and published V_p/V_s relationships (Castagna et al., 1993). Model 2 is a blocked model of the original logs. The bedding geometry has been simplified. The layer thickness averages about 8 ft (2.4 m). The total number of layers is 22. Model 3 is a simplified smooth model with the shale streak preserved. In this model we try to accommodate the coarsening upward and downward of two sand bodies. Model 4 is a smoothed model without the shale streak. Model 5 is a single gas sand layer embedded in shale. Model 6 is equivalent to a shale over gas sand half space model. The synthetics were generated with a 35 Hz Ricker wavelet.

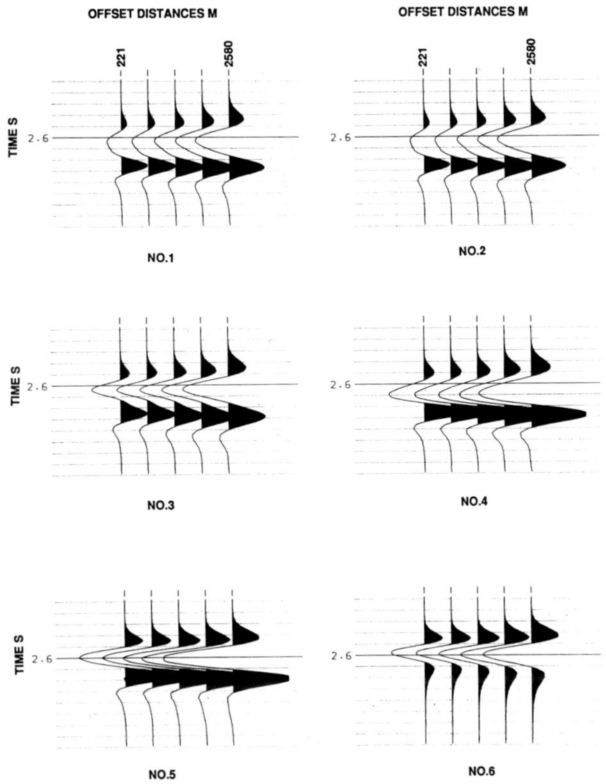

Fig. 12. Prestack synthetic seismograms from the six models shown in Figure 11.

The elastic prestack synthetics were corrected for spherical divergence and normal moveout (Figure 12). Attenuation due to overburden and *Q* effects was not removed. The near offset was 221 m and the far offset, 2849 m. The original synthetics had 50 traces per gather. Every tenth trace is displayed for all models. A single gain factor was applied to the data for display purposes.

The waveforms for Models 1 and 2 are almost identical. The waveform for Model 3 is slightly different from Model 1. Without the shale streak, Model 4 produced a much stronger reflection than Model 3, although the waveforms were similar. The waveforms for Models 5 and 6 were quite different from those produced by the original model. The waveform is sensitive to the reflection strengths and the distribution of the reflection coefficients (phase of the synthetic seismogram).

We then estimated two AVO coefficients, *A* and *B*, for all the synthetics by computing the rms amplitude for a 40 ms window versus offset and using $A + B \sin^2(\theta)$ (Richards and Frasier, 1976) to fit the rms amplitude as a function of the incident angle. Table 1 lists the computed *A* and *B* coefficients and their percent of differences from the original model (1), and the comparison of the waveforms to that of Model 1. *A* is the intercept of the rms amplitude at zero offset. *B* is the slope term. The results exhibit the sensitivity of the AVO parameters to the bedding geometry of the models.

A, the rms amplitude of the zero offset trace, is sensitive to bedding geometries. Only Model 2, the detailed blocked log model, had a value for *A* in close agreement with the original model. All other simplified models failed to reproduce the original model. Even though the shale streak is thin, it is important in terms

Table 1. Computed A and B coefficients and their percent differences from the original model, as well as waveform comparison to the original model.

Model No.	A	B	$(A-A_0)/A_0$ (%)	$(B-B_0)/B_0$ (%)	Waveform
1	0.447	0.907	0	0	
2	0.443	0.911	−0.7	0.4	same
3	0.585	0.965	30.9	6.4	similar
4	0.883	1.061	97.5	17.0	similar
5	1.032	1.275	130.9	40.6	different
6	0.543	0.862	21.5	−5.0	different

of modeling the amplitude (compare Models 3 and 4). The worst case is the simple three layer model (Model 5). The detailed bedding geometry and impedance distribution are important in modeling the zero offset amplitude and waveform.

B, the slope term, is less sensitive to the bedding geometry than *A*. Even the single interface model produces a reasonable *B* value. The percentage differences are all small (less than 10 percent) except for Models 4 and 5. Model 5 again is the worst model. This is good news, because this study has demonstrated that the B term is less sensitive to the internal bedding geometry of the target, but is sensitive to the elastic parameter contrasts.

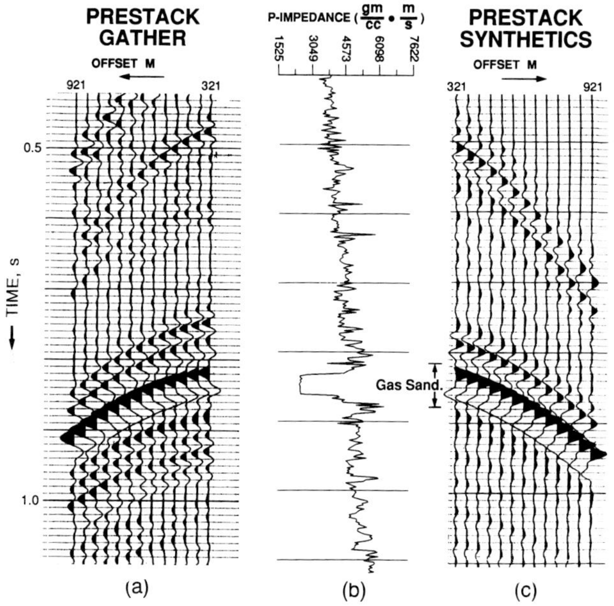

Fig. 13. Prestack seismic, impedance log, and prestack synthetic seismograms of well S gas sand model.

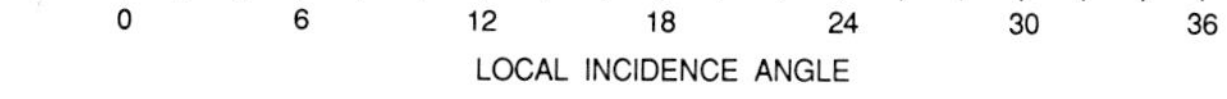

Fig. 14. AVO function of real and synthetic prestack data of well S gas sand. Dashed line: AVO fit to synthetic data. Dots: rms amplitude for real data. Solid line: AVO fit to real data.

AVO of a Gas Sand Target

This section shows a modeling study of a relatively shallow Gulf Coast gas sand target. We did not anticipate strong overburden effects. The sonic log in the target zone is incorrect, the sand V_p and density are corrected by fluid substitution (Castagna et al., 1993) in brine sand. The task for this case study was to model the V_p and V_s velocities and density of this gas sand target. Figure 13(a) shows a prestack gather of a CDP point close to well *S*. The gas sand, 30.5 m thick, is embedded in shale and has simple internal structure. Figure 13(b) shows the acoustic impedance. Not shown is the synthetic *S*-wave sonic log, used for synthetic generation. The *S*-wave sonic values were generated, using the mudstone V_p/V_s trend (Castagna et al., 1985). In the gas sand layer, the V_p/V_s ratio is assumed to be 1.5. The elastic prestack synthetic seismogram is shown in Figure 13(c). The match between the real and the synthetic seismogram is excellent. There are some minor differences between the real and the synthetic seismograms. The possible causes are: (1) a 1-D model was used to generate this synthetic seismogram, and the real seismic trace is the summation of the seismic wave scattering from the 3-D sedimentary rocks around the borehole; (2) we were unable to simulate the complicated seismic acquisition and processing effects in the prestack synthetic seismic generation process; (3) the sonic and density logs do not honestly represent the acoustic parameters of the earth; and (4) the synthetic shear sonic log represents a global V_p/V_s relationship of the sedimentary rocks, and the local V_p/V_s variation was not available.

In Figure 14, we show the AVO responses of the real and synthetic prestack seismograms. The dots are the rms amplitudes of the individual channels of the windowed real seismogram; the solid line is the best fit AVO function. The agreement is excellent. The small discrepancies at the large incidence angle could be attributed to the overburden factors or to any or all of the four factors already mentioned.

CONCLUSION

We have demonstrated the usefulness of the reflectivity method in modeling a finely layered elastic earth. By comparing the differences between acoustic and elastic earth models, we may estimate the importance of converted wave interferences and the elastic transmission losses. The transmission losses are shown to significantly affect AVO of the target. We modeled a realistic target interval with five other models with different bedding geometries to examine the sensitivities of the AVO parameters. The zero offset amplitude and waveform are sensitive to the bedding geometry. The B term was found to be less sensitive to the bedding geometry and to be more sensitive to the elastic parameters of the medium. In exploring a petroleum target using AVO, this full waveform modeling technique provides a means to examine not only the elastic parameters, but also the detailed bedding geometry of the reservoir target. We have illustrated the importance of adequate sampling of the model in prestack synthetic seismogram generation.

The simple interface reflection coefficient is only a starting point toward understanding offset dependent reflectivity. A stratigraphic earth model with realistic elastic parameters is required for meaningful modeling of prestack synthetic seismograms. This kind of synthetic seismogram is an effective tool for understanding some complex AVO issues.

ACKNOWLEDGMENTS

We extend our appreciation to Drs. Dennis Corrigan, Charles Sicking, Tim Keho, and John Castagna for their comments and suggestions. Appreciation is also extended to ARCO for permission to release those results.

REFERENCES

Backus, M. M., Nepomuceno, F., and Cao, J., 1982, The reflection seismogram in layered earth: Presented at the 52nd Ann. Internat. Mtg., Soc. Expl. Geophys.

Castagna, J. P., Batzle, M. L., and Eastwood, R. L., 1985, Relationships between compressional and shear-wave velocity in clastic silicate rocks: Geophysics, **50**, 571–581.

Castagna, J. P., Batzle, M. L., and Kan, T. K., 1993, Rock Physics: The link between rock properties and AVO response, *in* Castagna, J. P., and Backus, M. M., Eds., Offset-dependent reflectivity—Theory and Practice of AVO analysis: Soc. Expl. Geophys.

Ewing, M., Jardetzky, W., and Press, F., 1957, Elastic waves in layered media: McGraw-Hill.

Kennett, B. L. N., 1974, Reflections, rays and reverberations: Bull. Seis. Soc. Am., **64**, 1685–1696.

Ostrander, W. J., 1982, Plane wave reflection coefficients for gas sands at nonnormal angles of incidence: Presented at the 52nd Ann. Internat. Mtg., Soc. Expl. Geophys.

Ostrander, W. J., 1984, Plane wave reflection coefficients for gas sands at nonnormal angles of incidence: Geophysics, **49**, 1637–1648.

Richards, P. G., and Frasier, C. W., 1976, Scattering of elastic waves from depth-dependent inhomogeneities: Geophysics, **41**, 441–458.

Tatham, R. H., and Stoffa, P. L., 1976, V_p/V_s—A potential hydrocarbon indicator: Geophysics, **41**, 837–849.

Wenzel, F., Stoffa, P. L., and Buhl, P., 1982, Seismic modeling in the domain of intercept time and ray parameter: IEEE trans. on Acoustic, Speech, and Signal Processing, **30**, 406–423.

Young, C. Y., Jarrard, R. D., and Kan, T. K., 1983, Prestack synthetic seismogram of elastic layered earth: Presented at the 53rd Ann. Internat. Mtg., Soc. Expl. Geophys.

Weak Anisotropic Reflections

*Leon Thomsen**

INTRODUCTION

As discussed elsewhere in this volume, the angular dependence of the P-wave reflection coefficient $R_p(\theta)$ between planar isotropic media involves the S-wave properties of both media. This situation has raised the hopes of the petroleum industry that S-wave information about subsurface formations may be found cheaply, through the range dependence of P-wave reflection amplitudes.

However, the subsurface formations are invariably anisotropic, and it is intuitively obvious that the angular dependence of elastic wave velocities should modify the angular dependence of the quasi-P-wave reflection coefficient. This intuition is verified in numerical studies reported in Wright (1984, 1987), who demonstrated that the effect is nontrivial and may substantially interfere with the deduction of S-wave properties. Banik (1987) verified Wright's results analytically, with a closed analytical expression for the anisotropic reflection coefficient, in the limit of weak anisotropy and small angles. The present work generalizes these analytic results to larger angles, and discusses the extension of understanding beyond the numerical examples.

A very brief account of weak isotropic reflections is followed by a very brief account of weak anisotropic body wave propagation. The application to P-wave reflections between weakly anisotropic media follows, with derivations given in an Appendix.

Weak Elastic Reflections

The exact solution for the oblique P-wave reflection at a planar interface between two isotropic media is well-known; a good account is presented in Aki and Richards (1980, chapter 5). However, the exact result is so complex algebraically that it is difficult to grasp intuitively the physics contained in it. Furthermore, it depends upon compressional velocity (V_p), shear velocity (V_s), and density (ρ) in both media, which means that a numerical study necessarily involves the exploration of a six-parameter model space. Not even the choice of parameters is obvious, and inappropriate choices [e.g., Poisson's ratio, cf. Koefoed (1955)] have received much discussion.

Fortunately, in most contexts in exploration seismology, an appropriate simplification is available. Since at most reflecting interfaces, the contrast in elastic properties is small, it is appropriate to linearize the exact solution in the small quantities:

$$\left|\frac{\Delta V_p}{\bar{V}_p}\right| \ll 1,$$

$$\left|\frac{\Delta V_s}{\bar{V}_s}\right| \ll 1,$$

$$\left|\frac{\Delta \rho}{\bar{\rho}}\right| \ll 1,$$

where the bar denotes an average of properties, above and below the interface. Chapman (1976) showed that the linearized reflection coefficient is (cf. also Aki and Richards, 1980):

$$R_p(\theta) = \frac{1}{2}\left[\frac{\Delta Z_p}{\bar{Z}_p}\right] + \frac{1}{2}\left[\frac{\Delta V_p}{\bar{V}_p} - \left(\frac{2\bar{V}_s}{\bar{V}_p}\right)^2 \frac{\Delta G}{\bar{G}}\right] \sin^2\theta + \frac{1}{2}\left[\frac{\Delta V_p}{\bar{V}_p}\right] \tan^2\theta \sin^2\theta \qquad (1)$$

where θ is the incidence angle, $Z_p = \rho V_p$ is the P-wave impedance, and $G = \rho V_s^2$ is the shear modulus. Equation 1 is equivalent to, but algebraically simpler than, the linearized formulation in Shuey (1985) and differs from an expression in Wiggins et al., (1985) in not being restricted to small angles or to particular assumptions on $\bar{V}_p/\bar{V}_s$, and more importantly, in the sign of the $(\Delta V_p/\bar{V}_p)\sin^2\theta$ term.

Figure 1 shows the accuracy of the linear approximation compared to the exact solution, for a particular case. Bortfeld (1961) has presented an approximation

*Amoco Production Company, Tulsa, OK.

for $R_p(\theta)$ which is partly linearized. In some cases, Bortfeld's approximation will be more numerically accurate than equation (1), but less accurate in other cases. In any case, numerical accuracy is not the chief virtue of equation (1); all serious calculations should be done with the exact result. The chief virtue of equation (1) is that its simplicity of form allows one to understand the physics involved. Furthermore, equation (1) shows that a numerical study (fitting data to rock properties) need only explore a three-dimensional parameter space; the three parameters are the three combinations of differentials in brackets in equation (1) or combinations of these. We are reminded, once again, that reflections do not depend on absolute values of rock properties separately, but only on certain differences in properties. Absolute values of V_p, etc., may be found by integrating (through traveltime) the differences $\Delta V_p/\bar{V}_p$, etc., found at reflecting events. Further discussion of this isotropic problem is presented elsewhere in this volume; the minus-sign in the coefficient of the $\sin^2 \theta$ term in equation (1) is the crucial feature which makes this discussion interesting.

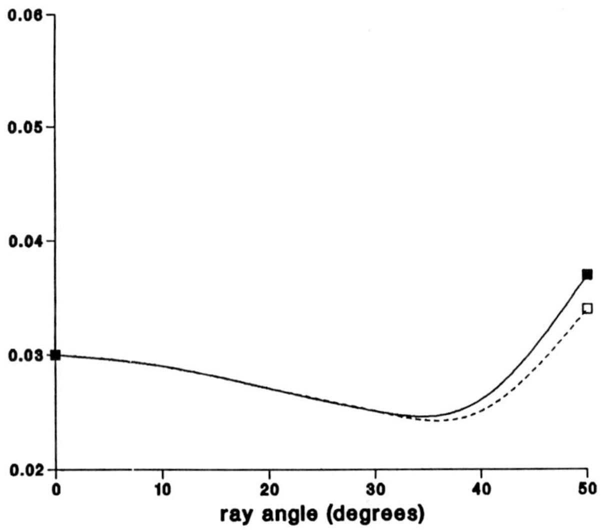

Fig. 1. Isotropic reflection coefficient variation with ray angle. Solid line—exact solution. Dashed line—weak elastic approximation [Equation (1)]. The model parameters are: incident V_p = 9500 ft/s (2895 m/s), incident V_s = 5800 ft/s (1768 m/s), incident density = 2.18 gm/cc, reflecting V_p = 10 000 ft/s (3048 m/s), reflecting V_s = 6000 ft/s (1829 m/s), reflecting density = 2.2 gm/cc.

Weak Elastic Anisotropy

Most rocks are weakly anisotropic, with elastic velocities depending both on angle of propagation (measured azimuthally and from the vertical) and on angle of polarization (for S-waves). The azimuthal dependence may be neglected, however, for P-wave propagation at small-to-moderate angles of incidence; cf. Thomsen (1988) and references cited therein. For body waves in an azimuthally isotropic rock, Thomsen (1986) showed that the various velocities normal to wavefronts are

$$V_p(\theta) = \alpha_0(1 + \delta \sin^2 \theta \cos^2 \theta + \varepsilon \sin^4 \theta) \quad (2)$$

$$V_{S\perp}(\theta) = \beta_0\left[1 + \left(\frac{\alpha_0}{\beta_0}\right)^2 (\varepsilon - \delta) \sin^2 \theta \cos^2 \theta\right] \quad (3)$$

$$V_{S\parallel}(\theta) = \beta_0(1 + \gamma \sin^2 \theta). \quad (4)$$

Here $S\perp$ and $S\parallel$ denote S-waves with polarization vectors that have a component perpendicular ($\perp$) to the plane of symmetry, or are parallel ($\parallel$) to the plane. If the symmetry plane is parallel to the ground surface, these waves are polarized in-line and cross-line, respectively. In a horizontally stratified isotropic context, these may be called SV- and SH-waves, respectively, but in anisotropic media, this labeling (with its irrelevant reference to the direction of gravity) may at times lead to confusion.

In equations (2, 3, and 4), α_0 and β_0 are P- and S-wave velocities, respectively, for propagation along the symmetry axis, and θ is the angle of propagation (trivially different from the wavefront-normal angle in this context). ε, δ, and γ are three independent anisotropic parameters, each assumed to be much less than one. ε is the "familiar" P-wave anisotropy; in terms of vertical and horizontal velocities

$$\varepsilon = \frac{V_p(90°) - \alpha_0}{\alpha_0}. \quad (5)$$

γ is the corresponding quantity for S-waves:

$$\gamma = \frac{V_{S\parallel}(90°) - \beta_0}{\beta_0}. \quad (6)$$

δ is the "strange" P-$S\perp$ anisotropic parameter, defined in terms of elastic moduli in Thomsen (1986):

$$\delta = \frac{(C_{13} + C_{44})^2 - (C_{33} - C_{44})^2}{2C_{33}(C_{33} - C_{44})}, \quad (7)$$

$$\simeq \frac{C_{13} + 2C_{44} - C_{33}}{C_{33}},$$

or alternatively in terms of P-wave velocity by

$$\delta \simeq 4[V_p(45°)/V_p(0°) - 1] - [V_p(90°)/V_p(0°) - 1]. \quad (8)$$

The parameter δ is intuitively inaccessible, and is tedious to measure in the laboratory, but Thomsen (1988) showed that δ is much more important in most exploration contexts than is the more familiar anisotropy parameter ε. The reason is that, for small angles θ, the ε-term in equation (2) is much smaller than the δ-term (unless in some instance $\delta \ll \varepsilon$), so that the δ-term usually dominates. Thomsen (1986) tabulated most published experimental values of δ, documenting that in general $\delta \neq \varepsilon$, and hence that P-wavefronts in most azimuthally isotropic rocks are not elliptical.

Other measures of anisotropy are possible, of course [cf. e.g., Banik (1987)]. However, the set (δ, ε γ) seems most convenient in discussions where near-vertical P-wave propagation is prominent, and hence where δ recurs as the leading anisotropy term in equations expressing P-wave angular dependence. If, in some other context, explicit introduction of a measure of $S\perp$ anisotropy were important, then the parameter

$$\sigma = \left(\frac{\alpha_0}{\beta_0}\right)^2 (\varepsilon - \delta)$$

is suggested by the form of equation (3). For near-vertical problems including $S\perp$-waves, the appropriate set of anisotropies is then (δ, σ, γ), although the redundant set (δ, σ, ε, γ) might be used if properly explained. Of course, these parameters are useful for characterizing the medium, and for intuitively understanding medium behavior, even when they are not small.

Weak Anisotropic Reflections

The solution to the general problem of P-wave reflections in anisotropic media was given in Keith and Crampin (1977). The specialization to a specific form of azimuthal anisotropy particularly relevant to seismic exploration was discussed in Thomsen (1988).

Here, we are concerned especially with angle-dependent effects due to P-wave reflections in azimuthally isotropic media, from an interface parallel to the symmetry planes of the media. The exact solution to this reflection problem was given in Daley and Hron (1977). Even more so than in the isotropic case, that solution is so algebraically complex that seeing the essential physics is difficult. Fortunately, an appropriate approximation is available, i.e., that the anisotropy is weak in both media, and that the contrasts in vertical elastic properties are also small. Specifically, the assumption is:

$$\left|\frac{\Delta\alpha_0}{\bar{\alpha}_0}\right| \ll 1 \qquad |\varepsilon_\mu| \ll 1,$$

$$\left|\frac{\Delta\beta_0}{\bar{\beta}_0}\right| \ll 1 \qquad |\delta_\mu| \ll 1, \quad (9)$$

$$\left|\frac{\Delta\rho}{\bar{\rho}}\right| \ll 1 \qquad |\gamma_\mu| \ll 1,$$

where subscript μ denotes either (1) incident medium or (2) reflecting medium.

The derivation of the quasi-P-wave reflection coefficient subject to these assumptions is given in the Appendix. The result is

$$R_p(\theta) = \frac{1}{2}\left[\frac{\Delta Z_0}{\bar{Z}_0}\right] + \frac{1}{2}\left[\frac{\Delta\alpha_0}{\bar{\alpha}_0} - \left(\frac{2\bar{\beta}_0}{\bar{\alpha}_0}\right)^2 \frac{\Delta G_0}{\bar{G}_0} + (\delta_2 - \delta_1)\right]\sin^2\theta + \frac{1}{2}\left[\frac{\Delta\alpha_0}{\bar{\alpha}_0} - (\delta_2 - \delta_1 - \varepsilon_2 + \varepsilon_1)\right]\tan^2\theta\sin^2\theta \quad (10)$$

where $Z_0 = \rho\alpha_0$ and $G_0 = \rho\beta_0^2$ are the vertical P-wave impedance and S-wave modulus. The corresponding transmission coefficient, and the equivalent quantities for S-waves are given in the Appendix. Equation (10) reduces to Banik's (1987) result in the limit of small angles. Figure 2 shows the accuracy of the linear approximation compared to the exact solution, for a particular case. Equation (10) is directly comparable to equation (1), the isotropic case. One sees immediately that the isotropic case may be recovered by assigning zero anisotropies in equation (10). At normal incidence ($\theta = 0$), only the first term is nonzero. This first term is the same in both equations because each anisotropic medium has been parameterized in terms of two *vertical* velocities (plus three anisotropies), instead of five elastic moduli.

The second term of both equations (in $\sin^2\theta$) gives the low-order angular variation. The coefficients of these terms involve contrasts in the vertical velocities and also contrasts in the anisotropies. The coefficient of the $\sin^2 \theta$ term,

$$\frac{1}{2}\left[\frac{\Delta\alpha_0}{\bar{\alpha}_0} - \left(\frac{2\bar{\beta}_0}{\bar{\alpha}_0}\right)^2 \frac{\Delta G_0}{\bar{G}_0} + (\delta_2 - \delta_1)\right], \tag{11}$$

illustrates the fundamental conclusion of this work: the anisotropy is a *first-order* effect in the range dependence of $R_p(\theta)$. That is, the anisotropy enters into terms like equation (11) as simple additions (of the same magnitude) to the isotropic terms ($\Delta\alpha_0/\bar{\alpha}_0$, etc.). Therefore, the anisotropy makes a nonnegligible (first-order) contribution, even when it is weak. This happens because, while an anisotropy of 10 percent is small compared to 1, and hence negligible in many contexts, it is not negligible when the leading term is also of order 10 percent, or less, as in the present discussion, e.g., equation (11). Another example of this situation was given by Thomsen (1986) in the Discussion concluding that work.

Another feature of anisotropic reflections is that the anisotropy which appears in the lowest angular order [cf. equation (11)] is the strange anisotropy δ, rather than the familiar anisotropy ε. This is another instance of the conclusion of Thomsen (1986) that the quantity δ is more important than ε in exploration seismology. In fact, when particular examples, such as those of Wright (1984, 1987), are considered for the effects of anisotropic elastic moduli, it is primarily the combination of moduli composing δ (rather than the individual moduli) which determines the result.

When the exact equations (Daley and Hron, 1977) are implemented, inputting five elastic moduli for each medium, it is difficult to rationally specify modulus values to yield values of δ appropriate for sedimentary rocks. A better approach is to first select anisotropy parameters, δ, ε, and γ, perhaps using Table 1 from Thomsen (1986) as a guide. These values for intrinsic anisotropy must usually be modified by considerations of effective long-wavelength (extrinsic) anisotropy, due to stratification (Backus, 1962). Note that for δ (in contrast to ε), both positive and negative values are plausible. If the selected anisotropies δ and ε are small, one may use equation (10) directly to calculate reflectivity. A better procedure is to first calculate elastic moduli using equations (5–8), and then calculate reflectivity using the exact formulas in Daley and Hron (1977). Whether or not the anisotropies are small, equation (10) provides a guide for iterative adjustment of values for forward modeling purposes.

A third feature of the linearized anisotropic reflection coefficient, equation (10), is that the anisotropy appears only as a *difference* in anisotropies across the reflecting boundary. Hence, the isotropic case may be recovered if both media are equally anisotropic, so that differences such as $\delta_2 - \delta_1$ vanish. This is possible in some cases, but implausible in general. In fact, many reflection scenarios involve a sandstone (very weak anisotropy) on one side of the interface and a shale (weak to moderate anisotropy) on the other. In these cases, a suitable approximation is to neglect the sandstone anisotropy altogether, leaving only terms in δ_{shale} in the equations, along with the isotropic terms $\Delta\alpha_0/\bar{\alpha}_0$, etc.

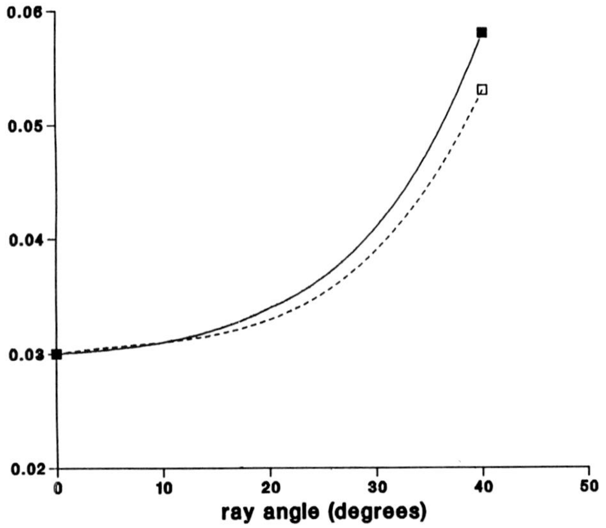

Fig. 2. Anisotropic reflection coefficient variation with ray angle. Solid line—exact solution. Dashed line—weak elastic and weak anisotropy approximation [equation (10)]. The model parameters are the same vertical velocities and densities as those of Fig. 1 and in addition incident delta (δ) = 0., incident epsilon (ε) = 0., reflecting delta (δ) = .10, and reflecting epsilon (ε) = .05.

Examples

The linearized equation (10) analytically verifies the conclusions of Wright (1984, 1987), determined by numerical computations using the exact solutions. In Wright's examples, one may verify that his assumed anisotropies were, in fact, plausible for sedimentary examples, by calculating

$$\varepsilon_1 = 0 \quad \delta_1 = 0$$

$$\varepsilon_2 = .26 \quad \delta_2 = .40$$

for the Model 2 shale in his Figure 5. This is a large (but not implausible) value, cf. Thomsen (1986); Wright used this value in a case where $\Delta\alpha_0/\bar{\alpha}_0$ was also large.

Wright's point may be made directly using the present analysis. The previous Figures 1 and 2, presenting isotropic and anisotropic calculations respectively, were done using the same vertical velocities (P and S) and densities in the two cases. The two calculations differ only in the non-zero anisotropies used in Figure 2. The reflecting horizon represents a normal lithologic change, i.e., one with velocity and density jumps leading to a decreasing $R_p(\theta)$, as calculated with isotropic theory (Figure 1). However, if the same normal lithologic change is associated with a plausible change in anisotropy, then $R_p(\theta)$ *increases* with angle (Figure 2) instead of decreasing. By isotropic theory, this increase could be interpreted as indicative of gas; here a similar effect is caused by anisotropy.

Banik (1987) has provided further examples. However, for developing insight it is more efficient to simply examine equation (10). One essential point is the size and sign of $\delta_2 - \delta_1$, relative to $\Delta\alpha_0/\bar{\alpha}_0$. If Wright had chosen an example with smaller $\Delta\alpha_0/\bar{\alpha}_0$, he could have shown the same anisotropic effect with a smaller δ_2. In fact, in any specific context, it is usually not clear what values of δ (and ε) are appropriate. Hence, in forward modeling, the anisotropy must often be allowed to assume a wide range of values. Therefore, in fitting the equation to data, the uncertainty in the other terms is correspondingly large, and a conclusion (such as the presence of gas in one medium) is correspondingly uncertain.

However, from equation (10), it is clear that, if the isotropic terms (e.g., $\Delta\alpha_0/\bar{\alpha}_0$) are large, then any plausible anisotropy will yield only quantitative (not qualitative) effects. Hence, among bright reflections, one may have more confidence in an isotropic interpretation than among less bright reflections, especially in a qualitative sense.

Nonetheless, *most* reflections are not bright (by definition), and so most have angular variations which are quantitatively affected by anisotropy. Hence, attempts to recover absolute values of ρ, V_s, etc., by integration over traveltime, will have to correct for these anisotropic contributions.

Implications for Exploration

It is shown here analytically that anisotropy constitutes a first-order effect in the angular dependence of P-wave reflection amplitude. Therefore, in the general case, it may not be neglected for quantitative work. In other words, the angular dependence of $R_p(\theta)$ contains information, not only about S-wave properties of the media, but also about their anisotropies. The two effects are plausibly of comparable size, and not easy to separate. Hence, the S-wave information may not be deduced from this angular dependence, without corrections or assumptions regarding the anisotropy. The assumption of zero anisotropy is probably not realistic.

The anisotropic parameter which primarily controls the effect is the "strange" anisotropy parameter δ, rather than the familiar parameter ε. Very little is known about appropriate values for δ. Thomsen (1986) tabulates most published laboratory and field measurements, but in any field context, there will be additional contributions to the effective value of δ from thin-stratification. In fact, it appears that the angular variation in $R_p(\theta)$, in conjunction with independent information on $\Delta\bar{V}_s/\bar{V}_s$, etc., may constitute one of the best ways to estimate anisotropy in situ.

Because anisotropy appears as a first-order contribution to the angular dependence of $R_p(\theta)$, it follows that exploration success, obtained by neglecting the effect in one context, may not presage similar success in other contexts. For example, the effect may be negligible (due to very small anisotropy or very similar anisotropies) in rocks in a certain basin, or of a certain age, yielding accurate predictions of gas by using the isotropic interpretation. Equation (10) shows that such success may not be confidently generalized to rocks of a different basin or a different age, where the anisotropies may not be so obliging. If the isotropic approximation is to be used, it must be justified separately in every context.

Similarly, although neglect of the anisotropic contributions may lead to qualitative success of the isotropic interpretation for bright events, this success may not extend to less bright events. Further, deduction of absolute values for in-situ S-wave velocity, etc., will be quantitatively in error, even when the isotropic theory is qualitatively adequate at selected horizons.

Therefore, it appears that the hope of obtaining cheap S-wave information from the range dependence of P-wave reflection amplitudes is complicated, in an essential way, by anisotropy. The extent of the complication should become clearer, within a few years, as the industry acquires a better understanding of the effective in-situ values of anisotropy.

APPENDIX A—DERIVATIONS

The most straightforward way of linearizing the reflection coefficient is to differentiate the exact expression for $R_p(\theta)$ [cf. Daley and Hron (1977), equation (17)] in each of the seven small quantities in equation (9) (not including γ_μ, which enters the $S\|$ problem only). Denoting these by d_n, $n = 1 \cdots 7$, the answer is

$$R_P(\theta) \simeq \sum_n \frac{\partial R_P(\theta)}{\partial d_n} d_n \tag{A-1}$$

where the summation is over the repeated indices. Although the algebra is tedious, it may be performed by symbolic manipulation software, such as REDUCE or Mathematica.

In practice, this straightforward approach overwhelmed the computer memory resources available at the time of this research (1982). However, substantially less memory is required by the following approach: Instead of linearizing the exact answer, we linearize the exact equations, and solve the resulting simpler problems.

The reflection coefficient $R_p(\theta)$ is found by solving the boundary conditions for continuity of displacement and stress at the interface. As in the isotropic problem, the incident P-wave excites only outgoing P- and $S\perp$-waves, with indices:

$\nu = 0$ incident quasi-P-wave
$\nu = 1$ reflected quasi-P-wave
$\nu = 2$ transmitted quasi-P-wave
$\nu = 3$ reflected quasi-$S\perp$-wave
$\nu = 4$ transmitted quasi-$S\perp$-wave

The displacement of the νth wave, propagating in the sagittal $(\hat{x} - \hat{z})$ plane, is

$$\mathbf{u}^\nu(x, z) = U^\nu \mathbf{g}_\nu e^{i(\omega t - \mathbf{k}_\nu \cdot \mathbf{x})}, \tag{A-2}$$

where U^ν is a scalar amplitude, $\mathbf{g}_\nu$ is the νth polarization vector, ω is circular frequency, and $\mathbf{k}_\nu$ is the νth propagation vector.

The propagation vector is

$$\mathbf{k}_\nu(\theta) = \frac{\omega}{V_\nu(\theta_\nu)} [\sin\theta_\nu \hat{x} + \cos\theta_\nu \hat{z}] \tag{A-3}$$

where V_ν is the wavefront velocity, given by equations (2) and (3) in the main text, and θ_ν is the angle between the wavefront normal and the symmetry axis. If the anisotropy is weak, the angle θ_ν differs negligibly from the ray angle. The polarization vector is

$$\mathbf{g}_\nu = (1 + \Delta\ell_\nu)\sin\theta_\nu \hat{x} + (1 + \Delta m_\nu)\cos\theta_\nu \hat{z}, \tag{A-4}$$

where

$$\Delta\ell_\nu = [\delta_\mu - 2(\delta_\mu - \varepsilon_\mu)\sin^2\theta_\nu]\cos^2\theta_\nu, \tag{A-5}$$

$$\Delta m_\nu = -[\delta_\mu - 2(\delta_\mu - \varepsilon_\mu)\sin^2\theta_\nu]\sin^2\theta_\nu, \tag{A-6}$$

with index $\mu = 1$ for the incident medium ($\nu = 0, 2, 4$), and $\mu = 2$ for the reflecting medium ($\nu = 1, 3$). Using these linearized solutions of the wave equations, the boundary condition for continuity of two components of displacement at the interface ($z = 0$) may be written

$$\sum_{\nu=1}^{4} (-1)^\nu u_i^\nu = u_i^0 \quad i = 1 \text{ or } 3. \tag{A-7}$$

The indexing scheme thus lends itself to collection of the outgoing waves on the left, and the incoming (source) wave on the right.

The corresponding equations for continuity of two components of stress, on the $z = 0$ plane, are

$$\sum_{\nu=1}^{4} (-1)^\nu \sigma_{i3}^\nu = \sigma_{i3}^0 \quad i = 1 \text{ or } 3 \tag{A-8}$$

where the stress corresponding to the νth displacement field is

$$\sigma_{ij}^\nu(x, z) = \frac{1}{2}\sum_k \sum_\ell C_{ijk\ell}^\mu \left(\frac{\partial u_k^\nu}{\partial x_\ell} + \frac{\partial u_\ell^\nu}{\partial x_k}\right). \tag{A-9}$$

It is easily established that no solution to equations (A-7, 8) is valid at all times, unless ω is the same for all ν, as shown. Similarly, no solution is valid at all x unless the "wavefront parameter"

$$p = \frac{\sin\theta_\nu}{V_\nu(\theta_\nu)}, \tag{A-10}$$

is the same for all ν. This conclusion is the anisotropic generalization of Snell's law.

This set of four equations (A-7, A-8) may be written compactly following Daley and Hron (1977) (but without their interchange of rows 2 and 3) in terms of matrices:

$$\underset{\sim}{\mathbf{A}}\mathbf{U} = \mathbf{B}, \tag{A-11}$$

where $\mathbf{U}$ is a column-vector of length four, containing the unknowns:

$$\mathbf{U} = \begin{bmatrix} U^1 \\ U^2 \\ U^3 \\ U^4 \end{bmatrix}, \tag{A-12}$$

$\mathbf{B}$ is a column-vector containing the source terms:

$$\mathbf{B} = U_0^0 \begin{bmatrix} -pV_0(\theta_0) \\ C_{55}^1 \sin\theta_0 \cos\theta_0 \\ \cos\theta_0 \\ E_1 \end{bmatrix}. \quad \text{(A-13)}$$

The mapping matrix $\underset{\sim}{\mathbf{A}}$ has the following form [from Daley and Hron (1977)]:

$$\underset{\sim}{\mathbf{A}} = \begin{bmatrix} pV_0 & -\dfrac{1+\Delta\ell_2}{1+\Delta\ell_1} pV_2 & \dfrac{1+\Delta m_3}{1+\Delta\ell_1}\cos\theta_3 & -\dfrac{1+\Delta m_4}{1+\Delta\ell_1}\cos\theta_4 \\ C_{55}^1 V_0 \cos\theta_0 & C_{55}^2 LpV_0 \cos\theta_2 & C_{55}^1 W_1 & C_{55}^2 W_2 \\ \cos\theta_0 & \dfrac{1+\Delta m_2}{1+\Delta m_1}\cos\theta_2 & -\dfrac{1+\Delta\ell_3}{1+\Delta m_1} pV_3 & -\dfrac{1+\Delta\ell_4}{1+\Delta m_1} pV_4 \\ -E_1 & E_2 & D_1 pV_0 \cos\theta_3 & -D_2 pV_0 \cos\theta_4 \end{bmatrix}, \quad \text{(A-14)}$$

where the following symbols have been used:

$$D_1 = (1+\Delta\ell_3)C_{33}^1 - (1+\Delta m_3)C_{13}^1,$$

$$D_2 = (1+\Delta\ell_4)C_{33}^2 - (1+\Delta m_4)C_{13}^2,$$

$$E_1 = (1+\Delta\ell_1)C_{13}^1 \sin^2\theta_0 + (1+\Delta m_1)C_{33}^1 \cos^2\theta_0,$$

$$E_2 = \frac{V_1(\theta_1)}{V_2(\theta_2)}[(1+\Delta\ell_2)C_{13}^2 \sin^2\theta_2 + (1+\Delta m_2)C_{33}^2 \cos^2\theta_2],$$

$$W_1 = \frac{V_1(\theta_1)}{V_3(\theta_3)} \frac{1}{2+\Delta\ell_1+\Delta m_1} \times [(1+\Delta m_3)\cos^2\theta_3 - (1+\Delta\ell_3)\sin^2\theta_3],$$

$$W_2 = \frac{V_1(\theta_1)}{V_4(\theta_4)} \frac{1}{2+\Delta\ell_1+\Delta m_1} \times [(1+\Delta m_4)\cos^2\theta_4 - (1+\Delta\ell_4)\sin^2\theta_4],$$

$$L = \frac{2+\Delta\ell_2+\Delta m_2}{2+\Delta\ell_1+\Delta m_1}.$$

It is easy to formally write the solution to equation (A-11) as

$$\mathbf{U} = \underset{\sim}{\mathbf{A}}^{-1}\mathbf{B}. \quad \text{(A-15)}$$

Further, it is easy to rewrite the entire system in terms of the degenerate set of equations (i.e., the set with all small quantities $d_n = 0$ identically), plus small departures from this degenerate set. Denoting the degenerate set with subscript 0, the degenerate version of equation (A-11) reads

$$\underset{\sim}{\mathbf{A}}_0 \mathbf{U}_0 = \mathbf{B}_0. \quad \text{(A-16)}$$

This equation describes the propagation of a plane P-wave through a phantom "interface" between identical, isotropic media. The solution is

$$\mathbf{U}_0 = \begin{bmatrix} 0 \\ U^0 \\ 0 \\ 0 \end{bmatrix} = \underset{\sim}{\mathbf{A}}_0^{-1}\mathbf{B}_0, \quad \text{(A-17)}$$

corresponding to a transmitted P-wave with unchanged amplitude U^0.

Denoting departures from this case with Δ, the general case, equation (A-11), may be rewritten as

$$(\underset{\sim}{\mathbf{A}}_0 + \Delta\underset{\sim}{\mathbf{A}})(\mathbf{U}_0 + \Delta\mathbf{U}) = \mathbf{B}_0 + \Delta\mathbf{B}.$$

Linearizing this expression in Δ, and using equation (A-16), the difference equation is

$$\underset{\sim}{\mathbf{A}}_0 \Delta\mathbf{U} + \Delta\underset{\sim}{\mathbf{A}}\mathbf{U}_0 = \Delta\mathbf{B},$$

whose solution is

$$\Delta\mathbf{U} = \underset{\sim}{\mathbf{A}}_0^{-1}[\Delta\mathbf{B} - \Delta\underset{\sim}{\mathbf{A}}\mathbf{U}_0]. \quad \text{(A-18a)}$$

With indices explicit, this is

$$\Delta U_\nu = \sum_\kappa (A_0^{-1})_{\nu\kappa}[\Delta B_\kappa - \Delta A_{\kappa 2} U^0];$$

$$\nu = 1, 2, 3, 4. \quad \text{(A-18b)}$$

For the incident-P problem, the only nonzero element in $\mathbf{U}_0$ is the second one, [cf. equation (A-17)], which means that only the second column of differentials $\Delta A_{\kappa 2}$ is required. The first component of $\Delta\mathbf{U}$ will be $U^1(\theta)$, the scalar amplitude of the reflected P-wave; $R_p(\theta_0) = U^1(\theta)/U_0$ is the reflection coefficient.

Equation (A-18) is much easier to solve analytically than is the exact result [equation (A-15)] since $\underset{\sim}{\mathbf{A}}^{-1}$ is very complex, whereas $\underset{\sim}{\mathbf{A}}_0^{-1}$ may be found in simple closed form because of the high degree of symmetry in $\underset{\sim}{\mathbf{A}}_0$:

$$\underset{\sim}{\mathbf{A}}_0 = \begin{bmatrix} pV_0 & -A_0(1,1) & \cos\theta_3 & -A_0(1,3) \\ C_{55}^1 p\alpha_0 \cos\theta_1 & A_0(2,1) & \dfrac{\alpha_0 C_{55}^1(\cos^2\theta_3 - p^2\beta_0^2)}{2\beta_0} & A_0(2,3) \\ \cos\theta_1 & A_0(3,1) & -pV_3 & A_0(3,3) \\ -C_{33}^1(1-2p^2\beta_0^2) & -A_0(4,1) & 2C_{55}^1 p\alpha_0\cos\theta_3 & -A_0(4,3) \end{bmatrix}, \tag{A-19}$$

which leads to

$$\underset{\sim}{\mathbf{A}}_0^{-1} = \begin{bmatrix} \left(\dfrac{\beta_0}{\alpha_0}\right)^2 \sin\theta_1 & \dfrac{p}{\rho\alpha_0\cos\theta_1} & \dfrac{1-2p^2\beta_0^2}{2\cos\theta_1} & \dfrac{-1}{2\rho\alpha_0^2} \\ -A_0^{-1}(1,1) & A_0^{-1}(1,2) & A_0^{-1}(1,3) & -A_0^{-1}(1,4) \\ \dfrac{(1-2p^2V_{s0}^2)}{2\cos\theta_3} & \dfrac{1}{\rho V_{p0}V_{s0}} & -\dfrac{V_{s0}}{V_{p0}}\sin\theta_1 & \dfrac{p}{2\rho V_{p0}\cos\theta_3} \\ -A_0^{-1}(3,1) & A_0^{-1}(3,2) & A_0^{-1}(3,3) & -A_0(3,4) \end{bmatrix} \tag{A-20}$$

This result was found using the facilities of REDUCE (Hearn, 1978). The differentials $\Delta\mathbf{B}$ and $\Delta\underset{\sim}{\mathbf{A}}$, required in equation (A-18), are straightforward, for example:

$$\Delta B_\kappa = \sum_n \frac{\partial B_\kappa}{\partial d_n} d_n\ ;\ \kappa = 1, 2, 3, 4, \tag{A-21}$$

and similarly for $\Delta A_{\kappa 2}$. Putting these together, the answer for the reflection coefficient is given as equation (10) of the main text. Equation (10) is valid at any angle, not too close to critical, for reflections between two similar, weakly anisotropic media, with symmetry planes parallel to the interface.

The corresponding expression for the transmission coefficient is

$$T_p(\theta) = 1 - \frac{1}{2}\frac{\Delta Z_0}{\bar{Z}_0} + \frac{1}{2}\left[\frac{\Delta\alpha_0}{\bar{\alpha}_0} + (\delta_2 - \delta_1)\right]\sin^2\theta + \frac{1}{2}\left[\frac{\Delta\alpha_0}{\bar{\alpha}_0} + (\varepsilon_2 - \varepsilon_1)\right]\sin^2\theta\tan^2\theta - (\delta_2 - \delta_1 - \varepsilon_2 + \varepsilon_1)\sin^4\theta \tag{A-22}$$

Notice that $T_p \neq 1 - R_p$ except at normal incidence. Equation A-22 reduces to Banik's (1987) expression at small angles.

The corresponding expressions for $S\perp$-waves may be derived using the same methods. The results are:

$$R_{S\perp}(\theta) = -\frac{1}{2}\left(\frac{\Delta\rho}{\bar{\rho}} + \frac{\Delta\beta_0}{\bar{\beta}_0}\right) - \frac{1}{2}\frac{\Delta\beta_0}{\bar{\beta}_0}\tan^2\theta + 2\left(\frac{\Delta\rho}{\bar{\rho}} + 2\frac{\Delta\beta_0}{\bar{\beta}_0}\right)\sin^2\theta - \frac{1}{2}\frac{\bar{\alpha}_0^2}{\bar{\beta}_0^2} \times (\delta_2 - \delta_1 - \varepsilon_2 + \varepsilon_1)\sin^2\theta, \tag{A-23}$$

$$T_{S\perp}(\theta) = 1 - \frac{1}{2}\left(\frac{\Delta\rho}{\bar{\rho}} + \frac{\Delta\beta_0}{\bar{\beta}_0}\right) + \frac{1}{2}\frac{\Delta\beta_0}{\bar{\beta}_0}\tan^2\theta - \frac{1}{2}\frac{\bar{\alpha}_0^2}{\bar{\beta}_0^2}(\delta_2 - \delta_1 - \varepsilon_2 + \varepsilon_1) \times \sin^2\theta(1 - 2\sin^2\theta). \tag{A-24}$$

Shear-Wave Velocity Control

John P. Castagna, Michael L. Batzle,‡ Kenneth M. Tubman,‡ James E. Gaiser,§ and Michael D. Burnett***

INTRODUCTION

Quantitative calibration of amplitude variation with offset data by seismic modeling requires *S*-wave velocity information. The purpose of this paper is to review the means of acquiring *S*-wave velocities in boreholes by: (1) laboratory measurements on core samples, (2) pseudo-shear velocity logs derived from empirical trends and conventional well log data, (3) conventional full-waveform sonic logs, (4) direct *S*-wave sonic logs, and (5) multicomponent vertical seismic profiles with "point" or horizontal sources.

Each technique has limited accuracy. However, when applied with sufficient caution and skepticism, each provides valuable information needed to understand seismic amplitude variation near a borehole. Laboratory measurements are useful for studying the systematics of velocity variations, but due to sampling, scale, and frequency considerations, do not provide direct velocity control for surface seismic data. *S*-wave sonic logs of various kinds give a high resolution measure of vertical velocity variation but suffer from processing errors, borehole damage, and limited lateral investigation. Vertical seismic profiles are comparable to surface seismic data in frequency and scale, but have poor vertical resolution in comparison to well logs.

This paper provides a review of borehole *S*-wave velocity acquisition methods, with some examples and comparisons between techniques in Gulf Coast clastics and Devonian hard shales.

LABORATORY MEASUREMENTS

S-wave velocities have been measured in the laboratory for over half a century. Many of the techniques used for rocks were developed first in materials science to study the properties of alloys, ceramics, composites, etc. The most common methods of measurement are (1) wave propagation experiments usually made at ultrasonic frequencies between 0.2 and 2 MHz, and (2) measurement of the modulus of rigidity and subsequent calculation of the *S*-wave velocity by resonating the sample at a frequency between 500 Hz and 40 KHz.

The complexity of the particular experimental procedure is determined by the objective of the measurement; from a simple beam resonance in an isotropic homogeneous material to a complex, multipath pulse propagation to derive the complete stiffness tensor. The equipment available to measure *S*-wave properties is quite sophisticated, and the accuracy is usually high in competent materials. The largest uncertainties result from (1) poor signal strength in soft rocks, (2) unrepresentative, inhomogeneous, or altered samples, or (3) inappropriately applied experimental conditions.

The laboratory provides numerous advantages and disadvantages for the investigation of *S*-wave properties. In the laboratory, a known, controlled environment can be applied to the sample including confining pressure or inhomogeneous stress, pore pressure, temperature, and the specific pore fluid composition. Velocities can be determined for multiple directions of propagation and wave polarization. The mineralogy and fabric of the specimen can be ascertained accurately. Finally, porosity, density, electrical conductivity, fluid permeability, etc. can be measured for use in theoretical models or empirical correlations. Laboratory methods also impose restrictions that can limit the usefulness of the results or invalidate them entirely. Normally, the frequency of the measurement is high, narrow band, and out of the range of surface seismic or borehole data. All of the samples measured in the laboratory are disturbed to some extent. For soft or clay-rich rocks, this disturbance can alter the measured properties substantially. One problem concerns scale and statistics: in situ measurement techniques

*ARCO Oil and Gas Company, Houston, TX
‡ARCO Exploration and Production Technology, Plano, TX
§Western Geophysical, Denver, CO
**Integrated Seismology, Garland, TX

average distances of meters to hundreds of meters while the laboratory scale is centimeters. As a consequence, laboratory measurements need to be averaged, however, the data generally are coarsely sampled (aliased) and the statistics poor. The major benefit of laboratory data lies in the development of theoretical relationships and empirical trends (Castagna et al., 1993, this volume, pages 135–171).

Constraints are placed on laboratory velocity measurements due to the ultimate application of the data and the nature of the materials being tested. These constraints, and the ancillary equipment required, make S-wave measurement techniques distinct. These measurements are usually made to ascertain in situ characteristics. At depth, rocks are exposed to elevated pressures and temperatures. A "lithostatic" stress is carried by the mineral frame of the rock and fluid pressure exists in the pore spaces. The mechanical properties of a rock, including the S-wave velocity, are sensitive to these pressures. Hence, most measurements are made inside pressure vessels where both lithostatic and pore pressures can be maintained. Space is limited and access to the test specimen is restricted. On the other hand, the influence of temperature is second order and most measurements are made under ambient temperatures except when great depths are simulated or a particular phenomenon is examined (e.g., steam flooding). Environmental conditions are controlled by a hydraulic system and heating coils (Figure 1). Often, the S-wave measurement is simple compared to the simulation of in situ environmental conditions.

The majority of laboratory S-wave velocity data are collected by propagating an S-wave pulse (pulse transmission). Velocities are derived from the observed time required for the pulse to travel from a transmitting transducer to a receiving transducer. A generalized experimental circuit is shown in Figure 2. A typical waveform recorded using this configuration is shown in Figure 3. In this figure, the shear arrival is distinct, but interfering P-wave arrivals are seen clearly in front of the S-wave. Alternatively, a single transducer can be used and the traveltime for the echo recorded. Propagation techniques are robust, flexible, and relatively simple to perform.

Propagation and polarization in several directions allow measurement of anisotropy. However, these measurements are restricted to frequencies above the seismic and sonic logging bands, and ambiguity is introduced due to differences in scaling. Signals can be weak or noisy due to poor coupling between the transducers and the sample, particularly when the

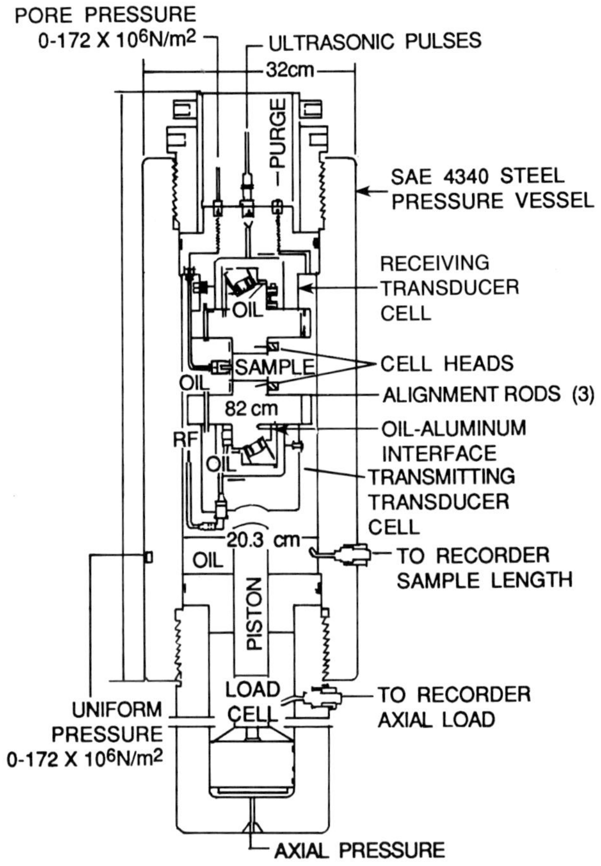

Fig. 1. Dual-mode ultrasonic apparatus for triaxial sample loading (after Gregory and Podio, 1970).

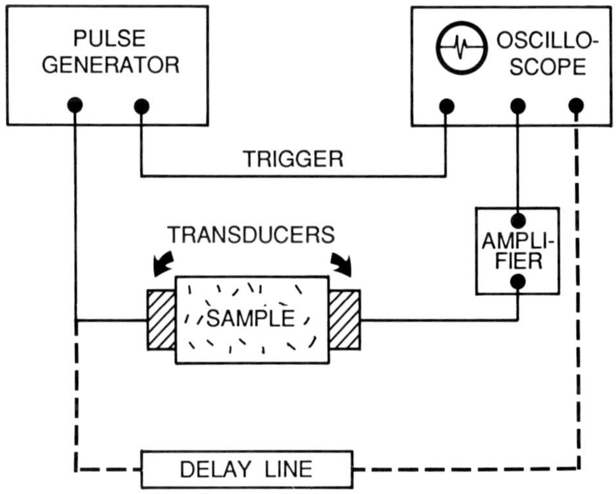

Fig. 2. Generalized schematic of the electronic components used in ultrasonic shear-wave velocity measurements.

sample is poorly consolidated. An excellent, and still applicable, review of these methods was provided in Simmons (1965).

The sources and receivers for these propagation techniques are piezoelectric crystals or poled ceramic aggregates. Transducer technology has undergone intensive development for nondestructive materials testing, biomedical imaging, sonar, etc. The requirements for velocity measurements on rock samples are crude and the details of impedance matching and backing materials are of more concern when attenuation experiments are performed. The frequencies used are constrained by the mineral grain size and overall size of the sample. The signal wavelength should be on the order of ten times the grain size or scattering will attenuate the signal and inhomogeneities dominate the travel path. Low frequencies are hard to excite and if the wavelength is too large, unwanted modes will be excited, near-field effects will dominate the signal, and arrival times will be difficult to determine accurately. Under laboratory conditions and for common samples, these criteria limit frequencies from several hundred kilohertz to 2 MHz. Figure 4 shows various transducer configurations. By using interference techniques of composite bending elements, as shown in Figure 4d, measurements have been made in the kilohertz range (Shirley and Hampton, 1978). Higher frequencies can be generated, and are useful for measuring the elastic constants of single crystals. Typical results of propagation measurements on a rock sample are shown in Figure 5.

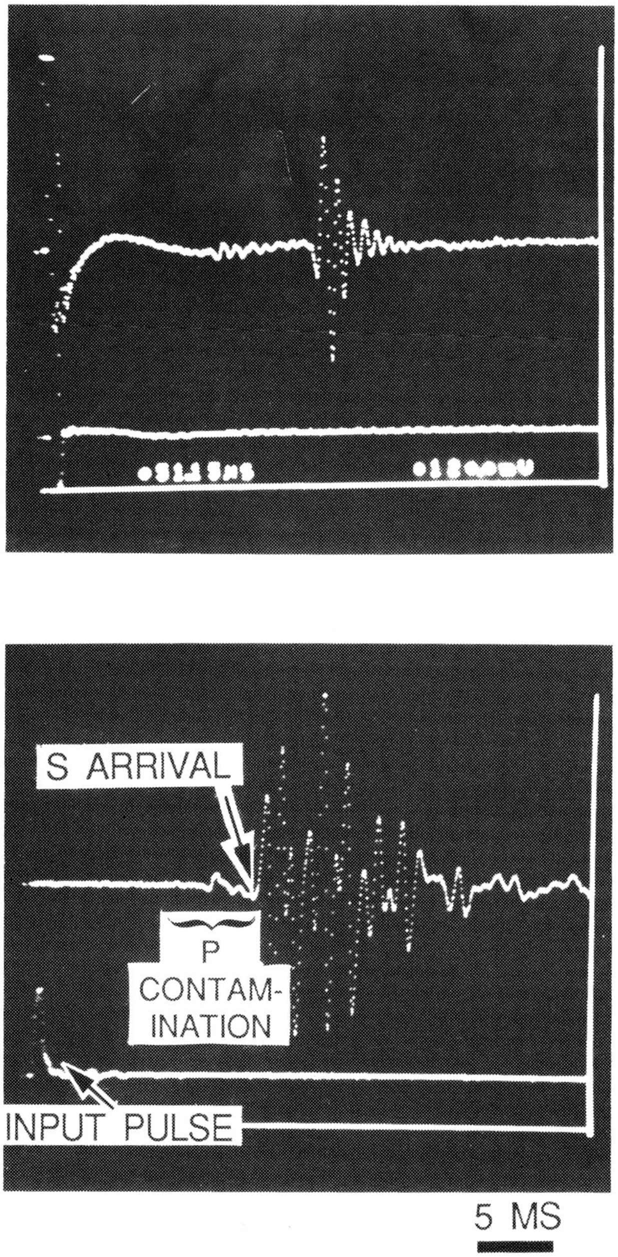

Fig. 3. Oscilloscope pictures of ultrasonic pulse transmission shear waveforms.

The transmission and reflection amplitudes as a function of angle are used to measure S-wave velocity. In this technique, a pulse is transmitted through a liquid to the face of a disk-shaped sample. The resulting transmitted wave is recorded as a function of angle and position. A schematic of the equipment is shown in Figure 6. In this case, the S-wave velocity is derived from the compressional energy minimum according to the relation

$$\ell = \frac{1}{2} L \text{ s } (n\phi) \tag{1}$$

where ℓ is the optimum thickness of the sample, L is the shear wavelength, ϕ the critical angle, and n a positive odd integer (amplitude variation with offset analysis in the laboratory). King and Fatt (1962), Wyllie et al. (1962), and Gregory (1963) adapted this method for rocks at elevated pressures. Hosten et al. (1987) have developed the technique sufficiently to derive the complex stiffness tensor for transversely isotropic materials. The advantage of this transmission-refraction technique is that numerous samples can be run quickly at room pressure. Unfortunately, the complexity of the signals does result in large errors in the derived S-wave velocities and it is difficult to apply confining and pore pressures.

Alternative methods measure the rock shear modulus from which the S-wave velocity can be calculated.

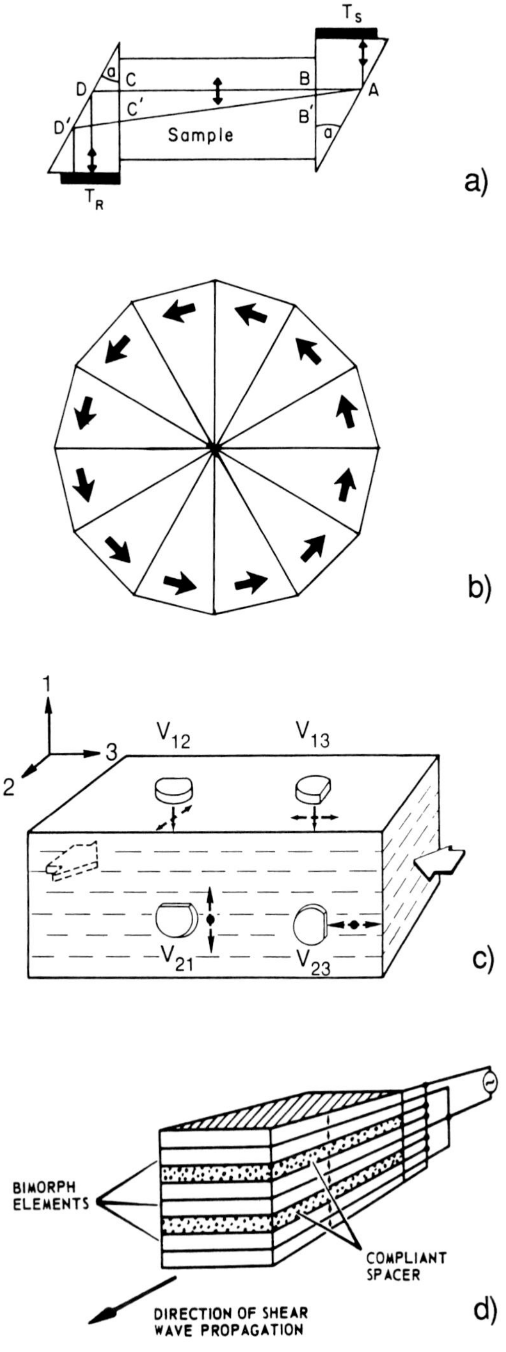

Fig. 4. Transducer configurations: (a) from Jamieson and Hoskins (1963): a glass wedge is used to generate shear waves from compressional transducers (side view), (b) from Rafavich et al. (1984): a composite transducer made from polarized wedges to generate a torsional wave (top view), (c) from Rai and Hanson (1988): placement and orientation of transducers to measure specific shear polarizations (projected view), (d) from Shirley and Hampton (1978): "Bender" element probe for use in soft sediments.

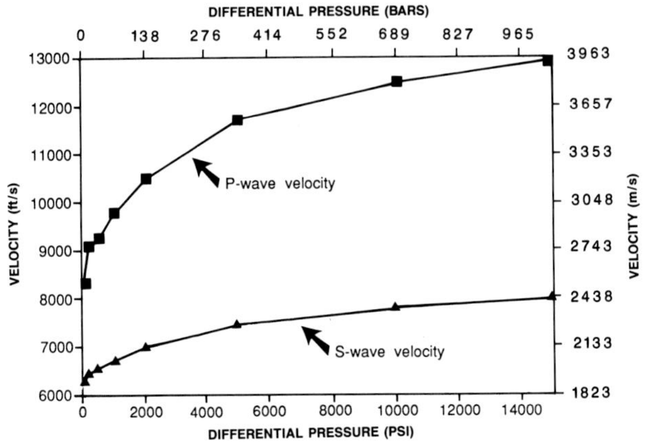

Fig. 5. Compressional and shear-wave velocity versus differential pressure for a sandstone sample.

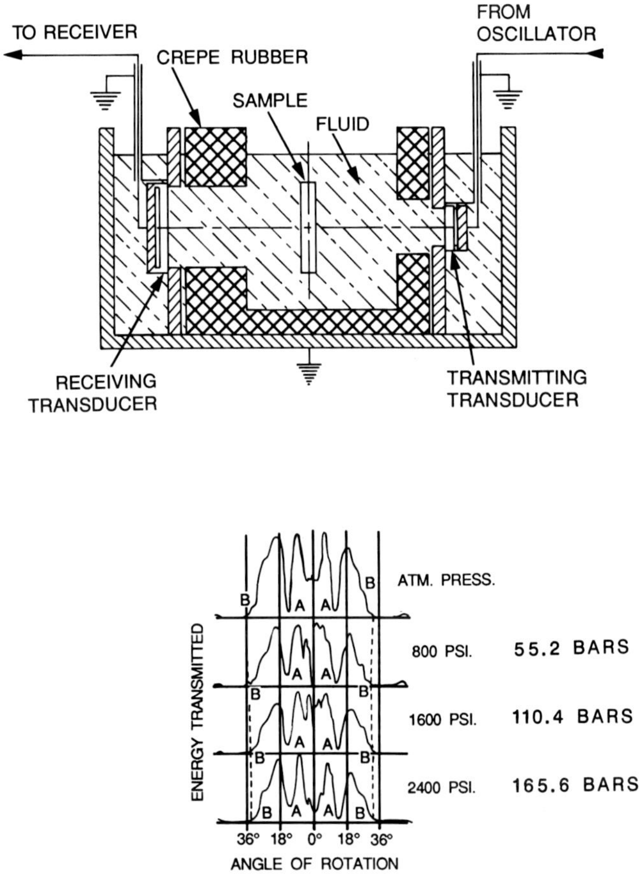

Fig. 6. (a) Section through bath. (b) Energy transmitted versus angle of rotation for Solenhofen limestone. From King and Fatt (1962).

The common rock mechanics approach of modulus determination from macroscopic stress-strain curves is inapplicable. Such large deformation measurements, often with strains greater than 10^{-4}, give low values for the moduli and velocities. This discrepancy is in excess of the minor correction for isothermal stress-strain to adiabatic dynamic wave propagation conditions (Simmons and Brace, 1965). For most rocks, the strain amplitude must stay below 10^{-6} for a valid velocity determination. The resonant beam and the subresonant, low-amplitude, stress-strain techniques have been successfully applied to velocity determination. These methods have the advantage of providing data closer to, or within, the seismic frequency band. These techniques often require cumbersome equipment such that the proper pressure and temperature conditions are difficult to apply. Samples need to be homogeneous and sufficiently indurated for precise shaping. Preparing a sample and mounting it in the experimental apparatus is an exacting and tedious task.

The resonant beam or vibrating bar provides a relatively robust measure of the shear modulus. The most common procedure is to force a long bar of rock into torsional resonance using an electromagnetic or other driver at one end of the sample. A sensor at the other end records the strain amplitude. The driving frequency is swept to maximize the strain amplitude. The shear modulus, μ, is calculated from the resonance frequency, f_n, by:

$$\mu = (2Lf_n/n)^2 \tag{2}$$

where, L is the length of the sample and n the mode number. Odd numbered modes have a node at the center of the sample where it can be supported and pore fluids injected during the experiment. An apparatus with supports at the ends is shown in Figure 7. Typical results of the measurements are shown in Figure 8. Shear velocity is determined at various amplitudes at several of the low mode frequencies, usually in the tens of kilohertz range. Alternatively, by adding fixed weights to the sample assembly, a broader range of frequencies is obtained (Tittman and Curnow, 1976).

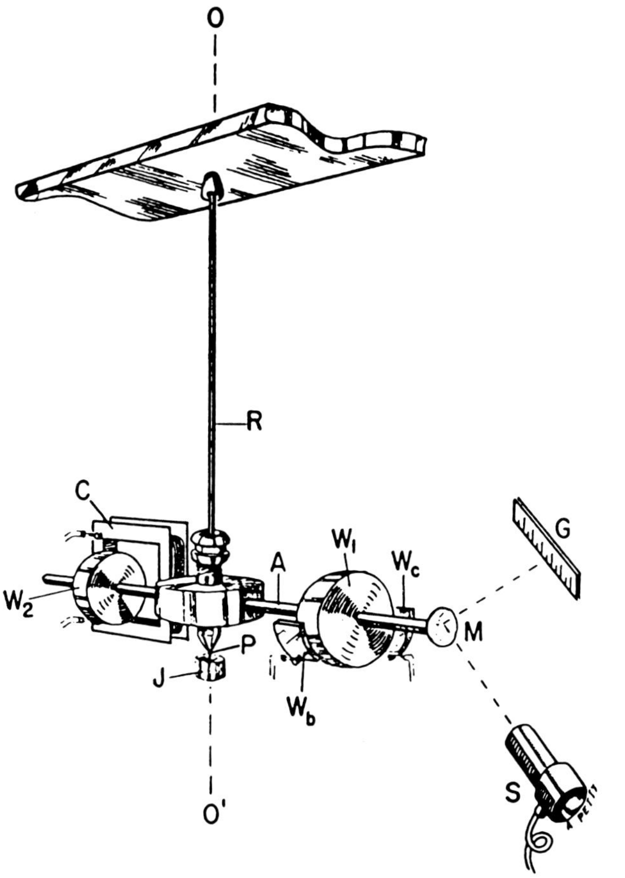

Fig. 7. Perspective drawing of the torsion pendulum. 00′, axis of torsional oscillation; R, specimen having length L between chucks; C, air core coil; W_1 and W_2 soft-iron inertia weights; A, stainless-steel threaded inertia arm; J, sapphire V jewel bearing; P, carpet sewing (from Peselnick and Outerbridge, 1961).

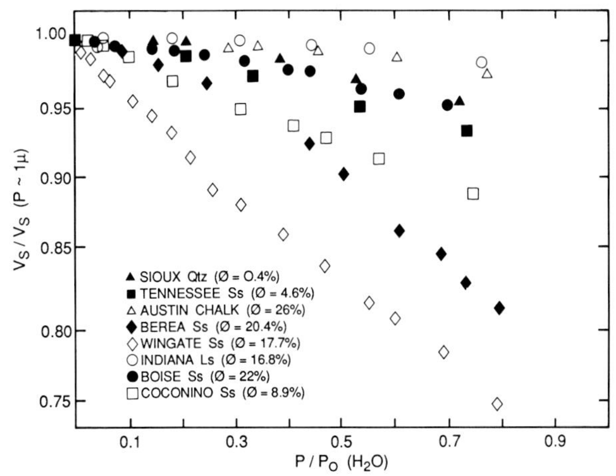

Fig. 8. The shear velocity, normalized to the shear velocity measured in a vacuum of ≈1 μ, (1 microtorr = 1.3×10^{-9} bars) versus the relative partial pressure of water vapor ($P_0(H_2O)$ = partial pressure of water vapor at STP, P = partial pressure of water vapor) for various rocks (from Clark, 1980).

Direct modulus measurements need not be made only at the resonance frequency of the rock sample assembly. The shear modulus is measured by determining the ratio of shear stress to shear strain. Jackson et al., 1984 (Figure 9) apply a low amplitude periodic torsional force to measure the resulting shear strain.

Laboratory measurements generally are not tied directly to seismic data, but rather are used to establish the systematics of velocity variation. Continuous measurements of S-wave velocity as a function of increasing depth in a borehole are generally required for calibration of the seismic response.

PSEUDO-SHEAR VELOCITY LOGS

In practice, the explorationist may not have access to an S-wave velocity log. Indeed, he may be fortunate to have a conventional sonic log of acceptable quality. He then must make a "best guess" of the S-wave velocity. Given a reliable sonic log and volumetric analysis from conventional well logs, it may be possible to estimate the S-wave velocity with a precision of about 5 percent (Greenberg and Castagna, 1992). The ability to predict S-wave velocity follows from: (1) the "global" $V_p - V_s$ trend curves for different lithologies (Pickett, 1963; Castagna et al., 1985; Castagna et al., 1993), (2) the applicability to first order of the Biot-Gassmann equations to real rocks, and (3) the similar mechanical properties of sedimentary rock-forming minerals which allow first-order assumption of mixing laws in multimineralic rocks.

We discuss the case of 100 percent water-saturated rocks with the understanding that partially saturated rocks can be handled in the same way with straightforward application of Gassmann's equations (Gassmann, 1951). We begin with $V_p - V_s$ trend curves for pure lithologies. These are similar to those reported in Castagna et al. (1993) and may be expressed as polynomials (or other empirical fits):

$$V_{si} = a_i V_p^2 + b_i V_p + c_i \tag{3}$$

where V_p is the measured P-wave velocity, V_{si} is the predicted S-wave velocity for lithology i, and the coefficients a_i, b_i, and c_i are "global" for each lithology. Considering well-consolidated rocks only, first order equations are adequate. However, as S-wave velocity logs are most uncommon in poorly consolidated low-velocity sediments, all three coefficients may be necessary to adequately handle low velocities.

For a monomineralic rock, the S-wave velocity is given by equation (3) without porosity, effective stress, or velocity versus depth information. For a mixed lithology, however, some form of mixing law must be assumed. In well log analysis, a time-average mixing is usually assumed. In fact, a variety of mixing laws fall within theoretical limits (Hashin-Shtrikman bounds). For example, a mixing law linear in velocity would yield:

$$V_{sm} = \sum_{i=1}^{n} X_i V_{si} \tag{4}$$

where

n = number of porous end-member lithologies,
X_i = volume fraction of lithology i,
V_{si} = S-wave velocity of lithology i,

and

V_{sm} = S-wave velocity for the mixed lithology

with

$$\sum_{i=1}^{n} X_i = 1$$

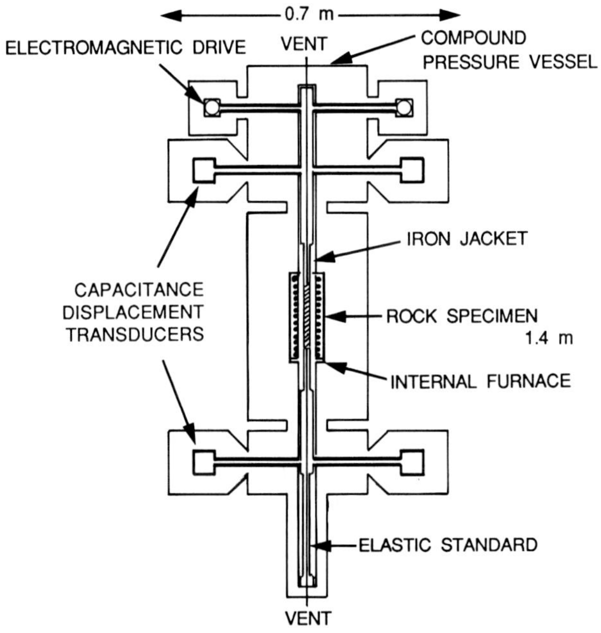

Fig. 9. Experimental configuration for the study of rock anelasticity ultimately at simultaneous high pressure (to 700 MPa) and temperature (to 1400°C), seismic frequencies (10^{-3} – 1 Hz), low strain amplitudes, ($<10^{-6}$) and controlled pore pressure of volatiles (from Jackson et al; 1984).

As the S-wave velocities of the pure end-member components are similar, the exact form of the mixing law is not crucial. A possible exception occurs for clay-rich rocks (Han et al., 1986).

Equations (3) and (4) may be used to predict S-wave velocities in mixed lithologies by assuming the measured P-wave velocity of the composite is in the neighborhood of the hypothetical P-wave velocity of each end-member lithology. Equation (3) is then utilized to produce fictitious S-wave velocities of the end-members. These are mixed using equation (4) which yields the predicted S-wave velocity of the composite. Figure 10 shows a comparison between pseudo and full-waveform sonic measured S-wave velocity logs. We have found that S-wave velocities can be predicted with about 5 percent precision providing: (1) the original log data are high quality, (2) the volumetric analysis is correct, and (3) fluid saturation is known and properly corrected for. Unfortunately, these three assumptions are frequently not valid. It is desirable, therefore, to calibrate with direct measurements whenever possible.

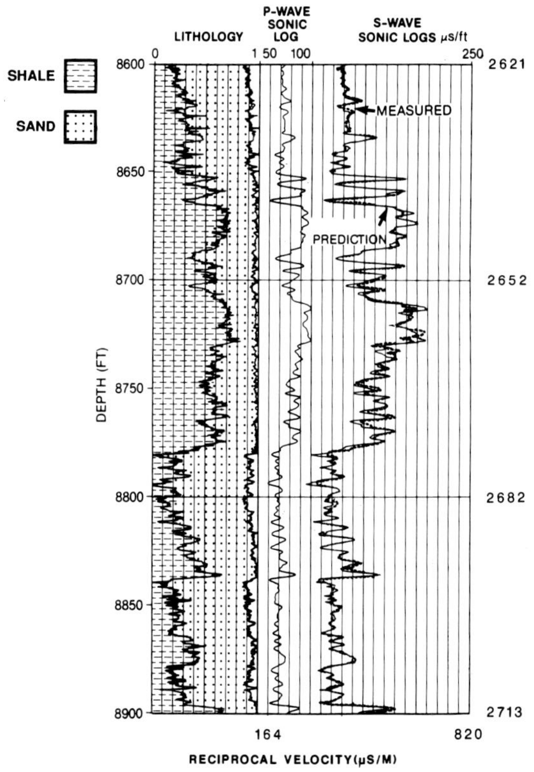

Fig. 10. Field verification of pseudo-shear velocity calculation in a mixed lithology.

FULL WAVEFORM SONIC LOGS

Conventional sonic logging signals contain S-wave velocity information (Sommers and Broding, 1952; Vogel, 1952; Biot, 1952; Pickett, 1963). Digital recording of the full sonic waveform became available in the 1970s and a considerable volume of literature has been published (for example Cheng and Toksoz, 1981; White and Tongtaow, 1981; Siegfried and Castagna, 1982; Paillet and White, 1982; Tubman et al., 1984; Williams et al., 1984; Schmitt, 1988; and many others).

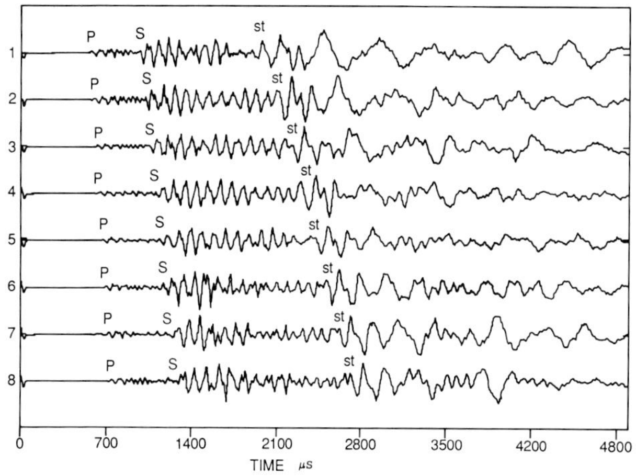

Fig. 11. Conventional full-waveform sonic data acquired with an array tool. The receiver spacing is .15 m. P = P-wave, S = S-wave, st = Stoneley wave.

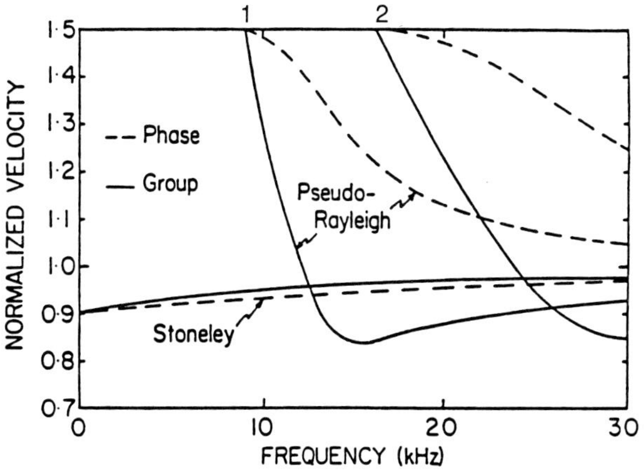

Fig. 12. Dispersion curves for conventional full-waveform sonic data in a "hard" formation. The velocities are normalized by fluid velocity. The first two pseudo-Rayleigh wave modes are shown (from Cheng and Toksöz, 1981).

At acoustic logging frequencies the various modes of propagation along the borehole have wavelengths the same order of magnitude as the borehole diameter. Consequently, guided waves are the dominant form of wave propagation. Figure 11 shows sonic waveforms with increasing source-receiver offset in a "hard" formation (defined as a formation where the *S*-wave velocity is greater than the borehole fluid velocity). The first arrival is the compressional head wave, followed by more compressional energy in the form of leaky-*P* modes (which at high frequencies can be thought of as borehole reverberations). Theoretically, these should be followed by a shear head wave. However, the shear head wave may be relatively weak (Paillet and Cheng, 1986) and inseparable from the following pseudo-Rayleigh wave. The pseudo-Rayleigh wave has been referred to as the *S*-wave, because its initial arrivals travel at or near the *S*-wave velocity. The pseudo-Rayleigh wave is highly dispersive and has a low-frequency cutoff, with phase velocity bounded by the formation *S*-wave velocity from above and the borehole fluid velocity from below (Figure 12). Higher order modes occur, but only one or two pseudo-Rayleigh modes are significant for typical frequencies and borehole geometries. The Stoneley-wave is an interface wave which travels along the borehole wall. As shown in Figure 12, the wave is dispersive (often exhibiting reverse dispersion) with phase and group velocities slightly less than the borehole fluid velocity in "hard" formations.

Successful *S*-wave logging with full-waveform sonic logs depends on the relationship between formation *S*-wave velocity and borehole fluid velocity (velocity of a *P*-wave in the infinite fluid medium).

Case 1: Vs ≫ Vf ("Hard" formations such as limestone, dolomite, and tight sandstone; see Figure 13)

S-wave velocities are obtained from pseudo-Rayleigh wave velocities. *P*-waves are weak compared to pseudo-Rayleigh waves, and due to the corresponding high signal to noise ratio (S/N), excellent results can be obtained with conventional short or long-spaced tools. Excepting fractured, washed out, highly interbedded or gas-saturated intervals, *S*-wave transit-times should have better precision than 4 μs/ft (13 μs/m) providing the waveforms have been carefully processed to avoid cycle skipping.

Case 2: Vs > Vf ("Medium-Hard" formations such as porous consolidated sandstones and some shales; see Figure 14)

S-wave velocities are obtained from pseudo-Rayleigh wave velocities. However, in contrast to Case 1, late arriving compressional energy is stronger and pseudo-Rayleigh waves are weaker. Source-generated noise results in poor *S*-wave S/N, so the redundancy provided by array tools is desirable. Figure 15 shows histograms of the difference between transit-times obtained over the same interval with short and long-spaced sonic tools in a sand-shale sequence at depths between 16100 ft (4907 m) and 16460 ft (5017 m). Depth shift and cable stretch errors were corrected prior to differencing. The repeatability of the *P*-wave transit-time is about 2 μs/ft (6.6 μs/m), whereas the *S*-wave repeatability is significantly poorer at about 6 μs/ft (19.7 μs/m). This is a "best case" scenario as S/N was

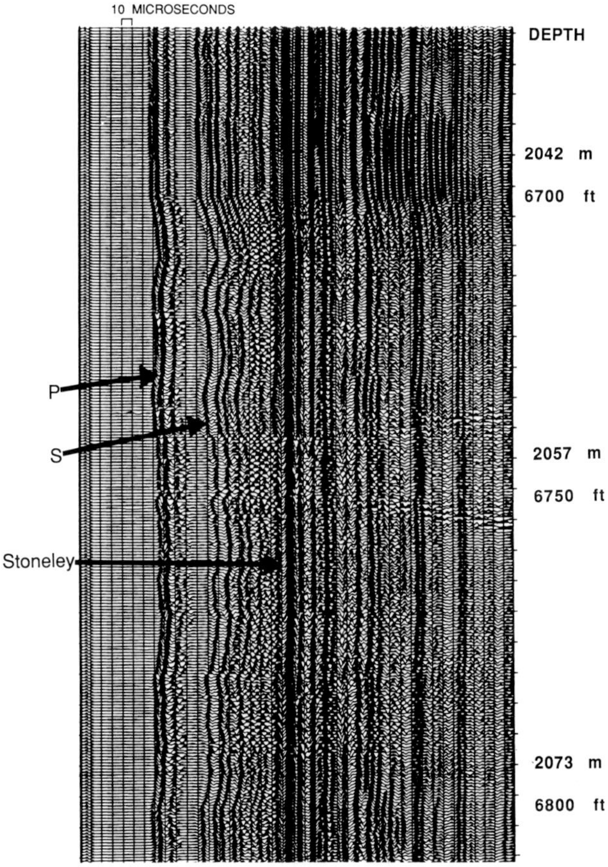

Fig. 13. Full-waveform sonic log in a "hard" formation. Source-receiver spacing equals 10 ft. The Airy phase of the pseudo-Rayleigh wave is more prominent than the Stoneley wave in this frequency band.

significantly enhanced by filtering and the velocities were picked very carefully.

Case 3: Vs < Vf ("Slow" formation such as poorly consolidated sandstones and shales, see Figure 16)

Pseudo-Rayleigh waves are not observed, so *S*-wave velocity is determined by other techniques. An estimate can be extracted from Stoneley-wave analysis, but this method has not proven to be robust. Array tools with good low frequency content (to better excite the Stoneley waves) and nonrigid bodies are advantageous. Borehole rugosity and alteration, as well as formation permeability variations (Williams et al., 1984) and anisotropy, complicate the analysis. The accuracy of the method is dependent on the borehole fluid properties, which may not be well known. *S*-wave velocities are best determined with an asymmetric (dipole or quadrapole) acoustic logging sonde.

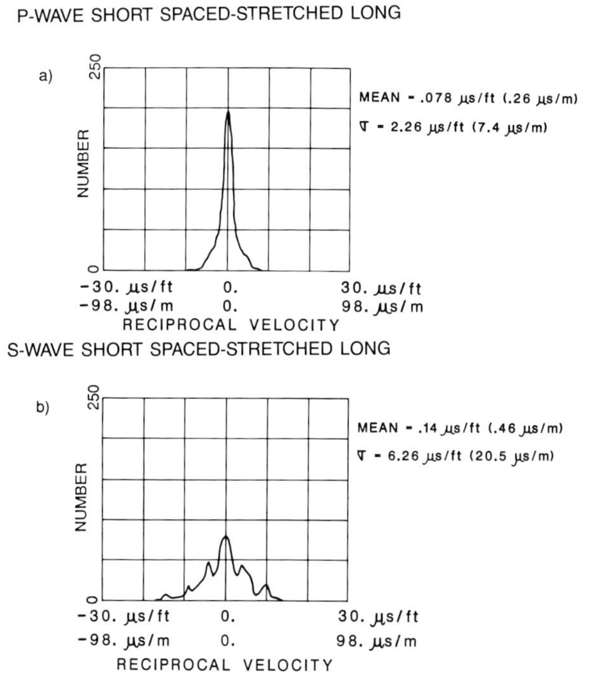

Fig. 15. Histograms of difference between short-spaced and long-spaced *P*- and *S*-wave sonic logs.

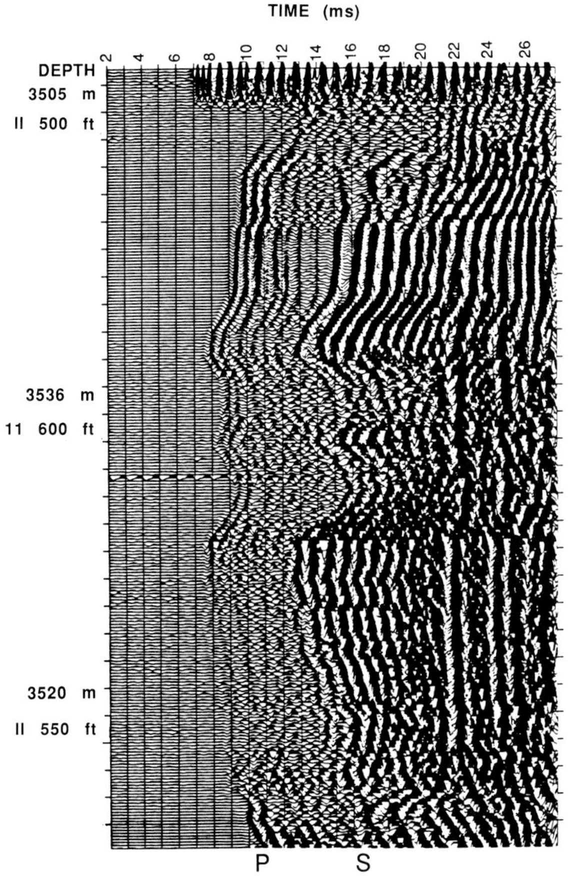

Fig. 14. Full-waveform sonic log in a "medium-hard" formation. Source-receiver spacing equals 10 ft.

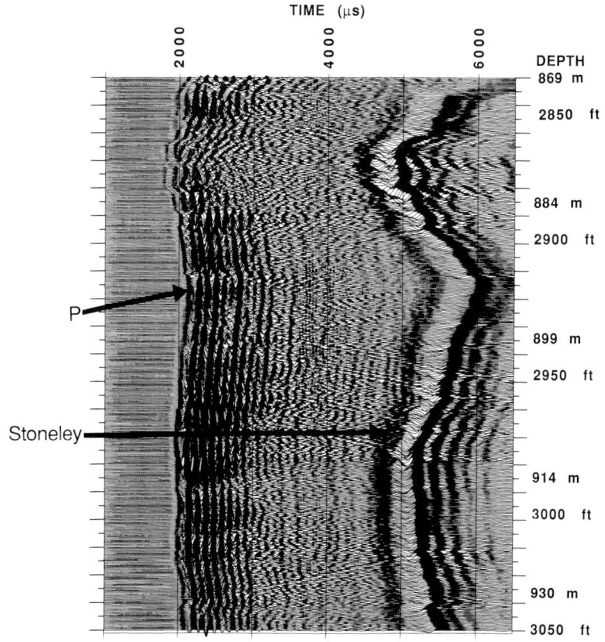

Fig. 16. Full-waveform sonic log in a "slow" formation. Source-receiver spacing equals 10 ft.

Case 4: Vs ≪ Vf ("Very Slow" formations such as unconsolidated sands and shales, see Figure 17)

The Stoneley-wave cutoff frequency moves to higher frequencies, excitation is poorer, and the wave is highly attenuated. Theoretically, *S*-wave velocity can be extracted from full-waveform inversion of the leaky-*P* modes. However, this has proven to be very difficult to accomplish with real waveforms. An asymmetric tool is recommended.

In all cases, velocity analysis methods vary depending on the number of receivers on the logging sonde (2 to 48 or more). For arrays of receivers, frequency-wavenumber methods (Ellefsen et al., 1989) or correlation-semblance methods (Aron et al., 1978; Goldberg and Gant, 1988) can be applied. For two-receiver systems, velocity analysis is more challenging (Siegfried and Castagna, 1982) and various signal enhancement techniques followed by stacking and mixing of correlation curves can improve the results.

Regardless of the method used, source-generated noise should be separated to the greatest extent pos-

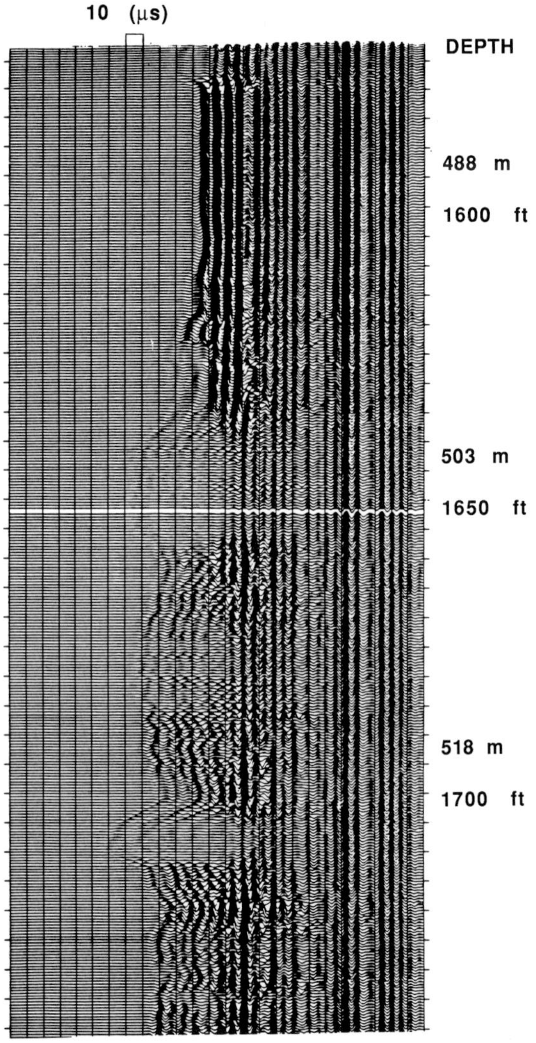

Fig. 17. Full-waveform sonic log in a "very slow" formation. Source-receiver spacing equals 10 ft.

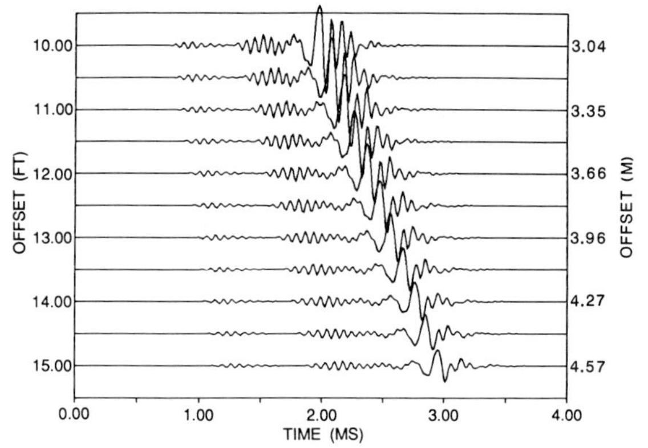

Radius (ft)	Radius (m)	VP (ft/s)	VP (m/s)	VS (ft/s)	VS (m/s)	Density (gm/cc)	QP	QS
.154167	..0470	5500	1676	0	0	1.20	20	0
.187500	.0571	20000	6096	11000	3353	7.50	1000	1000
.229167	.0698	5500	1676	0	0	1.20	20	0
.333333	.1016	9259	2822	5670	1728	1.92	40	30
0.	0.	16000	4877	8530	2600	2.16	60	60

Fig. 18(a). Synthetic full-waveform sonic data in a well-bonded cased borehole. Model parameters are given as a function of outer radius (*R*) from the center of the borehole.

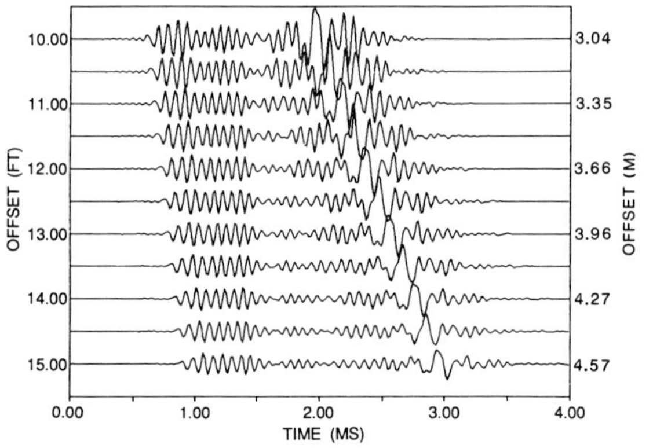

Radius (ft)	Radius (m)	VP (ft/s)	VP (m/s)	VS (ft/s)	VS (m/s)	Density (gm/cc)	QP	QS
.154167	.0470	5500	1676	0	0	1.20	20	0
.187500	.0571	20000	6096	11000	3353	7.50	1000	1000
.229167	.0698	5500	1676	0	0	1.20	20	0
.333333	.1016	9259	2822	5670	1728	1.92	40	30
0.	0.	16000	4877	8530	2600	2.16	60	60

Fig. 18(b). Synthetic full-waveform sonic data in a poorly bonded cased borehole. Model parameters are given as a function of outer radius (*R*) from the center of the borehole.

sible. The slowness-time-coherency method of Aron et al. (1978) uses a moving window and searches the slowness-time plane for maximum coherency within physical limits. The method is susceptible to cycle skipping if sufficient receivers are not available. The maximum likelihood method (Block et al., 1986) has the advantage of examining separate frequencies in the analysis. A priori knowledge (such as P-wave velocity and arrival times, expected spatial variation of arrival times, and empirical $V_p - V_s$ relationships) can be applied to constrain the window locations and coherency searches, thereby reducing cycle skipping.

With sonic waveforms, it is sometimes possible to log through casing. Figure 18 compares synthetic waveforms in well-bonded and poor-bonded casing in a "hard" formation. In the well-bonded casing, the formation P-waves and S-waves are evident. However, in the poor bonded casing, the casing reverberations swamp out the formation arrivals. The Stoneley-wave properties are dominated by the casing when the bond is poor and yield little rock information.

MULTIPOLE SONIC WAVEFORMS

The best way to determine S-wave velocity when the S-wave velocity of the formation is less than or about equal to the borehole fluid velocity is with

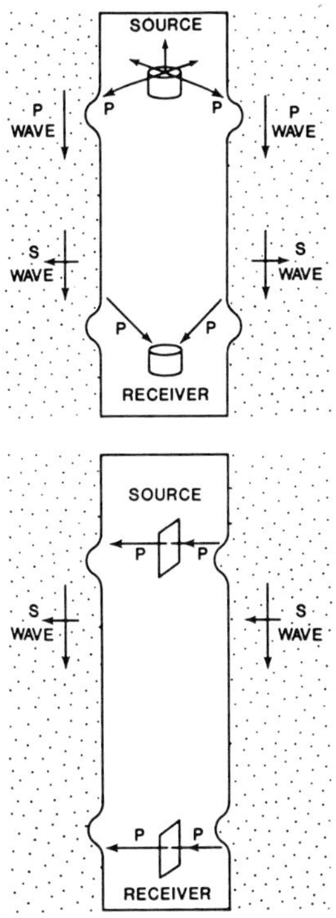

Fig. 20. Particle motions for monopole and dipole sonic logs (from Williams et al., 1984).

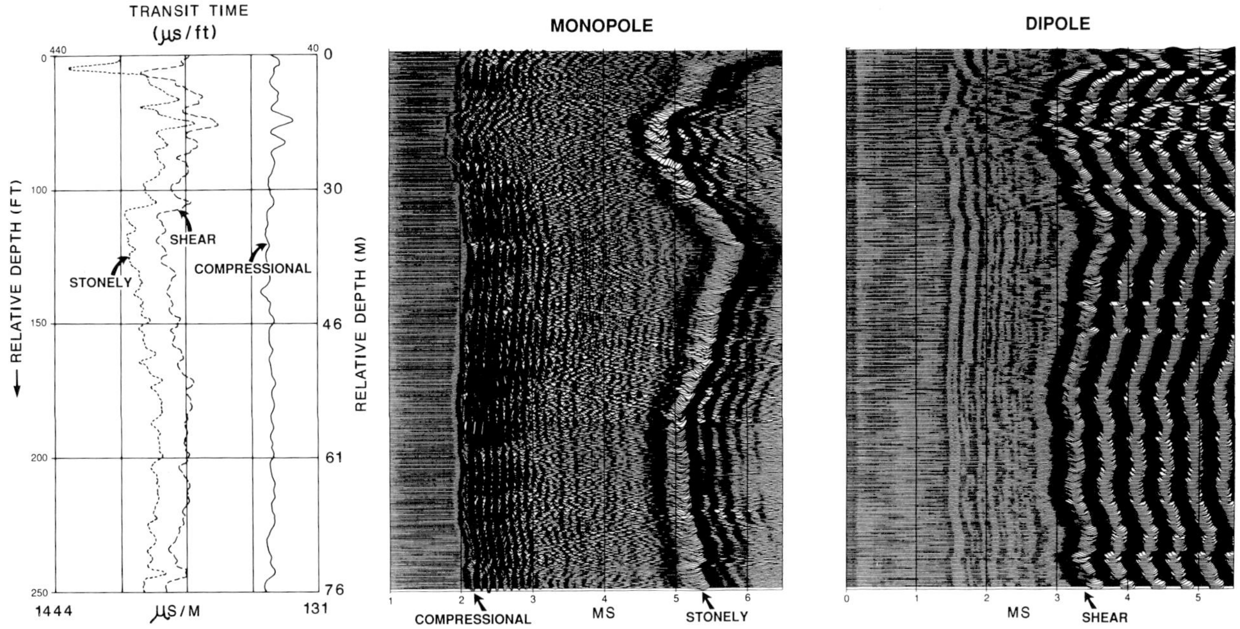

Fig. 19. Long-spaced full-waveform sonic log and dipole shear-wave sonic log in a soft formation. Source-receiver spacing equals 10 ft. Note that the digitizing delay is different for monopole and dipole waveforms.

multipole sonic logging devices. Axially asymmetric (multipole) sources set up asymmetric excitations which generate modes such as the flexural (dipole) and screw (quadrapole) waves. Formation S-wave velocity can be readily determined from the velocities of such asymmetric modes of propagation.

White (1967) proposed the use of asymmetrical sources with the "hula" log, named for the analogous movement between a hula dancer and the borehole flexural mode. The tool concept is a direct contact device, which does not allow continuous velocity logging. Kitsunezaki (1980) developed a dipole sonde with excitation by a dipole solenoid and impedance matched (floating) horizontal geophones for receivers. This device is practical for civil engineering purposes in shallow boreholes. However, neutral buoyancy requirements and S/N considerations prevent continuous logging. Williams et al. (1984) demonstrated the first practical dipole tool for petroleum industry applications. Figure 19 (Katahara, 1989, unpublished data) compares waveforms acquired with a monopole and a dipole sonde in a "slow" formation.

A dipole source excites a flexural mode in the borehole (see Figure 20). The flexural mode velocity depends strongly on the S-wave velocity of the formation. It is a dispersive mode (Figure 21) with the low-frequency portion most sensitive to the formation S-wave velocity. The high-frequency limit acts like the Stoneley wave from a monopole source. There is no theoretical cut-off frequency (Kurkjian, 1986), but at low frequencies the flexural mode is indistinguishable from the S-wave. The higher modes established by a dipole source have low-frequency cutoffs. These modes are only excited in "fast" formations.

While the fundamental mode excited by a dipole is a flexural mode, the fundamental mode excited by a quadrupole is a screw mode (Thurston, 1978; Kumar and Ram, 1969). This is another asymmetric mode

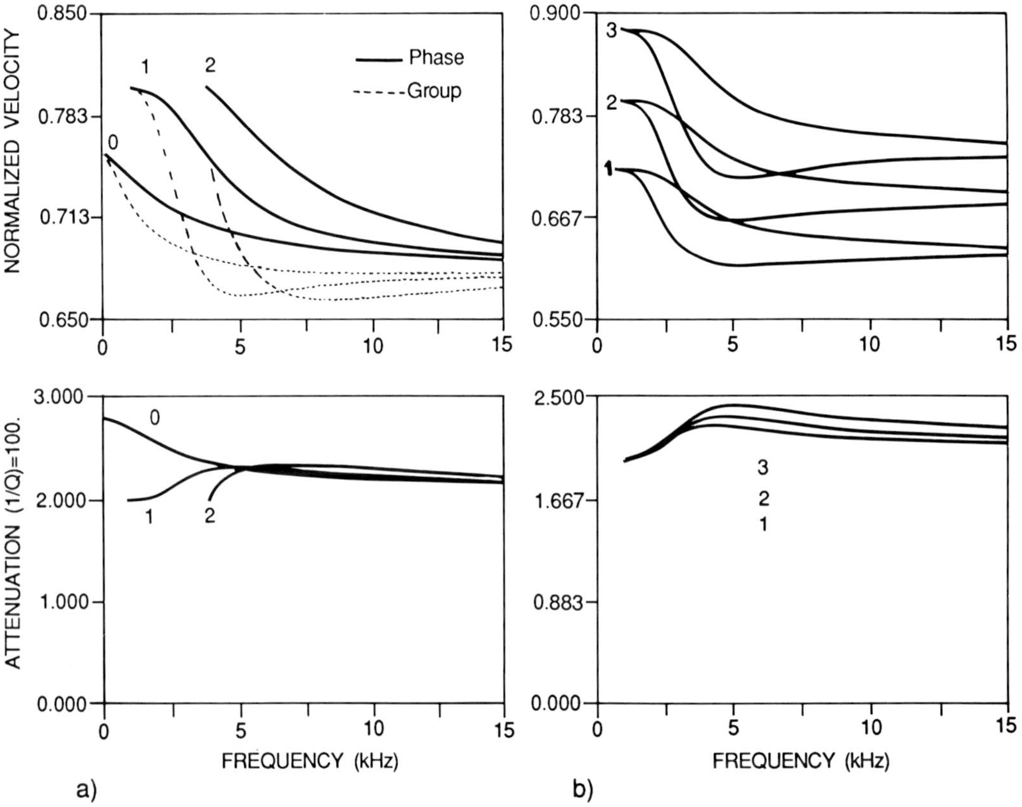

Fig. 21. Slow sandstone. (a) Dispersion and attenuation of the Stoneley wave (O), the flexural (1) and screw (2) modes excited by a monopole, dipole, and quadrupole source respectively. (b) Dispersion and attenuation of the flexural mode generated in the presence of a slow sandstone (2) and when the shear wave velocity is increased by 10 percent (3) and decreased by 10 percent (1). The velocities are normalized with respect to the borehole fluid velocity (from Schmidt, 1988).

from which the S-wave velocity can be determined. The useful energy of the screw mode is at a higher frequency than the flexural mode. As with the dipole, the lower portion of the frequency range is useful for determining the formation S-wave velocity. The higher frequency range behaves like the monopole Stoneley wave. Also similar to the dipole, higher modes are excited in fast formations only. These modes exhibit behavior similar to the pseudo-Rayleigh modes from a monopole and the higher modes of a dipole. Chen (1984) presented the first practical quadrupole device.

VERTICAL SEISMIC PROFILES

The elementary wavefield properties used to measure velocity from a vertical seismic profile (VSP) are absolute transit-time, traveltime differences, displacement amplitudes of body waves, and borehole tube waves (low-frequency Stoneley waves). An additional property important for S-waves is polarization direction for propagation in anisotropic media. Many inversion techniques have been developed to exploit one or more of these properties. Traveltime inversion methods, for example, patterned after the reciprocal problem of earthquake location (e.g., Aki and Lee, 1976), have been applied to P-wave and S-wave VSP data.

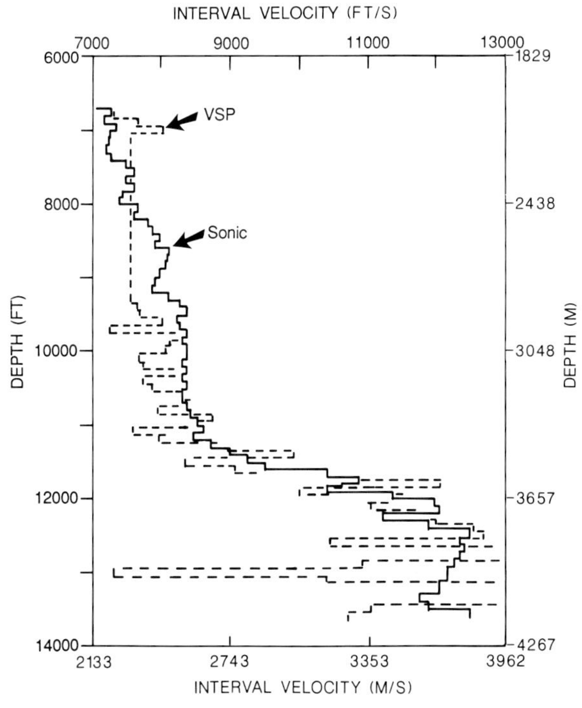

Fig. 22. Comparison of sonic log and VSP S-wave interval velocities.

S-wave vertical seismic profiles are acquired with a multicomponent sonde and (1) an S-wave source (usually a horizontal vibrator or inclined impactor) or (2) converted S-waves from a conventional P-wave source. In addition, as with sonic waveforms in slow formations, S-wave velocities can be calculated from tube-wave velocities.

An important aspect of determining S-wave velocities is that a consistent receiver coordinate frame be established to separate the two polarized S-waves. In an isotropic or a transversely isotropic medium, this can be done with independent gyroscopic information or by orienting the radial component in the direction of the P-wave.

DiSiena et al. (1984) computed S-wave velocities by iteratively inverting direct arrival traveltimes for laterally homogeneous layers between receivers. Interval velocities were varied until "ray" theoretical traveltimes agreed with observed traveltimes to within a specified amount.

Stewart (1984) developed a Levenberg-Marquardt damped least-squares technique using a forward ray-tracing scheme in a one-dimensional medium to invert VSP traveltimes. In a velocity sensitivity study, he showed that the precision of VSP interval velocity measurements is poor at short receiver spacings under 100 ft (30 m). The same technique was applied to invert S-wave traveltimes in an offset VSP to delineate an induced fracture zone (Stewart et al., 1984).

Figure 22 compares full-waveform sonic (12-channel array tool) S-wave interval velocities with those from a near-offset S-wave VSP (acquired with horizontal vibrators). The sonic log transit-times were block averaged to conform to the VSP station spacing. Note that the sonic log interval velocities have smaller variance and are geologically more reasonable than the VSP interval velocities.

Alternative methods of traveltime inversion that apply in principle to any wavefield have been developed. Using an iterative least-squares inversion in a "layer stripping" technique, Lines et al. (1984) inverted for velocities and local dip in the vicinity of the borehole from direct arrival and primary reflection times. Pujol et al. (1985) extended the Levenberg-Marquardt algorithm to simultaneously estimate velocity and ray parameters. This formulation is based on a Taylor series expansion of traveltimes and horizontal source-receiver distances as functions of velocity and ray parameter in laterally homogeneous media. Ahmed (1987) showed that the traveltimes of converted S-waves can be used to determine S-wave velocity.

The methods described invert for "ray" velocity or energy transport paths as opposed to "phase" velocity in connection with anisotropic and boundary-wave propagation. Using two closely spaced wells and multiple offset compressional (P), vertically polarized shear (SV), and horizontally polarized shear (SH) sources, White et al. (1983) determined transversely isotropic phase velocities for P-waves and S-waves in the Pierre shale. He employed a least-squares technique to solve a linearized form of the dispersion relation. Gaiser (1992) adapted this approach to a single borehole using multiple offset VSPs to obtain independent estimates of vertical and horizontal slownesses for compensating phase velocity. Figure 23 shows the results from six P-wave vertical seismic profiles. Least-squares estimates were made for vertical and horizontal P-wave velocity and the vertically polarized S-wave velocity at vertical and 45 degree phase angles of incidence. A near offset SH-wave VSP was used to constrain the vertical S-wave velocities. There is good correlation of the apparent velocity anisotropy with the different lithologies. Anisotropy is lowest in the sandstones and chalk and higher in the marl.

In "slow" formations, Biot (1952) and White (1965) have shown that S-wave velocity may be obtained from tube-wave velocity at zero frequency by:

$$V_s = V_f V_T[\rho_f/(\rho(V_f^2 - V_T^2)]^{-1/2} \tag{5}$$

where

V_f = borehole fluid velocity,
V_T = tube-wave velocity,
V_s = formation S-wave velocity,
ρ_f = borehole fluid density,

and

ρ = formation bulk density.

Figure 24 shows the relationship between zero-frequency tube-wave velocity and formation S-wave velocity. The reader is referred to Cheng and Toksoz (1984) for an excellent discussion; however, beware of typographical errors in equations (3) and (16).

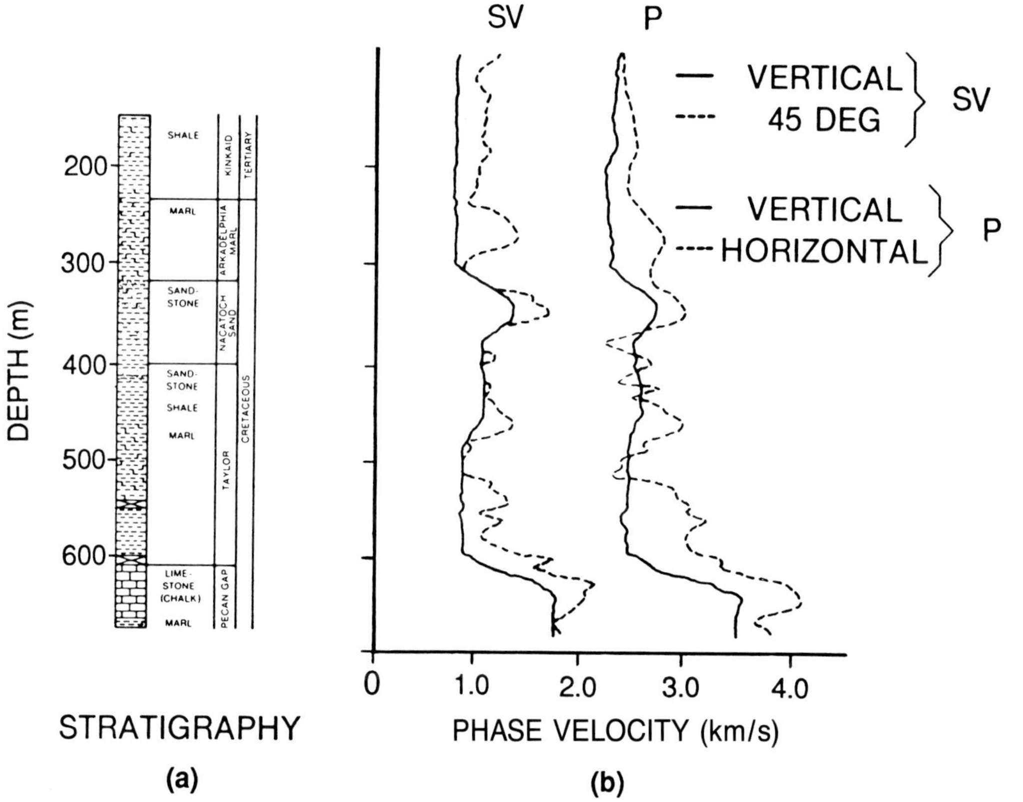

Fig. 23. Measured compressional (P) and shear (SV) anisotropy in a shallow test well.

Short of acquiring a full multicomponent VSP, S-wave checkshots can be used to calibrate derived or measured S-wave sonic logs for modeling purposes.

Example 1—Texas Gulf Coast Clastics

Example 1 compares S-wave VSP velocities acquired with a near-offset airgun source (which is apparently generating strong S-waves at or near the surface) to expected S-wave velocities when we assume brine- or gas-saturated rocks.

The VSP was acquired with a three-component borehole geophone tool with downhole signal conditioning and digitizing circuitry and clamping force to tool weight ratio better than two. The source was a 200 cubic inch (3279 cc) airgun charged to 2000 PSI (138 bars) and suspended in a water pit dug 250 ft (76.2 m) from the wellhead. There was no unusual configuration in the pit other than the beveled sides to prevent pit collapse. Pit dimensions were 15 ft (4.57 m) square on top, 10 ft (3.05 m) square on the bottom, and 10 ft (3.05 m) to 12 ft (3.66 m) deep. A monitor hydrophone was hung 5 ft (1.52 m) below the gun. The VSP was run in open hole (17.8 cm bit size with no washouts over 5.1 cm) from 1000 ft (305 m) to 5000 ft (1524 m) measured depth in 50 ft (15.2 m) increments. Five to seven good records were recorded at each depth.

Noisy traces were deleted and static corrections from the hydrophone monitor were applied. The traces were stacked in the common depth mode using a median stacking process. A 5–65 Hz bandpass filter was applied and times were picked interactively for the vertical channel. A coordinate system rotation was performed on the horizontal components level-by-

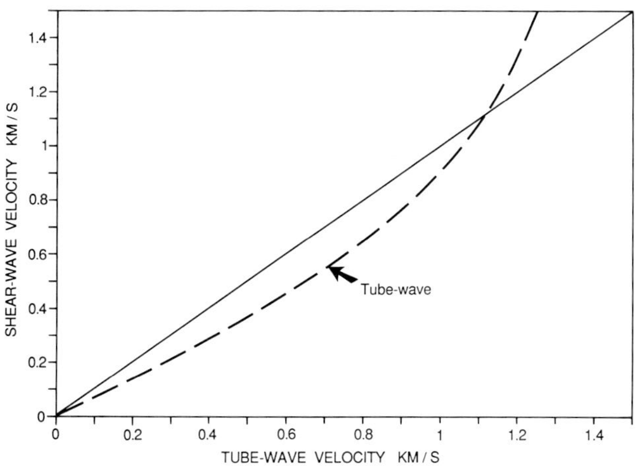

Fig. 24. S-wave velocity as a function of tube-wave velocity.

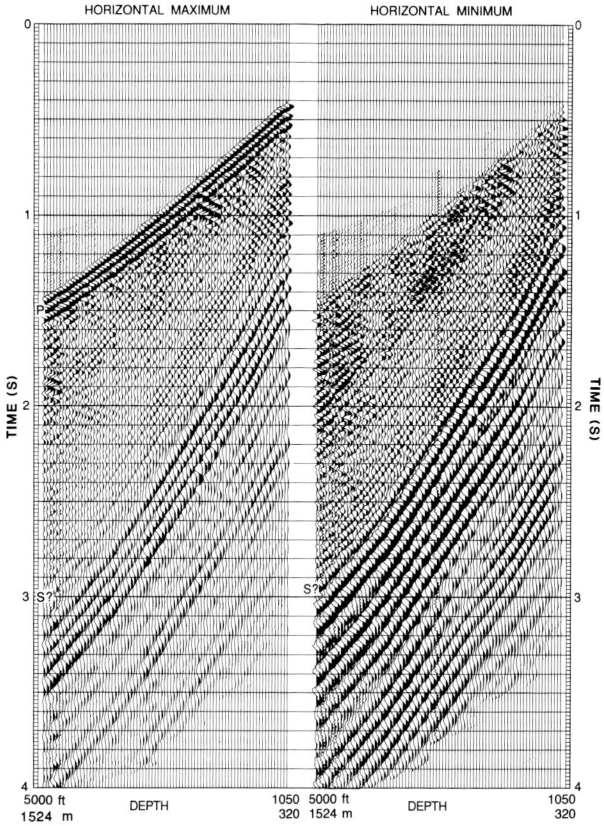

Fig. 25. Gulf Coast vertical seismic profile with rotated horizontal components.

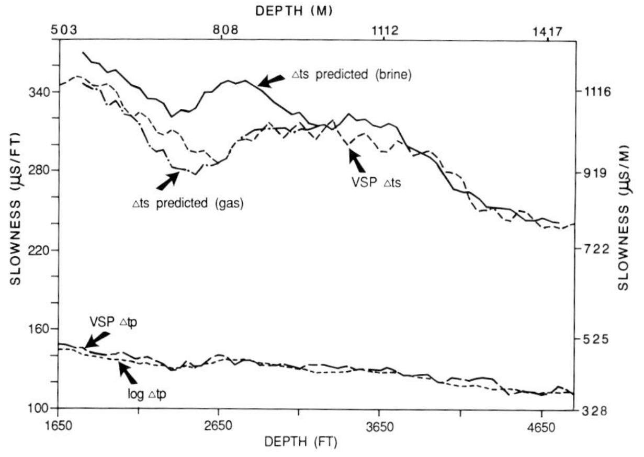

Fig. 26. VSP and log P-wave slowness and VSP and predicted S-wave slowness. Above 960 m S-wave slowness is predicted for both fully brine-saturated sands (solid) and partially gas-saturated sands (dot-dash).

level using the hodogram method (DiSiena et al., 1981). The analysis window was −10 ms to +45 ms about the picked breaktimes, yielding tangential and orthogonal projections. The tangential and vertical components were projected to isolate the *P*-wave energy from horizontal particle motion in the source-well plane. Figure 25 shows the resulting horizontal component seismograms. Shear breaktimes were picked on a constant phase and smoothed over five levels (200 ft or 61 m). The identification of the *S*-wave on Figure 25 is ambiguous. Elastic modeling indicates that strong converted *S*-waves are not expected at the near offset of this experiment. On the other hand, the main "shear" event does not strongly reflect from the bottom of the hole as would be expected for a tube wave. Furthermore, as will be shown, the velocities are consistent with the expected *S*-wave velocities. There is significant upgoing and downgoing "shear" energy other than the "main" arrival. The amplitudes of these events are greater than predicted by elastic modeling of converted waves.

Figure 26 compares the VSP interval slowness (tran-

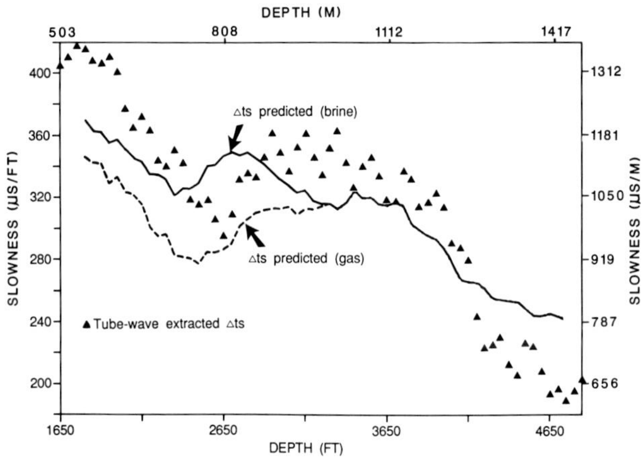

Fig. 27. Comparison of estimated *S*-wave slowness assuming the slower traveling wave is a tube wave (triangles) with predicted *S*-wave slownesses from lithological information. The mismatch suggests that the slow wave is not a tube wave.

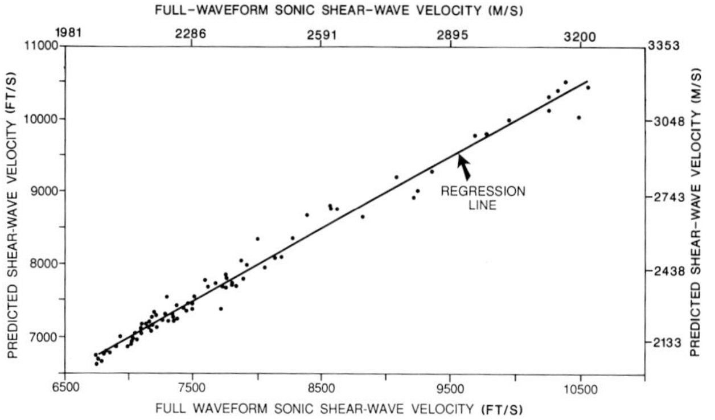

Fig. 29. Crossplot of predicted and measured *S*-wave velocity logs.

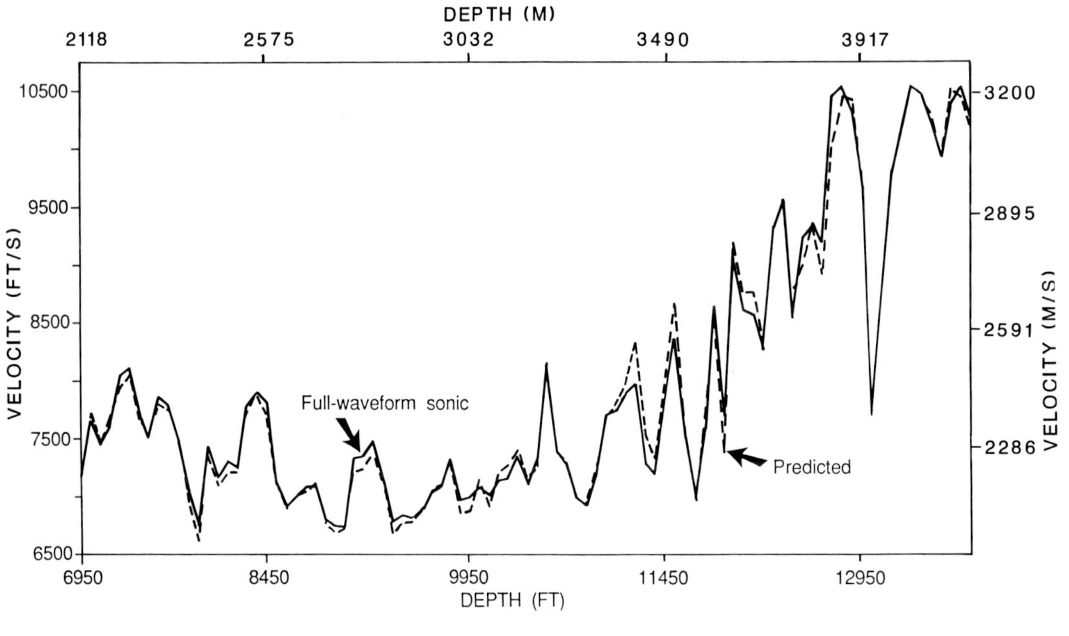

Fig. 28. Comparison of predicted and measured *S*-wave velocity logs in a Paleozoic interval.

sit time) to a smoothed conventional P-wave sonic log and predicted S-transit times using trend curves. The Schlumberger volumetric analysis (VOLAN) indicates primarily brine-saturated sands and shales. Given this information and the conventional sonic log, a brine-saturated S-wave transit-time log was estimated. Below 3150 ft (960 m) the VSP and predicted S-wave transit times are in agreement. However, above 3150 ft (960 m) the VSP S-wave transit times are faster than the predictions. Thus, the observed V_p/V_s ratios are lower than expected for brine-saturated rocks. Minor amounts of gas, not obvious on the resistivity logs, could be responsible for the anomalously low V_p/V_s ratios. This hypothesis was tested by assuming 10 percent gas saturation in all the sands above 3150 ft (960 m) and predicting new S-wave slownesses. In Figure 26, this results in a curve which overlays or, with the brine-saturated predictions, brackets the observed S-wave transit times.

To verify identification of the S-wave, the hypothesis was forwarded that the event in question was a tube wave. Assuming that the tube-wave slowness was measured, the corresponding S-wave velocities were computed using Equation 5. The result (Figure 27) is that the extracted S-wave transit times are in poor agreement with our expectations. The tube-wave hypothesis is, therefore, less satisfactory to us, than the assumption that S-wave slowness was directly measured with the VSP.

In conclusion, integrated interpretation of VSP and log information indicates low gas-saturation in the sands above 3150 ft (960 m) and full water-saturation below this depth. This information was not apparent from, nor within the reliability of conventional well log analysis. The implications for AVO analysis in this vicinity are (1) sands above 3150 ft (960 m) may exhibit "false" AVO responses whether they contain commercial hydrocarbons or not, and (2) the shallow gas-bearing sands may confound the AVO response for deeper horizons.

Example 2—Devonian Hard Shales

Example 2, compares an S-wave VSP acquired with horizontal vibrators at two different azimuths, a conventional full-waveform sonic log, and a pseudo-shear velocity log in hard, fractured Devonian shales.

The VSP was acquired with vertical and horizontal (inline and crossline) vibrators at two locations 2000 ft north and west of the wellhead. Horizontal components were oriented by maximizing the P-wave particle motion. Interval velocities were determined from principle component traveltimes. All VSP and well log slownesses were averaged over a span of 200 ft.

A well log lithology analysis was used to construct a pseudo-shear velocity log. This prediction is in excellent agreement with the full-waveform sonic log (see Figures 28 and 29). The standard error is 122 ft/s (37.2 m/s). The smoothed (5 station) VSP velocities are in agreement for the two azimuths, although some azimuthal anisotropy is evident (Figures 30 and 31). The standard errors are 273 ft/s (83.2 m/s) for P waves and 346 ft/s (105.4 m/s) for S waves. For the VSP and full-waveform sonic S-wave velocity comparison (Fig-

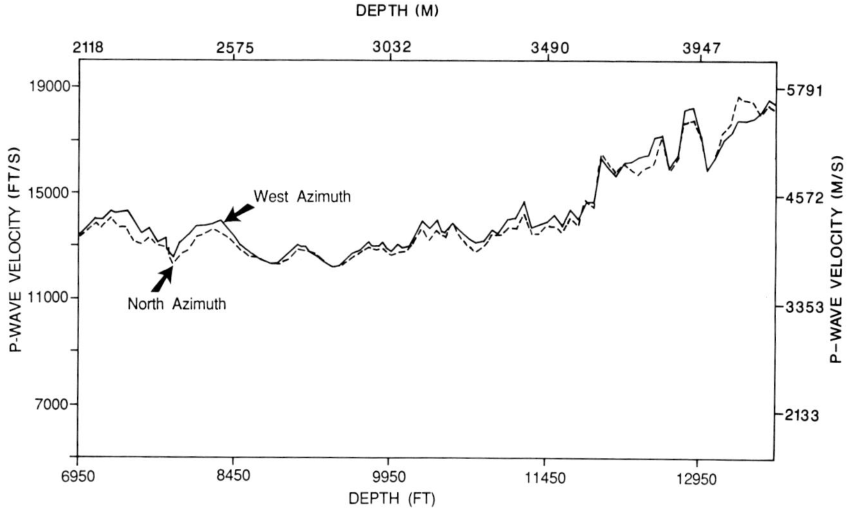

Fig. 30(a). Compressional VSP velocity-depth profiles for sources with North and West azimuths.

ures 32 and 33), the standard error is high (921 ft/s or 280.7 m/s) with no significant bias. In conclusion, the full-waveform and VSP *S*-wave velocities are in substantial agreement with expected velocities. The full-waveform *S*-wave velocities appear to be highly reliable (on the order of 100 ft/s or 30.5 m/s if the predicted velocities are assumed to be correct) when heavily smoothed with a 200 ft (61 m) operator (400 samples). Presumably, random error has been greatly suppressed by the smoothing operation. The VSP measurements for the different azimuths are in better agreement with each other, than with the smoothed sonic waveform results. This may be due to repeatable VSP picking errors caused by interference of early arrivals. The apparent azimuthal anisotropy is presumably due to fractures.

CONCLUSIONS AND DISCUSSION

Obtaining *S*-wave "ground truth" from boreholes involves the integrated interpretation of laboratory measurements, pseudo-velocity logs, conventional full-waveform sonic logs, dipole or quadrupole logs, and vertical seismic profiles of various varieties. Each method has inherent advantages and limitations.

Laboratory measurements are performed on a biased, undersampled representation of the subsurface. Sample sizes and signal wavelengths are a minute fraction of seismic wavelengths, so results are often unrepresentative. However, it is possible in the laboratory to control environmental conditions, to study the systematics of velocity variations, and to measure intrinsic anisotropy directly.

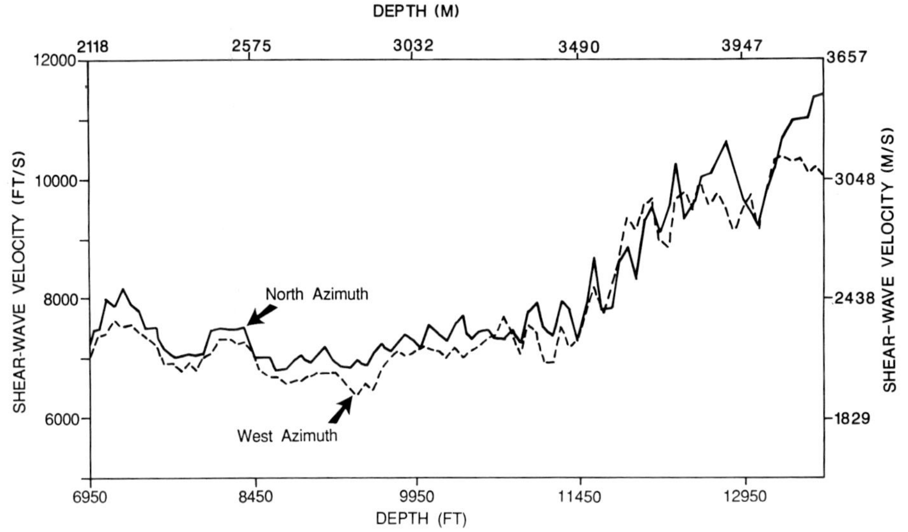

Fig. 30(b). Shear VSP velocity-depth profiles for sources with North and West azimuths.

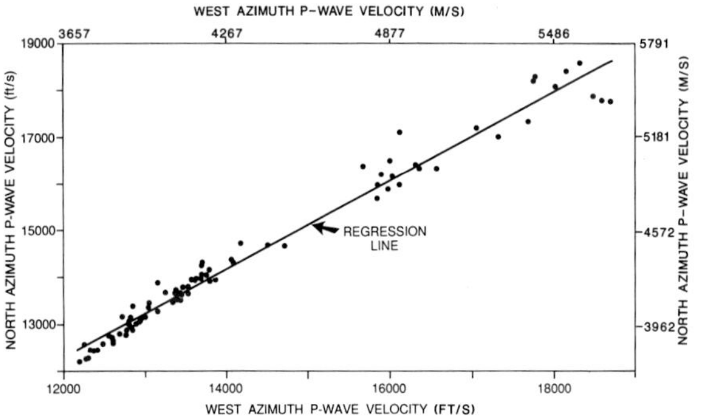

Fig. 31(a). Crossplot of *P*-wave velocities for north and west azimuths.

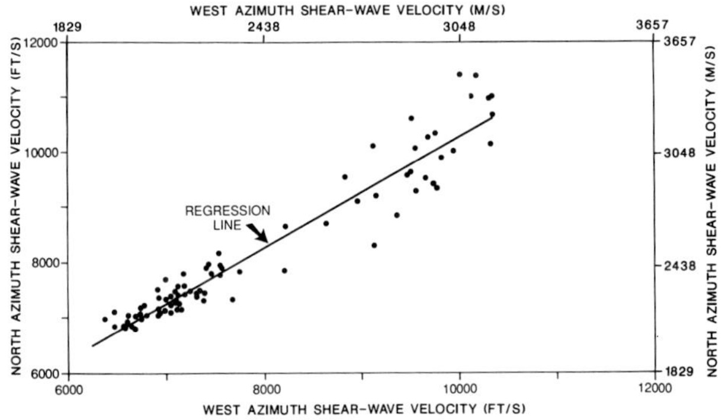

Fig. 31(b). Crossplot of *S*-wave velocities from north and west azimuths.

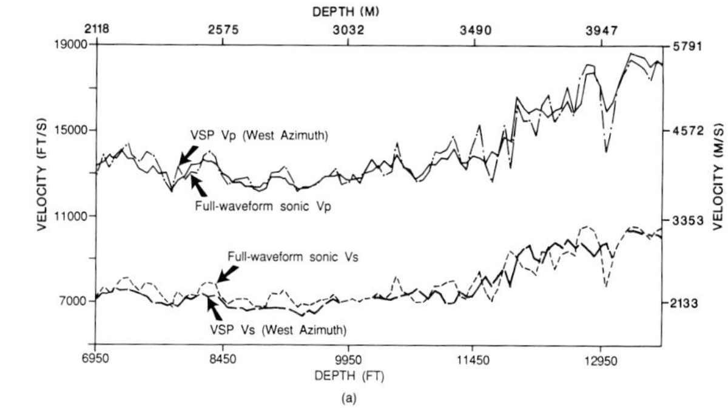

Fig. 32(a). West azimuth VSP and full-waveform sonic velocity-depth profiles.

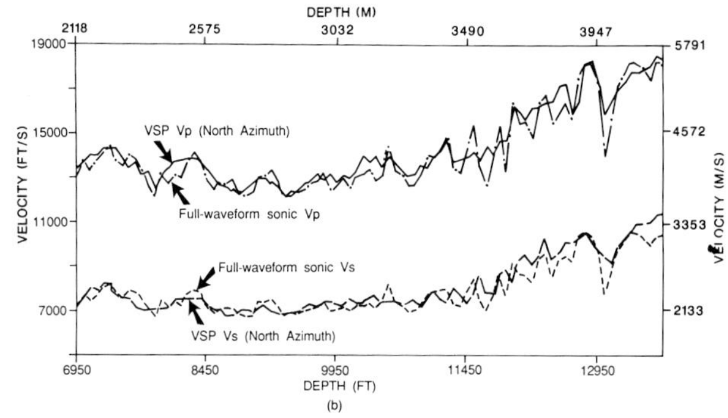

Fig. 32(b). North azimuth VSP and full-waveform sonic velocity-depth profiles.

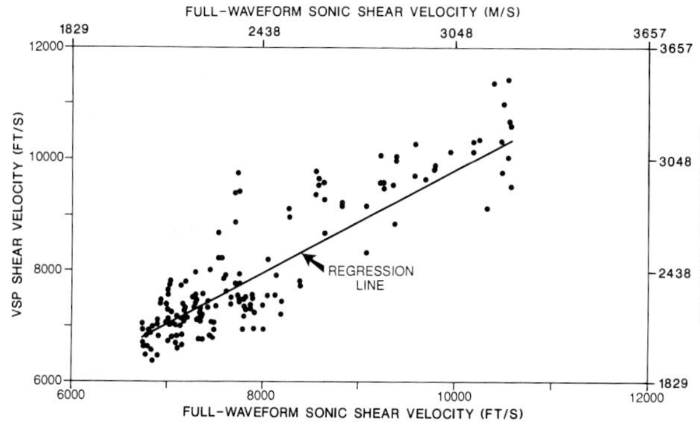

Fig. 33. Crossplot of VSP and full-waveform sonic S-wave velocities.

Well log measurements are subject to processing errors and the perturbing effects of the borehole. If processed correctly, they provide a reasonably precise, high-resolution measure of vertical variations in velocity. However, the lateral variation of velocity is sampled only at the wellbore, and may or may not adequately represent the subsurface properties averaged over the Fresnel zone of surface seismic data. Pseudo-shear-velocity logs from empirical trends are a useful supplement when direct shear-wave logs are unavailable or of questionable reliability.

VSP measurements are most comparable to surface seismic data in frequency and scale. However, interval velocities must be heavily smoothed to achieve the precision necessary for AVO modeling which limits the resolution of the method. For interval velocity determination, VSP's are best used as a means of calibrating well log velocities.

ACKNOWLEDGMENTS

Thanks to ARCO Oil and Gas Company for permission to publish this paper. Special thanks to M. Popa, K. Katahara, B. Smith, B. Phillips, S. Gronemeier, and G. M. Kolb for their contributions.

REFERENCES

Ahmed, H., 1987, Mode converted shear waves in 3-component VSP data: Presented at the 57th Ann. Internat. Mtg., Soc. Expl. Geophys.

Aki, K., and Lee, W. H., 1976, Determination of three dimensional velocity anomalies under a seismic array using first *P* arrivals from local earthquakes: J. Geophys. Res., **81**, 4381–4399.

Aron, J., Murray, J., and Seemans, B., 1978, Formation compressional and shear transit-time measurement by means of long-spacing and digital techniques: Soc. Petr. Eng. 7746.

Biot, M. A., 1952, Propagation of elastic waves in a cylindrical bore containing fluid: J. Applied Physics, **23**, 997–1005.

Block, L. V., Cheng, C. H., and Duckworth, G. L., 1986, Velocity analysis of multireceiver full-waveform acoustic logging data in open and cased holes: Presented at 56th Ann. Internat. Mtg., Soc. Expl. Geophys.

Castagna, J. P., Batzle, M. L., and Eastwood, R. L., 1985, Relationships between compressional and shear-wave velocities in clastic silicate rocks: Geophysics, **50**, 571–581.

Castagna, J. P., Batzle, M. L., and Kan, T. K., 1993, Rock physics: The link between rock properties and amplitude variation with offset response, *in* Castagna, J. P., and Backus, M. M., Eds., Offset-Dependent-Reflectivity: Soc. Expl. Geophys., 124–157.

Chen, S. T., 1984, Experimental studies of on-line shear wave logging: Presented at 54th Ann. Internat. Mtg., Soc. Expl. Geophys.

Cheng, C. H., and Toksoz, M. N., 1984, Generation, propagation, and analysis of tube waves in a borehole, *in* Toksoz, M. H., and Stewart, R. R., Eds., Vertical seismic profiling, Part B: Advanced Concepts: Geophysical Press.

Cheng, C. H., and Toksoz, M. N., 1981, Elastic wave propagation in a fluid-filled borehole and synthetic acoustic logs: Geophysics, **46**, 1042–1053.

Clark, V. A., 1980, Effect of volatiles on seismic attenuation and velocity in sedimentary rocks: Ph.D. Thesis, Texas A&M Univ.

DiSiena, J. P., Gaiser, J. E., and Corrigan, D., 1981, Three-component vertical seismic profiles: Orientation of horizontal components for shear wave analysis: Presented at 51st Ann. Internat. Mtg., Soc. Expl. Geophys.

DiSiena, J. P., Gaiser, J. E., and Corrigan, D., 1984, Horizontal components and shear wave analysis of three component VSP data, *in* Toksoz, M. N., and Stewart, R. R., Eds., Vertical Seismic Profiling, Part B: Advanced Concepts: Geophysical Press.

Ellefsen, K. J., Cheng, C. H., and Tubman, K. M., 1989, Estimating phase velocity and attenuation of guided waves in acoustic logging data: Geophysics, **54**, 1054–1059.

Gaiser, J. E., 1992, Anisotropic phase velocity analysis from slowness estimates: J. Geophys. Res., **95**, 11 241–11 254.

Gassmann, F., 1951, Elastic waves through a packing of spheres: Geophysics, **16**, 673–685.

Goldberg, D., and Gant, W. T., 1988, Shear-wave processing of sonic log waveforms in a limestone reservoir: Geophysics, **53**, 668–676.

Greenberg, M. L., and Castagna, J. P., 1992, Shear-wave velocity estimation in porous rocks: Theoretical formulation, preliminary verification, and applications: Geophys. Prosp., **40**, 195–209.

Gregory, A. R., 1963, Shear wave velocity measurements of sedimentary rock samples under compression: Proc. 5th Rock Mech. Symp., Pergamon, 439–471.

Gregory, A. R., and Podio, A. L., 1970, Dual-mode apparatus for measuring compressional and shear wave velocities of rock samples: IEEE Trans. Sonics. Ultrason., **17**, 77–85.

Han, D. H., Nur, A., and Morgan, D., 1986, Effects of porosity and clay content on wave velocities in sandstones: Geophysics, **51**, 2093–2107.

Hosten, B., Deschamps, M., and Tittmann, B. R., 1987, Inhomogeneous wave generation and propagation in lossy anisotropic solids: Application to the characterization of viscoelastic composite materials: J. Acoust. Soc. Am., **82**, 1763–1770.

Jackson, I., Paterson, M. S., Niesler, H., Waterford, R. M., 1984, Rock anelastic measurements at high pressure, low strain amplitude and seismic frequency: Geophys. Res. Letters, **11**, 1235–1238.

Jamieson, J. C., and Hoskins, H., 1963, The measurement of shear-wave velocities in solids using axially polarized ceramic transducers: Geophysics, **28**, 87–90.

King, M. S., and Fatt, I., 1962, Ultrasonic shear-wave velocities in rocks subjected to simulated overburden pressure: Geophysics, **27**, 590–598.

Kitsunezaki, C., 1980, A new method for shear-wave logging: Geophysics, **45**, 1489–1506.

Kumar, R., and Ram, S., 1969, Flexural vibrations of a fluid-filled cylindrical cavity in an infinite solid medium: Acustica, **22**, 163–171.

Lines, L. R., Bourgeois, A., and Covey, J. D., 1984, Traveltime inversion of offset vertical seismic profiles—a feasibility study: Geophysics, **49**, 250–264.

Paillet, F., and Cheng, C. H., 1986, A numerical investigation of head waves and leaky modes in fluid-filled boreholes: Geophysics, **51**, 1438–1449.

Paillet, F., and White, J. E., 1982, Acoustic modes of propagation in the borehole and their relationship to rock properties: Geophysics, **47**, 1215–1228.

Peselnick, L., and Outerbridge, W. F., 1961, Internal friction in shear and shear modulus of Solenhofen Limestone over a frequency range of 10^7 cycles per second: J. Geophys. Res., **66**, 581–588.

Pickett, G. R., 1963, Acoustic character logs and their applications in formation evaluation: J. Petr. Tech., **15**, 569–667.

Pujol, J., Burridge, R., and Smithson, S. B., 1985, Velocity determination from offset vertical seismic profiling data: J. Geophys. Res., **90**, 1871–1880.

Rafavich, F., Kendall, C. H. St. C., and Todd, T. P., 1984, The relationship between acoustic properties and the petrographic character of rocks: Geophysics, **49**, 1622–1636.

Rai, C. S., and Hanson, K. E., 1988, Shear-wave velocity anisotropy in sedimentary rocks—a laboratory study: Geophysics, **53**, 800–806.

Schmidt, D. P., 1988, Shear-wave logging in elastic formations: J. Acoust. Soc. Am., **84**, 2215–2229.

Shirley, D. J., and Hampton, L. D., 1978, Shear-wave velocity measurements in laboratory sediments: J. Acoust. Soc. Am., **63**, 607–613.

Siegfried, R. W., and Castagna, J. P., 1982, Full-waveform sonic logging techniques: Presented at the 23rd Ann. Logging Symposium, Society of Professional Well Log Analysts.

Simmons, G., 1965, Ultrasonics in geology: Proc. IEEE, **53**, 1337–1345.

Simmons, G., and Brace, W. F., 1965, Comparison of static and dynamic measurements of compressibility of rocks: J. Geophys. Res., **70**, 5649–5656.

Stewart, R. R., 1984, VSP interval velocities from traveltime inversion: Geophys. Prosp., **32**, 608–628.

Stewart, R. R., Turpening, R. M., and Toksoz, M. N., 1984, Subsurface fracture zone delineation, *in* Toksoz, M. N., and Stewart, R. R., Eds., Vertical Seismic Profiling, Part B: Advanced Concepts: Geophysical Press.

Sommers, G. C., and Broding, R. A., 1952, Continuous velocity logging: Geophysics, **17**, 598–614.

Thurston, R. N., 1978, Elastic waves in rods and clad rods: J. Acoust. Soc. Am., **64**, 1–37.

Tittman, B. R., and Curnow, J. M., 1976, Apparatus for measuring internal friction Q factors in brittle materials: Rev. Sci. Instrum., **47**, 1516–1518.

Tubman, K. M., Cheng, C. H., and Toksoz, M. N., 1984, Synthetic full waveform acoustic logs in cased boreholes: Geophysics, **49**, 1051–1059.

Vogel, C. B., 1952, A seismic velocity logging method: Geophysics, **17**, 586–597.

White, J. E., 1965, Seismic waves: McGraw-Hill.

White, J. E., 1967, The HULA log, a proposed acoustic tool: Presented at the 8th Ann. Logging Symposium, Soc. Prof. Well Log Analy.

White, J. E., Martineau-Nicoletis, L., and Monash, C., 1983, Measured anisotropy in Pierre Shale: Geophys. Prosp., **31**, 709–725.

White, J. E., and Tongtaow, C., 1981, Cylindrical waves in a transversely isotropic medium: J. Acoust. Soc. Am., **70**, 1147–1155.

Williams, D. M., Zemanek, J., Angona, F. A., Dennis, C. L., and Caldwell, R. L., 1984, The long-spaced acoustic tool: Presented at the 25th Ann. Logging Symposium, Soc. Prof. Well Log Analysts.

Wyllie, M. R. J., Gardner, G. H. F., and Gregory, A. R., 1962, Studies of elastic wave attenuation in porous media: Geophysics, **27**, 569–589.

Rock Physics—The Link Between Rock Properties and AVO Response

J. P. Castagna, M. L. Batzle,‡ T. K. Kan‡*

INTRODUCTION

The variation of seismic reflection amplitude with offset is dependent on intrinsic rock parameters such as compressional-wave velocity (V_p), shear-wave velocity (V_s), density, anisotropy, and attenuation. An understanding of the interrelationships among these parameters and rock properties such as lithology, porosity, and pore fluid content is needed for the quantitative extraction of rock properties information by amplitude variation with offset (AVO) analysis. This paper provides the necessary rock physics framework for AVO analysis.

After briefly discussing some elementary rock physics concepts, we will (1) synthesize available P-wave velocity, S-wave velocity, and density information; (2) give empirical V_p versus V_s or V_p versus density trend curves for the common sedimentary lithologies; (3) discuss, in detail, the pressure, temperature, and composition dependence of the density and compressibility of pore fluid mixtures of various kinds; (4) demonstrate fluid substitution in rocks using the well known Biot-Gassmann equations and derive the rock skeleton properties needed for these calculations; (5) compile a variety of attenuation measurements including new vertical seismic profiling measurements in a young clastic section; and (6) provide a very basic anisotropy tutorial.

In the course of writing this paper, we have made extensive use of several review papers and books. We have found the following particularly helpful: Gardner et al. (1974), Gregory (1977), White (1983), Thomsen (1986), Simmons (1965), and Töksoz and Johnston (1981).

SOME PRELIMINARIES

Velocity and Elastic Moduli

P- and S-wave velocities for an isotropic, homogeneous, elastic material are given by

$$V_p = \left[\frac{k + 4/3\mu}{\rho}\right]^{1/2} \tag{1}$$

and

$$V_s = \left[\frac{\mu}{\rho}\right]^{1/2} \tag{2}$$

where

V_p = compressional-wave velocity,
V_s = shear-wave velocity,
k = rock bulk modulus = rock incompressibility,
μ = rock shear modulus = rock rigidity,

and

ρ = rock bulk density.

Equations (1) and (2) provide the fundamental link between seismic velocities and rock properties. For example, it has been demonstrated many times (e.g., Domenico, 1976, 1977) that the rock bulk modulus may be strongly dependent on the pore fluid bulk modulus while the rock shear modulus may be unaffected by the fluids. Hence, when a compressible free gas replaces liquids in the pore space, the rock P-wave velocity will decrease significantly, whereas the rock S-wave velocity will be slightly increased due to the decreasing bulk rock density. Consequently, the ratio of compressional-to-shear-wave velocity (V_p/V_s) is expected to be an excellent indicator of free gas in the pore space. This is the basis for direct hydrocarbon detection using AVO.

For a homogeneous isotropic material which is deforming elastically under uniaxial compression, Poisson's ratio (σ) is defined as the negative fractional change in width divided by the fractional change in length (ratio of transverse to axial strain). Poisson's ratio is directly related to the V_p/V_s ratio by

$$\sigma = \frac{.5(V_p/V_s)^2 - 1}{(V_p/V_s)^2 - 1}. \tag{3}$$

*ARCO Oil and Gas Company
‡ARCO Exploration and Production Technology

A Poisson's ratio of zero corresponds to a V_p/V_s ratio of 1.41. Fluids have a Poisson's ratio of .5 (V_p/V_s equals infinity).

Shear modulus is the ratio of shear stress to shear strain. Similarly, bulk modulus is the ratio of volumetric stress to volumetric strain; this is the reciprocal of the compressibility.

Effects of Pressure and Temperature

At a given pressure and temperature, the velocity of a sedimentary rock depends on porosity, lithology, pore fluid content, texture, and lithification. In order to relate velocities to these rock properties, we must first comprehend the environmental effects of pressure and temperature.

Laboratory measurements (for example, Simmons and Brace, 1965) have shown that V_p and V_s for a given rock specimen can vary greatly depending on the confining pressure (P_c) and the pore pressure (P_p).

Numerous laboratory measurements have shown that most rocks obey an "effective stress" law with the "effective" pressure (P_e) given by

$$P_e = P_c - nP_p \text{ (with } n \text{ usually equal to 1).} \quad (4)$$

When n is equal to 1, the effective pressure is equal to the "differential pressure" which is the difference between confining and pore pressure. (Terzaghi and Peck; 1948). As shown in Figure 1, the P- and S-wave velocities are functions of the effective pressure. The typical behavior of most porous rocks is for velocities to increase rapidly at low pressure and more slowly at higher pressure. (See Appendix A for pressure units and gradients).

The increase in velocity with pressure is usually attributed to the closing of thin crack-like pores and grain contacts within the rock. In solid materials and rocks without cracks, the velocities stay almost constant with pressure. Figure 2 shows the measured changes in rock volume with pressure which correspond to the velocity changes shown in Figure 1. Since the constituent minerals of the rock are relatively incompressible, the bulk volume change can be assumed to be primarily due to pore volume changes.

The dependence of velocity on effective pressure complicates the establishment of velocity-porosity relationships. Any velocity-porosity transform which does not explicitly include pressure as a variable must be used very cautiously, particularly at low effective pressures.

The direct effect of temperature on V_p and V_s is small when we remain within the ranges commonly encountered in hydrocarbon exploration. Timur (1977) has shown that even for an increase in temperature of 100 degrees Celsius, the velocities drop by only a few percent. According to Nur and Simmons (1969) and DeVilbiss et al. (1978) such changes in velocity are largely due to the effects of temperature on the pore fluids while the skeleton properties remain approximately constant. Larger changes can be expected when high temperatures cause a vapor phase (i.e., steam flooding).

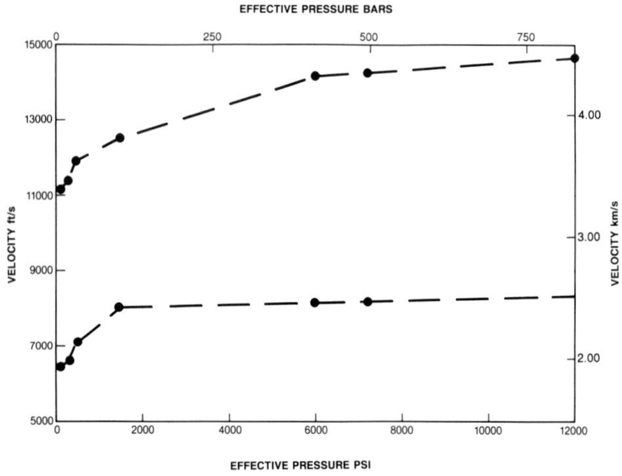

Fig. 1. Variation of P- and S-wave velocity with pressure for a water-saturated glauconitic sandstone with a porosity of 19.2 percent sampled from a depth of 11 833.5 ft (3606.7 m).

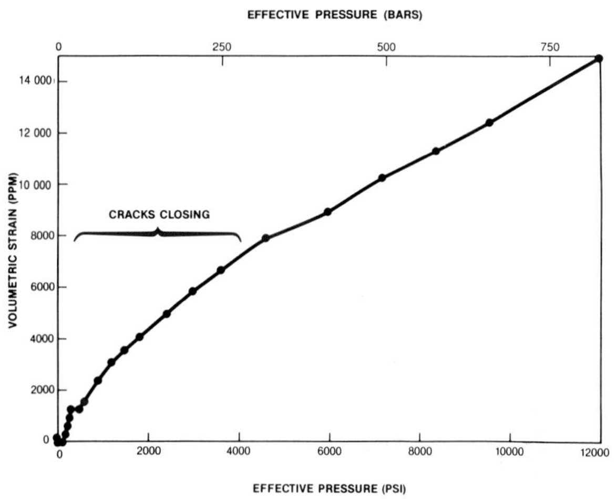

Fig. 2. Variation of volumetric strain with pressure for the sample described in Figure 1.

When we move from the laboratory into real in-situ conditions, the confining pressure becomes the pressure due to the weight of the overburden, which depends on depth and average rock bulk density and is often expressed as a pressure gradient (pressure divided by depth). The overburden (lithostatic or geostatic) pressure gradient is usually assumed to be about .0226 MPa/meter (1.0 psi/ft) for Tertiary deposits off the Texas-Louisiana Gulf Coast. This value corresponds to an average overburden bulk density of 2.31 gm/cc. As shown in Appendix A, the pressure gradient may deviate significantly from this value.

The hydrostatic pressure is the pressure due to the weight of the overlying fluid column. The hydrostatic pressure gradient is usually about .0105 MPa/meter (.465 psi/ft) for typical formation waters but can vary depending on the pore fluid salinities and temperature. In normally pressured rocks, pore pressure is equal to the hydrostatic pressure. However, in geopressured formations, the pore pressure is abnormally high, and effective pressure is abnormally low.

Geothermal gradients range from about 8 to 40 degrees Celsius per kilometer (0.4 to 2.2°F/100 ft.) and are typically around 20 degrees Celsius per kilometer (1.1°F/100 ft.) in many sedimentary basins. The gradients tend to be relatively low in young areas with rapid deposition. The reader is referred to Gretener (1981) for an excellent overview.

The indirect effects of pressure and temperature can be substantially larger than the direct effects discussed thus far. As a rock is buried, the increasing pressures and temperatures produce bulk porosity reduction, phase changes, cementation, and other diagenetic modifications. Bear in mind that exposing a rock to elevated pressures and temperatures in a laboratory does not produce the same result as when the rock is exposed to these conditions over geologic time.

V_P-V_S RELATIONSHIPS

Because rocks are aggregates of mineral grains, we expect the velocity of a highly lithified, low-porosity rock to be strongly dependent on the velocities of the grains. However, we expect the velocity of an unconsolidated rock to be only weakly dependent on the velocities of the grains due to the large pore volume and the influence of pore shape.

Because most sedimentary rock-forming minerals (such as quartz, calcite, and clays) are not isotropic, mineral velocity is taken to be that of a zero-porosity polycrystalline aggregate of randomly oriented mineral grains (Hill, 1952). In general, we expect the velocity of a monomineralic rock to approach the velocity of the pure mineral as porosity approaches zero.

Mineral velocity is generally referred to as the "matrix" velocity in log analysis and the "grain" or "solid" velocity in the rock physics literature. Table 1 gives P- and S-wave velocities for some minerals commonly found in sedimentary rocks. The S-wave velocity for dolomite is not well established. Conspicuously absent from this list are the clay minerals. Some information about clay velocities can be ascertained by extrapolation of measurements on mixed lithologies to 100 percent clay (see Table 2).

When V_p and V_s are crossplotted, we expect, for relatively "pure" (mono-mineralic) water-saturated rocks, the mineral velocity to constitute one endpoint of a distribution of points representing varying porosities and effective pressures. At the other extreme, as the rock becomes more fluid-like and the compressional velocity approaches that of water, we expect the S-wave velocity to go to zero. We show that this behavior is indeed exhibited by the common sedimen-

Table 1. Some reported mineral properties. Mineral velocities are averaged to represent zero-porosity isotropic aggregates.

Mineral	Density (gm/cc)	V_p (km/s)	V_s (km/s)	V_p/V_s	Reference*
Calcite	2.71	6.53	3.36	1.94	(1)
Calcite	2.71	6.26	3.24	1.92	(2)
Dolomite	2.87	7.05	4.16	1.70	(3)
Halite	2.16	4.50	2.59	1.74	(4)
Muscovite	2.79	5.78	3.33	1.74	(5)
Quartz	2.65	6.06	4.15	1.46	(2)
Quartz	2.65	6.05	4.09	1.48	(6)
Anhydrite	2.96	6.01	3.37	1.78	(7)

*(1) Dandekar (1968). (2) Anderson and Liebermann (1966), (3) Nur and Simmons (1969), (4) Birch (1966), (5) Aleksandrov and Ryzhova (1961), (6) McSkimin et al. (1965), (7) Rafavich et al. (1984).

Table 2. Some interpreted clay velocities. These data are extrapolations to 100 percent clay from mixed lithologies.

Description	V_p (km/s)	V_s (km/s)	V_p/V_s	Reference*
Mixed clays	3.40	1.60	2.13	(1)
Mixed clays	3.41	1.63	2.09	(2)
Montmorillonite/ illite mixture	3.60	1.85	1.95	(3)
Illite	4.32	2.54	1.70	(4)

*(1) Tosaya (1982), (2) Han et al. (1986), (3) Castagna et al. (1985), (4) Eastwood and Castagna (1986).

tary rocks by crossplotting V_p and V_s for different lithologies.

For the following V_p-V_s crossplots, we combined a variety of our own measurements with data from the literature. We attempted to plot data for only relatively pure lithologies; however, as compositional analyses are usually not reported with published velocity data, it is certain that a good deal of scatter is present due to compositional variations and associated uncertainties.

Figure 3, including the data of Pickett (1963) and Milholland et al. (1980) in addition to our own, shows the V_p-V_s relationship for limestone. For S-wave velocities greater than 1.5 km/s, Pickett's relation of $V_p = 1.9 V_s$ holds extremely well. However, as hypothesized, at low velocities, there is a substantial deviation from this trend as V_p approaches 1.5 km/s (water velocity) while V_s approaches zero. The least-squares polynomial fit to the data is

$$V_s \text{ (km/s)} = -.05509 \quad V_p^2 + 1.0168 \quad V_p - 1.0305. \tag{5}$$

Note that for V_s greater than 1.5 km/s, Pickett's linear trend is superior to equation (5) which is a better fit for the entire velocity range.

In contrast, V_s is nearly linearly related to V_p over the entire range of data for sandstones and shales (Figures 4 and 5). Included on these plots, in addition to our own data, are data reported in the compilations in Castagna et al. (1985) and Thomsen (1986). The V_p-V_s relationships are

$$V_s \text{ (km/s)} = .8042 \; V_p - .8559 \tag{6}$$

for sandstone, and

$$V_s \text{ (km/s)} = .7700 \; V_p - .8674 \tag{7}$$

for shale. These trends are similar to the mudrock line obtained exclusively from in-situ measurements in Castagna et al. (1985) given by

$$V_s \text{ (km/s)} = .8621 \; V_p - 1.1724, \tag{8}$$

and in close agreement with the results of Han et al. (1986) for 75 sandstone samples,

$$V_s \text{ (km/s)} = .7936 \; V_p - .7868. \tag{9}$$

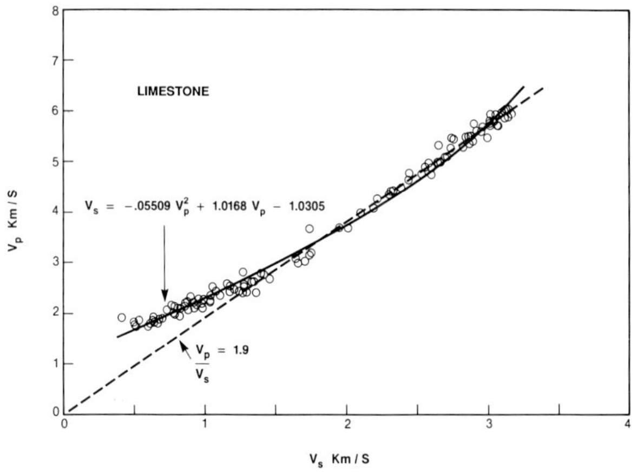

Fig. 3. Crossplot of V_p versus V_s for limestones. The dashed line is Pickett's (1963) trend. The solid line is a second-order polynomial fit to the experimental data.

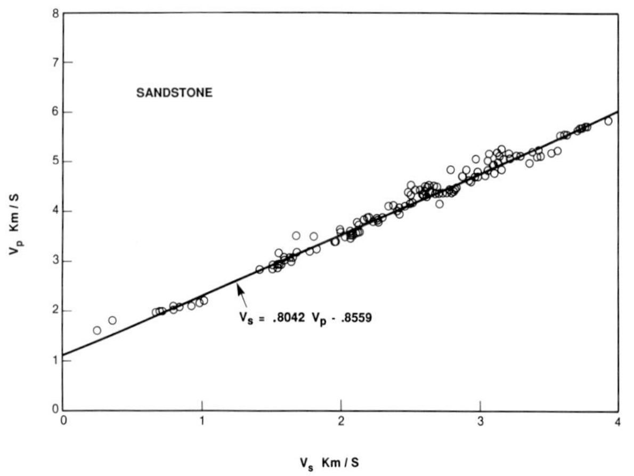

Fig. 4. Crossplot of V_p versus V_s for sandstones. The solid line is the linear regression fit to the experimental data.

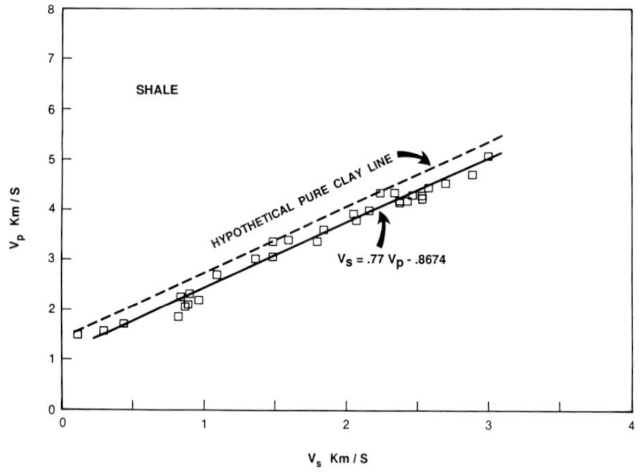

Fig. 5. Crossplot of V_p versus V_s for shales. The solid line is the linear regression fit to the experimental data. The dashed line represents a possible location for a hypothetical pure clay line.

Shale velocities reported in Johnston (1986) fall close to the shale trend of equation (7).

From equations (6) through (9), we may conclude that sandstones and shales exhibit very similar V_p-V_s relationships. If one assumes that the majority of the scatter about these trend curves is due to compositional variation in a binary quartz-clay system, it can be reasoned that a pure clay line would give a slightly higher V_p/V_s ratio than predicted by equation (7). Similarly, a pure quartz line would yield a slightly lower V_p/V_s ratio then given by equation (6). However, pure quartz sands and pure clay shales are scarce, so equations (6) and (7) are preferable for use when compositions are not precisely known; particularly prior to drilling. When estimating S-wave velocities from conventional sonic logs and well log volumetric analyses, slightly modified equations can be used to represent "pure" lithologies. In addition, our experience has shown that feldspathic sands tend to exhibit higher V_p/V_s than do quartzose sandstones.

According to Gregory (1977), gas sands have V_p/V_s ratios of about 1.5. In the laboratory, we have seen ratio variations from under 1.4 to over 1.8 with an average of about 1.5. The effect of gas saturation on V_p/V_s is discussed in greater detail below.

We can report velocity data for dolomites only over a limited range (see Figure 6). Once again, a linear fit is adequate for this range;

$$V_s \text{ (km/s)} = .5832\ V_p - .07776 \tag{10}$$

which is virtually the same as Pickett's trend $V_p = 1.8V_s$.

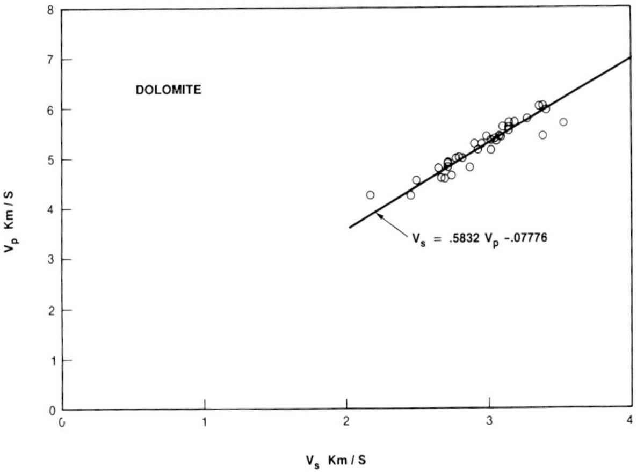

Fig. 6. Crossplot of V_p versus V_s for dolomites. The solid line is a linear regression fit to the experimental data.

Table 3 summarizes the V_p/V_s regressions for limestone, dolomite, sandstone, and shale.

Although our experience with V_p-V_s relationships in coals is limited, we have observed V_p-V_s ratios from about 1.9 to 2.2 in the laboratory. However, these results are biased toward competent samples, and we can expect higher V_p/V_s in low velocity coals (see Appendix B for more data). We have also found V_p/V_s to be about 1.8 for anhydrite and 1.7 for rock salt with relatively little variation due to small porosity variation for these rocks.

Figure 7 shows full-waveform sonic log velocities in a variety of rock types including mixed lithologies. We make the following observations for these data: (1) Clean limestones tend to fall along Pickett's limestone trend ($V_p = 1.9V_s$). (2) Sandstones and shales tend to fall along the mudrock line (equation 8). (3) Salt falls along a limited portion of the mudrock line with V_p about 4.4 km/s and V_p/V_s about 1.7. (4) Dolomites and mixed lithologies fall between the limestone and mudrock lines, and (5) Dolomites tend to have V_p/V_s ratios that are less than predicted by Pickett's dolomite trend ($V_p = 1.8V_s$). This variance may be due to (a) high silica content, or (b) pore shape distribution different from Pickett's dataset.

A composite plot of V_p/V_s relationships (Figure 8) allows us to draw certain conclusions about the use of V_p/V_s ratios to ascertain lithology. Lithology discrimination is best at high velocities where the rule of thumb that V_p/V_s is equal to 1.6 for sandstone, 1.8 for dolomite, and 1.9 for limestone is most nearly correct. In this region, however, it is difficult to distinguish high velocity shales from carbonates. Furthermore, the experimental error associated with extracting V_p/V_s ratios for seismic data will be large compared to the differences in V_p/V_s due to lithology variation. At high velocities, the difference in V_p/V_s between gas- and brine-saturated rocks will be relatively small. For low velocity rocks, lithology discrimination is also difficult; however, the difference in V_p/V_s between gas and brine saturated rocks will be relatively large. As a consequence, AVO analysis for hydrocarbons will be more robust for lower velocity targets.

VELOCITY-DENSITY RELATIONSHIPS

Bulk rock density (ρ_b) is directly related to mineral (matrix) density and porosity by

$$\rho_b = (1 - \phi)\rho_{ma} + \phi\rho_f \tag{11}$$

where

ϕ = porosity,

ρ_{ma} = matrix (grain) density,

and

ρ_f = pore fluid density.

The lithology affects bulk density through the matrix density parameter (ρ_{ma}). In mixed lithologies, this is taken to be the volume weighted average grain density.

Densities of minerals commonly found in sedimentary rocks are given in Table 4. We give the idealized formula for each mineral, and the density reported in Johnson and Olhoeft (1984) which represents the most commonly accepted density for each mineral. This density may not necessarily correspond to the exact chemical formula given. Furthermore, in recognition of variations due to cation substitution, non-stochiometric components, and impurities, we give a range of reported densities for which we have also drawn from Hurlbut (1971) and/or Judd and Shakoor (1981). In contrast, we also present clay mineral densities commonly used in well log analysis in Table 5 (Schlumberger, 1986). Note that these densities are consistently less than the clay densities given in Table 4. It is likely that the densities used in log analysis contain some contributions from excess water bound between clay layers or onto clay surfaces, and/or tied up in microporosity. For seismic modeling purposes, care must be taken in choosing the appropriate grain density; when total porosity is given (i.e., total water volume) Table 4 provides the correct grain densities (in practice, 2.65 gm/cc is a convenient choice for ''shale'' grain density). However, when effective porosity is used, Table 5 gives the appropriate grain density for clays.

Velocity-density relationships can be obtained from the velocity-porosity transforms which are used in well log analysis. The best known velocity-porosity transform is the Wyllie et al. (1956) time-average equation (WGG equation) given by

$$1/V_p = (1 - \phi)/V_{pma} + \phi/V_f \qquad (12)$$

where

V_p = compressional wave velocity,
V_{pma} = grain (''matrix'') compressional wave velocity, and
V_f = fluid compressional wave velocity.

Obviously, this relationship could also be written in terms of density if the grain and fluid densities were known.

The WGG equation is often used in well consoli-

Table 3. $V_p - V_s$ regression coefficients for various lithologies. Coefficients are for equations of the form: $V_s = aV_p^2 + bV_p + C$.

Lithology	Limestone	Dolomite	Sandstone	Shale
a	−0.05508	—	—	—
std err	0.00616	—	—	—
b	1.01677	0.58321	0.80416	0.76969
std err	0.04780	0.03736	0.00899	0.02039
c	−1.03049	−0.07775	−0.85588	−0.86735
std err of Y est	0.08783	0.10421	0.10603	0.11981
R squared	0.99096	0.87444	0.98352	0.97939
# observations	129	37	136	32

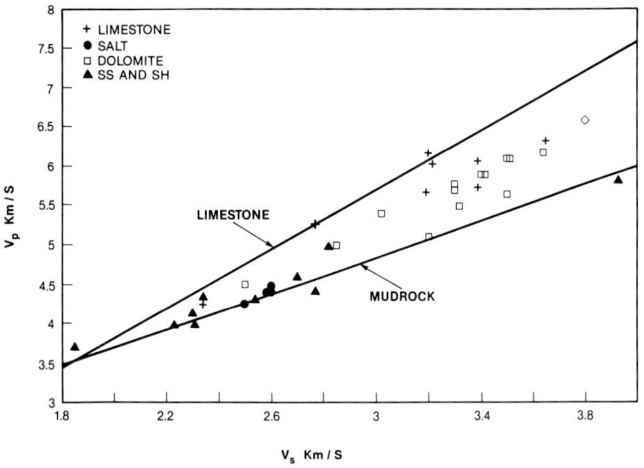

Fig. 7. Crossplot of V_p versus V_s for full waveform sonic log data. The mudrock line is from Castagna et al. (1985). The limestone line is from Pickett (1963).

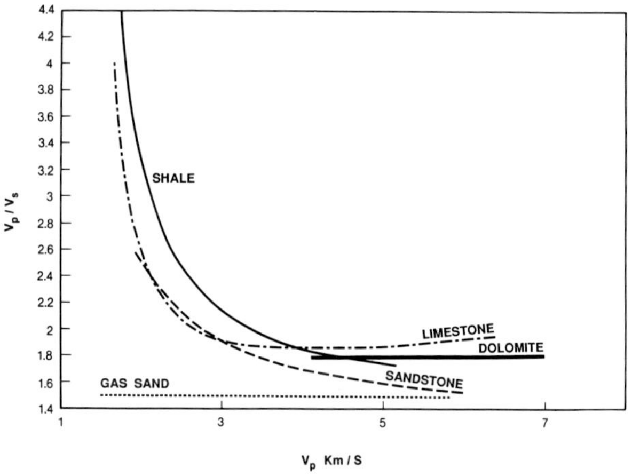

Fig. 8. V_p/V_s versus V_p trends for various lithologies. For water-saturated rocks V_p/V_s approaches infinity as V_p approaches fluid velocity.

dated sandstones for porosity determination from well logs. The equation can be easily derived by assuming that the wave propagates for a fraction of its time in the "matrix",

$$(1 - \phi)/V_{pma}$$

and a fraction of its time in the fluid,

$$\phi/V_f.$$

This is not a physically correct representation of wave propagation in porous media (as it is strictly valid only at wavelengths which are much smaller than the grain/pore size), but is a convenient means of deriving the form of the equation. One should resist the temptation to attach physical meaning to V_f, particularly in gas sands where the use of gas velocities can give erroneous results. This equation also tends to break down where the effective stress is low and where the rock is poorly consolidated. Under these circumstances a "compaction factor" is applied to correct the time-average porosity prediction.

An "improvement" to the time-average equation was proposed in Raymer et al. (1980). According to the Raymer-Hunt-Gardner (RHG) equation:

$$V_p = (1 - \phi)^2 V_{pma} + \phi V_f. \tag{13}$$

Raymer et al. (1980) claim that this equation for sands and sandstones is valid for porosities less than 37 percent. The RHG equation is also empirical in nature, and may not be applicable in all circumstances. As with the WGG equation, the RHG equation must be corrected for low effective stress and poor consolidation.

Table 4. Densities of minerals commonly found in sedimentary rocks. these data are commonly accepted values although variations exist due to cation substitution, non-stochiometric components, and impurities (from Johnson and Olhoeft, 1984; Hurlbut, 1971; and Judd and Shakoor, 1981).

Mineral	Formula	Density	Reported Range gm/cc
Kaolinite	$Al_2Si_2O_5(OH)_4$	2.594	(2.59–2.68)
Montmorillonite	$(Na, K, Mg, Ca)_{.33}(Al, Mg)_2Si_4O_{10}(OH)_2 \cdot nH_2O$	2.608	(2–3?)
Chlorite	$Mg_3(Si_4O_{10})(OH)_{12} \cdot Mg_3(OH)_6$	2.800	(2.6–3.3)
Illite	$(H, K)_3Al_8(Si, Al)_{16}O_{40}(OH)_8$	2.660	(2.6–2.9)
Glauconite	$K_{1.5}(Fe^{3+}, Mg, Al, Fe^{2+})_{4\text{–}6}(Si, Al)_8O_{20}(OH)_4$	2.3	highly variable
Muscovite	$KAl_3Si_3O_{10}(OH)_2$	2.831	(2.7–3.0)
Biotite	$K_2(Mg, Fe)_{4\text{–}6}(Si, Al)_8O_{20}(OH)_4$	2.900	(2.7–3.3)
Quartz	SiO_2	2.648	—
Feldspar	$KAlSi_3O_8$ (Microcline-Orthoclase)	2.56–2.57	(2.55–2.63)
	$NaAlSi_3O_8$ (Albite)	2.62	—
	$CaAl_2Si_2(OH)_8$ (Anorthite)	2.76	—
Calcite	$CaCO_3$	2.710	—
Dolomite	$CaMg(CO_3)_2$	2.866	(2.82–2.87)
Siderite	$FeCO_3$	3.944	—
Pyrite	FeS_2	5.011	(4.95–5.03)
Gypsum	$CaSO_4 \cdot 2H_2O$ (Alabaster)	2.305	—
Anhydrite	$CaSO_4$	2.963	—
Halite	NaCl	2.163	(2.04–2.17)
Iron Oxides		—	(5.200–5.275)
Iron Hydroxides		—	(3.1–4.0)
Zeolites		—	(2.0–2.4)

Table 5. Suggested matrix densities for well log interpretation (Schlumberger Log Interpretation Chartbook, 1986).

Mineral	Formula	Density (gm/cc)
Kaolinite	$Al_4Si_4O_{10}(OH)_8$	2.41
Chlorite	$(Mg, Fe, Al)_6(Si, Al)_4O_{10}(OH)_8$	2.76
Illite	$K_{1\text{–}1.5}Al_4(Si_{7\text{–}6.5}Al_{1\text{–}1.5})O_{20}(OH)_4$	2.52
Montmorillonite	$(Ca, Na)_7(Al, Mg, Fe)_4(Si, Al)_8O_{20}(OH)_4(H_2O)_N$	2.12

Castagna (1985) used Biot-Gassmann theory (Gassmann, 1951; Biot, 1956) to derive an S-wave RHG-like equation for sandstones:

$$V_s = (1 - \phi)^2 V_{sma} \tag{14}$$

where V_{sma} is the matrix S-wave velocity. It is satisfying to note that this equation can also be derived by replacing V_{pma} with V_{sma} and setting $V_f = 0$ in equation (13).

For unconsolidated sediments, Raymer et al. (1980) used a derivative of the Wood (1941) equation for suspensions. The Wood equation is

$$1/K_{su} = \sum_{i=1}^{n} \frac{X_i}{K_i} \tag{15}$$

where

K_{su} = suspension bulk modulus,
X_i = volume fraction of component i,
K_i = bulk modulus of component i, and
n = number of components including fluids and solids.

The Wood-like equation used in Raymer et al. (1980) accounts for some development of frame rigidity by using the plane-wave modulus rather than the bulk modulus of the solid:

$$1/M = (1 - \phi)/M_{ma} + \phi/K_f \tag{16}$$

where

M = bulk rock plane-wave modulus = $k + 4/3\mu$,
M_{ma} = solid grain plane-wave modulus, and
K_f = fluid bulk modulus.

One of the most important empirical reactions in seismic prospecting is the Gardner et al. (1974) equation which expresses density in terms of velocity for an average of all rock types:

$$\rho_b = 1.741\ V_p^{.25} \tag{17a}$$

where V_p is expressed in km/s, or

$$\rho_b = .23\ V_p^{.25} \tag{17b}$$

where V_p is expressed in ft/s, and density is given in gm/cc.

Unlike the WGG and RHG equations, the Gardner equation is fit to a very wide range of velocities and porosities and thus implicitly accounts for variations in consolidation and effective stress. Figure 9 compares the Gardner equation to laboratory and well log velocity and density data for sandstones and shales. The Gardner equation tends to overestimate density for sandstones and underestimate density for shales. Lithologic variation can be taken into account by using different Gardner equations for each lithology. These are given in Table 6 and were obtained by least-squares fits to digitized curves from Gardner et al. (1974), with the exception of the sandstone equation which was modified to approach the quartz point at low porosities.

In contrast to V_p/V_s relationships, V_p-density relationships are not well defined. This is amply demonstrated by Figure 10 which compares the different velocity-density relationships for sandstone including: (1) the WGG equation, (2) the RHG equation, (3) the Wood-like equation, (4) the Gardner equation, (5) the Gardner sandstone equation, and (6) the modified Gardner sandstone equation. Figure 11 is a crossplot of a sampling of laboratory and well log data for sandstones. The RHG and Wood-like equations tend to act as upper and lower limits for velocity at a given density, while the modified Gardner sandstone equation provides the best fit to the data; although this can be characterized as a rough fit at best. Similarly, the velocity-density crossplot for various shales (Figure 12) shows a great deal of scatter. When an arbitrary "pure clay" endpoint is selected, RHG and Wood-like equations again provide approximate upper and lower limits for velocity at a given density. The Gardner shale equation provides the most reasonable fit to the data, but deviations from this trend can be large. Similar conclusions may be drawn for the other lithologies; thus, when a rough velocity-density transform is required, we recommend the Gardner equations summarized in Table 6 and Figure 13. When more precise estimates are required, however, it will be necessary to explicitly consider effective stress and consolidation or to develop local empirical trends.

As velocity variation with pressure is typically greater than the porosity variation with pressure, velocity-porosity relationships should also vary with pressure. Domenico (1984) developed velocity-porosity relationships at different effective pressures for the sandstone and limestone measurements of Pickett (1963) and showed that the porosity dependence of velocity is greater at low effective stress. In the following discussion, for a wide range of data, we are unable to demonstrate any statistically significant relationship between pressure and the magnitude of the porosity dependence of velocity.

MIXED LITHOLOGIES

In the majority of situations, the explorationist must deal with lithologies that are mixed to some degree, rather than pure. Hence, we require methods for averaging the properties or relationships developed for pure rock types. Examples of averaging schemes include the Voigt-Reuss and Hashin-Shtrikman bounds (see Watt et al., 1976, for an excellent review). Unfortunately, these bounds may not be sufficiently tight for our applications. An alternative approach is composite media modeling assuming a solid with inclusions having specific shapes and elastic properties [for example, the Kuster-Töksoz (1974) and the O'Connell-Budiansky (1974) models]. Unfortunately, existing models require accurate and detailed information about the rock fabric before any calculations can be made. They also make assumptions (such as pore non-interaction) that have not been generally demonstrated to hold for sedimentary rocks. Consequently, it is necessary, once again, to resort to empirical equations.

A time-averaging approach is commonly taken in well log analysis by modifying the WGG equation to include the effects of composition. A "clay correc-

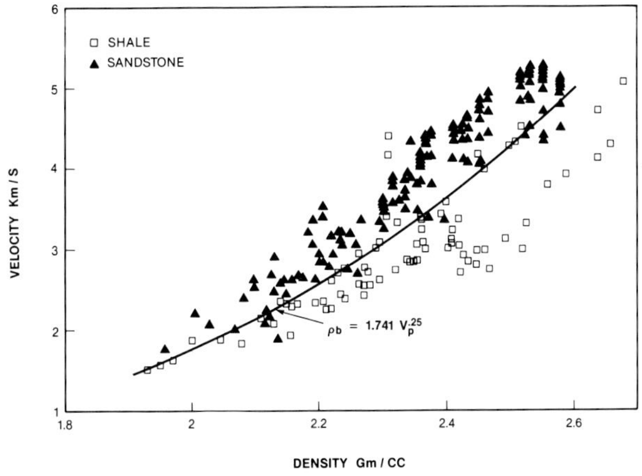

Fig. 9. Laboratory and log measurements for sands and shales showing Gardner's relation. (Gardner et al., 1974).

Table 6. Second-order polynomial fits to the Gardner et al. (1974) velocity-density relationships. The sandstone trend has been modified to approach the quartz point at low porosities. Units are km/s and gm/cc.

(a) Coefficients for the equation. $\rho = aV_p^2 + bV_p + c$

Lithology	a	b	c	V_p Range (Km/s)
Shale	−.0261	.373	1.458	1.5–5.0
Sandstone	−.0115	.261	1.515	1.5–6.0
Limestone	−.0296	.461	.963	3.5–6.4
Dolomite	−.0235	.390	1.242	4.5–7.1
Anhydrite	−.0203	.321	1.732	4.6–7.4

(b) Coefficients for the equation $\rho = dV_p^f$

Lithology	d	f	V_p Range (Km/s)
Shale	1.75	.265	1.5–5.0
Sandstone	1.66	.261	1.5–6.0
Limestone	1.50	.225	3.5–6.4
Dolomite	1.74	.252	4.5–7.1
Anhydrite	2.19	.160	4.6–7.4

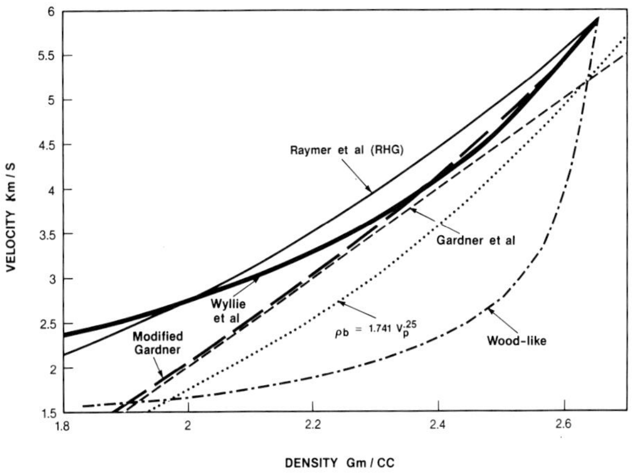

Fig. 10. Sandstone velocity-density trend curves. Light solid line—Raymer et al. (1980); heavy solid line—Wyllie et al. (1956); medium dashed line—modified Gardner (see text); dotted line—Gardner et al. (1974) for sedimentary rocks; dot-dashed line—Wood-like equation (see text).

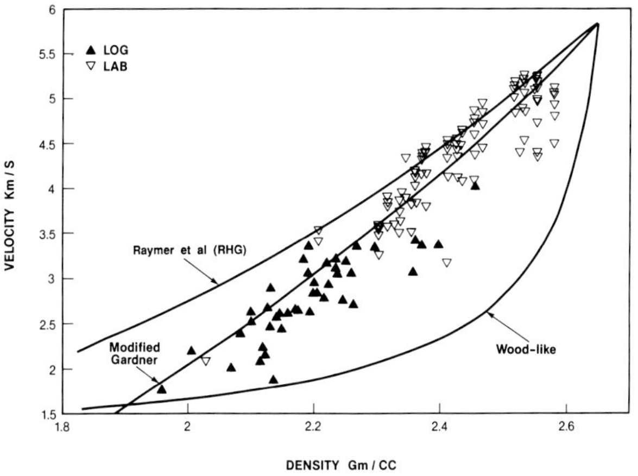

Fig. 11. Sandstone laboratory and log measurements.

tion'' utilizing the fractional volume of clay, ''X_{cl}'', and an assumed clay matrix velocity, ''V_{cl}'', (see Table 2) is often used. The modified WGG equation can be written:

$$1/V_p = X_{\mathrm{cl}}/V_{\mathrm{cl}} + X_{\mathrm{ss}}/V_{\mathrm{ss}} + \phi/V_f \qquad (18)$$

where

X_{ss} = volume of sandstone

and

V_{ss} = matrix velocity for sandstone

Similarly, terms may be added for other lithologic components.

A number of authors (mentioned in the following discussion) have presented similar equations which are, however, linear in velocity, rather than linear in the reciprocal of velocity (transit-time or slowness) as is the WGG equation.

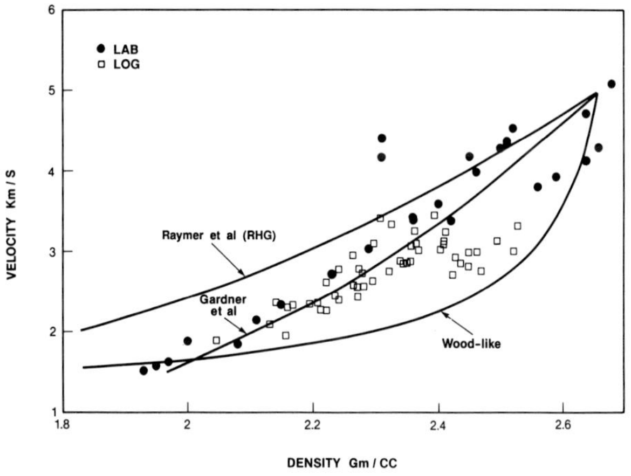

Fig. 12. Shale laboratory and log measurements.

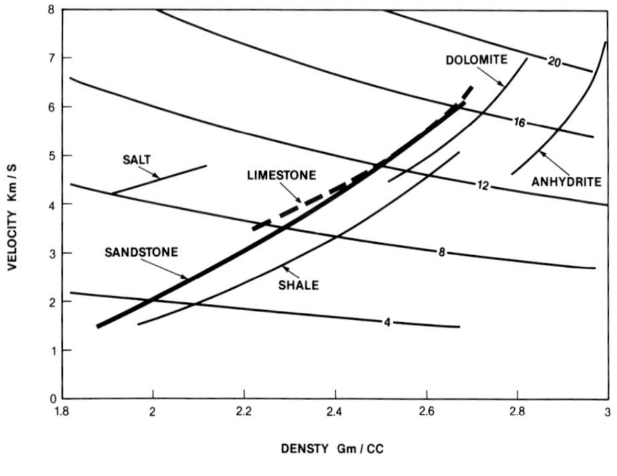

Fig. 13. Velocity-density trends from Gardner et al. (1974) and lines of constant impedance. Impedance units are (km-gm)/(cc-s).

Tosaya (1982) performed ultrasonic velocity measurements on brine-saturated sands and shales at 5800 psi. Clay contents were obtained by thin-section microscopy, X-ray diffraction, and elemental analysis. Multiple regression was used to derive the relationships between P- and S-wave velocity and porosity and clay content.

These equations are

$$V_p \text{ (km/s)} = 5.8 - 8.6\phi - 2.4X_{\mathrm{cl}} \qquad (19a)$$

and

$$V_s \text{ (km/s)} = 3.7 - 6.3\phi - 2.1X_{\mathrm{cl}} \qquad (19b)$$

where

V_p = P-wave velocity,
V_s = S-wave velocity,
ϕ = fractional porosity,

and

X_{cl} = fractional volume of clay.

Kowallis et al. (1984) performed ultrasonic velocity measurements on dry poorly consolidated sandstones. Clay contents were obtained by thin-section microscopy, X-ray diffraction, SEM observation, and elemental analysis. Multiple regression yielded:

$$V_p \text{ (km/s)} = 5.6 - 9.2\phi - 5.7X_{\mathrm{cl}}. \qquad (20)$$

Castagna et al. (1985) performed multiple regression on full-waveform sonic log measurements in the geopressured Frio formation at a depth of about 16 000 ft. Porosity and clay volume were determined by neutron-density crossplot calibrated by the X-ray diffraction data of Freed (1980). This yielded equations which are very similar to those of Tosaya (1982):

$$V_p \text{ (km/s)} = 5.81 - 9.42\phi - 2.21X_{\mathrm{cl}} \qquad (21a)$$

and

$$V_s \text{ (km/s)} = 3.89 - 7.07\phi - 2.04X_{\mathrm{cl}}. \qquad (21b)$$

Han et al. (1986) performed the most thorough examination of this kind for sandstones. They made ultrasonic velocity measurements on 75 different sandstones. Clay contents were obtained by point counting. Their equations for a number of different pressures are given in Table 7. Eberhardt-Phillips et al. (1989) fit these data with the following equations:

$$V_p = 5.77 - 6.94\phi - 1.73\sqrt{X_{cl}} + .446(P - e^{-16.7P}) \quad (22a)$$

$$V_s = 3.70 - 4.94\phi - 1.57\sqrt{X_{cl}} + .361(P - e^{-16.7P}) \quad (22b)$$

where P is the effective stress is kilobars.

Wilkens et al. (1984) and Rafavich (1984) performed lithological regressions on carbonates and evaporites. Rather than review their results separately, these data are incorporated into the following regression analysis.

In an attempt to unify the previously mentioned measurements and other data into one "global" set of equations, we have selected 972 measurements from the literature and our own database for analysis. These measurements were selected on the basis of the following criteria:

(1) specimen was fully water-saturated;
(2) composition was determined by x-ray diffraction, point count, or wet chemical analysis;
(3) both P- and S-wave velocity measurements were available; and
(4) composition was limited to the quaternary system quartz-clay-dolomite-calcite.

Velocities were assumed to be a composition-weighted linear average of the porous end-member lithologies:

Table 7. Sandstone velocity regressions by Han et al, (1986). Coefficients are for the equation $V = a + b\phi + cX_{cl}$ where ϕ equals porosity and X_{cl} equals clay volume fraction.

Effective Stress (bars)	P-Wave Velocity		
	a	b	c
400	5.59	−6.93	−2.18
300	5.55	−6.96	−2.18
200	5.49	−6.94	−2.17
100	5.39	−7.08	−2.13
50	5.76	−7.08	−2.02
Effective Stress (bars)	**S-Wave Velocity**		
	a	b	c
400	3.52	−4.91	−1.89
300	3.47	−4.84	−1.87
200	3.39	−4.73	−1.81
100	3.29	−4.73	−1.74

$$V = \sum X_i V_i = \sum X_i(A_i\phi + B_i \log P + C_i) \quad (23)$$

where P is the effective stress, X_i is the volume fraction of each component, i, V_i are the velocities of pure quartz, clay, calcite, and dolomite aggregates at the given porosity and pressure and

$$\sum X_i = 1. \quad (24)$$

Multiple linear regression yielded the coefficients given in Table 8. Residual analysis revealed no compelling nonlinearities or other obvious inadequacies in equation (23).

In particular, such improvements as (1) nonunity exponent for X_{cl}, (2) A_i a function of pressure, (3) B_i a function of porosity, and (4) pressure dependence different from log P (see Eberhardt-Phillips et al., 1989) did not significantly improve the residual error. This is presumably because the effects described are masked in this dataset by experimental error (primarily in composition). One should keep in mind that equation (23) is a significant improvement over the Tosaya-Han-Nur and Eberhardt-Phillips-Zoback forms in one respect: the porosity and pressure coefficients are lithology dependent.

Analysis of the coefficients given in Table 8 provides some very interesting insights.

(1) For compressional waves, the pressure dependence (partially described by B_i log P) is independent of lithology. For S-waves, the effect is lithology dependent. In both cases, the pressure dependence is the same for clay and quartz.
(2) For P-waves, the porosity dependence is the same for clay and quartz.
(3) For P-waves, items (1) and (2) indicate that the regressions presented by previous authors without lithology dependent porosity and pressure coefficients are valid for quartz-clay mixtures. However, for carbonates and for S-wave velocities, lithology dependence of the coefficients is required.
(4) Setting P = 500 bars and ϕ = 0, reasonable matrix velocities are obtained (see Table 9). Notable discrepancies from expected mineral velocities are (a) very low V_p/V_s(1.41) for zero-porosity clay, and (b) low V_p and V_s for dolomite. These deviations may indicate inadequacies in the regression model (equation 23) or our database.

Overall, the equations presented are consistent with our expectations. Velocity is most strongly dependent on porosity, with clay content being the next most important factor. When mineral properties are similar, we expect little variation with composition. Hence, the conclusion in Rafavich et al. (1984) that mineralogy only plays a secondary role.

For the equations presented in this section, the reader should be aware of a number of potential problems:

(1) The "independent variables" may be highly correlated, so the coefficients may be misleading.
(2) The equations are forced to be linear. Although this is generally adequate, problems may arise at times (e.g., according to Han et al. (1986), the effect of adding a small amount of clay to a sandstone can be nonlinear).
(3) The equations only apply to the specific effective pressure range for which the measurements were made.
(4) The use of a "volume clay" term ignores the variability of clay mineralogy and structural role.
(5) Variations in lithification, texture, and pore geometry are ignored.
(6) The pressure dependence is probably not adequately described.
(7) The effect of pressure may be exaggerated (at low effective stress) by sample damage.

Consequently, local independent confirmation of these equations with in-situ measurements (well logs) is desirable.

FLUID PROPERTIES

This section summarizes work which is more fully described in Batzle and Wang (1992). The pore fluids we are concerned with consist of three primary types: hydrocarbon gases, hydrocarbon liquids (oils), and

Table 8. Velocity regression coefficients for the system quartz-clay-calcite-dolomite: Equation $V = \Sigma X_i(A_i\phi + B_i \log P + C_i)$.

Coefficient	P	S
A_{quartz}	−7.636 ± .125	−4.691 ± .141
B_{quartz}	.192 ± .016	.166 ± .013
C_{quartz}	5.289 ± .0466	3.156 ± .035
A_{clay}	−7.636 ± .125	−12.655 ± 1.049
B_{clay}	.192 ± .016	.166 ± .013
C_{clay}	2.708 ± .059	1.844 ± .135
$A_{calcite}$	−10.136 ± .278	−5.429 ± .236
$B_{calcite}$	.192 ± .061	.068 ± .014
$C_{calcite}$	5.818 ± .039	3.156 ± .035
$A_{dolomite}$	−4.765 ± .444	−4.692 ± .141
$B_{dolomite}$	.192 ± .061	.137 ± .010
$C_{dolomite}$	5.524 ± .062	3.156 ± .035
Correlation Coefficient (R)	.969	.946
R^2	.940	.895
Standard Error	.220	.171
F	2143.687	1179.008
Signif F	.000	.000

Note: Coefficients are all significant (i.e., significance of $T \approx 0$). Velocities are in km/s and pressures are bars. Identical coefficients indicate that no statistically significant difference was found for an initial regression trial, and parameters were lumped for the final regression.

Table 9. Extrapolated velocities at zero porosity and 500 bars effective pressure. Units of velocity are km/s.

Mineral	Extrapolated V_p	Extrapolated V_s	Extrapolated V_p/V_s	Measured V_p	Measured V_s	Measured V_p/V_s
Quartz	5.81	3.60	1.61	6.06	4.12	1.47
Clay	3.23	2.29	1.41	—	—	—
Calcite	6.34	3.34	1.90	6.40	3.30	1.94
Dolomite	6.04	3.52	1.72	7.05	4.16	1.70

brines. The situation is complicated since we are dealing with substances of variable compositions. The hydrocarbons form a continuum of light to heavy compounds ranging from almost an ideal gas to almost solid organic residues. Brines can range from nearly pure water to saline solutions of nearly 50 percent salt. In addition, gases will be dissolved in both the oils and brines and will dramatically alter the velocities. Finally, we must also be concerned with mixtures of distinct phases since reservoirs usually have substantial brine saturations.

Gas

The gas phase is the easiest to characterize. The compounds are usually relatively simple and the thermodynamic properties have been thoroughly examined. Hydrocarbon gases usually consist of the light hydrocarbons of methane, butane, propane, etc. Additional gases, such as water vapor and heavier hydrocarbons, will occur in the gas depending on the pressure, temperature, and history of the deposit. The specific weight of these gases, as compared to air at standard temperature and pressure, will vary from about 0.6 for nearly pure methane to over 1.5 for gases with heavier components. Fortunately, when a rough idea of the gas weight is known, a fairly accurate estimate can be made of the gas properties at pressure and temperatures. Thomas et al. (1970) did a very complete analysis of the acoustic properties of natural gases and we will follow a similar analysis here.

The important seismic characteristics of a fluid (the bulk modulus, density, and sonic velocity) are all related to primary thermodynamic properties. Hence for gases we are obliged to start with the ideal gas law.

$$PV = nRT \tag{25}$$

where P is pressure, V is volume, n is the number of moles of the gas, R is the gas constant, and T the absolute temperature. This leads to a density ρ, of

$$\rho = \frac{nM}{V} = \frac{MP}{RT} \tag{26}$$

where M is the molecular weight. The isothermal compressibility β_T is

$$\beta_T = \frac{-1}{V}\left[\frac{\partial V}{\partial P}\right]_T \tag{27}$$

for compressibility defined as a positive number.

If we calculate the "isothermal" velocity V_T we find

$$V_T^2 = \frac{1}{\beta_T \rho} = \frac{RT}{M}. \tag{28}$$

Hence, velocity would go up with temperature and be independent of pressure.

A pair of mitigating factors must now be considered to bring the relationship closer to reality. First, since there are rapid temperature changes associated with the passage of an acoustic wave, we must use the adiabatic compressibility, β_s, rather than the isothermal compressibility.

$$\gamma \cdot \beta_S = \beta_T. \tag{29}$$

Here, γ is the ratio of heat capacity at constant pressure to the heat capacity at constant volume. In most solid materials, the difference between the isothermal and adiabatic compressibilities is negligible. However, in fluid phases, particularly gases, the isothermal compressibility can be twice the adiabatic value.

The second, more obvious factor stems from the inadequacies of the ideal gas law, equation (25). The gas law can be corrected by adding a compressibility factor (z). The relationships are thus modified

$$PV = znRT, \tag{30}$$

$$\rho = MP/zRT, \tag{31}$$

and

$$\beta_S = \beta_T/\gamma = M/(\gamma\rho zRT). \tag{32}$$

The heat capacity ratio can itself be derived if the equations of state of the material are known. The seismic characteristics of the gas can, therefore, be described if we have an adequate description of z with pressure, temperature, and composition.

Thomas et al. (1970), made use of the Benedict-Webb-Rubin (BWR) equation to define the gas behavior. The BWR equation of state is a rational equation with numerous constants based on the behavior of natural gas mixtures. These gas mixtures range in gravity G (relative to air) from about 0.5 to 1.8. The results of the density calculations are shown in Figure 14. As would be expected, the gas densities increase with pressure and decrease with temperature. However, the densities are also strongly dependent on the gas gravity which is composition dependent.

The adiabatic gas modulus K_s (the inverse of β_s) is also strongly dependent on the composition as well as the pressure and temperature conditions. Figure 15 shows the calculated modulus from the Thomas relationships. Again, the modulus increases with pressure and decreases with temperature but the relationship is not as linear. The impact of variable composition (gravity) is again obvious.

Oil

Crude oils can be mixtures of extremely complex organic compounds. Natural oils range from the lightest condensate liquids of low carbon number to very heavy tars. At the heavy extreme are bitumen and kerogen which may be denser than water and act essentially like solids. At the light extreme are condensates which may become gas with decreasing pressure. Oils can absorb large quantities of hydrocarbon gases under pressure, thus significantly decreasing the moduli. Under room conditions, the densities can vary from 0.5 gm/cc to greater than 1 gm/cc with most produced oils in the 0.7 to 0.8 gm/cc range. The American Petroleum Institute (API) number N_{API}, is defined as

$$N_{API} = \frac{141.5}{\rho} - 131.5. \tag{33}$$

This results in API numbers of about 5 for very heavy oils to near 100 for light condensates. The extreme variations in composition and ability to absorb gases produce greater variations in the seismic properties of oils.

If we have a general equation of state for oils we could calculate the moduli and densities as we did for the gases. Such equations abound in the petroleum engineering literature. Unfortunately, the equations are almost always strongly dependent on the exact composition of a given oil. In exploration, we normally only have a rough idea of what the oils may be like. In some reservoirs, individual yet adjacent zones will have quite distinct oil types. We will, therefore, proceed along rather empirical lines based on the density of the oil.

The acoustic properties of numerous organic fluids have been studied as a function of pressure or temperature (see, for example, Rao and Rao, 1959). Generally, if we stay away from phase boundaries, the velocities, densities, and moduli are quite linear with pressure and temperature. In organic fluids typical of crude oils, the moduli decrease with increasing temperature and increase with increasing pressure. Wang and Nur (1986) did an extensive study of several light alkanes, alkenes, and cycloparaffins and found simple relationships among the density, moduli, temperature, and carbon number or molecular weight. For velocity they found

$$V_T = V_o - b\Delta T \tag{34}$$

where V_o is the initial velocity, V_T the velocity at temperature T, ΔT is the temperature change and b is a constant for each compound of molecular weight M:

$$b = 0.306 - 7.6/M. \tag{35}$$

Similarly, the velocities are related in molecular weight by

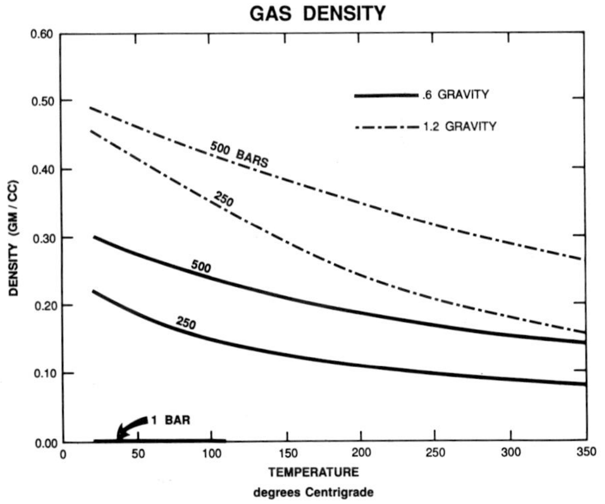

Fig. 14. Hydrocarbon gas density as a function of temperature pressure and composition. Densities are plotted for a light gas (0.6 gravity) and heavy gas (1.2 gravity) at 1, 250, and 500 bars. The two lines overlay at 1 bar.

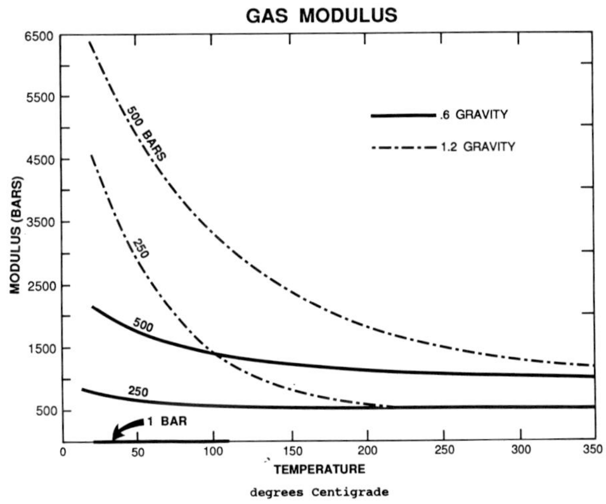

Fig. 15. Gas modulus at pressure and temperature. As in Figure 14, the values for the light and heavy gasses overlay at 1 bar.

$$V_{TM} = V_{TOMO} - b\Delta T - a_m\left[\frac{1}{M} - \frac{1}{M_o}\right] \quad (36)$$

where V_{TM} is the velocity of oil of weight M and V_{TOMO} is the velocity of a reference oil of weight M_o at temperature T_o. The variable a_m is a positive function of temperature. We can see from the right-most term in equation 36 that the velocity of the fluid will increase with increasing molecular weight. When compounds are mixed, Wang and Nur (1986) found that the resulting velocity is a simple fractional average of the end components. This is roughly equivalent to a fractional average of the bulk moduli of the end components. Pure simple hydrocarbons, therefore, behave in a simple predictable way. We need to extend this analysis to include crude oils which are generally much heavier and have more complex compositions. The influence of pressure must also be determined. In the petroleum engineering literature, broad empirical relationships are available. By empirically fitting equations to these data we can get density as functions of initial density (or API number), temperature, and pressure

$$\rho = F(\rho_0, T, P). \quad (37)$$

These densities are shown in Figure 16.

By differentiating equation (37) we obtain the isothermal compressibility β_I

$$\beta_I = \frac{1}{\rho}\left[\frac{\partial\rho}{\partial P}\right]_T. \quad (38)$$

If we assume a reasonable and constant heat capacity ratio γ, of 1.15 (Wang, personal communication) we obtain the adiabatic bulk moduli K_s,

$$K_S = \frac{\gamma}{\beta_I} = \frac{\gamma\rho}{\left[\frac{\partial\rho}{\partial P}\right]_T} \quad (39)$$

The ultrasonic velocities of a variety of crude oils measured recently are reported in Wang et al. (1988). A general relationship of oil velocity was derived.

$$V = 15\,454(77.1 + \text{API})^{-1/2} - 1.14T + 0.462P$$
$$- (1.146 \times 10^{-3} - 4.13 \times 10^{-4}\,\text{API}^{1/2})TP \quad (40)$$

where V is in m/s, T in °C, P in bars, and API is the API degree of the oil, or

$$V = 215 + 50700(77.1 + \text{API})^{-1/2}$$
$$- 6.72T + (.1096 - 0.00166\,\text{API}^{1/2})P$$
$$- (1.44 \times 10^{-4} - 5.18 \times 10^{-5}\,\text{API}^{1/2})TP \quad (41)$$

for V in ft/s, T in degrees F, and P in psi.

Using these velocities and the densities as shown in Figure 16, we get the moduli shown in Figure 17.

Very large amounts of gas or light hydrocarbons can go into solution in crude oils. In fact, the lighter crudes are condensates from the gas phase. We would expect the "live" or gas-saturated oils to have significantly different properties than the "dead" or gas-free oils commonly available and measured. The amount of gas that can be dissolved is a function of pressure, temperature, and the composition of both the gas and the oil (Standing, 1962).

$$R = 0.1266G[P \cdot \exp(.02878N_{\text{API}} - 0.00378T)]^{1.205} \quad (42)$$

where R is the gas-oil ratio in liters/liter (1 liter/liter = 5.615 cu ft/BBL) at atmospheric pressure and at 15.5°C and G is the gas gravity. Equation (42) indicates that much larger amounts of gas can go into the light (high API number) oils. In fact, heavy oils may precipitate heavy compounds if much gas goes into solution.

The effect of this gas in solution on the oil acoustic properties has not been well documented. Sergeev (1948) noted that gas in solution will reduce both oil and brine velocities. He calculated that this mix would change some reservoir reflection coefficients by more than a factor of two. A rough estimate of this dissolved gas effect can be made by assuming that the relationship in equation (40) remains valid and by adjusting the oil density to include the gas component. We are assuming here that the gas is a liquid component with its own volume and density and that the result is an ideal liquid mixture. The simple additive relations found in Wang and Nur (1986) support this concept. The estimated density becomes

$$\rho_G = \rho_0(0.972 + 0.000147F^{1.175}) \quad (43)$$

where ρ_0 is the dead oil density and ρ_G is the gas saturated live oil density. The factor F is derived from the gas/oil ratio

$$F = R\left[\frac{G}{\rho_0}\right]^{1/2} + 75. \quad (44)$$

Figure 18 shows the live and dead oil velocities measured in Wang et al. (1988) along with the estimates using equations (40), (42), and (43).

Brines

The great bulk of the pore fluids consists of brines. Their composition can range from almost pure water to saturated saline solutions. Figure 19 shows the salt concentrations found in brines from several wells from Arkansas and Louisiana. Gulf of Mexico area brines often have rapid increases in concentration with increasing depth. In other areas, California for example, the concentrations are usually lower but can vary drastically between adjacent fields.

The thermodynamic properties of aqueous solutions have been studied in detail. Kennan et al. (1969) give a relation for pure water that can be iteratively solved to give densities at pressure and temperature. Helgeson and Kirkham (1974) use this and other data to calculate a wide variety of water properties over an extensive temperature and pressure range. One obvious effect of salinity is to increase the density of the fluid. Rowe and Chou (1970) presented a polynomial to calculate both specific volume and compressibility of various salt solutions at pressure over a limited temperature

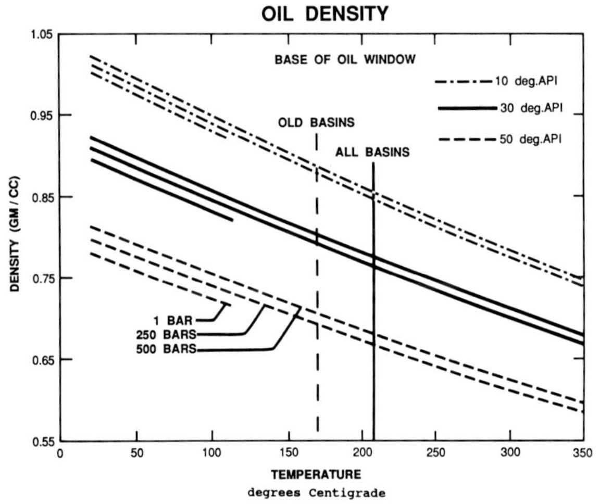

Fig. 16. Oil density as a function of pressure, temperature, and initial API gravity. Three constant-pressure lines are plotted for each oil weight. The 1 bar line is restricted because of phase instability at low pressures and high temperatures. The base of the oil window is represented for old basins (dashed vertical line) and all basins (solid vertical line).

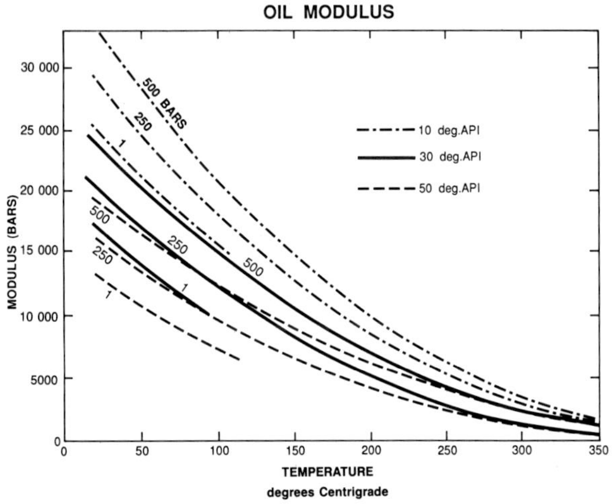

Fig. 17. Oil modulus as a function of pressure temperature, and initial gravity.

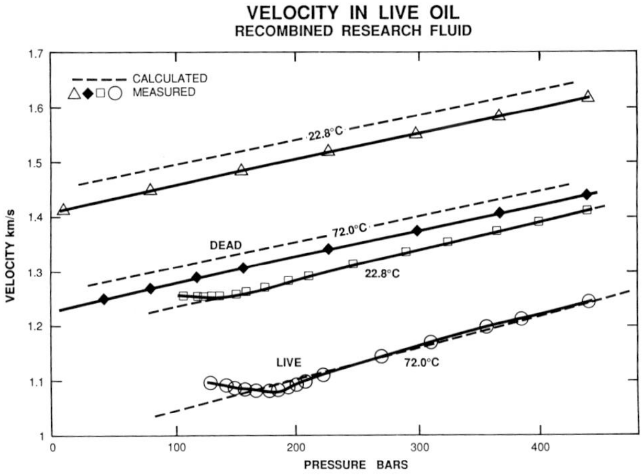

Fig. 18. Measured acoustic velocity of oil both "live" or gas saturated, and "dead" without gas. Measurements were made at 22.8 and 72.0 degrees centigrade. At about 200 bars, gas begins to come out of solution from the live oil. The dashed lines are the calculated velocities.

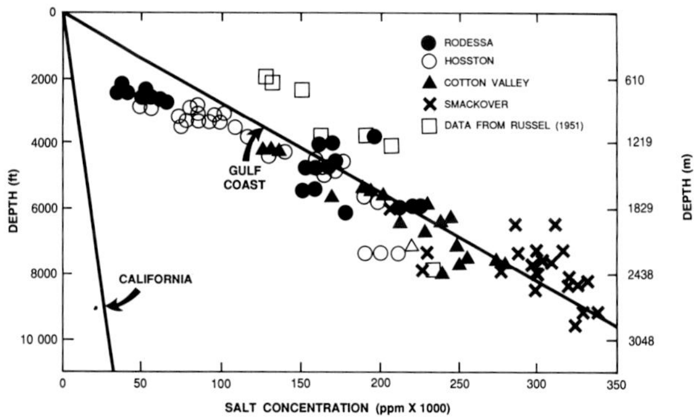

Fig. 19. Salt concentration in sand waters versus depth in southern Arkansas and northern Louisiana after Dickey (1986).

range. Extensive additional data on sodium chloride solutions is provided in Zarembo and Fedorov (1975) and Potter and Brown (1977). Utilizing all these data, a simple polynomial can be constructed that will adequately calculate the density of sodium chloride solutions:

$$\rho_W = \sum_{i=0}^{4} \sum_{J=0}^{2} a_{ij} T^i P^j \tag{45a}$$

and

$$\rho_B = \rho_W + \sum_{i=0}^{2} \sum_{j=0}^{2} \sum_{k=1}^{2} b_{ijk} T^i P^j x^k. \tag{45b}$$

Here T and P are in degrees centigrade and bars respectively, x is the weight fraction of sodium chloride, ρ_B is the density of the brine in gm/cm^3. The constants a_{ij} are given in Table 10. The calculated brine densities, along with selected data from Zarembo and Federov, 1975 are plotted in Figure 20. The accuracy of this relationship is limited largely to the extent that other mineral salts, particularly divalent ions, are in solution.

A vast amount of acoustic data is available for brines, but generally for pressure, temperature, and salinity expected under oceanic conditions. Wilson (1959) provides a relationship for the velocity V_w, of pure water to 100°C and about 1000 bars

$$V_W = \sum_{i=0}^{4} \sum_{j=0}^{3} W_{ij} T^i P^j \tag{46}$$

where the constants W_{ij} are given in Table 10. Millero et al. (1977) and Chen et al. (1978) give additional factors to be added to the velocity of water to calculate

Table 10. Coefficients for water density and modulus computation.

Water Density		Brine Density		Brine Velocity	
a00	0.9999786094	b001	6.683493 E−01	c00	1174.617 E+00
a10	4.463057 E−06	b101	8.981443 E−05	c10	−9.5739 E+00
a20	−5.666686 E−06	b201	2.616639 E−07	c20	5.9847 E−02
a30	1.732672 E−08	b002	4.367547 E−01	c30	−1.1431 E−04
a40	−2.891463 E−11	b102	−3.489720 E−03	c40	4.6486 E−08
a01	4.886269 E−05	b202	1.193570 E−05	c01	2.6337 E−01
a11	−2.217865 E−05	b011	6.864324 E−05	c11	−2.8892 E−04
a21	3.121667 E−09	b111	−2.034837 E−06	c21	−1.7427 E−07
a31	−1.264349 E−11	b211	8.432142 E−09	c30	0
a41	2.325326 E−14	b012	−2.423939 E−04	c40	0
a02	−3.334526 E−09	b112	4.656608 E−04		
a12	−2.404481 E−11	b212	−2.222059 E−09	c12	7.0642 E−09
		b022	−5.509494 E−08	c22	−2.3951 E−11
		b122	1.242326 E−09	c32	0
		b222	−5.091158 E−12	c42	0
Water Velocity				c03	5.2362 E−09
w00	1402.8505963			c13 to c43	0
w10	4.871385 E+00				
w20	−4.782846 E−02			d00	7.8760 E+02
w30	1.487149 E−04			d20	−7.5164 E−04
w40	−2.197453 E−07			d10	−4.2250 E−01
w01	1.523600 E−01				
w02	3.436930 E−05			d11	1.2715 E−04
w03	−1.186670 E−08			d21	−1.0739 E−07
w11	−1.110884 E−03			d02	1.6102 E−03
w21	2.746744 E−05			d12	0
w31	−6.503402 E−08			d22	0
w41	7.986792 E−11				
w12	1.739250 E−06			e0	−8.24262 E+02
w22	−2.135126 E−08			e1	9.5602 E−01
w32	−1.454887 E−10			e2	−1.5012 E−03
w42	5.230233 E−13				
w13	−1.628460 E−09				
w23	1.237155 E−11				
w33	1.326619 E−13				
w43	−4.613582 E−16				

the effects of salinity. Their corrections, unfortunately, are limited to 55°C and 1 molal ionic strength (55 000 ppm). We can extend their results by utilizing the data of Wyllie et al. (1956) to 100°C and 150 000 ppm NaCl. Still, this leaves the high temperature and pressure region with no data. Here we can use the isothermal modulus calculated from equation (45) to estimate the adiabatic moduli. We can also utilize the velocity function provided in Chen et al. (1978) but with the constants modified to fit the additional data. The heat capacity ratio for the brine can be estimated from the PVT relationship in equation (45) and estimates of the isobaric heat capacity from Helgeson and Kirkham (1974):

$$B = m \sum_{i=0}^{4} \sum_{j=0}^{3} c_{ij} T^i P^j + m1.5 \sum_{i=0}^{2} \sum_{J=0}^{2} d_{ij} T^i P^j + m^2 \sum_{i=0}^{2} e_i P^i \quad (47a)$$

and

$$V_B = V_W + B. \quad (47b)$$

In this equation, m is the molal salt concentration and c_{ij}, d_{ij}, and e_i are constants given in Table 10.

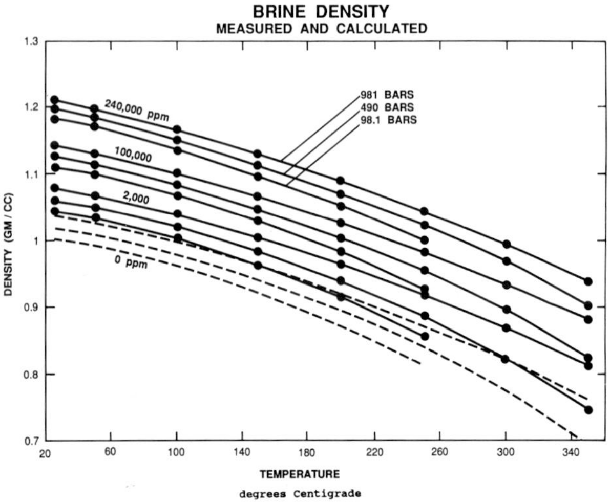

Fig. 20. Brine density as a function of pressure, temperature, and salinity. Symbols are data of Zarembo and Fedorov, 1975; Solid lines are regression fits to these data; Dashed lines are regression fits for pure water.

The calculated moduli using equations (45) and (47) are shown in Figure 21.

Gas can also go into solution with the brine. The amount of gas in solution is substantially less than for the light oils. However, sufficient gas can be in solution in some deep brines for them to be considered an energy resource. Culberson and McKetta, 1951; Sultanov et al., 1972; Brill and Beggs, 1977; and others have shown that the amount of gas that will go into solution increases with pressure and decreases with salinity. The effect of temperature is less substantial. Dodson and Standing (1945) found that the solution's isothermal modulus K_G, decreased almost linearly with gas content:

$$K_G = K_B/(1 + 0.0494\ g) \quad (48)$$

where g is the gas/water ratio at standard pressure and temperature. Equation (48) shows that for a reasonable gas/water ratio of 10 liters/liter (at room pressure and temperature) the isothermal modulus will be reduced by a third. We presume that the adiabatic modulus will be similarly affected. Assuming this to be true, brine modulus versus gas-water ratio plots are provided in Appendix C. The effect of dissolved gas on density is negligible.

GASSMANN'S EQUATIONS

A variety of theoretical models have been developed to predict the velocities of porous media. In broad terms, the models developed for isotropic rocks fall into two categories: (1) pore geometry specific, and (2) global. Well known examples of geometry specific models include the O'Connell-Budiansky (1974) and

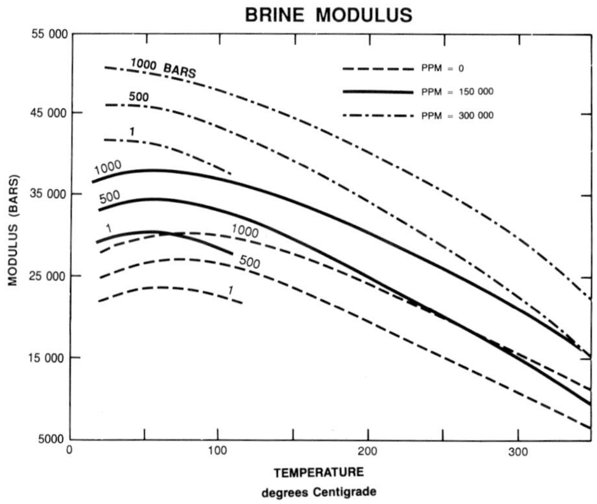

Fig. 21. Calculated brine modulus as a function of pressure, temperature, and salinity.

Kuster-Toksoz (1974) models. These models require the assumption of dilute distributions of idealized pores and have not been shown to be generally applicable to sedimentary rocks. Mavko and Nur (1977) demonstrated that the results can be strongly dependent on the assumed pore shape. Application of these models using the observed pore characteristics has proved difficult at best (Hadley, 1976).

More applicable results have been obtained using the global or bulk properties of the rocks and fluids without referring to any specific pore geometry. Gassmann (1951) and Biot (1956) produced fundamental and yet relatively simple relationships where the most significant unknown parameters are the bulk and shear moduli of the rock framework (skeleton). Gassmann's equations are valid only at low (seismic) frequencies (strictly valid at zero frequency). These are (Domenico, 1976)

$$k_w = k_d + (1 - b)^2/[(1 - \phi - b)/k_{ma} + \phi/k_f] \quad (49)$$

and

$$\mu_w = \mu_d \quad (50)$$

where

k_w = bulk modulus of the fluid filled rock,
k_d = bulk modulus of rock skeleton (aggregate of grains),
k_{ma} = bulk modulus of matrix (individual grains),
k_f = bulk modulus of pore fluid,
b = k_d/k_{ma},
μ_w = shear modulus of the fluid filled rock, and
μ_d = shear modulus of rock skeleton.

Often, the skeleton bulk modulus is unknown but the bulk modulus of the fluid-filled rock is easily derived. For example, if V_p at 100 percent water saturation is known for a given lithology, then the V_p-V_s trend curves can be used to obtain V_s. Given V_p, V_s, and density, we obtain the fluid filled moduli from:

$$k_{100\%} = \rho_{100\%}\left(V_p^2 - \frac{4}{3}V_s^2\right) \quad (51)$$

and

$$\mu_{100\%} = \rho_{100\%}\, V_s^2 \quad (52)$$

where the density and moduli are for water saturation of 100 percent.

Once the fluid-filled bulk modulus is known, it is possible to compute the skeleton bulk modulus from:

$$k_d = bk_{ma} \quad (53)$$

where

$$b = (XY - 1)/(X + Y - 2), \quad (54)$$

$$Y = 1 + \phi(k_{ma}/k_f - 1), \quad (55)$$

and

$$X = k_{100\%}/k_{ma}. \quad (56)$$

Once the skeleton bulk modulus is known, it is possible to calculate the fluid-filled modulus for any combination of fluids. The bulk modulus of a fluid mixture will vary with saturation (or liquid fractions) according to Wood's equation. For example in a gas-water system

$$k_f = (k_{\text{water}}k_{\text{gas}})/(k_{\text{gas}}S_{\text{water}} + k_{\text{water}}S_{\text{gas}}) \quad (57)$$

where

k_{water} = water bulk modulus,
k_{gas} = gas bulk modulus,
S_{water} = fractional water saturation, and
S_{gas} = fractional gas saturation = $1 - S_{\text{water}}$.

Given the fluid-filled moduli from equations (51) and (52), all that is required to compute the fluid-filled velocity is the density. For a gas-water system, the density is given by:

$$\rho_b = (1 - \phi)\rho_{\text{ma}} + S_{\text{water}}\phi\rho_{\text{water}} + S_{\text{gas}}\phi\rho_{\text{gas}} \quad (58)$$

where ρ_{ma}, ρ_{water}, and ρ_{gas} are the grain (matrix), water, and gas densities. P- and S-wave velocities can then be obtained from equations (1) and (2).

Figure 22 shows example velocity-saturation curves for a well consolidated porous sandstone. For the case of gas-water mixtures, the introduction of a minor amount of gas causes a large drop in compressional velocity. This is the well known "bright spot" effect which is due to the large drop in pore fluid modulus when a small amount of gas is added. Increasing gas saturation further has the effect of increasing the velocity, because the fluid bulk modulus remains essentially unchanged and the density decreases. For the case of oils, the change in velocity is less dramatic. Light oils cause the velocity to decrease, while very heavy oils can result in a velocity increase.

There has been some debate in the literature concerning the applicability of Gassmann's equations to partially saturated sandstones. In particular, the laboratory measurements of Gregory (1976) and Domenico (1976) do not agree with Gassmann's equations. However, Murphy (1982) has shown that low-frequency laboratory measurements are in accord with Gassmann's equations for water saturations above a few percent. On the other hand, measurements made at sonic and ultrasonic frequencies deviate significantly from Gassmann's equations, exhibiting a much more gentle drop in compressional velocity as gas is added to a fully water saturated rock. Inhomogeneous distribution of gas bubbles within the pore space may also cause deviations from Gassman's equations.

Conventional sonic logs generally operate at a frequency of about 10 KHz. In light of Murphy's experimental data, we would expect sonic gas sand velocities to be higher than seismic velocities. Furthermore, the radius of investigation of the sonic tool may not extend beyond the invaded zone around a borehole, where gas is flushed back or dissolved by drilling fluid. Thus, the sonic log response to gas will depend on the amount of residual gas in the flushed zone. This residual can also result in an overestimation of the gas sand velocity.

Sandstone V_p-V_s trends and Gassmann's equations imply a distinct relationship between brine-sand compressional velocity and gas-sand compressional velocity. This issue is discussed in Appendix E.

FRAME MODULI

Ideally, rock frame (skeleton) moduli should be the same for fluid-saturated rocks and dry rocks. In practice, however, some difference could result from chemical and physical interaction between the pore fluid and the solid grains. At low pressures, frame moduli for moist sandstones are less than frame moduli for dry sandstones (e.g., Murphy, 1982). However, this effect has not been demonstrated to be generally significant at typical formation effective stresses. For shaly rocks,

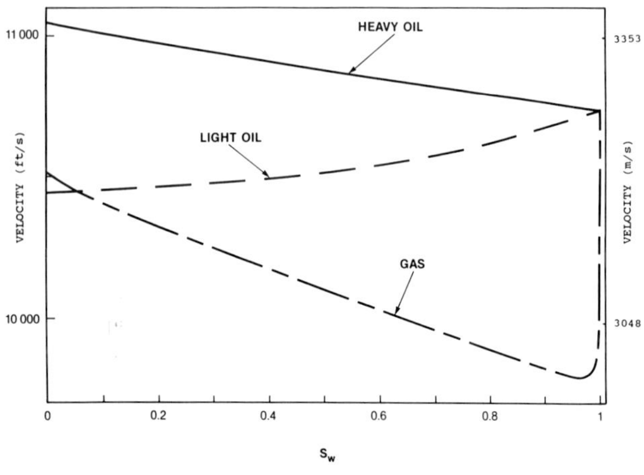

Fig. 22. Calculated velocity versus water-saturation for a typical well-consolidated sandstone.

there is the possibility of frame moduli modification due to interaction between pore fluids and clay minerals. This effect is also not yet well understood.

If there is a difference between dry and wetted frame moduli, the wetted frame moduli should be used for fluid substitution calculations. These can be obtained through Gassmann's equations as described in the previous section. However, if one wishes to calculate frame moduli for high-frequency laboratory data, Gassmann's equations are not adequate. Biot's theory includes the entire frequency range up to the point where the grain scattering becomes important and the rocks can no longer be considered homogeneous; Gassmann's equations are the low frequency limit of Biot's more general relationships. The infinite frequency Biot equations may be more applicable for ultrasonic measurements (see Domenico, 1976):

$$V_p = \left[K_D + \frac{4}{3}\mu + \frac{\dfrac{\phi\rho_b}{\kappa\rho_f} + (1-b)(1-b-2\phi/\kappa)}{(1-\phi-b)/K_{ma} + \phi/K_f} \cdot \frac{1}{\rho_b\left[1 - \dfrac{\rho_f\phi}{\rho_b\kappa}\right]} \right]^{1/2} \quad (59)$$

$$V_s = \left[\frac{\mu}{\rho_b\left[1 - \dfrac{\rho_f\phi}{\rho_b\kappa}\right]} \right]^{1/2} \quad (60)$$

where the new parameter, κ, has been introduced. This is the mass coupling factor which varies between 1 (no fluid-solid coupling) and infinity (perfect coupling). For the case of the perfect coupling these equations reduce to the zero frequency case [equations (1), (2), (49), and (50)].

We can study the relationships between saturated frame moduli by deriving bulk and shear moduli from fully water saturated laboratory data using equations (59) and (60). Figures 23a, 23b and 23c are bulk versus shear frame moduli crossplots for dry sandstones, fully-saturated sandstones assuming perfect coupling, and fully-saturated sandstones assuming no coupling. For the case of dry sandstones, dry bulk and shear moduli are about equal. This is consistent with observations in Gregory (1977) and Castagna et al. (1985). When perfect coupling is assumed, the fully saturated frame bulk modulus is greater than the frame shear modulus. However, when no coupling is assumed the frame bulk and shear moduli are again equal. At the ultrasonic frequencies of laboratory measurements, there should be relatively little fluid-solid coupling. We conclude that, for both dry and saturated laboratory sandstone data, frame bulk and shear moduli are about equal. This corresponds to a frame V_p/V_s ratio of about 1.5. Frame moduli relationships for various lithologies derived in this way are given in Table 11.

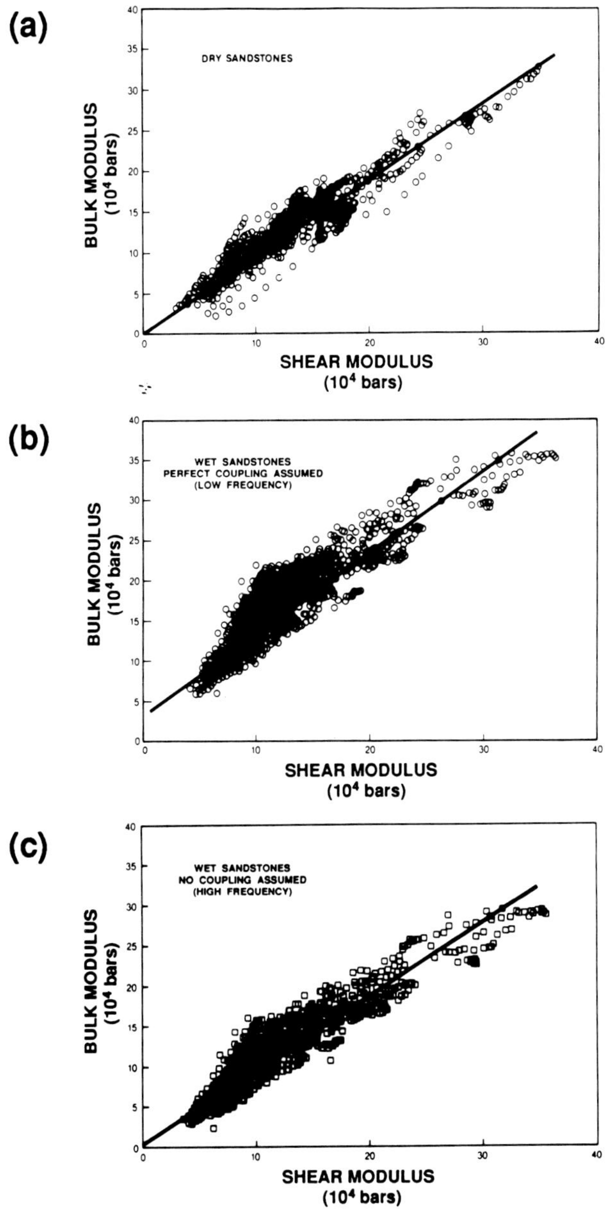

Fig. 23. Skeleton bulk modulus versus shear modulus for (a) dry sandstones, (b) wet sandstones with perfect rock-fluid coupling assumed, (c) wet sandstones with no rock-fluid coupling assumed.

ATTENUATION

Due to increasing path length with offset, overburden attenuation influences seismic amplitude variation with offset. This is most significant for highly attenuating overburdens resulting from poor consolidation or partial gas saturation. Extremely high attenuation in gas reservoirs may also affect AVO, although this effect is expected to be more subtle due to small variations in target zone path length with offset.

The attenuation of a seismic plane wave is described by

$$A(z) = A(0)e^{-\alpha z} \tag{61}$$

where

$A(z)$ = the amplitude of the signal as a function of position z,
$A(0)$ = initial amplitude, and
α = attenuation coefficient.

The attenuation coefficient is related to the quality factor (Q) by:

$$\alpha = \frac{\pi f}{QV} \tag{62}$$

where

f = frequency, and
V = velocity.

Typically, to measure the attenuation coefficient, amplitudes are measured at two locations. The ratio of these amplitudes eliminates the initial amplitude dependence.

$$A(z_1)/A(z_2) = e^{-\alpha(z_1/z_2)}. \tag{63}$$

Hence,

$$\alpha = \frac{\ln\,[A(z_1)/A(z_2)]}{(z_2 - z_1)} \tag{64}$$

It is generally assumed that, over the seismic frequency band, the quality factor (Q) is constant. This is generally supported for P-waves by spectral ratio measurements which indicate that the attenuation coefficient is linearly proportional to frequency (McDonal et al., 1958). This requires that the product Q times velocity be constant (see equation 62). Since velocity varies only slightly with frequency, Q is approximately constant over the seismic band.

Laboratory measurements above the normal seismic range (e.g., Spencer, 1981; Murphy, 1982) have shown, however, that over a broad frequency band, Q is strongly frequency dependent. In particular, attenuation peaks between seismic and sonic log frequencies (Figure 24). This effect is most pronounced for partial water saturations.

Figure 25 shows VSP attenuation measurements obtained with a spectral ratio method described in Kan et al. (1983) in a South Texas clastic section. One interval, corresponding to a known gas reservoir, exhibits anomalously high attenuation. This is consistent with laboratory observations (Murphy, 1982).

Table 12 summarizes various in-situ attenuation measurements made in clastic sediments and rocks. There appears to be some tendency toward higher attenuation in sands than in shales, however, the systematics are not at all clear at this time. For example, Anderson and Castagna (1984) observed higher full waveform sonic P-wave attenuation in Frio formation geopressured shales than in adjacent sandstones. The reader is cautioned that in situ measurements of Q tend to be unreliable. These measurements are commonly plagued by such factors as differential frequency dependent coupling, scattering, and narrow bandwidth.

Laboratory evidence indicates that microscopic fluid flow in sandstones is the most important attenuation mechanism. We have already seen how the Biot-Gassmann relationships are modified for the frequency-dependent fluid coupling factor. Winkler (1986) uses these relationships to compare low frequency velocities to ultrasonic laboratory data. Biot (1962), Jones and Nur (1983), Bulau et al. (1984), and Jones (1986) among others have examined how the porosity, permeability, and specific fluid properties will control attenuation and dispersion. Currently, this is a topic of intensive research.

ANISOTROPY

In anisotropic materials, the velocities of P- and S-waves are dependent on the direction of propagation. Furthermore, for S-waves, the velocity may also depend on the polarization of the wave. In AVO analysis, there is the concern that anisotropy of the rocks above and in the vicinity of the target (particularly in shales) will influence the AVO response.

Sedimentary rocks generally exhibit transverse isotropy for which two of the three orthogonal propagation directions are equivalent. For propagation perpendicular to bedding, transverse anisotropy exhibits one P-wave speed and one S-wave speed. For propagation parallel to bedding there are three wave speeds: compressional, horizontally polarized shear, and vertically polarized shear. The vertically polarized shear wave propagating parallel to bedding has the same speed as the vertically propagating shear-wave.

The P-wave anisotropy (E) is defined as:

$$E = (V_{\text{ppa}} - V_{\text{ppe}})/V_{\text{ppe}} \tag{65}$$

where

V_{ppa} = compressional velocity parallel to bedding, and

V_{ppe} = compressional velocity perpendicular to bedding.

Similarly, the shear-wave anisotropy G is defined as

$$G = (V_{\text{spa}} - V_{\text{spe}})/V_{\text{spe}} \tag{66}$$

where

V_{spa} = horizontally polarized shear-velocity parallel to bedding, and

V_{spe} = shear-velocity perpendicular to bedding.

Table 11. Frame moduli relationships derived from fully-saturated ultrasonic velocity measurement assuming no fluid-solid coupling and Biot's high frequency equations. Units are in bars.

Lithology	Equation
Limestone	$k = 2.29\mu - 120000.$
Dolomite	$k = 1.50\mu - 20000.$
Shale	$k = 1.16\mu + 10000.$
Sandstone	$k = \mu$

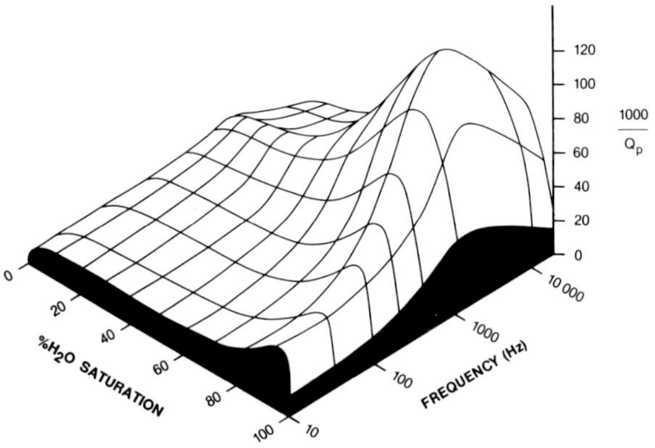

Fig. 24. Schematic representation of the dependence of compressional wave attenuation versus frequency and water-saturation (Murphy, 1982).

In reality, we expect anisotropy to be continuously frequency (and wavelength) dependent. As a convenience, however, we follow Thomsen (1986), in referring to intrinsic and extrinsic anisotropy. By intrinsic anisotropy we mean anisotropy observable on a core plug or hand specimen with dimensions on the order of a few inches. By extrinsic anisotropy we mean the difference between anisotropy observable at seismic wavelengths and ultrasonic wavelengths.

Intrinsic anisotropy may be due to preferred orientation of: (1) anisotropic mineral grains (most likely for micas and clay minerals), (2) grain shapes (such as flat-lying clay platelets of shell debris), and (3) pores or microfractures, or may be caused by fine laminations.

As a consequence of these factors, shales often have high intrinsic anisotropy. Many shales exhibit fissility as a result of preferred orientation of micas and clay minerals. Fissility decreases with increasing pressure and silica or carbonate content and increases with organic content (Pettijohn, 1975). Shales also tend to be laminated, exhibiting fine layering from .05 to 1 mm in thickness (Pettijohn, 1975). These laminations may be due to (1) alternations of silt and clay, (2) alternations of organic content, and (3) alternations of carbonate content.

Figure 26 compares *P*-wave and *S*-wave anisotropy for sandstone and shale measurements compiled in Thomsen (1986). The tendency for shales to exhibit higher anisotropy than sandstones is clear. However, some shales exhibit very low anisotropy, and there is considerable overlap with the sandstone data. Bioturbation and soft sediment deformation often destroy the initial preferred orientations and laminations in shales.

Table 13 gives bench top measurements for dry Sevier Shale samples from Tennessee. The non-fissile Rogersville sample exhibits weak compressional *P*- and *S*-wave anisotropy. In contrast, the fissile Kingsport samples exhibit extremely large anisotropy.

The effect of pressure on shale anisotropy is illustrated in Figure 27 for a well consolidated dry shale from a depth of over 19 000 ft. As pressure increases, anisotropy decreases for both *P*- and *S*-waves. The velocity increase with pressure is greatest for waves propagating perpendicular to bedding. This suggests that the anisotropy is due, in part, to preferred orientation of flat pores or cracks between clay platelets which close under pressure. The significant difference between *S*-wave velocity perpendicular to bedding and vertically polarized parallel to bedding indicates that the sample is not perfectly transversely isotropic (or perhaps not perfectly homogeneous). Similar behavior is observed for a well-consolidated, water-saturated shale from a depth of about 17 000 ft (Figure 28).

Figure 29 illustrates the behavior of an inhomogeneous water-saturated shale specimen. This specimen exhibits many features expected for transverse isotropy:

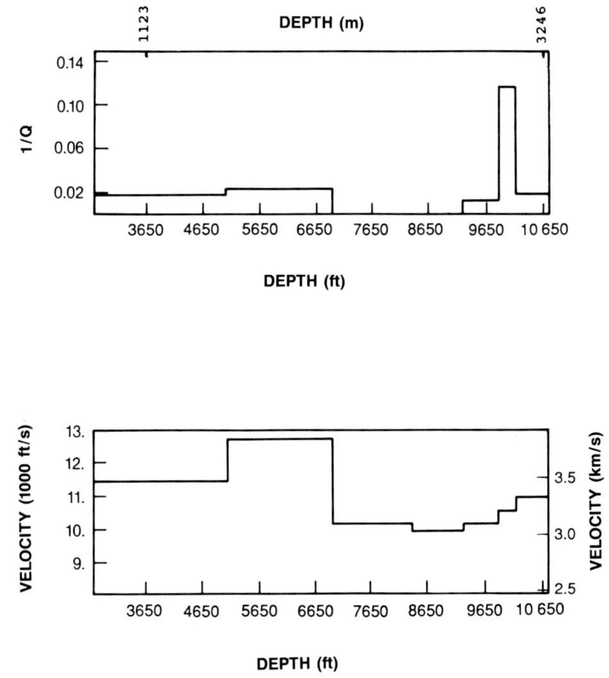

Fig. 25. Interval velocity and attenuation versus depth from a South Texas vertical seismic profile.

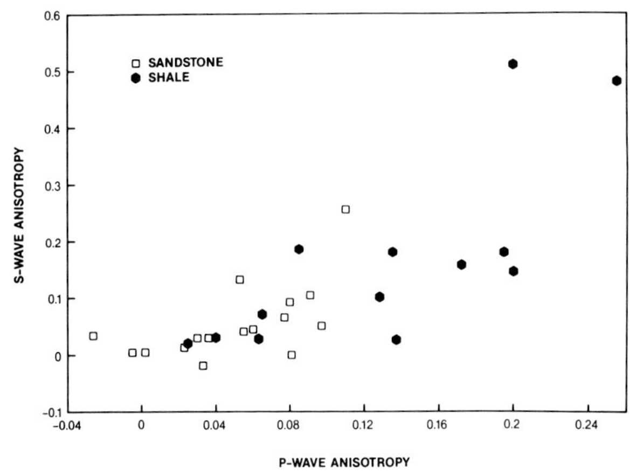

Fig. 26. *P*- and *S*-wave anisotropy for sandstones and shales for data compiled in Thomsen (1986).

Table 12. Reported in-situ attenuation measurements.

Location	Depth m	Hz	Q_p	$1/\alpha_p$	Method	Rock Type	References
Offshore Louisiana Pleistocene	2070–2850	≤125	31	.032	VSP	Sand and Clay	Hauge (1981)
Southeast Texas	660–1320	15–40	34	.027	VSP	Sand 45% and Clay	
Southeast Texas	1800–2100	≤80	37	.027	VSP	Sand 23% and Clay	
Southeast Texas	600–1560	≤80	46	.022	VSP	Sand 20% and Clay	
Offshore Louisiana Pleistocene	1170–1770	≤125	67	.015	VSP	Clay/Sand	
Southeast Texas	>1020	40–70	94	.0106	VSP	Sand 24% and Clay	
East Texas	317–402	≤200	38	.0263	VSP	Sand/Shale interbed	Kan et al (1983)
East Texas	402–610	≤200	—	$.004581f^{.35}$	VSP	Shale	
East Texas	235–371	≤200	—	$.004f^{.4}$	VSP	Shale	
East Texas	152–235	≤200	—	$.031f^{.4}$	VSP	Shale	
Beaufort Sea (Canada)	549–1193	125	43	.023	VSP		Ganley and Kanasewich (1980)
Beaufort Sea (Canada)	945–1311	425	67	.015	VSP		
Limon (Colorado)	0–225	50–450	32	.0313	VSP	Pierre Shale	McDonal et al (1958)
Gulf Coast		50–400	18	.0055	VSP	Sands and Shales	Tullos and Reid (1969)
130 km south of		50–400	75	.0133	VSP	Sandy Clay	
Houston		50–400	136	.0014	VSP	Clay/Sand	
Sarawak Basin	700	10–60	116–870		Sonobuoys	Marine Sediments	Stoll and Houtz (1983)
South China Sea	1800		137–324		Diving Wave		
San Diego	Shallow	14 KHz	31–32		Transducer	Sand	Hamilton (1972)
		14 KHz	111–437		Probes	Clay/Silt	
						Sand/Silt/Clay	
		25 KHz	44			Fine Sand	
		25 KHz	54			Sandy Silt	
		25 KHz	118			Clayey Silt	
		100 KHz	23–104			Sand/Silt/Clay	

Table 13. Velocities parallel and orthogonal to bedding, Sevier Shale, Tennessee.

			Shear 1	Shear 2
		V_p	$Vs1$	$Vs2$
NON-FISSILE SAMPLE	⊥	5.65	3.12	3.13
(Rogersville)	‖	6.03	3.29	3.31
FISSILE SAMPLE A	⊥	1.83	—	0.61
(Kingsport)	‖	4.32	—	2.66
FISSILE SAMPLE B	⊥	1.66	0.72	0.69
(Kingsport)	‖	4.21	—	2.69

Note: V-Velocity (km/s), ⊥-Propagation perpendicular to bedding, ‖-Propagation parallel to bedding, and —Signal too poor to make determination. For propagation parallel to bedding, shear 1 is polarized with particle motion perpendicular to bedding, shear 2 is polarized with motion parallel to bedding. No confining pressure (bench-top measurements).

(1) Compressional velocity and horizontally polarized S-wave velocity parallel to bedding are greater than the compressional P- and S-wave velocities perpendicular to bedding.
(2) The anisotropy decrease with increasing pressure is due primarily to velocity increases perpendicular to bedding.
(3) For propagation perpendicular to bedding, two orthogonal polarizations of the S-wave have virtually the same velocity.

However, there is one important difference: The vertically polarized S-wave travelling parallel to bedding has about the same S-wave velocity as the horizontally-polarized S-wave travelling parallel to bedding, and has a greater velocity than the S-wave travelling perpendicular to bedding. We suspect that this is caused by layering on a scale larger than about one-tenth of an inch, which is the approximate wavelength of the ultrasonic signal. Thus, we see that at laboratory frequencies, only very fine layering can be

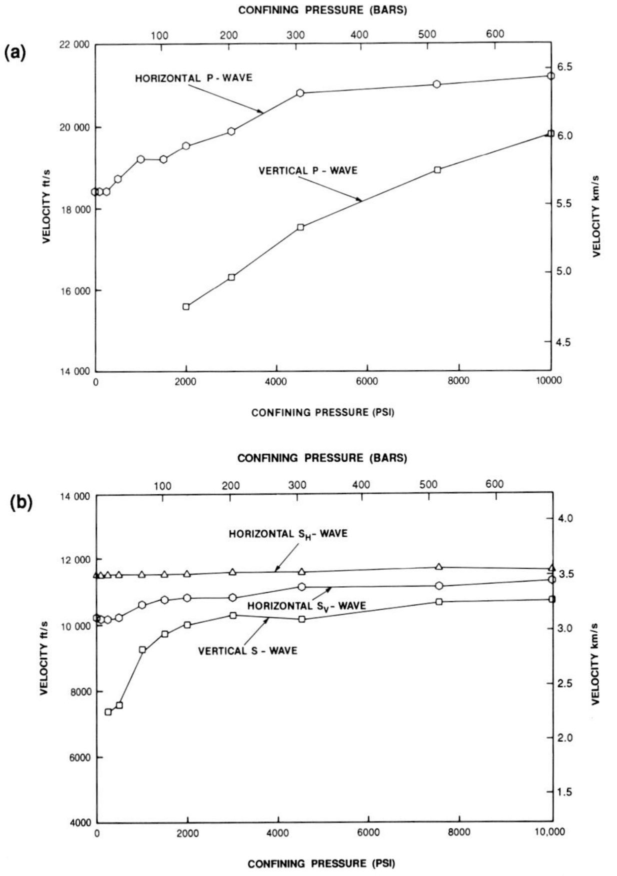

Fig. 27. Vertical and horizontal velocities versus confining pressure for a dry shale sampled from a depth of 19 454 ft (5929 m) (a) compressional velocities, (b) shear velocities.

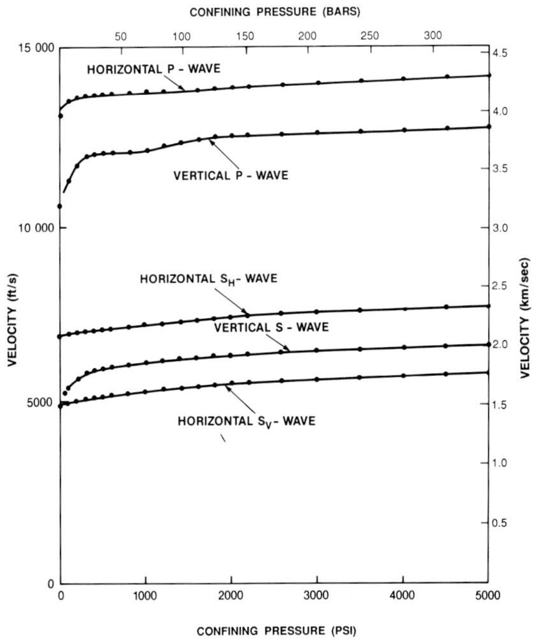

Fig. 28. Vertical and horizontal velocities for a bedded fully water-saturated shale sampled from a depth of 16 624 ft (5067 m).

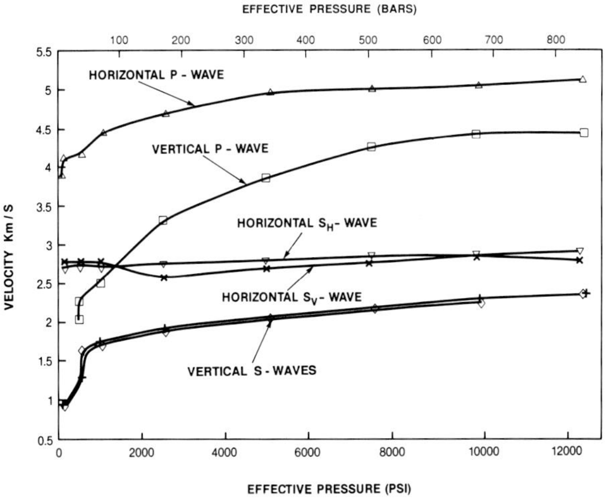

Fig. 29. Vertical and horizontal velocities for a water-saturated shale from a depth of about 17 000 ft (5181 m).

correctly characterized by bulk velocity measurements. On the other hand, we expect seismic waves to respond to anisotropy of all scales smaller than a seismic wavelength. Clearly then, laboratory measurements alone are insufficient to characterize anisotropy for seismic waves.

Extrinsic anisotropy is usually a result of geologic layering or preferred orientation of fractures. Assuming transverse isotropy, one can use the equations of Backus (1962) to average the elastic constants of layers which can be resolved by the sonic log (about 2 ft) over a seismic wavelength (hundreds of feet). Thus, given *P*- and *S*-wave velocity logs and a density log, one can compute the amount of anisotropy due to isotropic layers between 2 ft thick and some larger fraction of a seismic wavelength. We did this, choosing an arbitrary depth interval of 50 ft for each anisotropy calculation, for three different depositional settings: (1) shallow Gulf of Mexico Upper Miocene-Pliocene shales and sands, (2) shallow post-Eocene nonmarine shales and sands from the Sacramento Valley, and (3) North Slope interbedded carbonates and shales.

As shown in Table 14, for the massive Gulf of Mexico sands and shales, the mean extrinsic anisotropy is relatively small (about .5 percent for *P*-waves and 3 percent for *S*-waves). The Sacramento Valley interbedded sands and shales exhibit somewhat higher *P*- and *S*-wave anisotropy (2 and 10 percent, respectively). The interbedded North Slope carbonates and shales have the highest *P*-wave anisotropy (3 percent) but comparatively low *S*-wave anisotropy (4 percent). In the Gulf of Mexico example, calculated transverse isotropy is relatively small even though the section is predominantly shale because the shales and sands are massive and have a low degree of interbedding. In the Sacramento Valley example, there is a great deal of sand-shale interbedding and anisotropy is high. *S*-wave anisotropy is much greater than *P*-wave anisotropy because V_p/V_s is strongly correlated to *P*-wave slowness. In contrast, the interbedded North Slope shales and carbonates exhibit relatively high *P*-wave anisotropy due to the strong *P*-wave velocity contrasts, but only slightly greater *S*-wave anisotropy due to the relatively constant V_p/V_s ratios of the shales and carbonates.

It should be emphasized that these calculations predict only anisotropy induced by layers which can be resolved by the sonic log. Additional anisotropy can be expected from finer layers and intrinsic anisotropy. However, some conclusions of consequence can be drawn from these calculations: (1) The magnitudes of the predicted anisotropies in Table 14 are about the same as observed intrinsic anisotropies. Thus, we conclude that transverse layering can be as important as intrinsic anisotropy. (2) The degree of anisotropy is strongly dependent on the amount of interbedding and consequently the depositional environment. Knowledge of depositional environment may allow estimation of anisotropy which can be input into elastic modeling algorithms to aid AVO analysis. Conversely, anisotropy estimates extracted from the seismic data may be useful facies indicators. (3) Although the amount of anisotropy may not be easily related to lithology, comparing *P*- and *S*-wave anisotropy may provide some insight into the mix of lithologies. We have seen that interbedded isotropic sandstones and shales tend to have much higher *S*-wave anisotropy relative to the *P*-wave while interbedded isotropic carbonates and shales of similar V_p/V_s ratio have similar *P*- and *S*-wave anisotropy.

In practice, anisotropy is not necessarily transverse. Azimuthal anisotropy can be developed when there is structural complexity or the rocks are fractured. Numerous theories exist to calculate the effects of a population of oriented fractures with specified characteristics. Unfortunately, as with the geometry specific poroelastic models, information is required about the fracture shapes, densities, etc. Solutions are not unique, so measured anisotropy cannot unambiguously be related to the fracture characteristics. However, oriented microfractures have been demonstrated

Table 14. Calculated transverse anisotropy from compressional and shear wave velocity and density logs for 15 m thick intervals using the Backus (1962) elastic moduli averaging equations.

Area	Depth Range	Lithology	Parameter	Min	Max	Mean	Std. Dev
Gulf of Mexico	1025′–3775′	sh and ss	E	−.024	.021	.005	.005
			G	.001	.156	.032	.029
Sacramento Valley	1525′–3000′	sh and ss	E	.001	.161	.023	.024
			G	.003	.502	.103	.077
North Slope	8805′–10 075′	lms and sh	E	.000	.115	.030	.026
			G	.002	.165	.041	.037

Note: E = compressional-wave anisotropy, G = shear-wave anisotropy to have a substantial effect on ultrasonic velocities. For example, by applying a uniaxial stress we would expect to open fractures parallel to this stress. Hence, velocities should increase for propagations parallel to the stress and decrease normal to the stress. This effect has been reported in Volarovich et al. (1985). Alternatively, Engelder and Plumb (1984) attributed the measured azimuthal velocity anisotropy of core samples to microfractures opening from stress relief caused by the coring process.

CONCLUSIONS

Various key rock physics issues are directly relevant to amplitude variation with offset analysis. These include V_p-V_s, V_p-density, and velocity-lithology-porosity relations, pore fluid properties, fluid substitution, skeleton properties, attenuation, and anisotropy. We find that:

(1) V_p-V_s relationships are simply related to lithology and pore fluid content,
(2) Velocity-density relationships are highly variable,
(3) Pore fluid properties are highly variable and vary systematically,
(4) Skeleton moduli trends can be derived from water-saturated measurements,
(5) Gas-bearing rocks exhibit anomalously high attenuation, and
(6) Anisotropy due to transverse layering can be as important as intrinsic anisotropy.

These conclusions provide a framework for the interpretation of amplitude variation with offset data.

ACKNOWLEDGMENTS

Thanks to ARCO Oil and Gas Company for permission to publish this paper. Special thanks to M. Popa, B. Smith, C. Landisman and M. Backus for their contributions. If not for S. Gronemeier, this paper would never have been completed.

APPENDIX A—MORE PRESSURE CONCEPTS

Pressure is defined as normal force per unit area. Table A-1 provides conversion factors between various units of pressure. Overburden pressure (P_o) is the pressure due to the weight of fluid-saturated rock and is given by

$$P_o(z) = \bar{\rho}_b(z)gh = g\int_0^h \rho_b(z)\,dz \qquad \text{(A-1)}$$

where

$\bar{\rho}_b(z)$ = mean overburden fluid-saturated bulk density as a function of depth,
g = acceleration due to gravity,
h = total depth,
z = depth, and
$\rho_b(z)$ = fluid-saturated bulk density as a function of depth.

Similarly, hydrostatic pressure (P_h) is the pressure due to the weight of a column of fluid, and is given by:

$$P_h(z) = \bar{\rho}_f(z)gh = g\int_0^h \rho_f(z)\,dz \qquad \text{(A-2)}$$

where

$\bar{\rho}_f(z)$ = mean overburden pore-fluid density as a function of depth, and
$\rho_f(z)$ = pore fluid density as a function of depth.

In engineering practice, pressure gradients are defined as pressure divided by depth, rather than, dP/dz, which is the "true" pressure gradient. Thus, the overburden pressure gradient is P_o/z and the hydrostatic pressure gradient is P_h/z. Various pressure gradient units are given in Table A-2.

Table A-1. Conversion table for units of pressure (after Billings; 1972).

Pounds per Square Inch	Atmospheres	Bars	Mega Pascals (MPa)
1	0.0680458	0.0689474	.0069
14.6960	1	1.01325	.1013
14.5038	0.986924	1	.1
14.2234	0.967842	0.980665	.0981
.000014504	.000000986924	.000001	.0000001
145.038	9.86924	.10	.01

Table A-2. Densities and associated pressure gradients (after Gretener, 1979).

Oil:

g/cc	API° (60°F)	psi/ft	MPa/ft	MPa/m	bars/m	bars/ft
0.7	70.6	0.303	.00209	0.0069	.069	.0209
0.8	45.4	0.346	.00239	0.0078	.078	.0239
0.9	25.7	0.390	.00269	0.0088	.088	.0269
1.0	10.0	0.433	.00299	0.0098	.098	.0299

Formation Waters:

g/cc	total solids ppm	psi/ft	MPa/ft	MPa/m	bars/m	bars/ft
	0	0.433	.00299	0.0098	.098	.0209
1.00	28,000	0.441	.00305	0.0100	.100	.0305
1.02	55,000	0.450	.0031	0.0102	.102	.0311
1.04	84,000	0.459	.00317	0.0104	.104	.0317
1.06	113,000	0.467	.00323	0.0106	.106	.0323
1.08	144,000	0.476	.00329	0.0108	.108	.0329
1.10	176,000	0.485	.00335	0.0110	.110	.0335
1.12	210,000	0.493	.00341	0.0112	.112	.0341
sea water:						
1.025	35,000	0.443		0.0100	.100	

Rocks, Drilling Mud:

g/cc	Lb/ft^3	Lb/gal	psi/ft	MPa/ft	MPa/m	bars/m	bars/ft
1.0	62.4	8.35	0.433	.00299	0.0098	.098	.0299
1.2	74.9	10.02	0.520	.00359	0.0118	.118	.0359
1.4	87.4	11.69	0.607	.00419	0.0137	.137	.0419
1.6	99.8	13.36	0.693	.00478	0.0157	.157	.0478
1.8	112.3	15.03	0.780	.00538	0.0176	.176	.0538
2.0	124.8	16.70	0.867	.00598	0.0196	.196	.0598
2.2	137.3	18.37	0.953	.00658	0.0216	.216	.0658
2.4	149.8	20.04	1.040	.00718	0.0235	.235	.0718
2.6	162.2	21.71	1.126	.00777	0.0255	.255	.0777
2.8	174.7	23.38	1.213	.00837	0.0274	.274	.0837
3.0	187.2	25.05	1.300	.00897	0.0294	.294	.0897

*Strictly speaking, pressure gradients are related to the specific weight (γ) rather than the density (ρ). Note: $\gamma = \rho * g$, where g: 980.665 cm/s^2.

Table A-3. Example density-depth functions for various sedimentary sections world-wide. Functions may be for specific locations and are not necessarily statistically representative for a wide geographic area. Curve letters refer to Figure A-1. ρ_b = bulk rock density. z = depth.

	Curve area	Equation	Remarks
A	Kentucky	$\rho_b = 2.76 + .059z + .026z^2$	Borehole gravity; Chapin (ARCO data)
B	Michigan	$\rho_b = 2.63 + .049z$	Borehole gravity; Chapin (ARCO data)
C	Delaware Basin	$\rho_b = 2.52 + .029z$	Gravity model; Keller et al. (1985)
D	Western Ross Sea	$\rho_b = 2.00 + .254z - .021z^2$	Gravity model; Davey and Cooper (1987)
E	Santa Barbara Channel	$\rho_b = 2.05 + .160z - .020z^2$	Well data; Fertl (1976)
F	Deep Sea Terrigenous Sediments	$\rho_b = 1.53 + 1.39z - .617z^2$	Core; Hamilton (1976)
G	Gulf Coast Shales	$\rho_b = 2.14 + .505 \log z$	Well data; Dickinson (1953)
H	Offshore Morocco	$\rho_b = 1.75 + .200z - .010z^2$	Gravity model; Hinz et al. (1982)
I	Ventura Basin	$\rho_b = 2.19 + .035z$	Gravity model; Jachens and Griscom (1985)
J	North Sea	$\rho_b = 1.98 + .118z - .012z^2$	Well data; Fertl (1976)
K	Gulf Coast	$\rho_b = 1.95 + .108z - .009z^2$	Well data; Fertl (1976)

Figure A-1 shows typical average overburden density functions, $\bar{\rho}_b(z)$, and associated pressure gradients for various locations worldwide. The average density functions were obtained by integrating the density-depth functions given in Table A-3.

Figure A-2 shows overburden pressure gradients versus depth as a function of age used by drilling engineers for fracturing/completions calculations. As expected, older rocks tend to exhibit higher overburden pressure gradients.

Inspection of Figures A-1 and A-2 reveals that the common assumption that overburden pressure gradients are "about 1 psi/ft" can be grossly in error, particularly for young unconsolidated clastics and predominantly carbonate or Paleozoic age geologic columns.

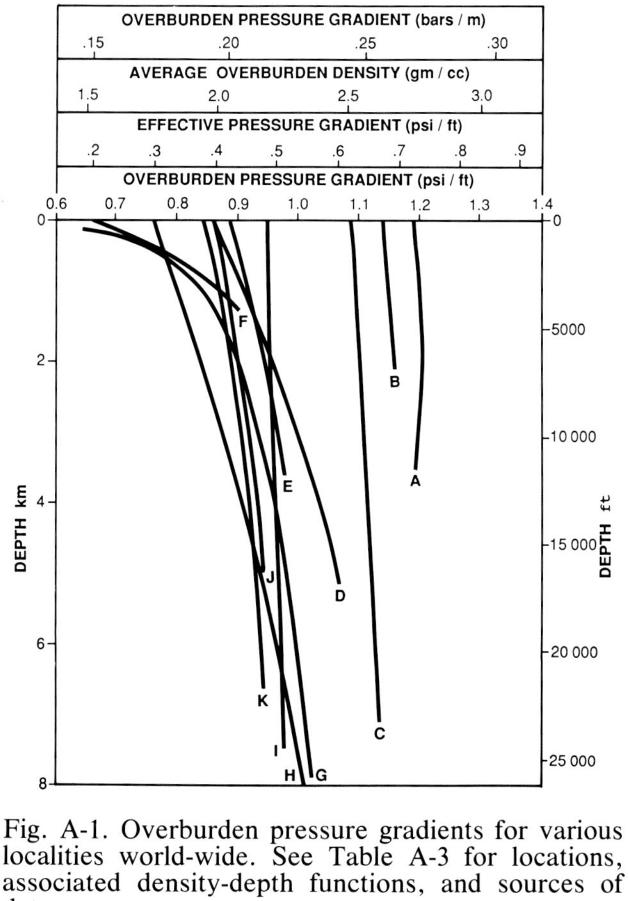

Fig. A-1. Overburden pressure gradients for various localities world-wide. See Table A-3 for locations, associated density-depth functions, and sources of data.

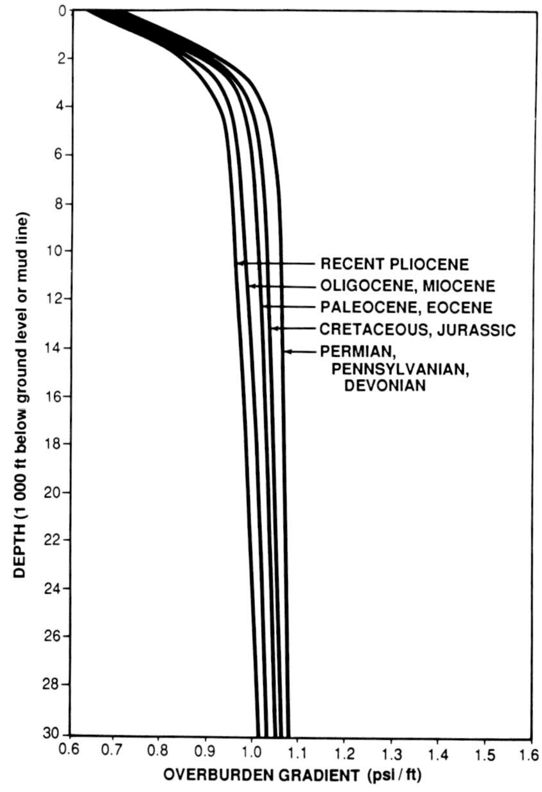

Fig. A-2. Overburden pressure gradients commonly used by engineers (from Matthews and Matthews; 1985).

APPENDIX B—COMPRESSIONAL- AND SHEAR-WAVE VELOCITY IN COAL

We can report only limited reliable S-wave velocity information for coals.

Table B-1 gives velocities of Permian bituminous coals from the Sydney Basin, New South Wales, presented by Greenhalgh and Emerson at the 1981 SEG Convention.

Figure B-1 shows velocity measurements on two samples of Ferron coal under wet and dry saturation conditions and for effective stresses from 34 to 1034 bars. There is no significant difference in the V_p-V_s trends of the dry and water-saturated data.

Figure B-2 combines the data from Table B-1 and Figure B-1 and shows the resulting second-order polynomial fit. Table B-2 gives the regression coefficients. The reader is cautioned that this trend is for a very limited dataset and that further data are required to establish a more universally applicable

Table B-1. Physical properties of Permian bituminous coals from the Sydney Basin, New South Wales (Greenhalgh and Emerson, 1981). Pressure and saturation conditions for velocity measurements were unspecified. Dry bulk densities: 1.34 to 1.45 gm/cc. Effective porosity: 1 to 9 percent, mean 4 percent.

Propagation direction	Mean V_p (km/s)	Mean V_s (km/s)	V_p/V_s
Perpendicular to bedding	≈2.0	.96	2.08
Parallel to bedding	≈2.3	1.08	2.13
Perpendicular to bedding	1.74	≈.8	2.18
Parallel to bedding	1.85	≈.8	2.31

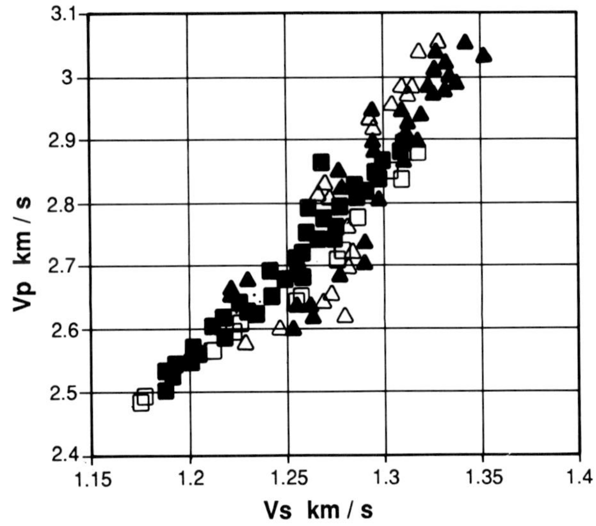

Fig. B-1. Velocity measurements on two samples of Ferron coal for effective stresses from 34 to 1034 bars. Solid symbols are fully water-saturated measurements. Open symbols are dry.

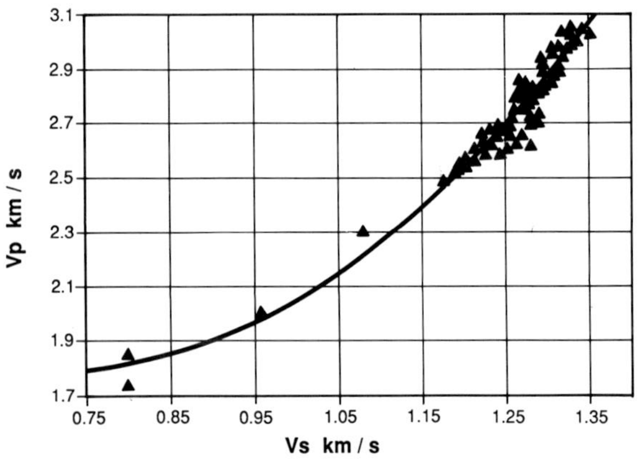

Fig. B-2. V_p-V_s relationship for coal combining data from Table B-1 and Figure B-1.

Table B-2. Regression coefficients for coal V_p-V_s trend (for velocity units of km/s).

(a) $V_p = aV_s^2 + bV_s + c$

Coefficient	Value	Standard Error
a	3.162	.296
b	−4.535	.664
c	3.420	.052
R-squared	.934	
Observations	143	

(b) $V_s = dV_p^2 + eV_p + f$

Coefficient	Value	Standard Error
d	−0.232	.012
e	1.542	.064
f	−1.214	.014
R-squared	.966	
Observations	143	

V_p-V_s relationship. In particular, the variation of V_p-V_s relationship with coal grade has yet to be reported.

APPENDIX C—EFFECT OF DISSOLVED GAS ON WATER MODULUS

Equation (48) provides the relationship between isothermal water modulus and gas-water ratio. For velocity calculations, the adiabatic modulus is required. If we assume equation (48) to hold for adiabatic conditions, we can calculate fluid modulus versus temperature, pressure, and gas-water ratio. Plots

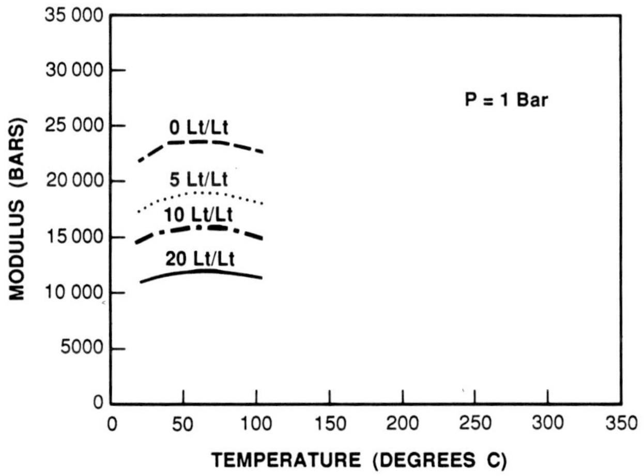

Fig. C-1. Tenuous water modulus versus temperature and gas-water ratio at a pressure of 1 bar.

of these calculations are provided as Appendix C rather than in the main body of this paper due to (1) the inadequate documentation provided in Dodson and Shanding (1948) regarding the empirical determination of equation (48) and, (2) the assumption that equation (48) does indeed apply for the adiabatic modulus. Thus, the modulus plots provided here (Figures C-1 through C-3) must be considered highly tenuous and subject to change as more data become available. Figures C-1 through C-3 have been included primarily to emphasize the importance of dissolved gas on water modulus, to illustrate the systematics of modulus variation, and to provide a rough estimate of the modulus only when better information is lacking.

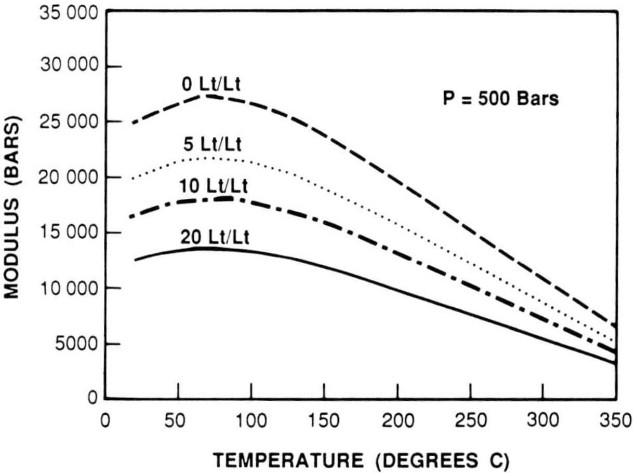

Fig. C-2. Tenuous water modulus versus temperature and gas-water ratio at a pressure of 500 bars.

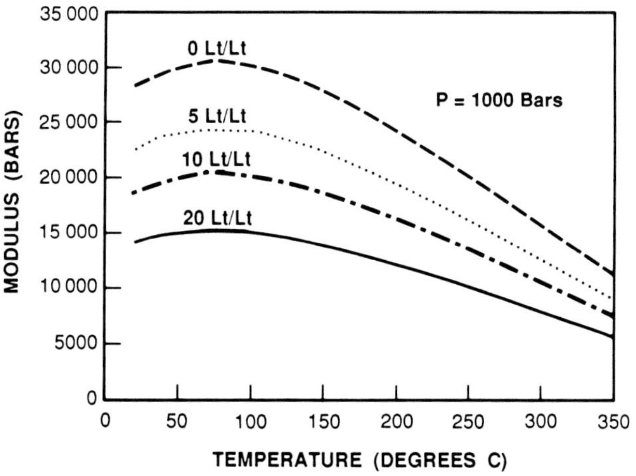

Fig. C-3. Tenuous water modulus versus temperature and gas-water ratio at a pressure of 1000 bars.

APPENDIX D—PROPERTIES OF SHALES

Although shale is the most abundant sedimentary rock component of the stratigraphic column worldwide, it is also the most poorly understood of the common sedimentary rocks. The physical, chemical, transport, electrical, and acoustic properties of shales are not well established. Even a definition of what constitutes a shale is generally unavailable. From the perspective of seismic explorationists, a knowledge of shale physical and acoustic properties is needed to interpret reflection amplitudes and to understand transmission effects above the target.

Rock physicists are faced with a number of difficult challenges when attempting to study shale properties:

(1) It is generally very difficult to obtain unaltered shale core samples and to perform velocity and physical properties measurements on them under in situ conditions. This is also compounded by the difficulty in economically justifying the costs of coring and analysis for nonreservoir rocks.
(2) Mineralogical analysis is greatly hindered by the extremely fine grain size of the clay particles. Mineral abundances can only be roughly estimated by painstaking X-ray diffraction analysis or by attempting to unravel bulk chemical and physical property analyses. This becomes very important in the study of diagenetic alteration which is required for an understanding of depth-dependent variations.
(3) Many shale properties are interrelated and it is often difficult to isolate them. For example, the relationship between velocity and effective stress may be difficult to separate from the relationship between porosity and effective stress.

While compiling data for shales, we came across three sources: (1) compaction measurements carried out on natural sediments and clay slurries, (2) in situ measurements based on wireline logs, and (3) laboratory measurements on core samples restored to in situ conditions. While laboratory measurements on clay-rich rocks are valuable, laboratory derived velocity-pressure curves may be questionable due to problems in achieving pore pressure equilibrium within the limited time available for laboratory measurements. Shales are known to exhibit long term "creep" phenomenon and, therefore, most of the data in the literature from such measurements must be inter-

Table D-1. Shale terminology.

Term	Definition
Silt	Particle between 1/16 and 1/256 of an inch in diameter.
Clay	Particle smaller than 1/256 of an inch.
Clay Minerals	A group of hydrous aluminum silicates exhibiting a sheet-like structure similar to mica. Clays usually result from the weathering of feldspars. For our purposes, we will also consider micas as clay minerals.
Mud	Silt and clay.
Mudrock	Lithified mud.
Fissility	The tendency of a rock to separate along smooth surfaces parallel to bedding.
Siltstone	A non-fissile mudrock composed of two-thirds silt or more.
Claystone	A non-fissile mudrock composed of two-thirds clay or more.
Shale (strict definition)	A fissile mudrock.
Shale (general usage)	A mudrock rich in clay minerals.

preted with caution. Notably, Johnston (1986) paid special attention to ensuring that his samples had sufficient time to equilibrate. Wireline measurements are valuable in establishing shale properties versus depth in a given area, but suffer from borehole alteration. They are also restrictive in that there are no means for controlling the independent variables such as effective stress.

Terminology for fine-grained clastic silicate rocks is not well established. We follow the terminology of Blatt et al. (1972) given in Table D-1. We use the term "shale" in the most general sense. It is common in logging to define "clean shales" based solely on the SP or gamma ray response. Hence, rocks often referred to as shales can actually be a host of materials.

Table D-2. Some reported mineral compositions of shales (percent).

Paleozoic and younger shales: Shaw and Weaver (1965)

Constituent	Range	Mean
Clay minerals	<20–90	61
Quartz	10–80	31
Feldspar	0–30	4.5
Carbonates		3.6
Iron oxides		<0.5
Miscellaneous		<2
Organic matter		1

Calcareous Devonian shales of Alberta: Campbell and Oliver (1968)

Constituent	Range
Illite	13–39
Quartz	3–26
Dolomite	1–36
Calcite	16–76

Shales are mixtures of clay-sized particles, consisting primarily of clay minerals, and silt particles, which are mostly quartz. Krynine (1948) estimated the "average" shale to be about 50 percent silt, while Pettijohn (1975) and others suggest that shales average about two-thirds silt. Mineralogically, shales tend to be composed of about half clay minerals; the rest being primarily quartz plus a few percent feldspar and calcite. Table D-2 shows shale compositions reported in Shaw and Weaver (1965) for 300 samples of Paleozoic and younger shales and by Campbell and Oliver (1968) for calcareous Devonian shales in Alberta. Note that quartz content can be as great as 80 percent, that clay content can be less than 20 percent, and that carbonate content forms a continuum between shale and limestone.

Figure D-1 from Pettijohn (1975) shows the structure of clay minerals. The basic building blocks of clay minerals are (a) silica tetrahedral layers, (b) aluminum and/or magnesium hydroxide octahedral layers, and (c) interlayer water molecules and cations. The simplest clay structure is that of kaolinite, which is essentially a stack of two-layer units, each containing one octahedral and one tetrahedral layer. Montmorillonite consists of three-layer units, each containing one octahedral and two tetrahedral layers. The units are bound to each other by water molecules and cations such as sodium, potassium, magnesium, and calcium. The number of water molecules is governed by volumetrics, rather than stochiometry, and the structure can expand to accommodate more water. The mica, muscovite, is also composed of three-layer units, however, the units are held together exclusively by potassium cations and there is substitution of aluminum for silicon in the tetrahedral layers. Illite is essentially hydrous muscovite, with a combination of water molecules and potassium cations bonding the three-layer units and less aluminum for silicon substitution. Chlorite consists of three-layer units bonded to each other by single magnesium hydroxide octahedral layers. Mixed-layer clays are stacks of alternating clay types such as kaolinite-illite or montmorillonite-illite. The alternations may be random or may have varying degrees of order. Cation substitution is common in clay minerals, so compositions generally vary greatly.

The distribution of clay minerals encountered in a sedimentary section depends on a number of factors including (a) source, (b) paleoclimate, (c) hydraulic sorting, and (d) diagenesis. Factors such as the chemical environment of deposition are believed to be of lesser importance. When shale/clay composition is plotted versus time (see Figure D-2) there is a clear tendency for an increase in chlorite and illite content at the expense of other clays with increasing age. This is presumably primarily a diagenetic effect. The irregularity of the curves can be explained by variations with time in factors (a) thru (c) presented earlier and variations in factors affecting diagenesis, such as geothermal gradient.

Upon deposition, clay sediment porosities may be as high as 90 percent. As a result of compaction, porosities decrease rapidly with depth. According to Rieke and Chilingarian (1974), shale porosity at 5000 ft varies from about 5 percent to 25 percent (see Figure D-3). Hedburg (1936) gives 9 percent as average shale porosity at 6000 ft, although density log readings indicate that shales typically have much larger water contents by volume at this depth. The primary factor controlling shale porosity is effective stress, but infilling of pores by diagenesis is also important (Pettijohn, 1975). Judd and Shakoor (1981) report shale densities from .97 (loose clays) to 2.72 gm/cc with a mean of 2.34 and a standard deviation of .379.

As described in the main body of this paper, velocity-density relationships may be quite variable for shales. Figure D-4 shows velocity-density data for shales from the DOE Pleasant Bayou #1 and #2 wells in Brazoria Co., Texas. There are two distinct groups of data corresponding to shales shallower than 8000 ft. and deeper than 12 000 ft. Note that these groups fall below the Gardner curve and exhibit two distinctly different velocity-density trends. This change in be-

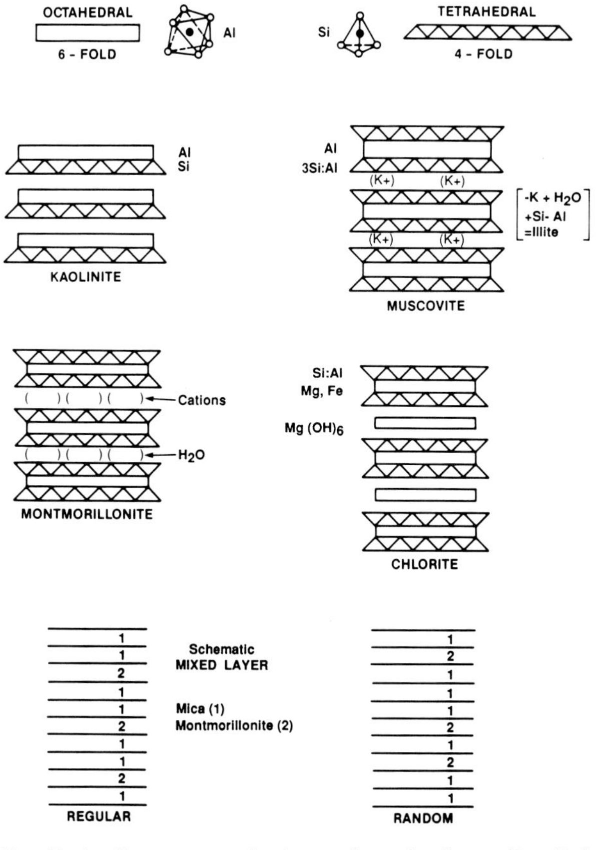

Fig. D-1. Structure of clay minerals from Pettijohn (1975).

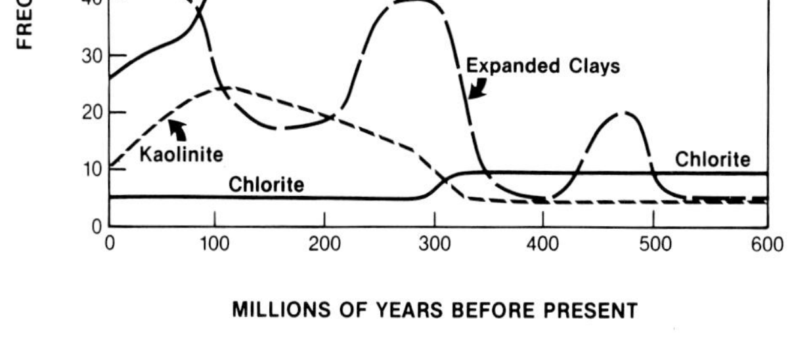

Fig. D-2. Shale clay composition as a function of geologic time from Blatt et al. (1972).

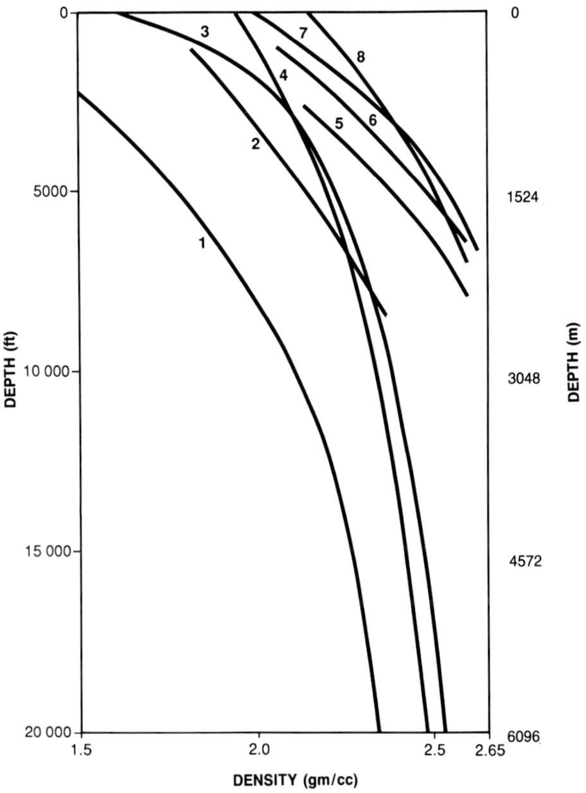

Fig. D-3. Variation of shale bulk densities with depth in sedimentary basins. 1 = methane-saturated clastic sedimentary rock (probable minimum density) (after McCulloh, 1967, p. A19); 2 = mudstone-Po Valley Basin, Italy, (after Storer, 1959); 3 = average Gulf Coast shale densities; values derived from geophysical data (after Dickinson, 1953, p. 427); 4 = average Gulf Coast shale densities derived from density logs and formation samples (after Eaton, 1969); 5 = Motatan-1-Maracaibo Basin, Venezuela (after Dallmus, 1958, p. 916); 6 = Gorgeteb No. 1—Hungary; calculated wet density values (after Skeels, 1943, unpublished material); 7 = Pennsylvanian and Permian dry shale density values, Oklahoma and Texas; Athy's adjusted curve (after Dallmus, 1958, p. 913); 8 = Las Ollas-1-eastern Venezuela (after Dallmus, 1958, p. 918). From Rieke and Chillingarian (1974).

havior is associated with the montmorillonite-illite transition which occurs at about 10 000 ft. The change in velocity-density behavior may be due directly to this change in mineralogy and/or may be associated with geopressuring.

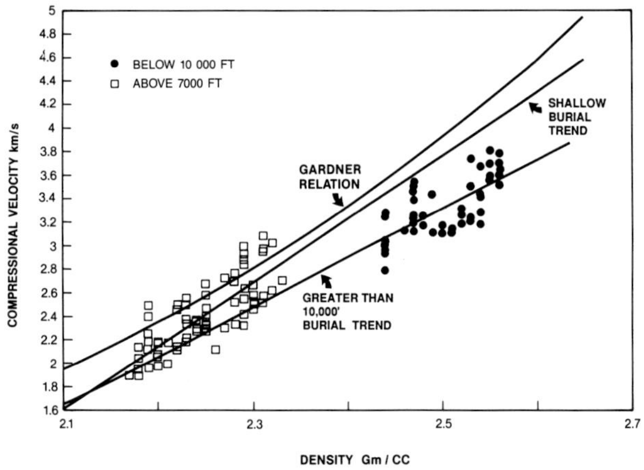

Fig. D-4. Velocity-density crossplot for shales from DOE Pleasant Bayou #1 and #2 wells in Brazoria, Co., Texas.

APPENDIX E—ROUGH APPROXIMATION RELATING GAS SAND AND BRINE SAND VELOCITIES

By assuming equality of sandstone frame bulk and shear moduli, Gassmann's equations may be used to compute gas sand compressional velocity as a function of fully saturated water sand velocity if the skeleton properties are assumed to be unchanged by the pore fluid. If the modified Gardner equation is used to estimate porosity, skeleton bulk modulus and shear modulus are assumed to be equal (Castagna, et al. 1985), and water saturation is taken to be 50 percent, gas sand compressional velocity (V_{pgas}) as given by (see Figure E-1):

$$V_{\text{pgas}} = .07V_p^2 + 1.67V_p - 1.74$$

where V_p = brine-saturated compressional velocity.

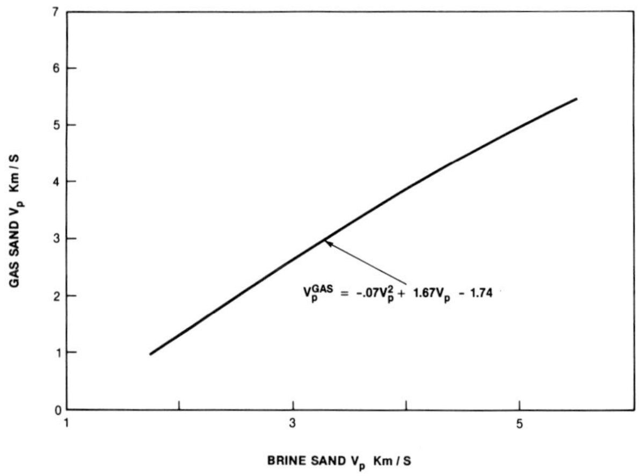

Fig. E-1. Rough relationship between brine-sand and gas-sand velocity using assumptions described in text.

REFERENCES

Aleksandrov, K. S., and Ryzhova, T. V., 1961, The elastic properties of crystals: Sov. Phys. Crystallgr., **6**, 228.

Anderson, O. L., and Lieberman, R. C., 1966, Sound velocities in rocks and minerals: VESIAC state-of-the-art report no. 7885-4-X Univ. of Michigan.

Anderson, R. G., and Castagna, J. P., 1984, Analysis of sonic log compressional wave amplitudes using borehole compensation techniques: Soc. Prof. Well Log Analysts 25th Annual Symposium, Transactions, paper K.

Backus, G. E., 1962, Long-wave elastic anisotropy produced by horizontal layering: J. Geophys. Res., **67**, 4427–4440.

Batzle, M. L., and Wang, Z. J., 1992, Seismic properties of pore fluids: Geophysics, **57**, 1396–1408.

Billings, M. P., 1972, Structural geology: Prentice-Hall.

Biot, M. A., 1956, The theory of propagation of elastic waves in a fluid-saturated solid, I lower frequency range, II higher frequency range: J. Acoust. Soc. Am., **28**, 179–191.

Biot, M. A., 1962, Mechanics of deformation and acoustic propagation in porous media: J. Appl. Phys., **23**, 1482–1498.

Birch, F., 1966, Compressibility; elastic constants, Clark, S. P., Jr., Ed., *in* Handbook of physical constants: Geol. Soc. Am., Memoir, **97**, 97–174.

Blatt, H., Middleton, G. V., and Murray, R. C., 1972, Origin of sedimentary rocks: Prentice-Hall.

Brill, J. P., and Beggs, H. D., 1977, Two phase flow in pipes: University of Tulsa.

Bulau, J. R., Tittman, B. R., Abdel-Gawad, M., and Salvado, D., 1984, The role of aqueous fluids in the internal friction of rock: J. Geophys. Res., **89**, 4207–4212.

Campbell, F. A., and Oliver, T. A., 1968, Mineralogic and chemical composition of Freton and Duvernay Formations Central Alberta: Bull. Canadian Pet. Geol., **16**, 40–63.

Castagna, J. P., Batzle, M. L., and Eastwood, R. L., 1985, Relationships between compressional-wave and shear-wave velocities in clastic silicate rocks: Geophysics, **50**, 571–581.

Castagna, J. P., 1985, Shear-wave time-average equation for sandstones: Presented at the 55th Ann. Internat. Mtg., Soc. Expl. Geophys.

Chen, C. T., Chen, L. S., and Millero, F. J., 1978, Speed of sound in NaCl, $MgCl_2$, Na_2So4, and $MgSO_4$ aqueous solutions as functions of concentration, temperature, and pressure: J. Acoust. Soc. Am., **63**, 1795–1800.

Culberson, O. L., and McKetta, J. J., Jr., 1951, The Solubility of methane in water at pressures to 10,000 PSIA: Petroleum Trans., AIME, **192**, 223–226.

Dallmus, K. F., 1958, Mechanics of basin evolution and its relation to the habitat of oil in the basin, *in* Weeks, L. G., Ed., Habitat of oil: Am. Assoc. Pet. Geologists, Mem. **36**, 2071–2174.

Dandekar, D. P., 1968, Pressure dependence of the elastic constants of calcite: Phys. Rev., **172**, 873.

Davey, F. J., and Cooper, A. K., 1987, Gravity studies of the Victoria Land Basin and Iselin Bank, *in* Cooper, A. K., and Davey, F. J., Eds., the Antarctic Continental Margin and Geophysics of the Western Ross Sea, CPCEMR Earth Science Series, V. J. B., Houston, Texas: Circum-Pacific Council for Energy and Mineral Resources.

DeVilbiss, J., Ito, H., and Nur, A., 1979, Measurement of compressional and shear wave velocities of water filled rocks during water-steam transition: Geophysics, **44**, 407.

Dickinson, G., 1953, Geological aspects of abnormal reservoir pressures in Gulf Coast Louisiana: Am. Assn. Petr. Geol., **37**, 410–432.

Dickey, P. A., 1966, Patterns of chemical composition in deep subsurface waters: Am. Assn. Petr. Geol., **50**, 2472–2478.

Dodson, C. R., and Standing, M. B., 1945, Pressure-volume-temperature and solubility relations for natural-gas-water mixtures, *in* Drilling and production practices, 1944: Am. Pet. Inst., 173–179.

Domenico, S. N., 1976, Effect of brine-gas mixture on velocity in an unconsolidated sand reservoir: Geophysics, **41**, 887–894.

Domenico, S. N., 1977, Elastic properties of unconsolidated porous sand reservoirs: Geophysics, **42**, 1339–1369.

Domenico, S. N., 1984, Rock lithology and porosity determination from shear and compressional wave velocity: Geophysics, **49**, 1188–1195.

Eastwood, R. L., and Castagna, J. P., 1986, Interpretation of V_p/V_s ratios from sonic logs, *in* Danbom, S. H., and Domenico, S. N., Eds., Shear wave exploration, Geophysical Developments No. 1: Soc. of Expl. Geophys.

Eaton, B. A., 1969, Fracture gradient prediction and its application to oil-field operations: J. Pet. Tech., **21**, 1353–1360.

Eberhart-Phillips, D., Han, D. H., and Zoback, M. D., 1989, Empirical relationships among seismic velocity, effective pressure, porosity, and clay content in sandstone: Geophysics, **54**, 82–89.

Engelder, T., and Plumb, R., 1984, The relation between ultrasonic properties and strain relaxation: Int. J. Rock Mech., Min. Sci. and Geomech. Abs., **21**, 74–83.

Fertl, W. H., 1976, Abnormal formation pressures: Elsevier.

Freed, R. L., 1980, Shale mineralogy of the No. 1 Pleasant Bayou Geothermal Test Well, a progress report, *in* Proceedings 4th United States Gulf Coast Geopressured-Geothermal Energy Conference, Research and Development, V.1: Center for Energy Studies, University of Texas—Austin.

Ganley, D. C., Kanasewich, E. R., 1980, Measurement of absorption and dispersion from check shot surveys: J. Geophys. Res., **85**, 5219–5226.

Gardner, G. H. F., Gardner, L. W., and Gregory, A. R., 1974, Formation velocity and density—The diagnostic basis for stratigraphic traps: Geophysics, **39**, 770–780.

Gassman, F., 1951, Elastic waves through a packing of spheres: Geophysics, **16**, 673–685.

Greenhalgh, S. A., and Emerson, D. W., 1981, Physical properties of Permian bituminous coals from the Sydney Basin, New South Wales: Presented at the 51st Ann. Internat. Mtg., Soc. Expl. Geophys.

Gregory, A. R., 1976, Fluid saturation effects on dynamic elastic properties of sedimentary rocks: Geophysics, **41**, 895–921.

Gregory, A. R., 1977, Aspects of rock physics from laboratory and log data that are important to seismic interpretation, *in* Seismic stratigraphy—Application to hydrocarbon exploration: Memoir 26, Am. Assn. Petr. Geol.

Gretener, P. E., 1979, Pore pressure: fundamentals, general ramifications, and implications for structural geology (revised): Education Course Notes Series No. 4, Am. Assn. Petr. Geol.

Gretener, P. E., 1981, Geothermics—using temperature in hydrocarbon exploration: Education Course Notes Series No. 17, Am. Assn. Petr. Geol.

Hadley, K., 1976, Comparison of calculated and observed crack densities and seismic velocities in Westerly Granite: J. Geophys. Res., **81**, 3484–3494.

Hamilton, E. L., 1972, Compressional-wave attenuation in marine sediments: Geophysics, **37**, 620–646.

Hamilton, E. L., 1976, Variations of density and porosity with depth in deep-sea sediments: Jour. Sed. Pet., **46**, 280–300.

Han, De-hua, Nur, A., and Morgan, D., 1986, Effects of porosity and clay content on wave velocities in sandstones: Geophysics, **51**, 2093–2107.

Hauge, P. S., 1981, Measurement of attenuation from vertical seismic profiles: Geophysics, **46**, 1548–1558.

Hedburg, H. D., 1936, Gravitational compaction of clays and shales: Am. J. Sci., **31**, 241–281.

Hegelson, H. C., and Kirkham, D. H., 1974, Theoretical prediction of the thermodynamic behavior of aqueous electrolytes: Am. J. Sci., **274**, 1089–1198.

Hill, R., 1952, The elastic behavior of a crystalline aggregate: Proc. Phys. Soc. London, Ser. A, **65**, 349–359.

Hinz, K., Dostmann, H., and Fritsch, J., 1982, The continental margin of Morocco: Seismic sequences, structural elements, and geological development, *in* von Rad, U., Jinz, K., Sarrthein, J. M., and Seibold, E., Eds., Geology of the northwest African continental margin: Springer-Verlag.

Hurlbut, C. S., 1971, Dana's manual of mineralogy: Wiley.

Jachens, R. C., and Griscom, A., 1985, An isostatic residual gravity map of California—a residual map for interpretation of anomalies from intracrustal sources, *in* Hinze, W. J., Ed., The utility of regional gravity and magnetic anomaly maps: Soc. of Expl. Geophys.

Johnson, G. R., and Olhoeft, G. R., 1984, Density of rocks and minerals, *in* Carmichael, R. S., Ed., CRC handbook of physical properties of rocks, Volume II: CRC Press, 2–37.

Johnston, D. H., 1986, Shale properties at temperature and pressure: Oil and Gas J., 60–55.

Jones, T. D., 1986, Pore fluids and frequency-dependent wave propagation in rocks: Geophysics, **51**, 1939–1953.

Jones, T. D., and Nur, A., 1983, Velocity and attenuation in sandstone at elevated temperatures and pressures: Geophys. Res. Lett., **10**, 140–143.

Judd, W. R., and Shakoor, A., 1981, Density, *in* Touloukian, Y. S., and Ho, C. Y., Eds., Physical properties of rocks and minerals: McGraw Hill Data Series on Material Properties, Volume II-2: McGraw-Hill.

Kan, T. K., Batzle, M. L., and Gaiser, J. E., 1983, Attenuation measured from VSP, evidence of frequency dependent *Q*: presented at 53rd Ann. Internat. Mtg., Soc. Expl. Geophys.

Keenan, J. H., Keyes, F. G., Hill, P. G., and Moore, J. G., 1969, Steam tables: John Wiley and Sons.

Keller, G. R., Smith, R. A., Hinze, W. J., and Aikens, C. L. V., 1985, Regional gravity and magnetic study of West Texas, *in* Hinze, W. J., Ed., The utility of regional gravity and magnetic anomaly maps: Soc. of Expl. Geophys.

Kowallis, B., Jones, L. E. A., and Wang, H. F., 1984, Velocity-porosity-clay content systematics of poorly consolidated sandstones: J. Geophys. Res., **89**, 10355–10364.

Krynine, P. D., 1948, The megascopic study of field classification of sedimentary rocks: Jour. Geol., **56**, 130–165.

Kuster, G. T., and Töksoz, N. M., 1974, Velocity and attenuation of seismic waves in two-phase media, I. Theoretical formulations: Geophysics, **39**, 607–618.

Matthews, J. C., and Matthews, W. R., 1985, Program calculates frac gradients for many basins: Oil and Gas Journal.

Mavko, G. M., and Nur, A., 1977, The effect of nonelliptical cracks on the compressibility of rocks: J. Geoph. Res., **84**, 4769–4776.

McCulloh, T. H., 1967, Mass properties of sedimentary rocks and gravimetric effects of petroleum and natural gas reservoirs: U.S. Geol. Surv., Prof. Pap., 528 A.

McDonal, F. J., Angona, F. A., Mills, R. L., Sengbush, R. L., Van Nostrand, R. G., and White, J. E., 1958, Attenuation of shear and compressional waves in Pierre Shale: Geophysics, **23**, 421–439.

McSkimin, H. J., Andreatch, P., Jr., and Thurston, R. N., 1965, Elastic moduli of quartz vs. hydrostatic pressure at 25° and −195.8°C: J. Appl. Phys., **36**, 1632.

Milholland, P., Manghnani, M. H., Schlanger, S. O., and Sutton, G. H., 1980, Geoacoustic modeling of deep-sea carbonate sediments: J. Acoust. Soc. Am., **68**, 1351–1360.

Millero, F. J., Ward, G. K., and Chetirkin, P. V., 1977, Relative sound velocities of sea salts at 25°C: J. Acoust. Soc. Am., **61**, 1492.

Murphy, W. F., III, 1982, Effects of microstructure and pore fluids on acoustic properties of granular sedimentary materials: Ph.D. thesis, Stanford Univ.

Nur, A., and Simmons, G., 1969, The effect of viscosity of a fluid phase on velocity in low porosity rocks: Earth and Planetary Sci. Lett., **7**, 99–108.

O'Connell, R. J., and Budiansky, B., 1974, Seismic velocities in dry and saturated cracked solids: J. Geophy. Res., **79**, 5412–5426.

Pettijohn, F. S., 1975, Sedimentary Rocks: Harper and Row.

Pickett, G. R., 1963, Acoustic character logs and their applications in formation evaluation: J. Petr. Tech., **15**, 650–667.

Potter, R. W., and Brown, D. L., 1977, The volumetric properties of sodium chloride solutions from 0 to 500°C at pressures up to 2000 bars based on a repression of available data in the literature: U.S. Geol. Surv. Bull. 1421-C.

Rafavich, F., Kendal, C. H., St. C., and Todd, T. P., 1984, The relationship between acoustic properties and the petrographic character of carbonate rocks: Geophysics, **49**, 1622–1636.

Rao, K. S., and Rao, B. R., 1959, Study of temperature variation of ultrasonic velocities in some organic liquids by modified fixed-path interferometer method: J. Acoust. Soc. Am., **31**, 439–441.

Raymer, L. L., Hunt, E. R., Gardner, J. S., 1980, An improved sonic transit-time-porosity transform: Soc. Prof. Well Log Analysts, 21st Annual Logging Symposium Transactions, Paper P.

Rieke, H. H., and Chilingarian, G. V., 1974, Compaction of argillaceous sediments: Elsevier.

Rowe, A. M., and Chou, J. C. S., 1970, Pressure-volume-temperature-concentration relation of aqueous NaCl solutions: J. Chem. Eng. Data, **15**, 61–66.

Russell, W. L., 1951, Principles of petroleum geology: McGraw-Hill.

Schlumberger Inc., 1986, Log interpretation charts: Schlumberger, Inc.

Sergeev, L. A., 1948, Ultrasonic velocities in methane saturated oils and water for estimating sound reflectivity of an oil layer: 4th All-Union Acoust. Conf. Izd. Nauk SSSR.

Shaw, D. B., and Weaver, C. E., 1965, The mineralogical composition of shales: J. Sed. Petrology, **35**, 213–222.

Simmons, G., 1965, Ultrasonics in geology: Proc. IEEE, **53**, 1337–1345.

Simmons, G., and Brace, W. F., 1965, Comparison of static and dynamic measurements of compressibility of rocks: J. Geophys. Res., **70**, 5649–5656.

Spencer, J. W., 1981, Stress relaxation at low frequencies in fluid saturated rocks: attenuation and modulus dispersion: J. Geophys. Res., **86B-3**, 1803–1812.

Standing, M. B., 1962, Oil-system correlations, *in* Frick, T. C., Ed., Petroleum production handbook Volume II, Part 19: McGraw-Hill.

Stoll, R. D., and Houtz, R. E., 1983, Attenuation measurements from sonoboys: J. Acoust. Soc. Am., **73**, 163–172.

Storer, D., 1959, Compaction of the argillaceous sediments in the Padano Basin, *in* The gasiferous deposits of Western Europe, 2: Acad. Nazl. dei Lincei, Roma, 519–536.

Sultanov, R. G., Skripka, V. G., and Namiot, A. Y., 1972, Solubility of methane in water at high temperatures and pressures: Gazovaia Promyshlenmost, 17 (May), 6–7.

Terzaghi, K., and Peck, R. B., 1948, Soil mechanics in engineering practice: Wiley.

Thomas, L. K., Hankinson, R. W., and Phillips, K. A., 1970, Determination of acoustic velocities for natural gas: J. Pet. Tech., **22**, 889–892.

Thomsen, L., 1986, Weak elastic anisotropy: Geophysics, **51**, 1954–66.

Timur, A., 1977, Temperature dependence of compressional and shear wave velocities in rocks: Geophysics, **42**, 950–956.

Töksoz, M. N., and Johnston, D. H., 1981, Seismic wave attenuation: Soc. Expl. Geophys.

Tosaya, C. A., 1982, Acoustical properties of clay-bearing rocks: Ph.D. Thesis, Stanford Univ.

Tullos, F. N., and Reid, A. C., 1969, Seismic attenuation of Gulf Coast sediments: Geophysics, **34**, 516–528.

Volarovich, M. P., Bayuk, E. I., and D'Yaur, J. J., 1985, Change in the velocity of longitudinal waves during deformation of samples: Izvestiya, Earth Physics, **21**, 302–304.

Wang, Z. J., and Nur, A., 1986, The effect of temperature on the seismic wave velocities in rocks saturated with hydrocarbons: SPE paper 15646, Proc. 61 Mtg. Soc. Pet. Eng.

Wang, Z. J., Nur, A. M., and Batzle, M. L., 1988, Acoustic velocities in petroleum oils: SPE paper 18163, Proc. 63 Mtg. Soc. Pet. Eng., 571–584.

Watt, J. P., Davies, G. F., and O'Connell, R. J., 1976, The elastic properties of composite materials: Rev. Geophys. and Space Physics, **14**, 541–563.

White, J. E., 1983, Underground sound: Application of seismic waves: Elsevier.

Wilkins, R., Simmons, G., and Caruso, L., 1984, The ratio V_p/V_s as a discriminant of composition for siliceous limestones: Geophysics, **49**, 1850–1860.

Wilson, W. D., 1959, Speed of sound in distilled water as a function of temperature and pressure: J. Acoust. Soc. Am., **31**, 1067–1072.

Winkler, K. W., 1986, Estimates of velocity dispersion between seismic and ultrasonic frequencies: Geophysics, **51**, 183–189.

Wood, A. B., 1941, A textbook of sound: MacMillan.

Wyllie, M. R. J., Gregory, A. R., and Gardner, L. W., 1956, Elastic wave velocities in heterogeneous and porous media: Geophysics, **21**, 41–70.

Zarembo, V. I., and Fedorov, M. K., 1975, Density of sodium chloride solutions in the temperature range 25-350°C at pressures up to 1000 kg/cm^2: J. Appl. Chem. USSR, **48**, 1949–1953, (English translation).

Chapter III

Processing

Seismic Data Processing for AVO and AVA Analysis

*Jeffrey R. Resnick**

INTRODUCTION

Clues to the lithology and fluid content of an exploration target may be revealed by analyzing variations of seismic reflectivity with source-receiver offset or reflection angle. Field recordings of seismic data, however, do not directly indicate target reflectivity, because numerous other factors (for example, geometric spreading, variations in shot strength or receiver coupling, multiple reflections, ground roll, transmission loss, dissipation, and noise) affect seismic amplitude (Ostrander, 1984; Dey-Sarkar, et al., 1986).

The contaminating influence of many of these factors can be ameliorated through data processing. Regrettably, the benefits of data processing cannot be realized without some risks: removing noise or other unwanted influences from seismic data may also distort or damage the underlying signal. Therefore, when data are processed for amplitude variation with offset (AVO) or amplitude variation with angle (AVA) analysis, the interpreter must know how the processing has affected the final results. To minimize processing artifacts and aid the interpreter's understanding of the processing strategy, displays of the data at the completion of processing and at several intermediate stages of the data processing flow are especially helpful.

This paper describes the processing flow that prepared one specific East Texas seismic line for AVO and AVA analysis. My intent is not to prove that a particular combination of parameters or algorithms is optimal; indeed, it is likely that many other processing schemes could have been effective. Instead, primarily with the interpreter in mind, my purpose is to provide a step-by-step case history that illustrates how the data evolved as AVO/AVA processing progressed.

The methods and results presented draw heavily on a number of references that have appeared in the AVO literature (Ostrander, 1984; Yu, 1985; Todd, 1986; Dey-Sarkar, et al., 1986; Chiburis, 1987). A common theme in all these papers is that the objective of processing is to reveal the "relative" amplitude, though not necessarily the "true" absolute amplitude, of primary reflections. To quote Ostrander, "Because one has great difficulty in separating out 'true' reflection amplitudes, the interpreter typically must rely on relative changes, concentrating on anomalous behavior of the amplitude." A similar philosophy is adopted here. The processing removes many, but not all, undesired influences; consequently, though true amplitude is not recovered, amplitude anomalies are effectively highlighted.

This is a case history of data collected on land. Although data collected at sea are often of higher quality, AVO analysis is not restricted to marine data sets. While some land surveys suffer from irregular coverage and missing offsets so extensively that AVO analysis is impossible, others can be good candidates for such work. Land and marine processing differ in some important respects; for example, marine processing need not remove ground roll, but might require attenuation of water-bottom multiples. Land and marine processing also have much in common, and many of the processes described are applicable in either environment. For this reason, both land and marine interpreters will find this case history of interest.

DATA PROCESSING

The data for this study consist of a land line shot with buried dynamite charges and with 120 receivers arranged in a split-spread configuration. The acquisition parameters for these data were: shot-station spacing, 150 m; receiver-station spacing, 50 m; near offset, 100 m; and far offset, 3050 m. The final stack of the line (Figure 1) reveals the target zone, which is a bright reflector located in a sand-shale environment at a two-way traveltime of about 2.0 s (Figures 1 and 2). The line was processed as part of an exploration program, and at the time of the original processing, no wells penetrated the target zone. Since that time, two

*Acuson, 1220 Charleston Rd., P. O. Box 7393, Mountain View, CA 94039-7393. Formerly Western Geophysical, Houston, TX.

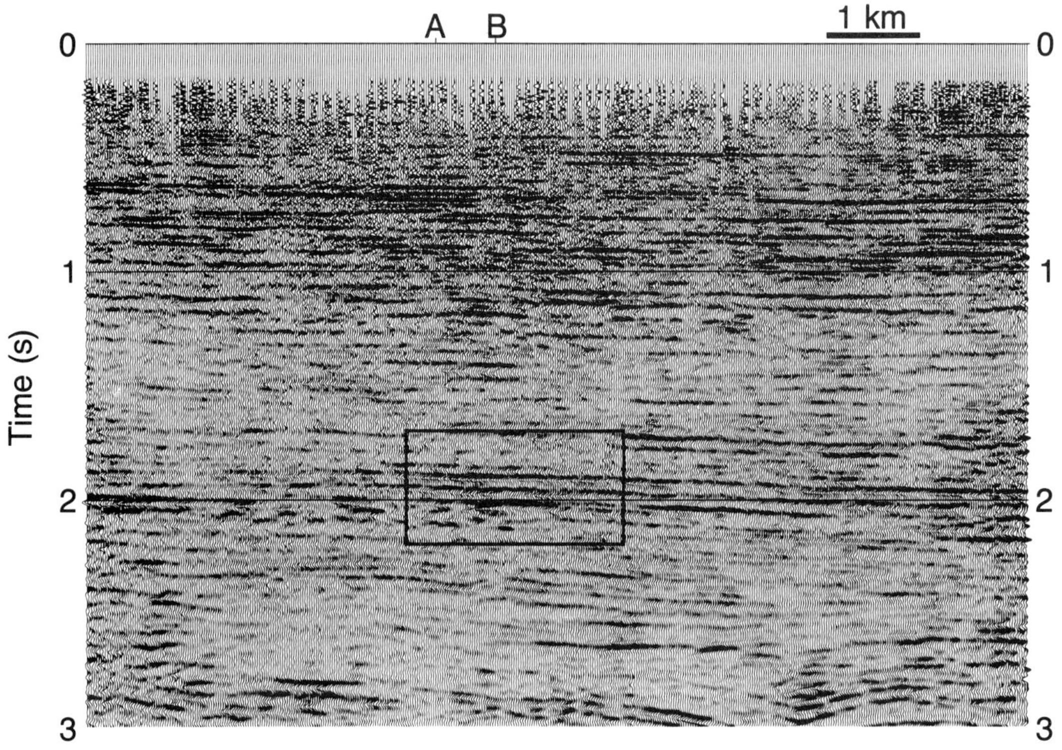

Fig. 1. Final stack. Rectangle encloses the zone of interest.

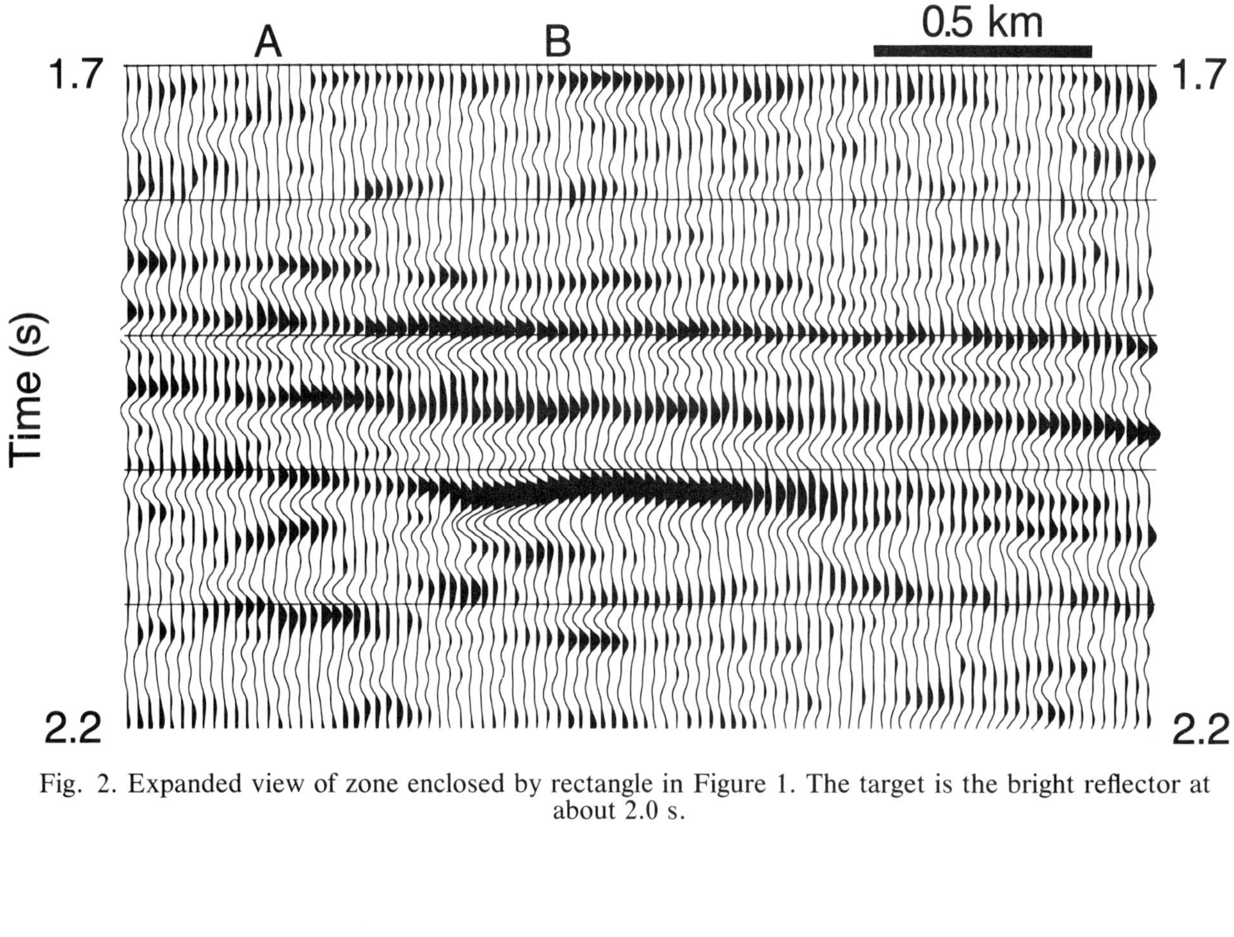

Fig. 2. Expanded view of zone enclosed by rectangle in Figure 1. The target is the bright reflector at about 2.0 s.

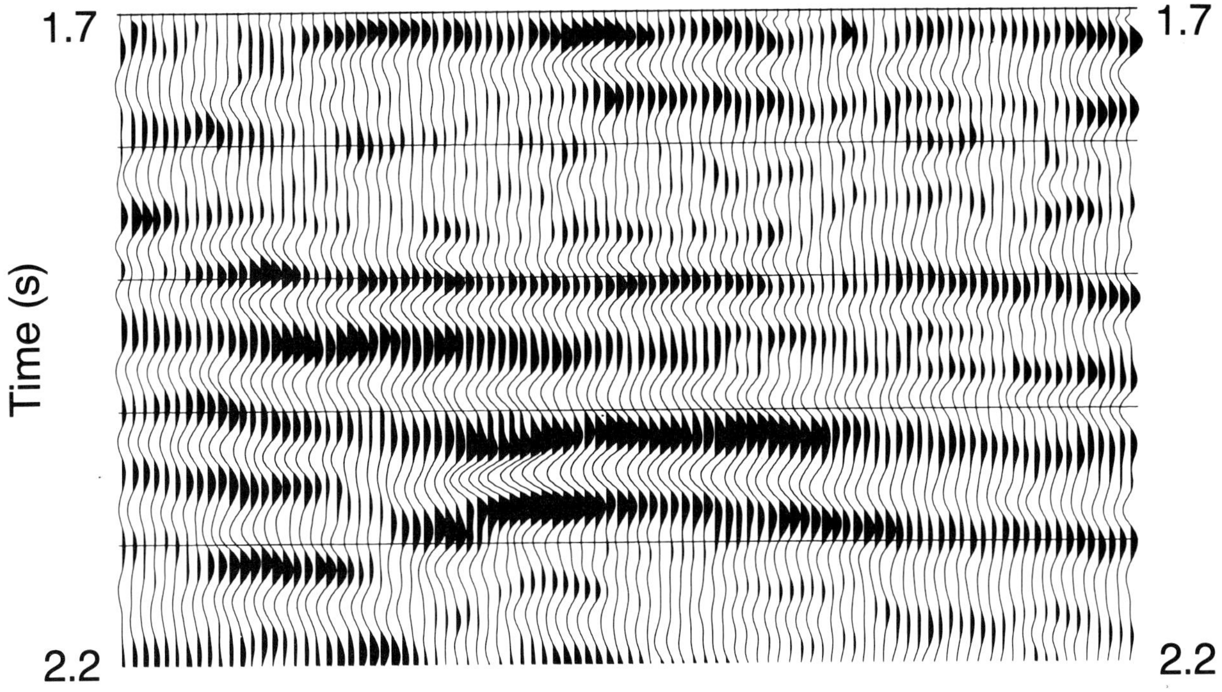

Fig. 7. Stack after *f-k* filtering the shot gathers. Continuity of the reflections is much improved; compare with Figure 5.

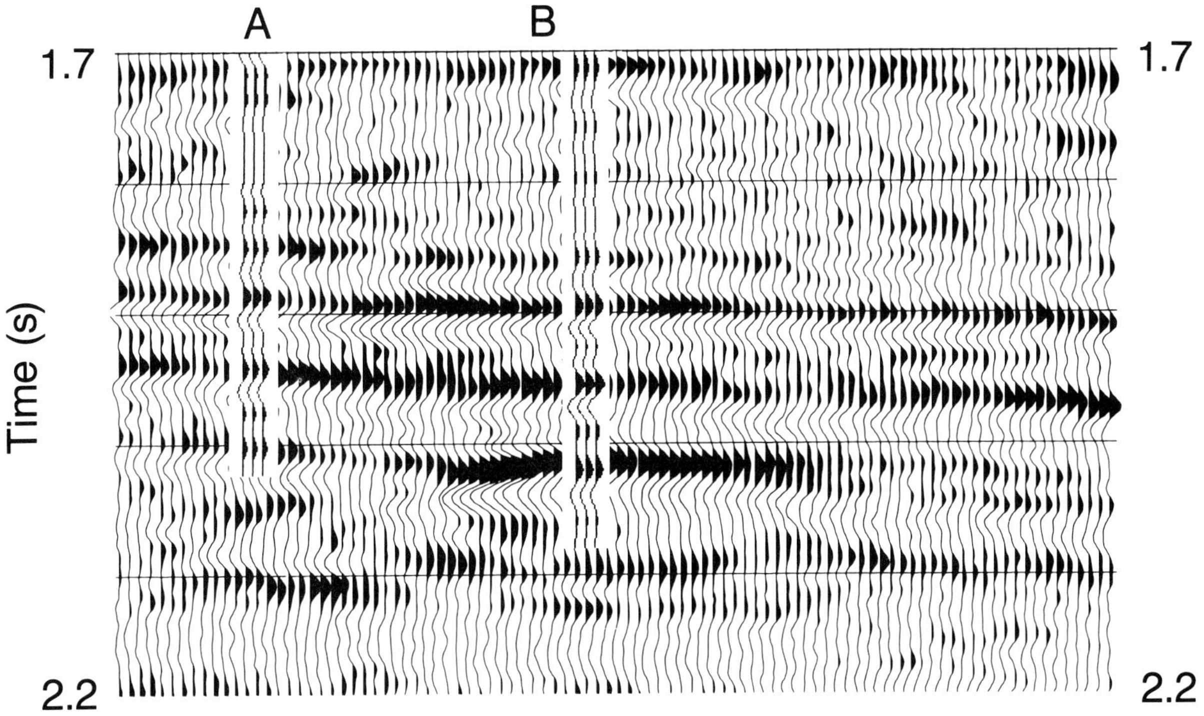

Fig. 8. Stack following surface-consistent deconvolution, with overlay of zero-phase synthetic traces at the well locations. The deconvolution compresses and shifts the phase of the effective wavelet in the data, which makes it possible to tie with the well over a broad bandwidth.

uniformly between the minimum of 100 m and the maximum of 3050 m, except that, because of the split-spread geometry, some adjacent traces have identical offsets; these "reciprocal" traces are indicated by the small dark rectangles beneath the bars. Differences in amplitude on reciprocal traces are due to random noise or variability of source waveform, shot strength, or receiver coupling.

The bar graphs in Figure 9 permit a comparison between changes of amplitude with offset in the different windows. Because so many undesired effects remain in the data at this early stage of processing, it is difficult to distinguish between noise and definite trends in AVO. Perhaps a tendency to increase with offset is more evident in the lower window at B than elsewhere, but it is impossible to be certain, because large spikes in the bar plots are found on a number of traces, including in some cases one member of a pair of reciprocal traces.

How do the CMP gathers at A and B change with each stage of processing? The *f-k* filter attenuates linear noise trains in the shot gathers, and reduces the amplitude of spikes in the bar plots (Figure 10). Surface-consistent deconvolution compresses the wavelet and further reduces the scatter in the heights of the bars (Figure 11); notice in particular the greater consistency of amplitudes on reciprocal traces. (The windows in Figure 11 are 10 ms earlier than those in Figure 10 to account for phase shift induced by the deconvolution process.) Application of static corrections to these data has relatively little effect on the bar graphs. Bandpass filtering yields some additional noise rejection, primarily by reducing the amplitude of high-frequency noise that was boosted by the whitening action of deconvolution. Finally, residual amplitude compensation affects the pattern of AVO on this line only marginally (Figure 12), indicating the absence of any strong average trend in AVO across the line at this stage in the processing.

A comparison of the data at the beginning of the processing flow (Figure 9) and near the end (Figure 12) shows the overall effect of the processing on the data. Noise is attenuated, amplitude differences between reciprocal traces are diminished, and the contrast between the AVO behavior in the target window and in the other windows is enhanced. After processing (Figure 12), the reflections in the lower window at B exhibit a trend of amplitude increase with offset out to about 2000 m, while reflection amplitude in the other windows tends to remain the same or to decay across the entire range of offsets.

ANGLE-GATHERS

Seismic reflectivity at an interface is a function of reflection *angle*. A seismic trace, however, is recorded at a fixed *offset*, and reflection angle changes with time. The final stage of data processing on this line was to convert from offset to reflection angle through the construction of angle-traces.

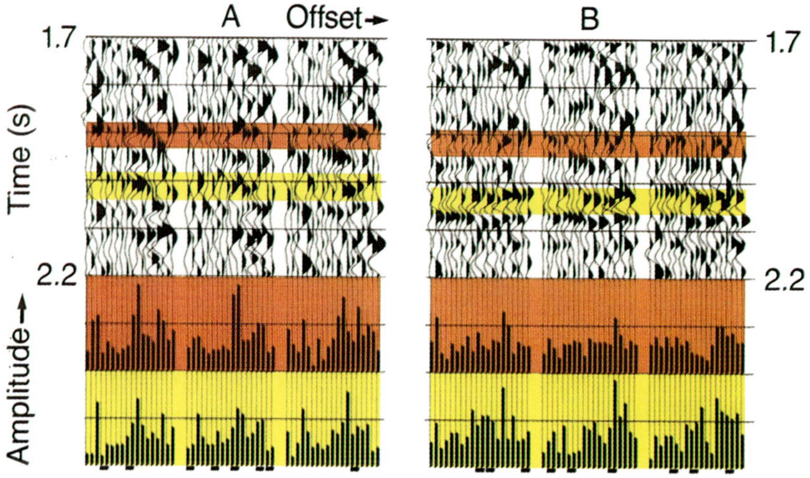

Fig. 9. Moveout-corrected CMP gathers at locations A and B (Figure 2) after application of a geometric spreading compensation. The yellow and orange bars indicate rms amplitude within the yellow and orange time-windows, respectively. Offset varies from 100 m to 3050 m; small rectangles beneath the lower bars indicate "reciprocal" traces, which have identical offsets. Noise at this early stage obscures AVO trends in the data.

In simply layered geology, the pattern of amplitude variation at a single location is relatively insensitive to whether offset or angle is the independent variable, though each does have some benefits that the other lacks. For example, AVA is preferable to AVO analysis when comparing a deeper target with a shallower one, because the different angle ranges of the two targets can produce misleading results if angle is not explicitly taken into account. On the other hand, AVA analysis requires information about velocities and raypaths which is not needed for AVO analysis. In practice, both offset and angle displays may be helpful to the interpreter because each offers a different perspective on the data.

To construct an angle-trace, it is necessary first to estimate reflection angle as a function of time and offset. In this paper, the sine of the reflection angle is given by the expression

$$\left(\frac{v_{\text{int}}}{v_{\text{smooth}}}\right)\left\{\frac{x}{[x^2 + (v_{\text{smooth}}t_0)^2]^{1/2}}\right\}, \tag{1}$$

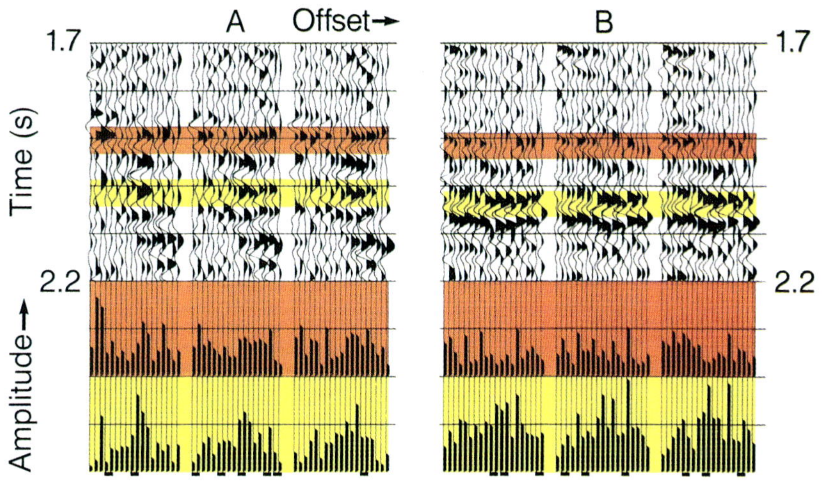

Fig. 10. CMP gathers at A and B after *f-k* filtering. The bars show less variability in amplitude than in Figure 9, indicating a reduction in noise level.

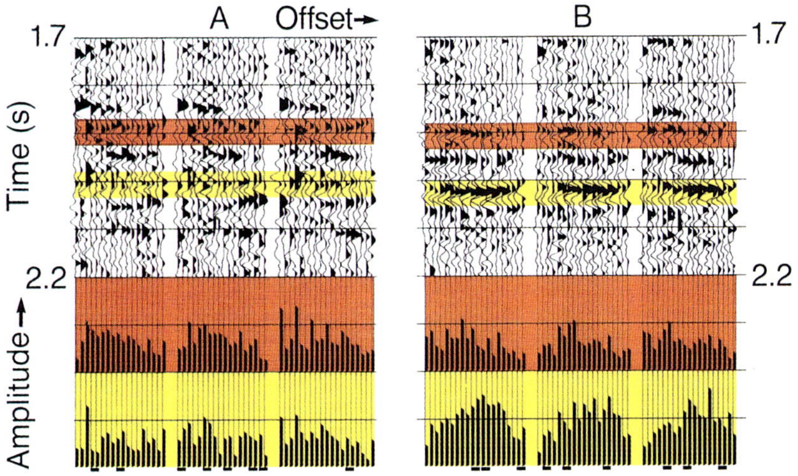

Fig. 11. CMP gathers at A and B following surface-consistent deconvolution and scaling. Surface-consistent compression of the wavelet improves the contrast between the AVO behavior in the target zone (yellow window at B) and that in the other windows, and decreases amplitude variability on reciprocal traces.

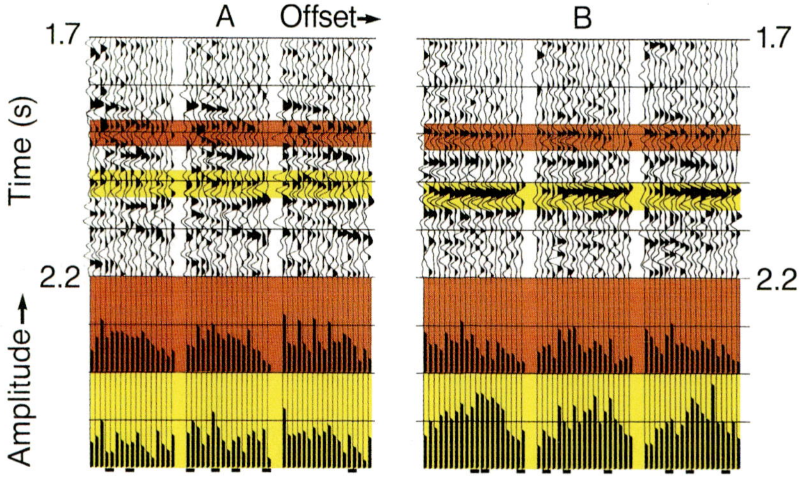

Fig. 12. CMP gathers at A and B following static corrections, bandpass filtering, and residual amplitude compensation. Compared with the data in Figure 11, some additional improvement in signal-to-noise ratio is evident.

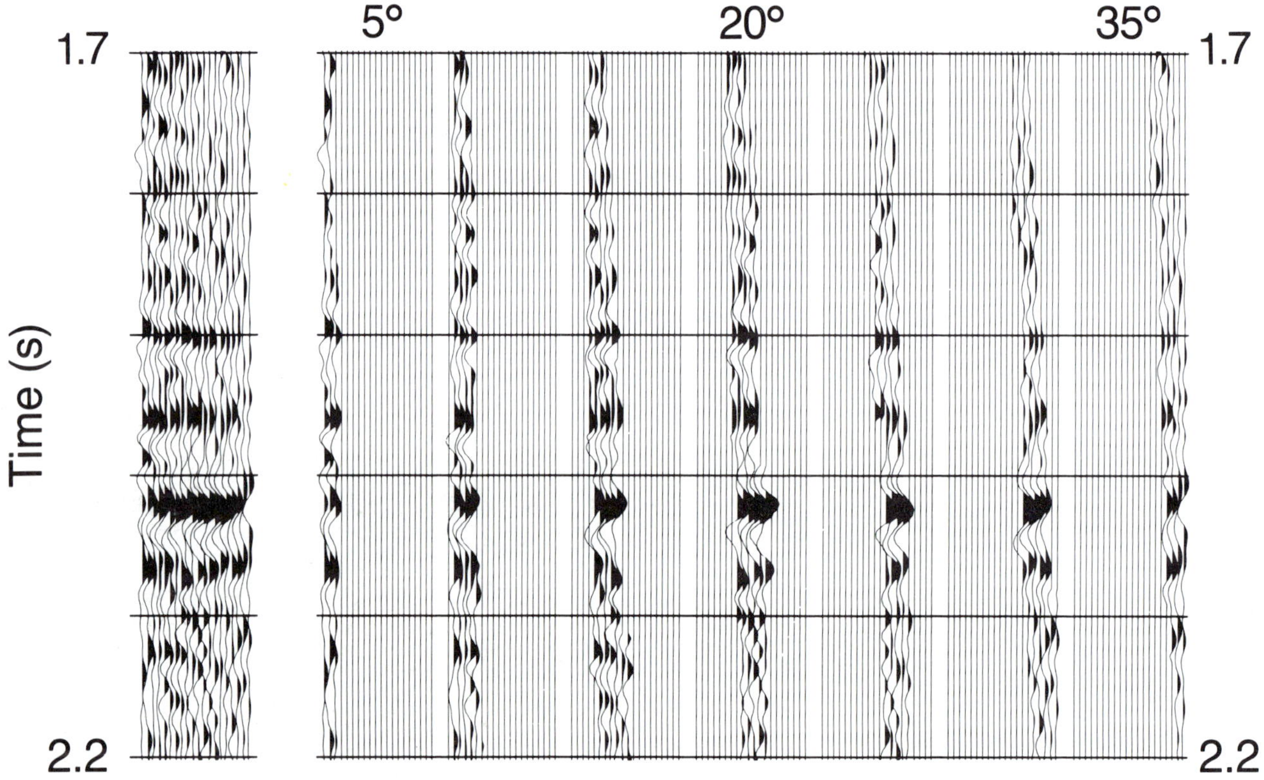

Fig. 13. Method of converting from offset to angle. Left panel is a single CMP gather. Other panels show angle-mute patterns. Each angle-mute pattern spans a range of plus and minus 4 degrees. The central angle in successive panels increases in 5-degree increments from 5 to 35 degrees. Stacking each panel yields an angle-trace.

where x is offset, v_{smooth} is a spatially varying velocity function derived by smoothing the stacking velocities over a cable length, v_{int} is the interval velocity obtained from v_{smooth} using the Dix equation, and t_0 is zero-offset two-way traveltime. The second factor in this expression is the sine of the angle in the absence of refraction. The first factor is an approximate correction for raybending. For these data, the raybending correction at 2.0 s amounts to only about 5 degrees at the far offset. Though this correction is included here for completeness, for a qualitative analysis of the data, such a correction could have been safely ignored.

The method of assembling angle-traces is illustrated in Figure 13. The left panel of Figure 13 consists of a single CMP gather, while the other panels to the right show the same gather after the application of different "angle-mute patterns." Each angle-mute pattern retains energy reflected with a particular central angle, plus or minus 4 degrees; the seven mute patterns displayed in Figure 13 differ in their central angle, which increases in 5 degree increments from 5 to 35 degrees. Stacking together the traces within a panel yields an angle-trace. Thus, a "5-degree trace" consists of a partial stack of reflections that range in angle from 1 to 9 degrees.

Notice in Figure 13 that the fold of each angle-mute pattern varies in time. Many of the angle-mute patterns in Figure 13 have a minimum of 4 "live" traces, yet portions of both the 5-degree and 35-degree angle-mute patterns exhibit a minimum fold of 2 traces. By examining angle-mute patterns at several CMP positions, a range of angles with a consistent fold across the line can be chosen, and this range facilitates comparisons of the AVA behavior at different locations. For these data angle-traces were constructed increasing in central angle from 7 degrees to 33 degrees in 2-degree increments.

Angle-gathers at locations A and B (Figure 14) are much less noisy than the data prior to conversion to angle (Figure 12), because each angle-trace is itself a partial stack over all the offsets in the corresponding angle-mute pattern. In addition, each angle-gather in Figure 14 is obtained from a running average of data from 3 adjacent CMP locations. Despite the smoother appearance, a strong contrast remains between the AVA behavior of the data in the lower window at B and that in the other windows.

OTHER AVA DISPLAYS

Figures 9, 10, 11, 12, and 14 illustrate how six CMP gathers evolved through the processing flow, but do not show the results of processing at other locations. To summarize the AVO or AVA behavior across the entire line, other displays are needed. In this case, the data were organized into a cube with axes of time, midpoint, and angle, and were displayed on a workstation. Different portions of the data were then easily viewed by slicing the cube in different directions. For example, a slice through the cube at 9 degrees (Figure 15) yields a display analogous to a CMP stack, but showing only near angle information. The bar plots in Figure 14 suggest the target reflector is strongest in amplitude between 21 and 25 degrees. Comparing the

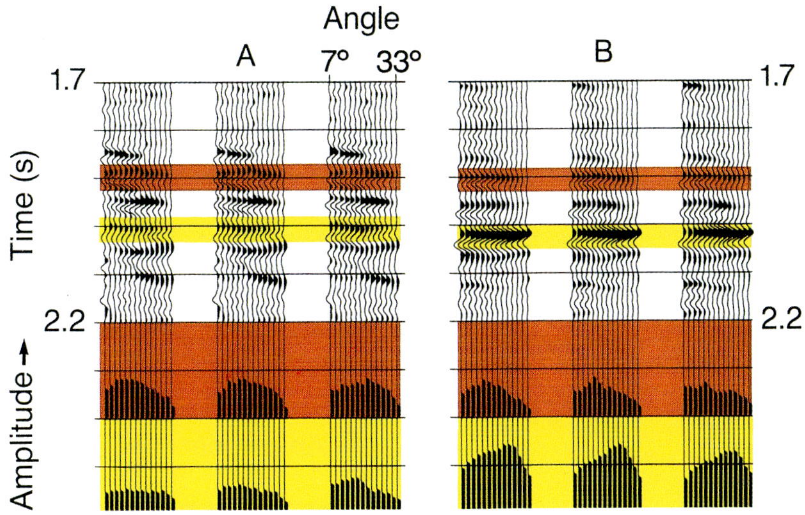

Fig. 14. CMP angle-gathers at A and B, with angle ranging from 7 to 33 degrees. Because of the partial stacking and mixing used to create the angle-traces, the data are much smoother than in Figure 12, yet strong contrast remains between the AVO behavior in the lower window at B and in the other windows.

9-degree stack to a 23-degree stack (Figure 16) reveals the location of the AVA anomaly on both the peak and trough of the bright reflector.

Comparing Figures 15 and 16, other zones where amplitude increases with angle of incidence also are evident, including the reflection peak at 1.95 s at location A. Is it possible that this event, or even the stronger reflection at B, is a processing artifact? By looking only at Figures 15 and 16, it is very difficult to answer such a question. The evolution of both events, however, can be traced through the entire processing flow by reviewing the earlier figures. (The reflection at 1.95 s, though not monitored by bar plots, appears in the figures at A between the upper and lower windows.) With the hindsight of the final results, the precursors of both anomalies can be found in the data at the earliest stage (Figure 9); each successive processing step improves the visibility of these reflections, but none creates an anomaly that was absent before. Thus, the final and intermediate displays *together* indicate that the AVA increase seen in these reflections is real.

RESULTS FROM DRILLING

What is the meaning of the amplitude variations observed at A and B? After these data were processed, a well was drilled on the peak of the anticlinal feature 100 m to the right of location B. Though the interpreter was intrigued by the presence of the AVO anomaly, many other factors, both geologic and economic, affected the decision to drill. The well was completed as a gas well; tests showed an initial flow rate of 572 MCF from a sand at the level of the target reflector. A second well drilled updip at location A encountered gas shows in sands highly interbedded with shales; the target was judged too tight to be commercial, and the well was plugged. Subsequently, the well at B was also shut down when overpressure in the formation resulted in some unexpected problems in production, and the flow rate decayed. Since that time, other operators have successfully drilled a number of producing gas wells on AVO anomalies in this area.

In this example, information from nearby wells or modeling was unavailable at the time the line was

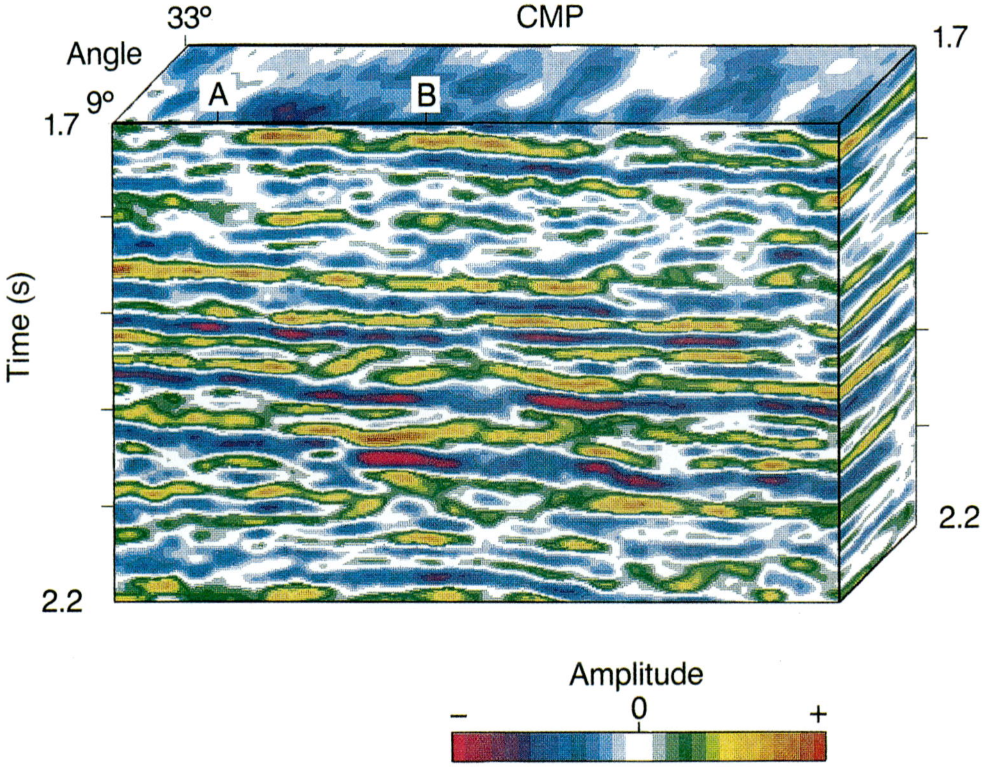

Fig. 15. Data cube. Axes are time, midpoint, and angle, and color indicates amplitude. The cube is sliced open to reveal the data at 9 degrees.

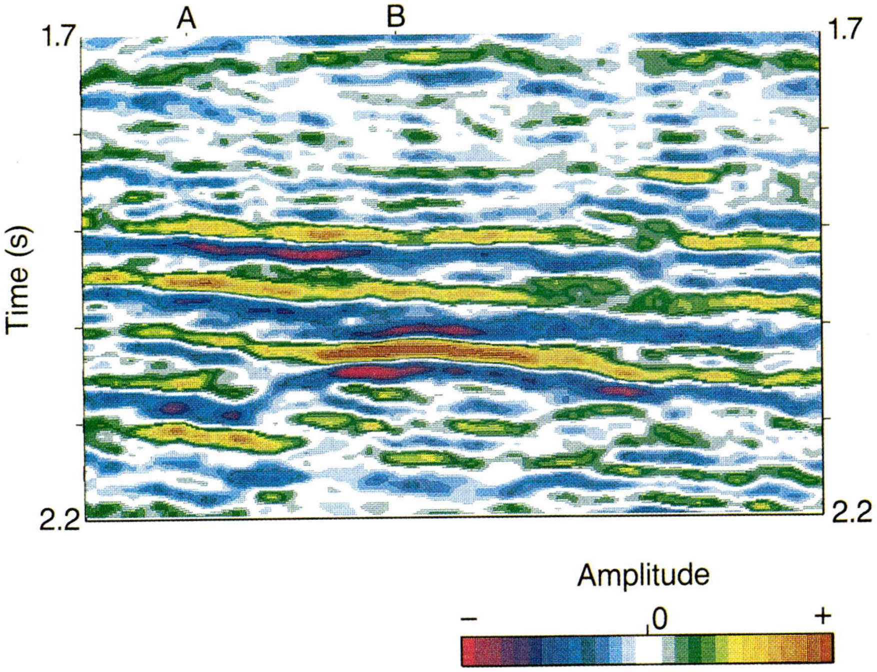

Fig. 16. A slice through the data cube at 23 degrees. The AVA anomaly on both the peak and the trough of the target reflector is clearly delineated.

interpreted. In the absence of such information, it was conjectured that gas would be expressed seismically as a bright spot with relative amplitude increasing with offset. That assumption was partly, but not entirely, correct. The reflection at 2.0 s at B satisfies both criteria (large amplitude, increasing AVO), and proved to be a gas reservoir. The same horizon at A meets neither criteria, and though it does contain some gas, the formations at A are much tighter than those at B. Moreover, a reflection above the target horizon at A at 1.95 s exhibits a trend of increasing amplitude with offset, yet it does not contain gas. Thus a unique association between a relative AVO increase and the presence of gas is not supported by the data.

The logs from the well near B reveal that the target is a 50 m shaly sand with about 15 m of gas pay. The corresponding sand at A exhibits greater shale content than at B. The velocity of the sand at both locations is approximately 20 percent greater than the overlying shale. If the gas at B were associated with an anomalously low Poisson ratio, as is often assumed for a gas-filled clean sand, simple modeling suggests that seismic reflections from the target should exhibit a strong decrease of amplitude with offset, quite contrary to what is observed. Indeed, to match the seismic data, the Poisson ratio of the shaly gas sand at B must be larger than that of the encasing shale. Unfortunately, logs which indicate shear-wave velocity or Poisson ratio were not recorded.

The mechanism for the AVO anomaly therefore remains uncertain. Perhaps the anomalous behavior reveals a change in lithology and porosity which is capable of trapping hydrocarbons, but which is not itself a direct indicator of gas. This would explain the absence of a one-to-one relation between amplitude increasing with offset and the presence of gas, while also explaining why gas is often associated with AVO anomalies in this area.*

*Editors note: This would require very unusual but not impossible geological circumstances.

DISCUSSION

These results illustrate both the benefits and risks of AVO and AVA interpretation. A relative increase of amplitude with offset is neither a necessary nor sufficient indicator of gas (Rutherford and Williams, 1989); moreover, levels of gas saturation too small to be commercial can produce strong AVO variations (Ostrander, 1984). Nevertheless, in some regions and some formations, the presence of gas is correlated with an anomalous change of amplitude with offset. In the absence of other information, data processing can at the very least help highlight these zones.

The potential benefits from AVO/AVA processing are greater when other information is available. For example, several workers (Chiburis, 1987; Burnett, 1989) have been successful at delineating the boundaries of a producing field by comparing the AVO and AVA response at different locations in the area to that at known producers and known dry wells, and then deciding which is the closer match. A second technique is to compare processed seismic data with synthetic data derived from modeling (Wrolstadt, 1986; Hilterman, 1989; Martens, et al., 1989). The modelers test various hypotheses about the target, contrasting, for instance, the seismic response of hydrocarbon- and water-bearing formations. Another method is inverse modeling, in which the Poisson's ratio, *P*-wave velocity, and density of a model are automatically adjusted so that the model data approximately match the processed field data. (Gassaway and Richgels, 1983; Gelfand, et al., 1986). Both forward and inverse modeling depend crucially on the modeler's knowledge of area geology, and the validity of the model's "known" parameters.

In this case study, a thorough analysis of model responses at the well locations was not undertaken. Such a study might lead to a better understanding of the mechanism responsible for the AVO anomaly, and remains a worthy topic for future investigations.

CONCLUSION

Processing for AVO and AVA analysis should be designed to attenuate noise and to reveal relative amplitude variations contained in the reflected signal. Striking the proper balance between these two objectives is essential. This case history illustrates one processing flow; other data will require different or additional processing, and various other processing schemes might be equally effective.

Interpreting AVO/AVA results is often difficult, especially when well logs or other geologic information are unavailable. The task is made more difficult if the interpreter is afflicted by uncertainty with the data processing. An interpreter who is acquainted with the processing strategy, who knows what decisions have been made in processing, and who looks at intermediate and final displays both before and after stack, is best positioned to understand the results.

ACKNOWLEDGMENTS

I extend special thanks to Bob Parker and Ladd Petroleum, through whose courtesy these data are made public. Thanks also to Glen Krajca-Radcliffe and Steve Checkles, former and current AVO processing analysts in Western's Land Processing Department, for helping to process the data in this study, and to Soraya Brombacher and Beth Linesch, for preparing the figures.

REFERENCES

Burnett, R. C., 1989, Seismic amplitude anomalies and AVO analyses at Mestena Grande Field, Jim Hogg Co., Texas: 59th Ann. Internat. Mtg., Soc. Expl. Geophys., Expanded Abstracts, 690–694.

Chiburis, E. F., 1987, Studies of amplitude versus offset in Saudi Arabia: 57th Ann. Internat. Mtg., Soc. Expl. Geophys., Expanded Abstracts, 614–616.

Dey-Sarkar, S. K., Svatek, S. A., and Crewe, O. W., 1986, Prestack analysis: Data processing: 56th Ann. Internat. Mtg., Soc. Expl. Geophys., Expanded Abstracts, 339–341.

Gassaway, G. S., and Richgels, H. J., 1983, SAMPLE: Seismic amplitude measurement for primary lithology estimation: 53rd Ann. Internat. Mtg., Soc. Expl. Geophys., Expanded Abstracts, 610–613.

Gelfand, V., Ng, P., Nguyen, H., and Larner, K., 1986, Seismic lithologic modeling of amplitude-versus-offset data: 56th Ann. Internat. Mtg., Soc. Expl. Geophys., Expanded Abstracts, 334–337.

Hilterman, F. J., 1989, Is AVO the seismic signature of rock properties?: 59th Ann. Internat. Mtg., Soc. Expl. Geophys., Expanded Abstracts, 559.

Hubral, P., 1978, Discussion on "Wavefront curvature in a layered medium" by Bjorn Ursin: Geophysics, **43**, 1551–1552.

Martens, K. N., Quale-Goranson, H., and Curts, B., 1989, Amplitude versus offset: A practical interpretation tool. Two case studies in South Central Alberta, Canada: Can. Soc. Expl. Geophys., Expanded Abstracts, 45.

Newman, P., 1973, Divergence effects in a layered Earth: Geophysics, **38**, 481–488.

Ostrander, W. J., 1984, Plane-wave reflection coefficients for gas sands at nonnormal angles of incidence: Geophysics, **49**, 1637–1648.

Rutherford, S. R., and Williams. R. H., 1989, Amplitude-versus-offset variations in gas sands: Geophysics, **54**, 680–688.

Taner, M. T., and Coburn, K. W., 1980, Surface consistent estimation of source-and-receiver response functions: Presented at the 50th Ann. Internat. Mtg., Soc. Expl. Geophys.

Taner, M. T., and Koehler, F., 1981, Surface consistent corrections: Geophysics, **46**, 17–22.
Taner, M. T., Koehler, F., and Alhilali, K. A., 1974, Estimation and correction of near-surface time anomalies: Geophysics, **39**, 441–463.
Todd, C. P., 1986, Isolation. display, and interpretation of offset dependent phenomena in seismic reflection data using offset to depth (ODR) range partial stacking: M.A. thesis, Univ. of Texas.
Todd, C. P., and Backus, M. M., 1985, Offset-dependent reflectivity in a structural context: 55th Ann. Internat. Mtg., Soc. Expl. Geophys., Expanded Abstracts, 586–588.
Ursin, B., 1978, Wavefront curvature in a layered medium: Geophysics, **43**, 1011–1013.
Ursin, B., 1989, Offset-dependent geometrical spreading in a layered medium: 59th Ann. Internat. Mtg., Soc. Expl. Geophys., Expanded Abstracts, 1156–1159.
Wiggins, R. A., Larner, K. L., and Wisecup, R. D., 1976, Residual statics analysis as a general linear inverse problem: Geophysics, **41**, 922–938.
Wrolstadt, K. H., 1986, Offset-dependent amplitude analysis by viscoelastic lithological modeling: 56th Ann. Internat. Mtg., Soc. Expl. Geophys., Expanded Abstracts, 341–346.
Yu, G., 1985, Offset-amplitude variation and controlled amplitude processing: Geophysics, **50**, 2697–2708.

Wavelet Attenuation and Bright-Spot Detection

*Peter C. Luh**

INTRODUCTION

A successful amplitude variation with offset (AVO) analysis coupled with standard exploration techniques can greatly improve the odds of finding hydrocarbons. In Tertiary clastic basins, AVO analysis not only augments bright-spot analysis, but helps delineate fluid contacts and reservoir facies. In carbonate provinces, AVO analysis can potentially aid in lithology discrimination and porosity prediction.

The success of any AVO analysis hinges on correctly compensating various effects that can corrupt the reflectivity-slope estimation from observed prestack seismograms. Among factors affecting the slope, Spratt (1987) and Swan (1988) have shown that a small normal-moveout (NMO) error, normally tolerable in the stacking process, can lead to a large error in the slope estimation. Since the slope term contains information about shear and density contrasts of subsurface reflectors, AVO analysis could be compromised if the ambiguity between moveout-velocity error and shear component of the AVO signal remained unresolved.

This paper examines the effect of an anelastic overburden on the AVO-slope term and shows that even in the absence of NMO errors a substantial slope change could be expected from wavelet attenuation and corresponding dispersion. The pair, phase dispersion and amplitude attenuation, are analytically related in the causal sense of Hilbert transform (Futterman, 1962; Wuenschel, 1965). Beside anelastic absorption, intrabed multiples also cause wavelet attenuation and dispersion. Intrabed multiples exhibit characteristics similar to absorption for a frequency-independent Q model (O'Doherty and Anstey, 1971).

The offset dependence of wavelet attenuation for any single reflector may be decomposed into vertical and horizontal components. The vertical component is the attenuation that the reflector's zero-offset arrival undergoes in its propagation, whereas the horizontal component is the additional attenuation that the wavelet must experience as it traverses obliquely to reach various finite-offset receivers. Numerical examples with no moveout errors demonstrate the effect of the horizontal attenuation on the AVO signal. AVO results with and without Q compensation are then shown for field data from the Gulf of Mexico.

*Oryx Energy Company, 13155 Noel Road, Dallas, Texas 75240-5067. Formerly ARCO Oil and Gas Company.

ATTENUATING OVERBURDEN

Anelastic absorption in a linear theory of wave propagation is described by the frequency-independent or constant-Q model (Kjartansson, 1979; Bickel and Natarajan, 1985). To first order in $1/Q_e$, the anelastic absorption transfer function $\hat{\mathbf{G}}(\omega, t)$ for a plane wave of angular frequency ω and propagation time t is

$$\hat{\mathbf{G}}(\omega, t) = \exp\left[-\frac{\omega t}{2Q_e} + i\,\frac{\omega t}{\pi Q_e}\ln\left(\frac{\omega}{\omega_0}\right)\right], \quad (1)$$

where Q_e is an effective quality factor of overburden along the wave propagation path and ω_0 is an angular reference frequency. Equation (1) reduces to unity in nonabsorbing media or $Q_e = \infty$. In absorbing media, the amplitude-decay factor is $\exp(-\omega t/2Q_e)$, and the minimum-phase dispersive delay is $(\omega t/\pi Q_e) \ln(\omega/\omega_0)$.

As suggested in Bickel and Natarajan (1985), it is convenient to set the reference frequency, ω_0, equal to the dominant frequency of the observed reflection. This approach yields zero phase shift at the dominant frequency in equation (1). In equation (1), $t = x/v$, where v is the phase velocity and x is the distance traveled. The Fourier transform of (1) yields the waveform of an impulse response after attenuation and dispersion by propagation along the distance x. The transfer function controlling the change in waveform with offset is given by equation (1) using normal moveout for a homogeneous overburden. Since the normal moveout is only a fraction of the zero offset time, some simplifications can be used in characterizing the offset dependence. We will consider the effect of a homogeneous, isotropic, attenuating overburden. On real data, variations in velocity and Q with depth will introduce complications. An anisotropic effect, both intrinsic (Hosten et al., 1987), and that related to fine layering, might be anticipated on real data. Since the offset dependent effect of Q is the attenuation in equation (1), we will focus on this attenuation.

HORIZONTAL WAVELET ATTENUATION AND DISPERSION

An AVO reflectivity function for a single horizontally flat reflector is usually expressed as $A + B \sin^2(\theta)$, where A and B are zero-offset P reflectivity and its slope estimate for small angles, respectively (Aki and Richards, 1980). Angle θ normally refers to an average reflection angle of two media, but here denotes an angle for the stacking or root-mean-square (rms) velocity of the reflector; i.e., θ subtends from normal to a straight line between common reflection point and receiver. Any conversion between the two angles follows from the Snell's law.

The reflectivity function is obtained by expanding the Zoeppritz coefficient of two media in Taylor series for jumps in density and compressional v_p and shear v_s velocities. The slope term B and higher-order expansion coefficients are valid for two nearly identical and adjacent media. This condition is violated for gas bright spots with large jumps in density and v_p.

A flat reflector's two-way group arrival time t at an offset can be expressed by θ as $t = \tau \sec(\theta)$, where τ is the reflector's zero-offset arrival time and "sec" stands for secant function. For small angles,

$$t \simeq \tau[1 + \sin^2(\theta)/2]. \quad (2)$$

Further, in isotropic absorbing media, Q_e for different offset arrivals is nearly the same, analogous to the near equality of the normal and oblique-incident rms velocities. Thus, substituting equation (2) into equation (1) at any specific reflector results in a product of vertical and horizontal attenuations. The vertical attenuation results from the first term of equation (2) and depends on the zero-offset arrival time τ. The horizontal attenuation comes from the second term $\tau \sin^2(\theta)/2$, which is the residual moveout time. Both attenuations obey the functional form $\exp[-\alpha\omega + i\beta\omega \ln(\omega/\omega_0)]$, where α is $\tau/2Q_e$ for vertical attenuation, but takes on $\tau \sin^2(\theta)/4Q_e$ for horizontal attenuation, and $\beta = 2\alpha/\pi$.

AVO analysis is a slight misnomer when directly analyzing prestack seismic data because the theory is based on reflection angles. For any prestack gather at a fixed finite offset, residual moveouts of deep events are smaller than those of shallow events. However, at a constant angle, deep events experience greater horizontal attenuation than shallow events because residual propagation path increases proportionally with depth.

It can be shown that the relative slope error, normalized by the zero-offset term, is approximately $f_1\tau/Q_e$, where f_1 is the dominant frequency of a narrow-band wavelet (Luh, 1988). For a 30-Hz wavelet ($f_1 = 30$) propagating through an overburden with $Q_e = 150$, about 20 percent error in AVO slope may be expected from the horizontal dispersion for a reflection target at $\tau = 1$ s.

NUMERICAL EXAMPLES

In the examples shown here, prestack synthetic seismograms at each offset are constructed with time arrivals static-shifted to their respective zero-offset arrival times (Figures 1 and 2). The wavelet attenuation and dispersion is computed using the exact transfer function $\hat{G}(\omega, t_0) = \exp\{-i\omega t_0[(i\omega/\omega_0)^{-\nu} - 1]\}$ where frequency-dependent phase velocity is $v(\omega) = v_0(\omega/\omega_0)^{\nu} \sec(\pi\nu/2)$, $Q = ½ \cot(\pi\nu/2)$, v_0 is v_p measured at ω_0, and $v_0 t_0$ is twice the layer thickness. The model used consists of three thick isotropic horizontal layers of constant density and v_p/v_s ratio of 2. The first and the third layer are identical while the second layer has slower v_p and v_s than those of the outer layers. The exact fit of $A + B \sin^2(\theta)$ to the Zoeppritz curve yields $A = -B = -0.171$.

Two isolated events at 1 and 2 s are shown with a 30-Hz Ricker wavelet for $Q = \infty$, 200, 100, and 50 (Figures 1a and b). Each gather is normalized with a different scale for ease of comparisons. No other corrections are applied. Except for the polarity reversal, the zero-offset arrivals at 1 s for $Q = 100$ and 50 in Figure 1a are the same as those for $Q = 200$ and 100 at 2 s in Figure 1b. The effect of the vertical dispersion is illustrated for Q of 50; viz., the wavelet for $Q = 50$ at 2 s undergoes a phase shift of nearly 90 degrees from that for $Q = \infty$ (Figure 1b). The long tail of the wavelet for $Q = 50$ is due to the slow decay (t^{-3}) of Hilbert transform of zero-phase Ricker wavelet (Ricker, 1953).

The AVO signals of the raw gathers are extracted by fitting $A + B \sin^2(\theta)$ in a 40-ms window with angles less than 25 degrees for both events. A, B, and the ratios of B over A are tabulated (Table 1). The normalized slope B/A (Chiburis, 1987) is used so relative slope change due to horizontal dispersion can be measured for varying Q values. The last column of

Table 1 tabulates $B/A - (B/A)_\infty$ values which may be approximated by $f_1 \tau / Q_e$. At the 1 s event, the relative slope changes due to horizontal dispersion of Q_e = 200, 100, and 50 are, respectively, 16, 31, and 46 percent when compared to the infinite-Q case.

The third set of synthetics shows the zero-offset time at 1.05 s, after the thickness of the second layer is decreased 20 fold (Figure 2). Except for the infinite-Q case, shown at left of Figure 2, a Q of 15 is used in the second layer for simulating a low-Q gas zone in the three finite-Q cases. The top reflection at 1 s is deliberately masked to exclude the thin-layer interference on the bottom reflection at 1.05 s. The resulting $B/A - (B/A)_\infty$ in Table 1 shows a greater horizontal dispersion at 1.05 s as a result of lower Q_e.

Table 2 examines the relative slope decay for three different zero-phase wavelets, 30 Hz Ricker, 14-56 Hz 8 s sweep Klauder, and 5-10-20-40 Hz bandpass. The

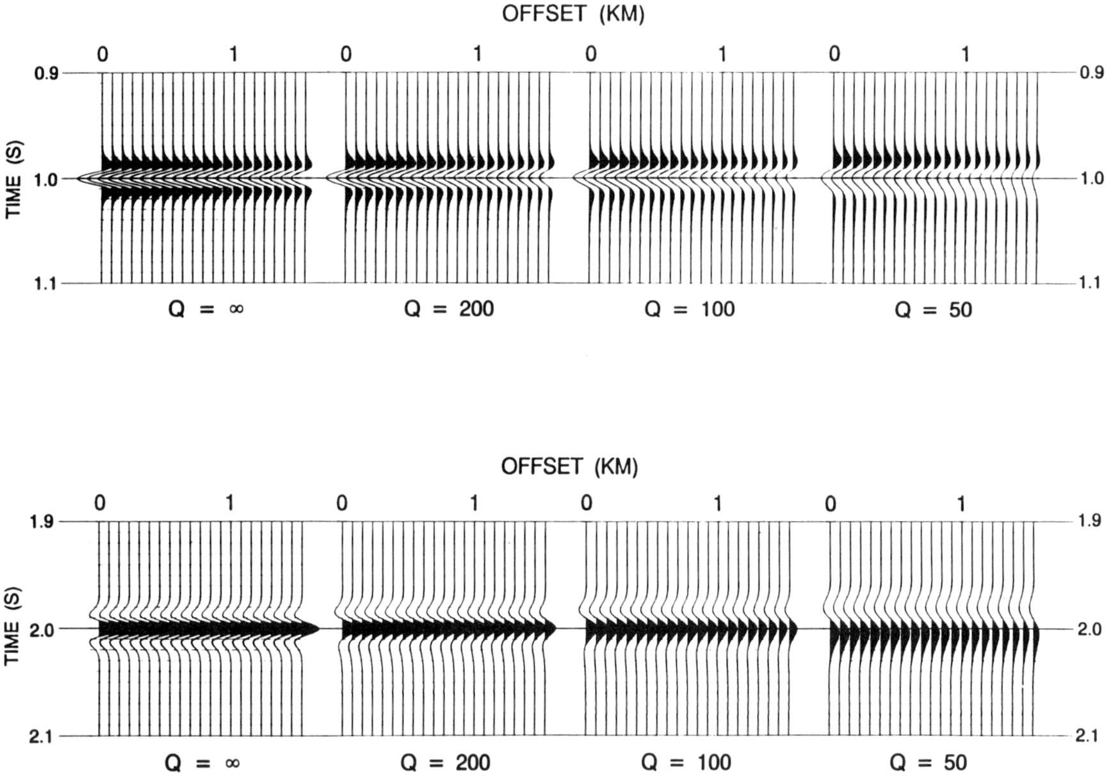

Fig. 1. Prestack synthetic seismograms at 1 and 2 s for $Q = \infty$, 200, 100, and 50 with a single normalized scale for each gather. The offset ranges from 0 to 1.6 km.

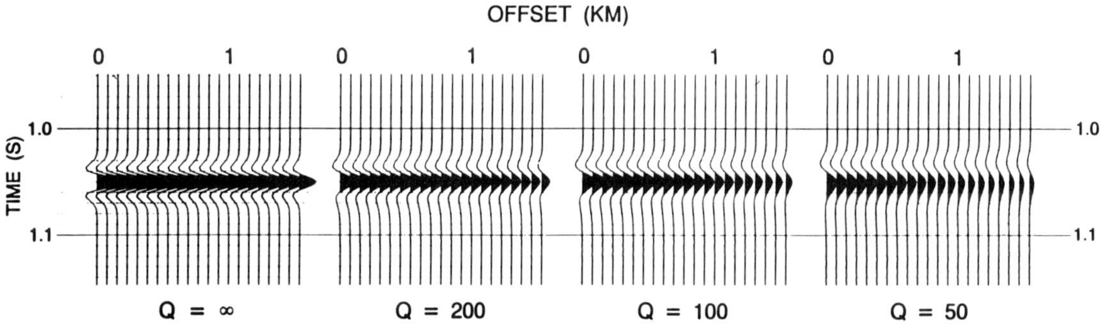

Fig. 2. The same as Figure 3 except for the deeper reflector now at 1.05 s and for Q = 15 in the second layer of the three finite-Q cases. The top reflection at 1 s is masked to exclude thin-layer interference.

first two agree well, but the slope corruptions of the bandpass wavelet are about half of the other two because the dominant frequency of the bandpass wavelet is half of the other two wavelets. These results confirm that in the high-Q approximation of $\alpha\omega_1 < 1$, $f_1\tau/Q_e$ approximates well the relative slope change of a typical seismic reflector at time τ and effective Q_e.

Additional synthetic modelings show that the valid range of τ (in s) in the high-Q approximation depends on the choice of Q_e. Further, $f_1\tau/Q_e$ fits best at around $Q_e/\tau = 100\ (\mathrm{s}^{-1})$. Slight underestimation and overestimation result for higher and lower Q_e/τ, respectively. Thus, for a relatively deep event at 3 s and Q_e of 150, the slope-change estimate will be high. However, a slight artificial boosting to the slope may be helpful in AVO interpretations since the deeper an event, the weaker will be its far-offset signals.

Table 1. Numerically estimated A and B for Figures 1 and 2. $B/A - (B/A)_\infty$ is the amount of relative slope discounting due to horizontal dispersion. Q_e for the last three rows are smaller due to a thin layer of $Q = 15$ between 1 and 1.05 s.

Time(s)	Q_e	A	B	B/A	$B/A - (B/A)_\infty$
1.0	∞	−0.171	0.171	−1	0
1.0	200	−0.104	0.121	−1.16	−0.16
1.0	100	−0.068	0.090	−1.31	−0.31
1.0	50	−0.030	0.044	−1.46	−0.46
2.0	50	−0.009	−0.016	−1.79	−0.79
1.05	∞	0.166	−0.173	−1.05	0
1.05	129	0.074	−0.099	−1.33	−0.28
1.05	79	0.049	−0.075	−1.54	−0.49
1.05	45	0.023	−0.041	−1.76	−0.71

Q COMPENSATION

The application of Q compensation is straightforward except for selection of the right Q profile. Since anelastic absorption depends on effective Q along the wavelet's propagation path, knowing approximate val-

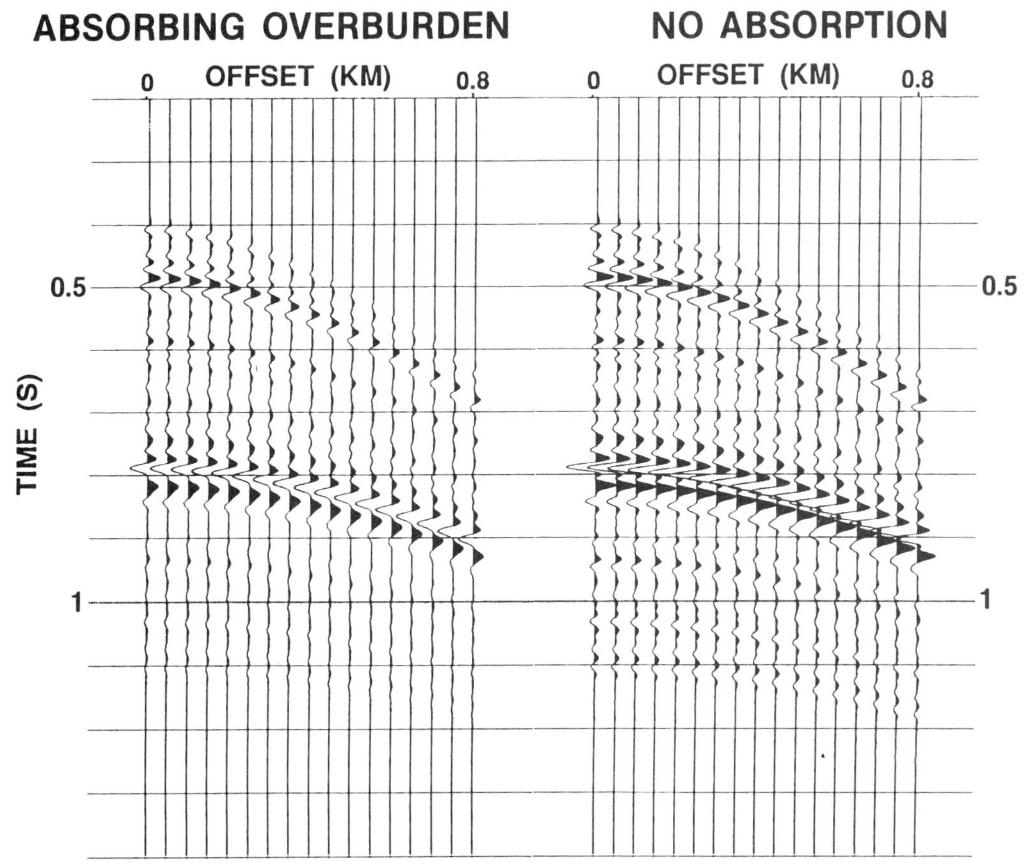

Fig. 3. Two primaries only ray trace prestack synthetic CDP gathers with and without anelastic absorption. The near-offset distance is 20 m with a 50 m trace interval.

ues of average Q suffices. The resulting Q compensation is generally robust to small errors in Q profiles. This robustness is analogous to that of spherical-divergence correction with small rms-velocity errors.

Figure 3 compares two primaries-only prestack synthetic common-depth-point (CDP) gathers with and without anelastic absorption, plotted in the same amplitude scale. The seismic model for the gathers is derived from elastic-moduli averaging (Backus, 1962) of a well in the Gulf of Mexico (Figure 4). Although the effect of transverse isotropy due to fine layering (see other references in this book) is excluded in the synthetic calculation. The Q profile is 150 throughout except for 15 at the 18 m gas zone at 0.8 s. The near-offset distance is 20 m with 50 m trace interval. Note that the wavelet at the bottom of the gas reflector at 0.82 s is conspicuously broader than that without absorption at the right.

An AVO analysis on the two gathers produces A, B, and AB traces. The AB traces are a filtered product of the A and B traces. The product accentuates any increase in absolute reflection amplitude as a function of offset (Figure 5). The difference in the A and B traces, aside from their magnitude, is largest at the bottom of gas or just below 0.8 s. The product traces show a large positive peak at the top of gas or slightly above 0.8 s.

Table 2. Numerically estimated B/A at 1 s for three zero-phase wavelets, 30-Hz Ricker, 14–56-Hz Klauder, and 5–10–20–40-Hz bandpass.

Time(s)	Q_e	B/A (Ricker)	B/A (Klauder)	B/A (bandpass)
1.0	∞	−1	−1	−1
1.0	200	−1.16	−1.19	−1.10
1.0	100	−1.31	−1.31	−1.17
1.0	50	−1.46	−1.48	−1.32

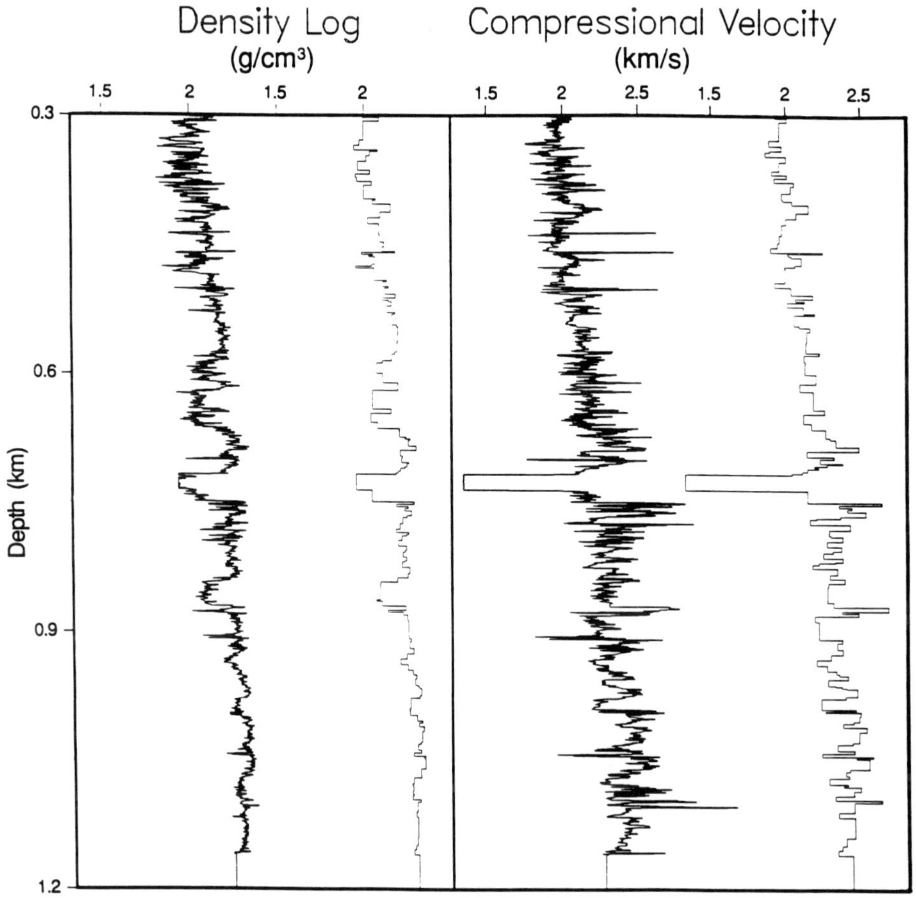

Fig. 4. The raw and blocked densities and compressional velocities from a density and P-wave sonic log in the Gulf of Mexico. The blocked velocities are derived from elastic-moduli averaging.

Q compensation with a uniform Q of 150 is applied to a seismic line from the Gulf of Mexico (Figure 6). This Q of 150 represents a compromise among various experimental studies, where measured values vary widely from 30 to 300 (McDonald et al., 1958; Hamilton, 1972). In general, shallow sediments have a lower Q, whereas the deep ones, a higher Q. The point, however, is whether the effect from a relatively high Q of 150 would be observable. If so, any lower Q will certainly affect observed seismic data in AVO analysis.

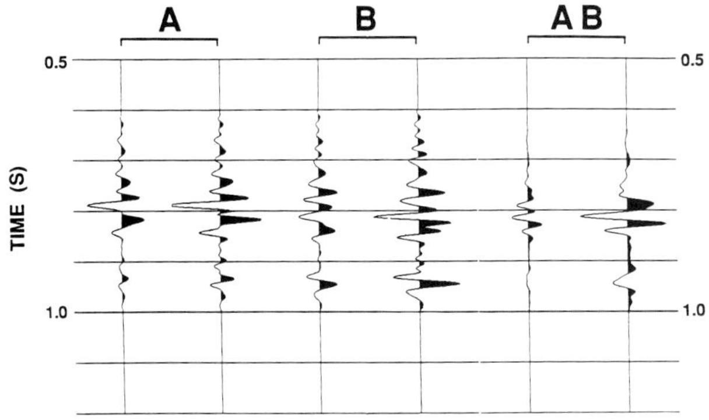

Fig. 5. Zero-offset reflectivity A, slope B, and their product AB traces of Figure 3.

A gas well is located near CDP 800. Of interest are the gas reservoir of the shallow Miocene at 0.8 s and a regional marker (top Oligocene) at 1.3 s. A slight sag in the Oligocene marker between CDP 800 and 930 is due to anomalously low gas velocities. Further, the Oligocene stacking amplitudes are weakened between CDP 800 and 1000 because the gas pay above has absorbed a large fraction of transmitted and reflected seismic energies. This example demonstrates the need for customized overburden corrections in multipay zones, especially for thick gas reservoirs having low Qs.

The A and AB traces centered about the two events are computed with and without Q compensation (Figures 7 and 8). The difference between the zero-offset reflectivity A traces show phase corrections and increased amplitudes at the Oligocene marker after Q compensation. Similarly, the product AB traces show improved signals at the Oligocene. Despite the simplistic choice of Q of 150, we have partially corrected the absorptive effect of the gas reservoir on seismic signals of underlying reflectors. A greater dynamic contrast in the AB traces is observed at the Oligocene after Q compensation. An obvious implication is that a judicious Q compensation can improve seismic and AVO signals of deeper pays in multiple-pay zones. As AVO technology matures, subtle nuances such as

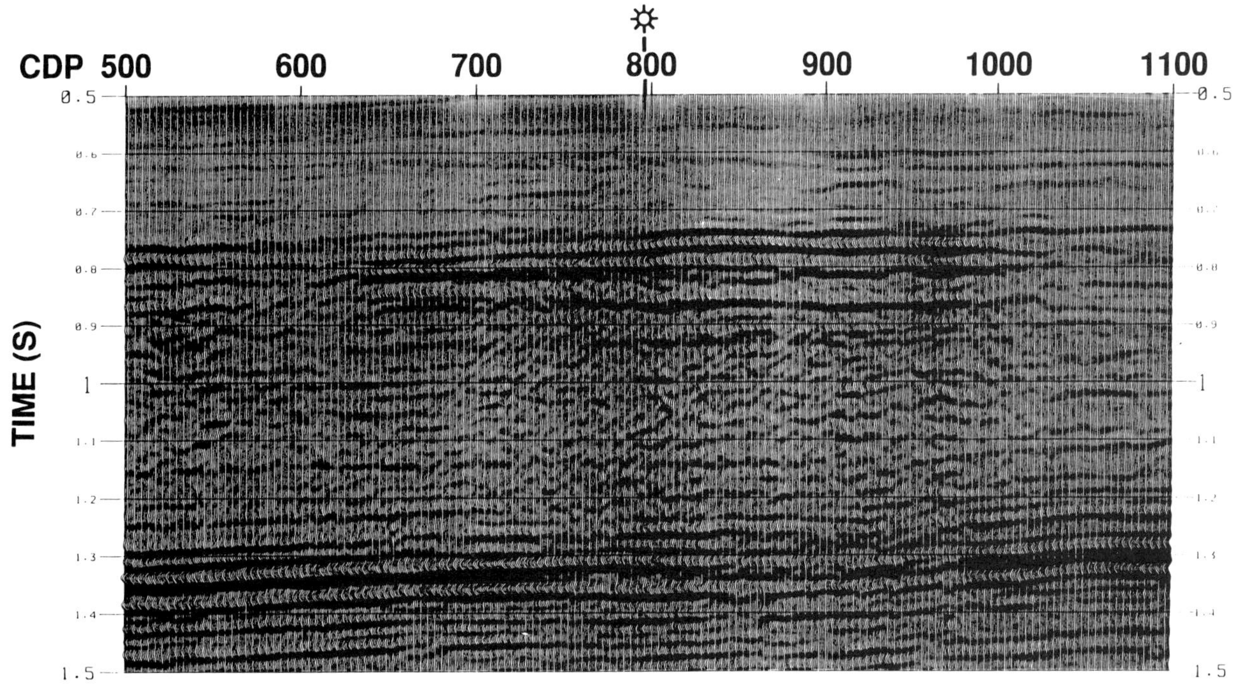

Fig. 6. A stacked section from the Gulf of Mexico with a 6.25 m trace interval.

these may be directly translated into lithology and pore-fluid content of potential target plays.

CONCLUSIONS

Since Ostrander (1984) showed that variations in seismic reflectivity as a function of offset can be used for bright-spot validation, overburden effect has been known to bias the estimation of AVO slope. A rule-of-thumb for estimating relative AVO-slope change due to an absorbing overburden in the high-Q approximation is $f_1 \tau / Q_e$, where τ is zero-offset arrival time of a reflector of interest, Q_e is the effective quality factor along its propagation path, and f_1 is the dominant seismic frequency of its wavelet. For example, AVO slope for a typical shallow gas-bearing sand at 1 s (or about 1-km depth) will be discounted by about 20 percent if a 30 Hz Ricker wavelet were to propagate through media with Q_e of 150. The synthetic examples show that such an estimate, however, is best for Q_e/τ at around 100 s^{-1}.

In AVO analysis, for any fixed angle, the deeper a reflector is, the greater will be its AVO slope change from the absorption effect because the residual travel path lengthens proportional to depth. The resulting slope change can be examined after the attenuation of reflection signals is decomposed into vertical and

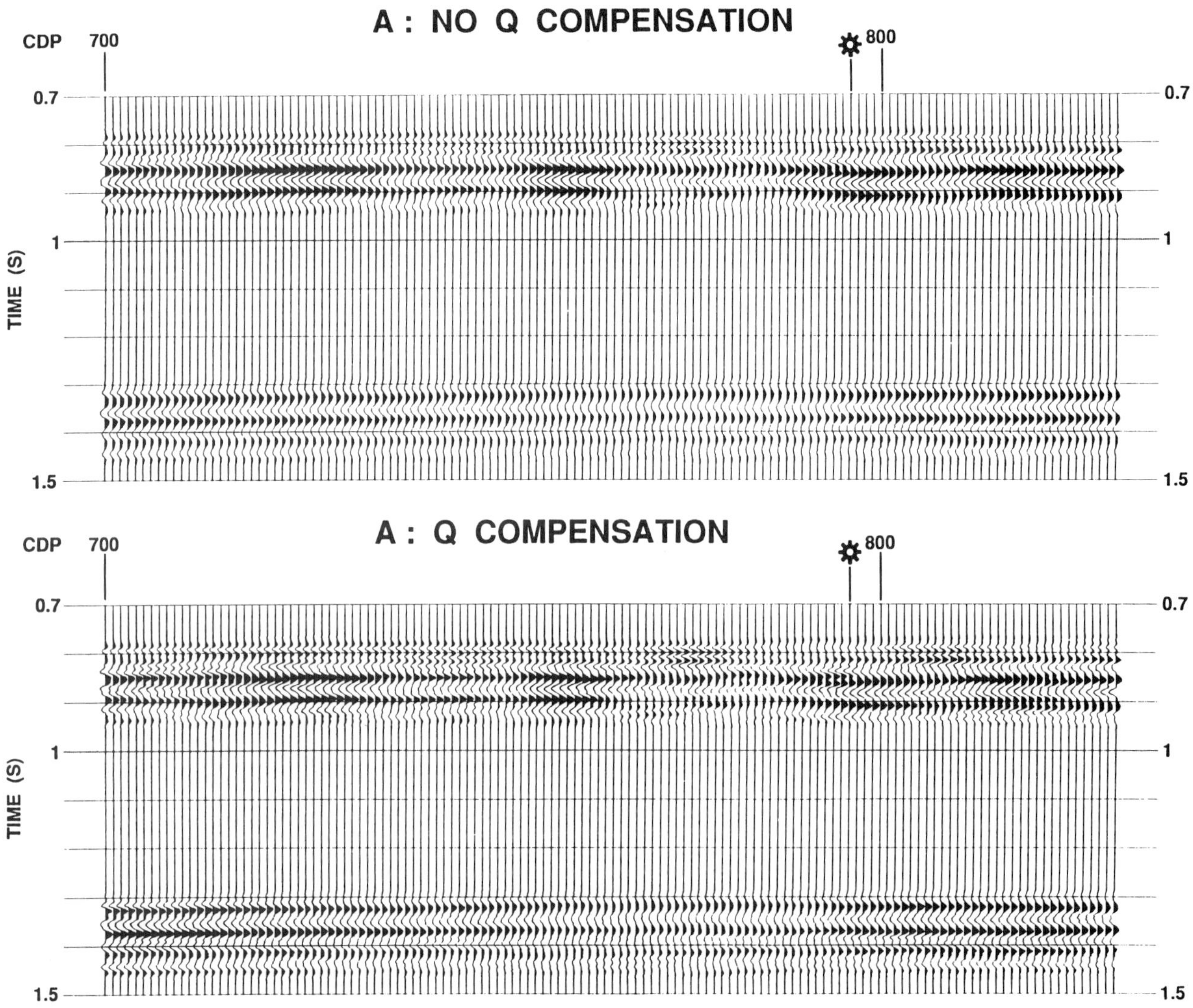

Fig. 7. Zero-offset reflectivity A traces with and without Q compensation of Q = 150. The gas well is near CDP 800. Only reflectivities centered about the two events are shown.

horizontal components. The vertical component depends on cumulative effect of overburden and disperses the wavelet more as its zero-offset arrival time increases. The horizontal component depends on residual moveout time along an oblique path and contributes to AVO-slope decay and dispersive phase delays.

Since wavelet dispersion exists even with perfect measurements, AVO signals for reflectors below a known gas reservoir require inverse Q filtering to compensate the dispersion effect. In practice, if a Q profile is known, Q compensation should be applied to prestack seismograms before normal moveout. Since inverse Q filtering is generally robust to small errors in Q profiles, any judicious compensation can improve seismic and AVO results of deeper pays in multiple-pay zones. A field-data example from the Gulf of Mexico shows improved AVO signals following Q compensation.

The pseudo-shear (or slope-error) term in the AVO-slope B depends on Q_e, zero-offset arrival time, dominant seismic frequency, and zero-offset reflectivity A. The slope change (except for $A = 0$) always causes a decrease in absolute amplitude as offset

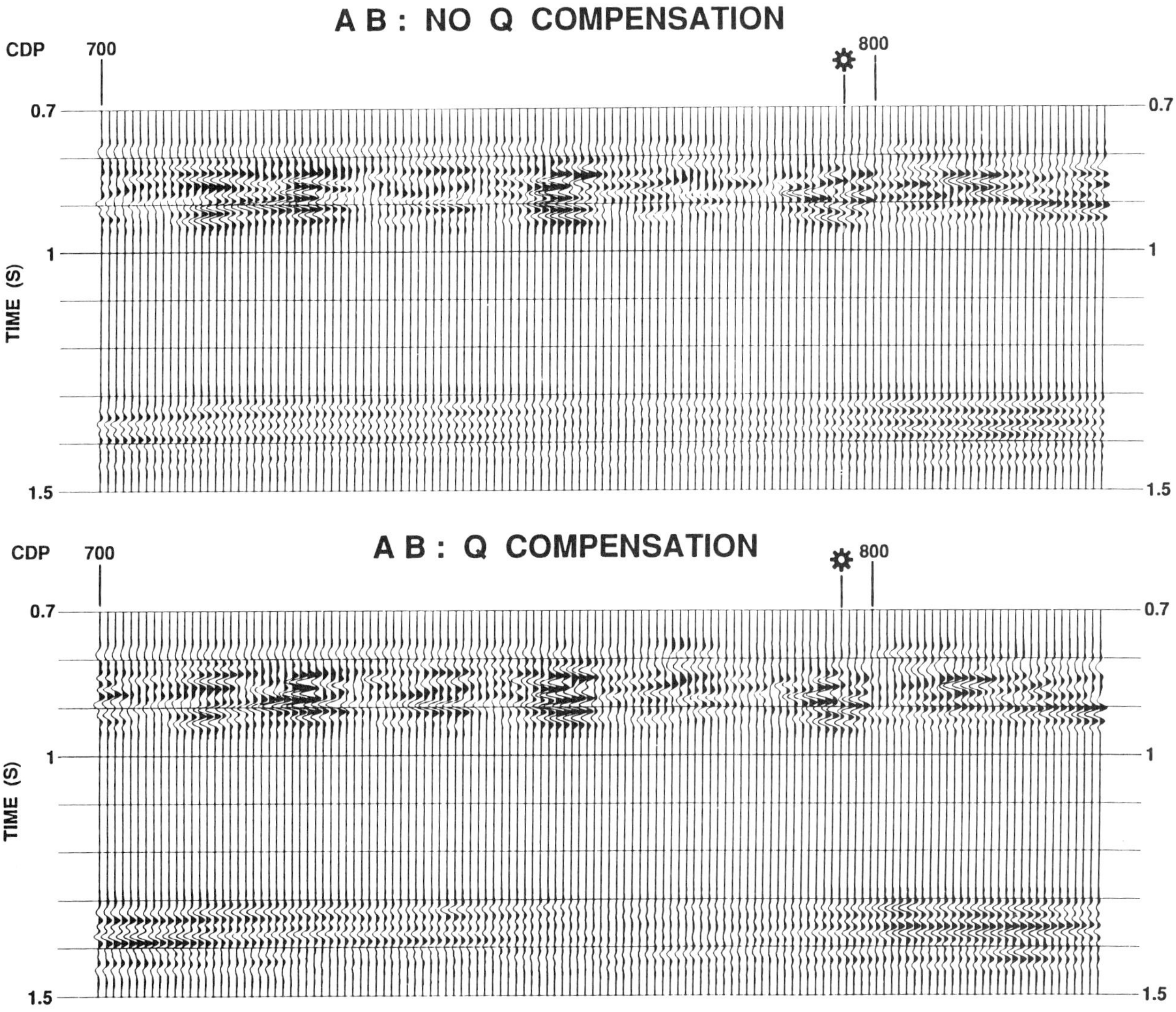

Fig. 8. Product AB traces with and without Q compensation of $Q = 150$. Only those AB values centered about the two events are shown.

increases. Further, the magnitude of the slope error is proportional to A.

ACKNOWLEDGMENT

Thanks to ARCO Oil and Gas Company for permission to publish; to Sam Bickel, John Castagna, and Bill Lamb for critical reviews of the first draft; and to Mike Batzle for discussion on experimental Q measurements.

REFERENCES

Aki, K., and Richards, P. G., 1980, Quantitative seismology: Theory and methods, Vol. 1: W. H. Freeman and Co.

Backus, G. E., 1962, Long-wave elastic anisotropy produced by horizontal layering: J. Geophys. Res., **67**, 4427–4440.

Bickel, S. H., and Natarajan, R. R., 1985, Plane-wave Q deconvolution: Geophysics, **50**, 1426–1439.

Chiburis, E. F., 1987, Studies of amplitude versus offset in Saudi Arabia: 57th Ann. Internat. Mtg., Soc. Expl. Geophys., Expanded Abstracts, 614–616.

Futterman, W. I., 1962, Dispersive body waves: J. Geophys. Res., **67**, 5279–5291.

Hamilton, E. L., 1972, Compressional-wave attenuation in marine sediments: Geophysics, **37**, 620–646.

Hosten, B., Deschamps, M., and Tittmann, B. R., 1987, Inhomogeneous wave generation and propagation in lossy anisotropic solids. Application to the characterization of viscoelastic composite materials: J. Acoust. Soc. Am., **82**, 1763–1770.

Kjartansson, E., 1979, Constant Q-wave propagation and attenuation: J. Geophys. Res., **84**, 4737–4748.

Luh, P. C., 1988, Wavelet dispersion and bright-spot detection: 58th Ann. Internat. Mtg., Soc. Expl. Geophys., Expanded Abstracts, 1217–1220.

McDonald, F. J., Angona, F. A., Mills, R. L., Sengbush, R. L., van Nostrand, R. G., and White, J. E., 1958, Attenuation of shear and compressional waves in Pierre Shale: Geophysics, **23**, 421–439.

O'Doherty, R. F., and Anstey, N. A., 1971, Reflections on amplitudes: Geophys. Prosp., **19**, 430–458.

Ostrander, W. J., 1984, Plane-wave reflection coefficients for gas sands at nonnormal angles of incidence: Geophysics, **49**, 1637–1648.

Ricker, N., 1953, The form and laws of propagation of seismic wavelets: Geophysics, **18**, 10–40.

Spratt, S., 1987, Effect of normal moveout errors on amplitude versus offset-derived shear reflectivity: 57th Ann. Internat. Mtg., Soc. Expl. Geophys., Expanded Abstracts, 617–620.

Swan, H. W., 1988, Amplitude versus offset analysis in a finely layered media: 58th Ann. Internat. Mtg., Soc. Expl. Geophys., Expanded Abstracts, 1195–1198.

Wuenschel, P. C., 1965, Dispersive body waves—an experimental study: Geophysics, **30**, 539–551.

Automated Extraction of AVA Information in the Presence of Structure

Z. Shang, J. A. McDonald*, and G. H. F. Gardner**

INTRODUCTION

Amplitude variation with offset (AVO) information has become one of the most important tools in geologic interpretation of seismic data. AVO information can be used to predict the elastic constants of rocks and to provide a valuable additional tool in stratigraphic interpretation. AVO information can also be an important hydrocarbon indicator. Often the study of AVO is based on common-midpoint (CMP) gathers because of the difficulties of extracting amplitude variation with angle of incidence (AVA) information from seismic data. However, AVO is only equivalent to AVA in the case of a horizontal, planar reflector. A data-smearing CMP gather is not a common-reflection point (CRP)[1] gather. Figure 1 shows the problems of AVO when a CMP gather is used. The first problem is the amplitude-smearing effect because the amplitudes on the traces at different offsets in the CMP gather are reflected from different points. The second problem is the inaccuracy in the calculation of the incident angle if the CMP gather is treated as a CRP gather. Detailed demonstrations of these problems can be found in the literature (Resnick et al., 1987; and Shang, 1990). To overcome these problems, we suggest: (1) sort the data into true CRP gathers and then, (2) compute the incident angle for each reflected signal. There are probably many ways to sort CRP gathers but we use the dip-moveout (DMO) concept. We demonstrate that, after the transformation of DMO plus NMO (Forel and Gardner, 1988), the signals reflected from a true CRP lie along the ridge line of a reflection time surface. We suggest two ways to use this property to extract the AVA information from a multifold seismic line, even when the reflectors are tilted or curved. The idea is to study traces which have a true CRP rather than a CMP. Assuming a constant medium velocity, we found that the DMO plus NMO transformation makes it possible to sort the data into CRP gathers. One way to extract the AVA information is to estimate the dip angle at the reflection point and then calculate the incident angle for each reflected signal. In this case the original measurements are used and there is no problem with amplitude and phase preservation. A second way is to stack the data along common-incident-angle contours in the CRP gather, scale the amplitude to remove the stacking effect, and apply a phase-shift filter to preserve the waveform. The location of the reflection point in either method can be determined relative to the zero offset trace for the CRP gather and hence its migrated position is known. With the second method suggested, the AVA information is automatically calculated for each reflection point and the output panel for each analysis point is organized in such a manner that the current techniques in AVO analysis can be used to obtain different kinds of AVA attribute sections.

The extraction of AVA information using this method can be incorporated into DMO plus NMO velocity analysis. There is virtually no additional computational cost except for a weight calculation for each point.

Since other DMO algorithms are equivalent to the definition employed in this paper, they probably can be used to do the same thing.

TRUE COMMON REFLECTION POINT GATHERING

In Figure 2, BC is a multifold seismic line with S being a source and R being a receiver. The solid arc represents a part of an arbitrary reflector. P is the reflection point and AP is a tangent to the reflector. A is the outcrop point of the tangent. A multifold seismic survey along BC can be represented by the 3-D data volume (h, m, t), where h is the half offset, m is the midpoint coordinate, and t is the reflection time. An analysis point O is selected and is considered the origin of the x and z axes. M is the midpoint of S and

Expanded version of the presentation at the Society of Exploration Geophysicists 59th Annual International Meeting.

*University of Houston, Allied Geophysical Laboratories.

[1]CRP in the SEG dictionary refers to *common receiver point* but is used for *common-reflection point* throughout this paper.

R. The half offset of R from S is defined as h and b is the distance from O to M.

The multifold seismic data can now be expressed in the 3-D domain (h, b, t). DMO plus NMO (Forel and Gardner, 1988) is then equivalent to transforming t to t_0 using h, b, and medium velocity v by:

$$t_0^2 = \frac{h^2 - b^2}{h^2} t^2 - 4 \frac{h^2 - b^2}{v^2}, \quad -h \le b \le h. \qquad (1)$$

This definition indicates that only when the source and receiver are on opposite sides of the analysis point will a trace contribute to the analysis point gather. For each analysis point, the effective data volume is limited by two boundaries as shown in Figure 3, where t is the reflection time. The arrival times shown in Figure 3 correspond to a planar reflector with slope $\alpha = \tan^{-1} 0.5$ and $AO = 2500$ m. The velocity above the reflector is 3000 m/s.

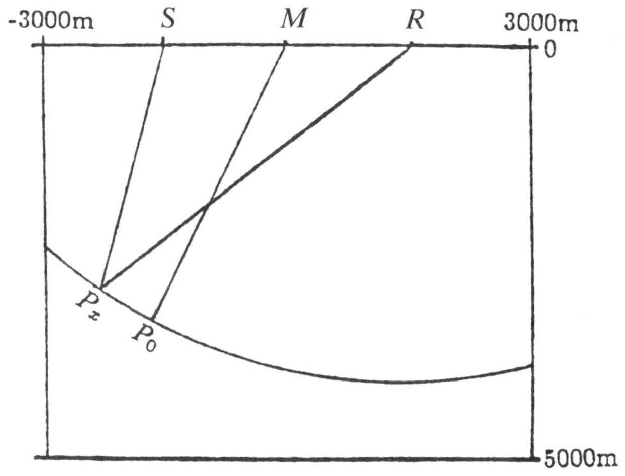

Fig. 1. Model showing the problems in AVO study. Even though M is the midpoint of S and R, the reflection point is at P_x, not P_0 as assumed in CMP gather AVO studies. This leads to amplitude smearing and inaccuracy in the computation of incident angle.

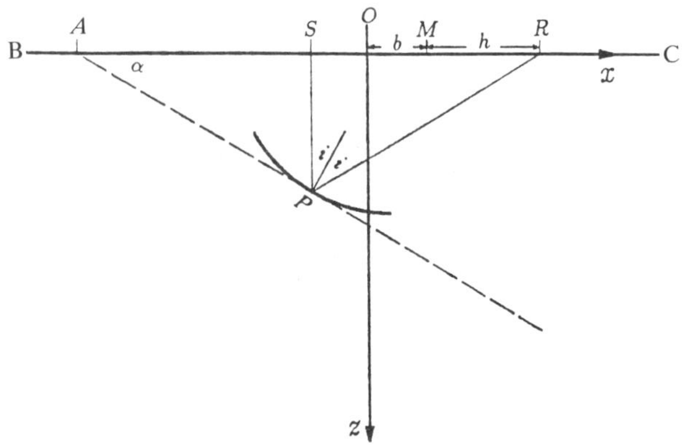

Fig. 2. Geologic model of a structured reflector. AP is tangential to the reflector (solid arc) at P. S and R are a source-receiver pair on the multifold seismic line BC. O is the analysis point and the origin of the coordinate system. h is the half offset and b indicates the location of midpoint M with respect to O.

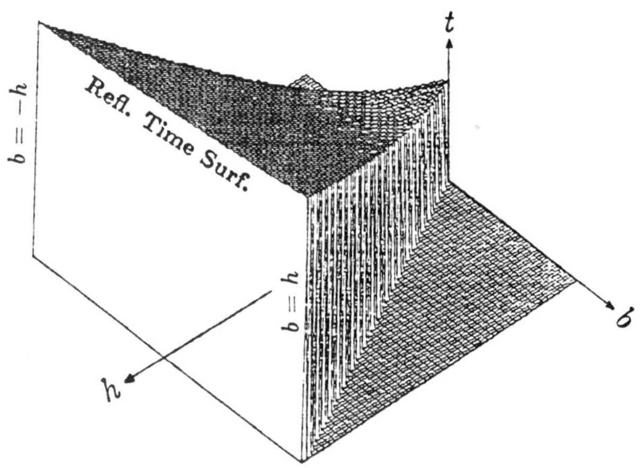

Fig. 3. Reflection time surface in the (h, b, t) domain for a dipping planar reflector. $b = -h$ and $b = h$ are the two boundaries beyond which the traces do not contribute to the DMO plus NMO corrected panel.

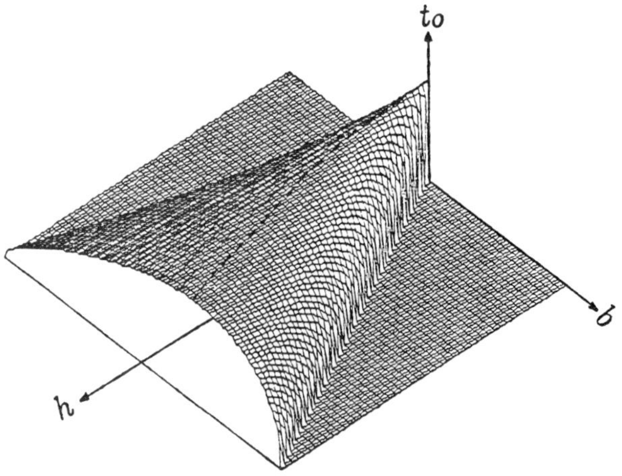

Fig. 4. The DMO + NMO transformed reflection time surface in the (h, b, t_0) domain. The transformed time t_0 vanishes on the two boundaries in Figure 3.

Mathematically, equation (1) transforms the original seismic data from the (h, b, t) domain into the (h, b, t_0) domain at each different analysis point (Figure 3 to Figure 4).

In Figure 2, once the reflection point P with coordinates (x, z), the incident angle i and the slope of the tangent AP are known, the half offset h and relative midpoint location b are given by

$$h = \frac{z}{2}\left[\tan(\alpha - i) + \tan(\alpha + i)\right] \tag{2}$$

and

$$b = \frac{z}{2}\left[\tan(\alpha + i) - \tan(\alpha - i)\right], \tag{3}$$

where

$$\alpha = \tan^{-1}\left(\frac{dz}{dx}\right). \tag{4}$$

Substituting equations (2), (3), and (4) into equation (1), we obtain the DMO plus NMO transformed reflection time surface for the reflector $z(x)$:

$$t_0^2 = -\frac{4[(x + zz')^2 - c^2(xz' - z)^2]}{v^2c^2(1 + z'^2)} \tag{5}$$

where $c = \tan(i)$, i is the incident angle, v is the velocity of the medium, and $z' = dz(x)/dx$. Figure 4 shows an example for a planar reflector.

We now show that the peak (the dashed line in Figure 4) of this reflection time surface is a true CRP gather. The derivative of equation (5) with respect to x is

$$\frac{dt_0}{dx} = -\frac{4(zz' + x)}{t_0 v^2 c^2}\left[1 + \frac{z''(1 + c^2)(z - xz')}{(1 + z'^2)^2}\right]. \tag{6}$$

Setting equation (6) to equal zero, we can determine the reflection point coordinates for the ridge line of the surface in Figure 4.

Putting equation (6) equal to zero produces two solutions. The first,

$$z' = -\frac{x}{z}, \tag{7}$$

means that the reflection point is such that the normal to the reflector at this point passes through the origin, i.e., the analysis point. That is, all the rays reflected at the zero-offset reflection point satisfy equation (7) regardless of the incident angle. Therefore, the ridge line of the surface in Figure 4 is a true CRP gather corresponding to the zero-offset reflection point.

The other solution obtained by setting equation (6) equal to zero is

$$z''(1 + c^2)(z - xz') + (1 + z'^2)^2 = 0, \tag{8}$$

Equation (8) represents a reflector such that for all rays incident upon it with a constant incident angle $i = \tan^{-1} c$, the transformed reflection time, t_0, will be constant. This will rarely occur in practice because it requires a special reflector and is for a particular incident angle only.

AVA DATA EXTRACTION ON TIME SLICE

Having proved that the peak line of the transformed reflection time surface is a CRP gather, we can cut a time slice from the (h, b, t_0) cube at the peak and extract the AVA information for the reflection point. The trajectory of the peak on the time slice can be derived as

$$h^2 = b^2 + \frac{vT_0}{2\sin(\alpha)}b \tag{9}$$

where T_0 is the maximum t_0. Combining equation (7) with equation (5) produces the maximum of t_0, T_0, in terms of z, z' and i. Using equations (2), (3), and (4) to remove z and z', we can obtain equation (9). Figure 5

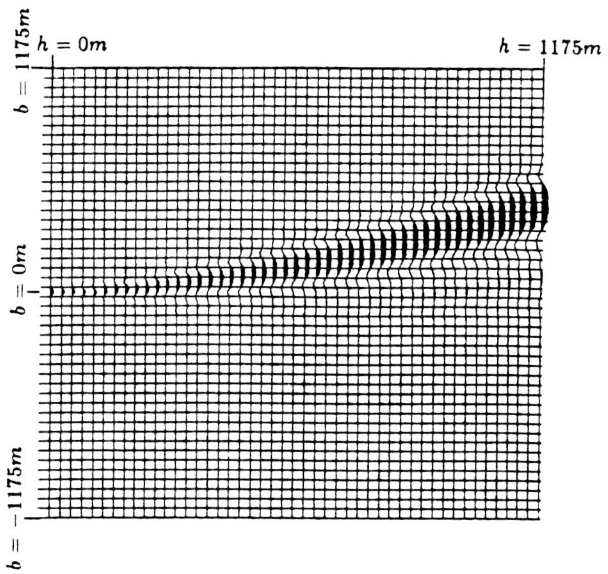

Fig. 5. A time slice cut at the top of the transformed reflection time surface. All the signals along the hyperbolic event are reflected from the reflection point. The characteristic of the hyperbola can be used to determine the dip of the reflector at the reflection point.

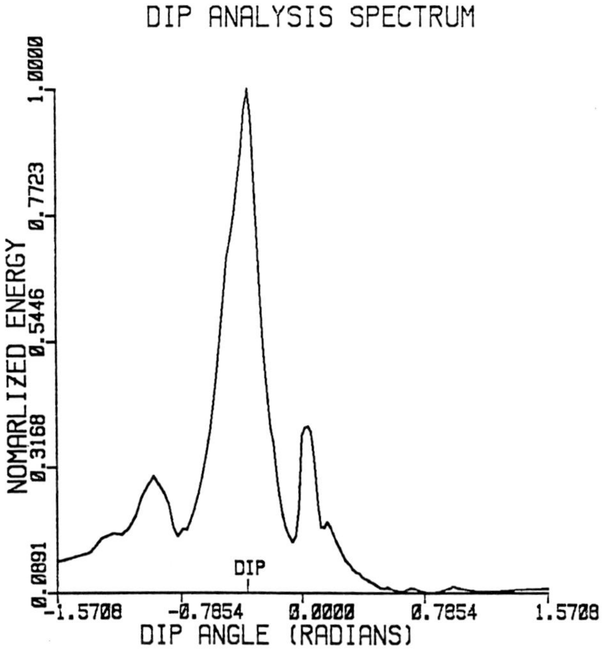

Fig. 6. A dip spectrum extracted from Figure 5. The dip angle is given by the maximum energy location. Negative dip value indicates the reflector is tilted toward the negative *b* direction.

is a time slice corresponding to the reflection time surface in Figure 4 but with wavelets included.

Equation (9) can be used to estimate the dip angle α of the reflector at the reflection point. For a given dip angle α, we can collect all the energy along the hyperbola defined by equation (9) on the time slice. The sum of the energy collected is a measurement of the possibility of this given dip being the true dip angle. This idea is not new but is similar to NMO velocity analysis.

Figure 6 is an example of a dip spectrum obtained in this manner where the true dip angle corresponds to the maximum of the total energy curve. After the dip is estimated, the incident angle of each reflection can be calculated by using

$$h = \left(\frac{vT_0}{2\sin(\alpha)} + b\right)\tan(\alpha)\tan(i) \qquad (10)$$

which is derived from the law of sines and the law of cosines. Figure 7 is an example of AVA relationship extracted along the CRP gather at $T_0 = 0.745$.

The advantage of this method is that all the data on the time slice are original measurements. However, a time value has to be given at which the AVA information is desired. Next we suggest one algorithm to overcome this in which the data will be processed time level by time level automatically.

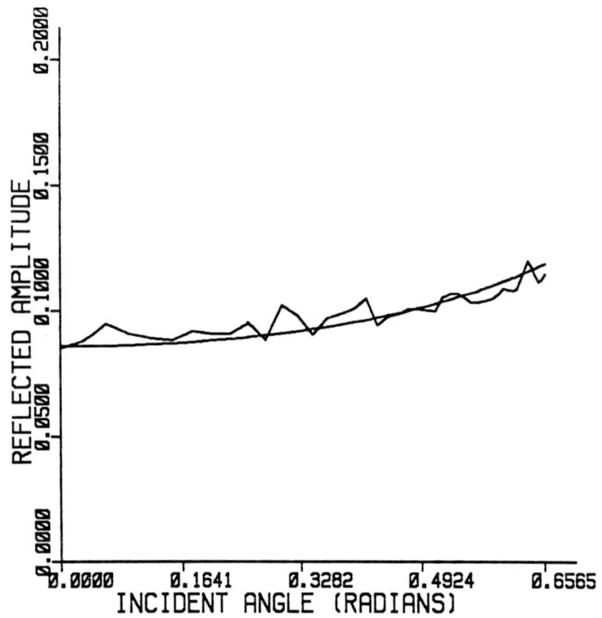

Fig. 7. Extracted AVA information by using equation (10) and the dip value in Figure 6.

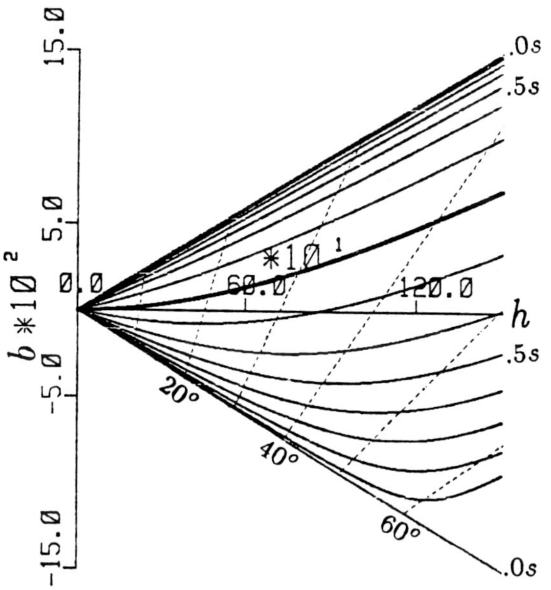

Fig. 8. DMO + NMO transformed reflection time contour lines (solid) and incident angle contour lines (dashed). The thick solid hyperbolic curve is the ridge line of the reflection time surface.

AVA DATA EXTRACTION THROUGH STACKING OF CONSTANT INCIDENT ANGLE CONTOURS

Incident Angle Contour Map For a Dipping Reflector

The incident angle contour map is made up of straight lines, h equals constant, parallel to the mid-point axis on a time slice (h, m) if the reflector is horizontal. However, equal offset does not mean equal incident angle on a time slice if the reflector dips. Equation (10) can be used to draw the constant incident angle lines. For given v, zero reflection time T_0 and α, each value of i produces an i contour. Figure 8 shows an example, where dashed lines are the incident angle contours, solid lines are the time t_0 contours, and the thick solid line is the ridge line of the surface (Figure 4). The reflector dips with slope $\alpha = \tan^{-1}$ (0.5) and the perpendicular depth is 1118 m.

If we cut the reflection time surface in 3-D (h, b, t_0) along an incident angle contour line, a section is obtained on which all the energy is reflected at the same angle. Figure 9 is such a section cut from Figure 8 with an incident angle of 40°. Stacking this section produces a stacked trace on which only the wavelets close to the apex will be preserved. Two problems occur here; (1) we do not know the dip angle in practice; and (2) how do we calibrate the amplitude on the stacked trace to represent the reflection coefficient?

Effective Incident Angle Contour Map For the Dip Spectrum

Since only the energy very close to the apex of the curved event contributes to the stacked trace, we need only consider the intersection points between the incident angle contours and the ridge line of the surface (Figure 8). That is, only one effective point needs to be considered for each incident angle at a given dip. We define this point as an effective point because, when the constant incident angle section is stacked, the signal at this point contributes to the stacked result most effectively. This effective point can be determined from equations (9) and (10) since, for a set of given values for α, i, v, and T_0, equations (9) and (10) yield a particular set of (h, b) which is the effective point corresponding to the apex on the constant incident angle section (Figure 9).

To obtain the trajectory of the effective point when α varies, we can eliminate α from equations (9) and (10), and then the effective constant i contour is given by

$$b^2 = h^2 - \frac{1}{2}\left[\sqrt{\left(\frac{vT_0}{2}\right)^4 + 4\left(\frac{vT_0}{2}\right)^2 h^2\, s^2(i)} - \left(\frac{vT_0}{2}\right)^2\right], \tag{11}$$

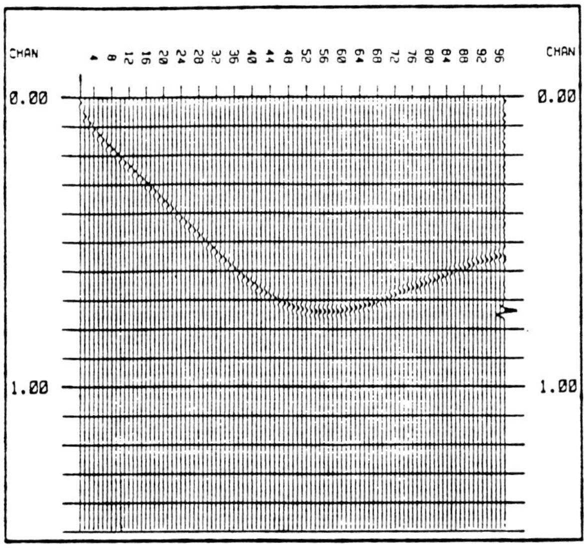

Fig. 9. Common incident angle (40° in Figure 8) section and the stacked trace (at the right). The amplitude depends on the amplitude and the curvature of the original curved event.

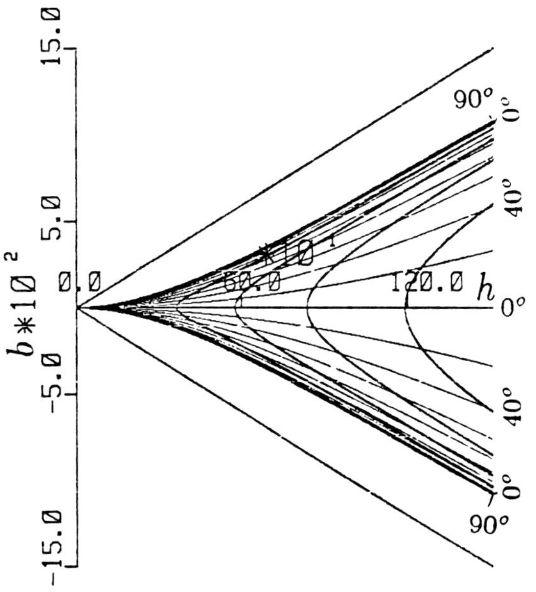

Fig. 10. Equal incident angle stacking trajectories (parabola-like, thick curves) and peak lines (thin and convergent) of the ridge surfaces for all the dip angles.

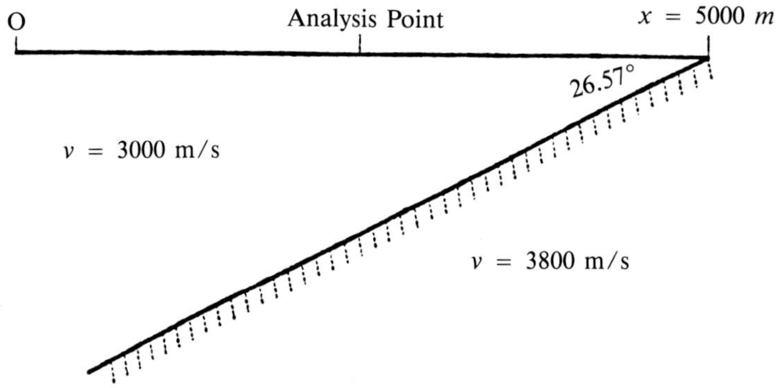

Fig. 11. A geologic model of a dipping reflector. The reflector outcrops at x = 5000 m and dips at 26.57°. The symbol v is P-wave velocity; the densities are 2.5 g/cm^3.

for all the possible dip angles. Figure 10 is an example of an effective incident angle contour map corresponding to the whole dip spectrum (0° to 90°). The hyperbola-like thick curves are effective-incident-angle contours and the thin convergent lines are the ridge lines of the surfaces corresponding to different dip angles. The incident angle contour lines in Figure 8 are tangential to the effective incident angle contours in Figure 10. Figure 10 corresponds to values of T_0 = 0.745 s and v = 3000 m/s. Since we do not know the dip angle in practice, we ought to stack along the effective incident angle contours. This solves the first problem of dip angle uncertainty mentioned in the last section.

Amplitude Calibration

In Figure 9, the number of wavelets arriving at the apex within a certain wavelet duration is inversely proportional to the square root of the second-order derivative of the arrival time curve with respect to b (Gardner and Forel, 1988). All these traces will contribute to the stacked trace, so the flatter the event, the stronger the stacked wavelet. Assuming that the reflection amplitude has been corrected for spherical divergence and for the absorption factor Q, the stacked amplitude multiplied by the square root of the second-order derivative of equation (5) with respect to b should be representative of the reflection coefficient at this incident angle. This second order derivative is given by

$$\frac{\delta^2 t_0}{\delta b^2} = -\frac{t_0}{h^2}\left[1 - 2\frac{(vt_0)^2 b^2}{h^2 - b^2} \times \frac{\tan^2(i) + 1}{8(h^2 - b^2) - (vt_0)^2 \tan^2(i)}\right]^2. \quad (12)$$

There is also a 45 degree phase shift due to the stacking process. This shift can be corrected by applying a phase shift filter.

Automated Extraction of AVA Information

The above extraction algorithm can be completely implemented on a computer without interaction assuming velocity analysis has been accomplished. The steps are:

(1) For a given seismic data set (h, m, t), the data are gathered into the (h, b, t) domain for each analysis point,

(2) DMO plus NMO is applied to the 3-D space (h, b, t) to obtain the 3-D space (h, b, t_0); the velocity values are obtained from velocity analysis,

(3) A matrix, $T_0H(It_0, Ih)$, can be obtained by stacking along effective incident angle contours with a weight for each point. In this matrix, the incident angle is determined from the grid point by $i = \tan^{-1}(h/vt)$

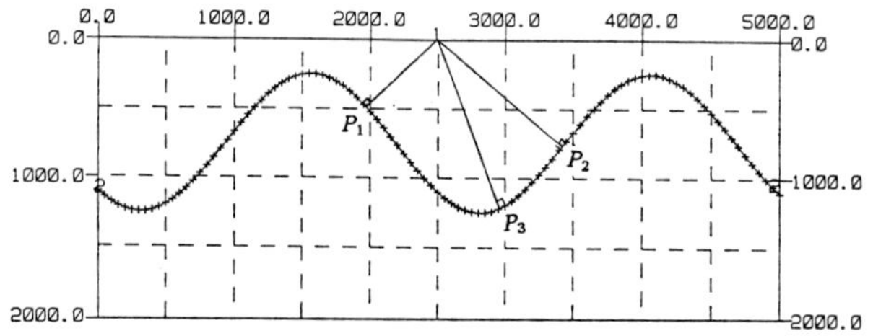

Fig. 12. A geologic model of a folded structure. Three reflections will be received for a zero offset shooting at analysis point X. The velocities above and below the reflector are 3000 m/s and 3300 m/s, respectively. The densities are 2.5 g/cm^3.

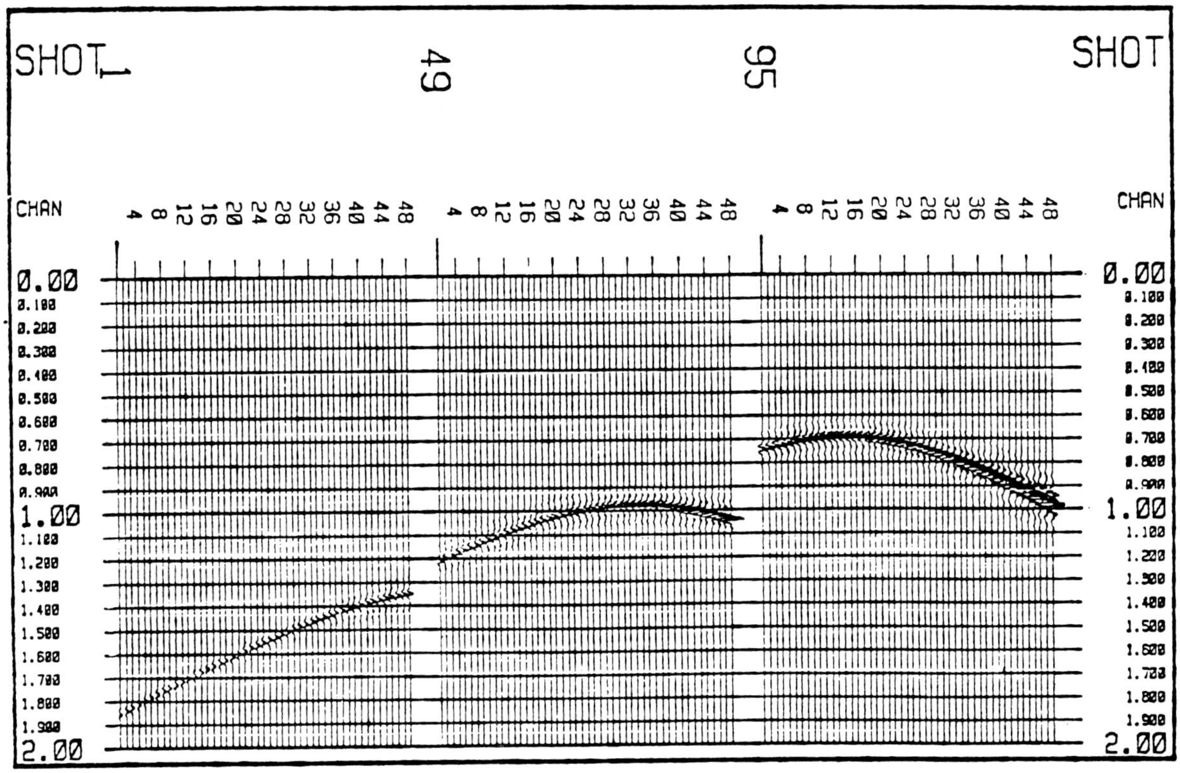

Fig. 13. Some shot gathers of a multifold seismic survey over the model in Figure 11. 95 shots are used with the first being at $x = 0$ and 25 m shot spacing. 48 channels are used for each shot with near offset 0 m and receiver spacing of 50 m.

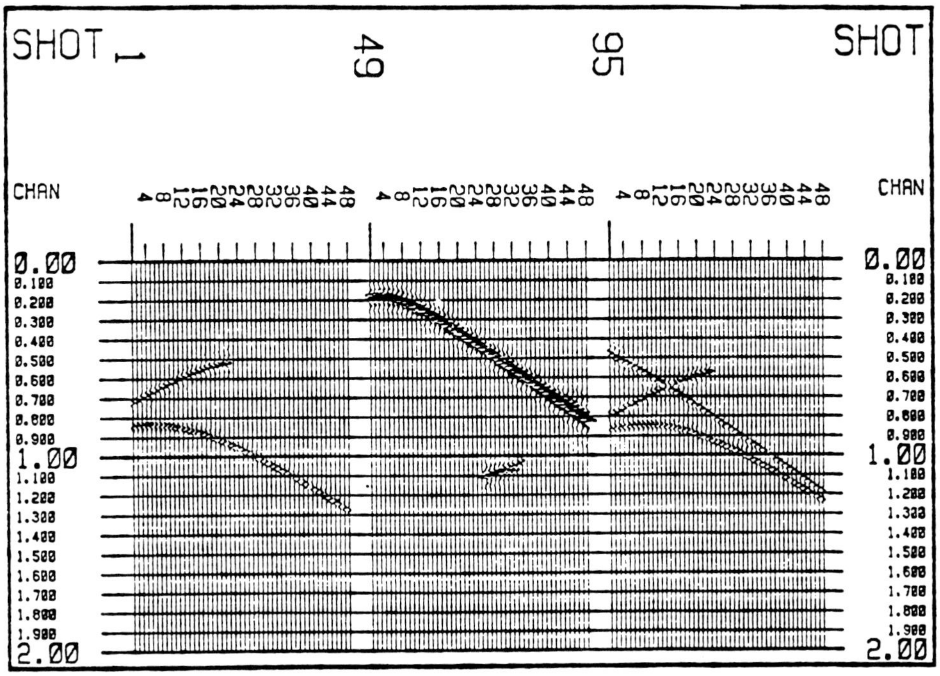

Fig. 14. Some shot gathers of a multifold seismic survey over the model in Figure 12. 95 shots are used with the first being at $x = 0$ and 25 m shot spacing. 48 channels are used for each shot with near offset 0 m and receiver spacing of 50 m.

so that the offset-angle relationship is the same as in the CMP panel of a horizontal reflector. The weight is given by equation (12). The trace is shifted by a 45 degree phase filter to correct the stacking effect on the phase.

EXAMPLES

To demonstrate the validity of the algorithm, two geologic models were used. The commercial software package AIMS by GeoQuest was used to generate the multifold seismic data sets. Figure 11 shows model 1 which is a simple dipping reflector. Figure 12 shows model 2 which is a folded structure. In both surveys, 95 shots were created with 48 channels per shot. The shot spacing is 25 m and receiver spacing is 50 m; the nearest offset is 0 m. A (t_0, h) panel at 2500 m is calculated for each data set.

Figures 13 and 14 show shot gathers from the models. Note that the abrupt truncation of the wave train is due to the interruption to the raypaths by the neighboring anticlines.

Figures 15 and 16 are the processed (t_0, h) panels for Figures 13 and 14, respectively.

Figures 17 and 18 are the AVA plot at $t_0 = 0.745$ s for model 1 and at $t_0 = 0.800$ s for model 2. The smooth curves are the theoretical relationship and the rough curves show the extracted relationship. As can be seen, there is a good fit between the theoretical calculations and automatically extracted values.

CONCLUSION

Two ways are presented by which we can extract amplitude variation with angle of incidence information even in the presence of structure. Important features of the methods are (1) the ability to extract AVA information without knowing the shape of a reflector and without interaction to determine the dip angle in the second method; (2) improved accuracy in incident angle calculation and removal of the smearing effect inherent in common midpoint AVO analysis since true CRP gathers, rather than CMP gathers, are used; and (3) low computational cost.

ACKNOWLEDGMENTS

This research was partially supported by the industrial sponsors of the Seismic Acoustics Laboratory at the University of Houston.

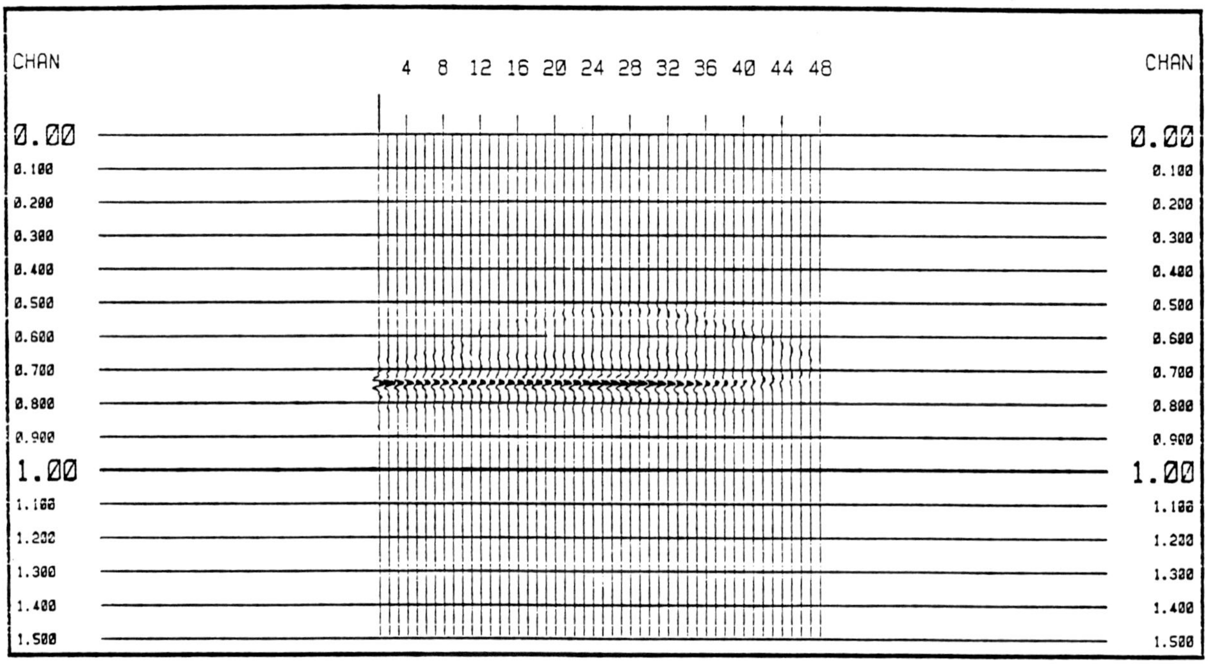

Fig. 15. The output CRP panel at the analysis point $x = 2500$ m. This panel can be used for velocity analysis and AVA analysis as if the reflector were horizontal since the dip effect has been removed. Note that the tail of the event is not reliable because of a limited offset range. See model Figure 11.

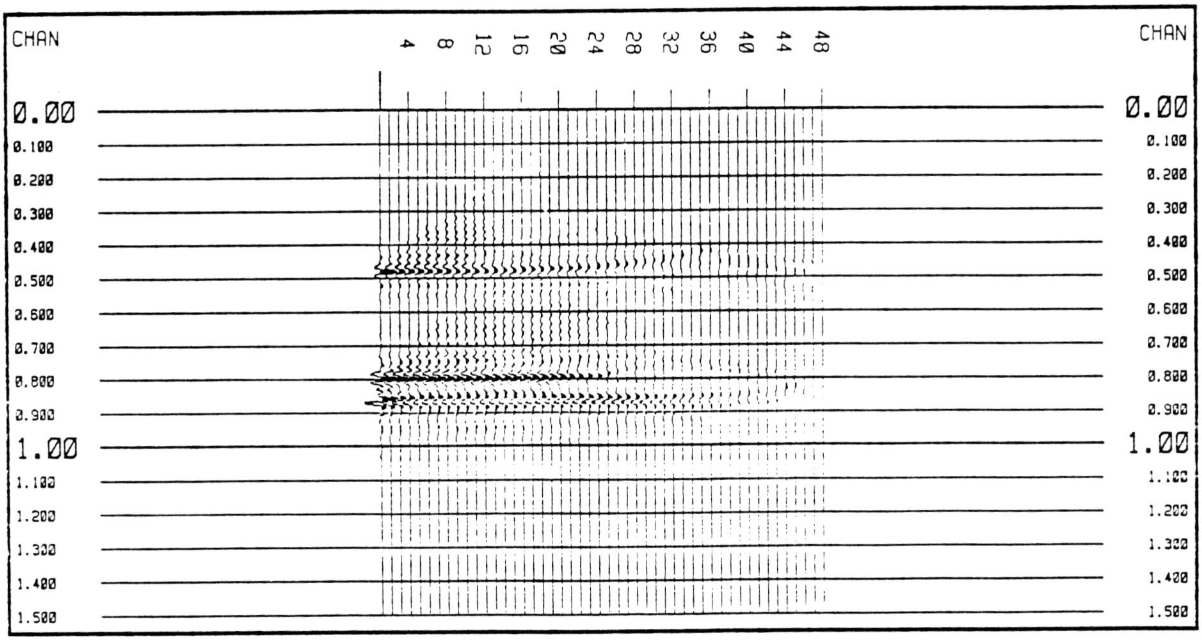

Fig. 16. The output panel at the analysis point $x = 2500$ m. This panel can be used for velocity analysis and AVA analysis as if the reflector were horizontal since the dip and structure effect has been removed. As we can see, the three reflections are correctly positioned. Note that the tail of each event is not reliable due to limited offset range and steep dip. See model Figure 12.

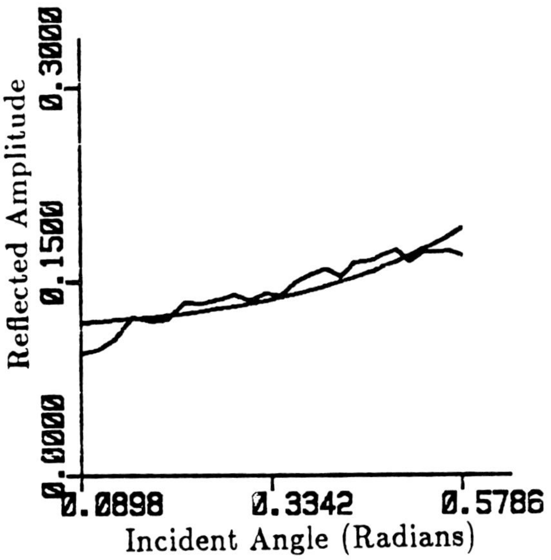

Fig. 17. The extracted AVA curve (rough) and the theoretical AVA curve (smooth) at $t_0 = 0.745$ s for the model in Figure 11.

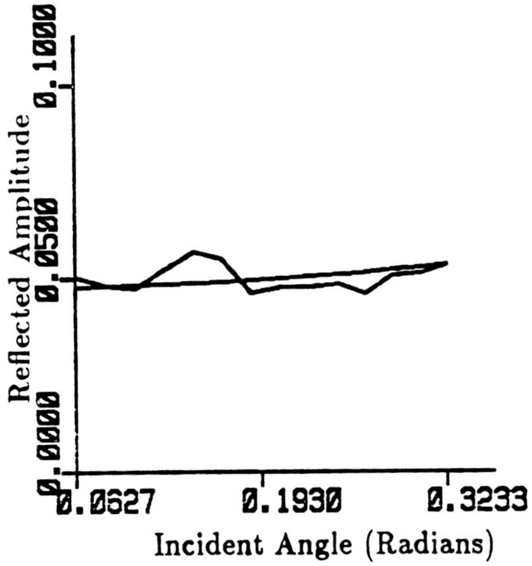

Fig. 18. The extracted AVA curve (rough) and the theoretical AVA curve (smooth) at $t_0 = 0.800$ s for the model in Figure 12.

REFERENCES

Forel, D., and Gardner, G. H. F., 1988, A three-dimensional perspective on two-dimensional dip moveout: Geophysics, **53**, 604–610.

Gardner, G. H. F., and Forel, D., 1990, Amplitude preservation equations for DMO: Geophysics, **55**, 485–487.

Resnick, J. R., Ng, P., and Larner, K., 1987, Amplitude versus offset analysis in the presence of dip: 57th Ann. Internat. Mtg., Soc. Expl. Geophys., Expanded Abstracts, 617–620.

Shang, Z., McDonald, J. A., and Gardner, G. H. F., 1989, The extraction of amplitude versus incident angle information in the presence of structure: 59th Ann. Internat. Mtg., Soc. Expl. Geophys., Expanded Abstracts, 713–715.

Shang, Z., 1990, The extraction of amplitude versus incident angle information in the presence of geologic structure: Ph.D. dissertation, University of Houston.

Chapter IV

Interpretation

AVO Applications in Saudi Arabia

*Edward F. Chiburis**

INTRODUCTION

The past several years have seen steps taken in the application of amplitude variation with offset (AVO) techniques to prestack reflection data. These steps were obviously prompted by an attempt to exploit information contained in the additional dimension and by knowing the theoretic amplitude changes which can occur due to different lithologies and contained fluids. Offsets of 2000 m for both land and marine programs provide sufficient sampling to detect AVO changes, which may be related to change in formation properties. Regardless of source-and receiver-array attenuation effects and of the use of vertical-component receivers on land, the offset dimension can provide more significant information relevant to reservoir and stratigraphic properties than is usually appreciated.

Ostrander (1984) was one of the earlier workers in exploration to point out the value of AVO analysis in distinguishing between "bright spots" produced by gas or by high-impedance formations such as basalt stringers. Gassaway and Richgels (1983) and Gassaway (1984) have utilized AVO in determining changes in formation properties (usually Poisson's ratio) as a function of depth in the vicinity of well control. Onstott et al. (1984) and Todd and Backus (1985) have utilized various display schemes in an attempt to detect amplitude anomalies in gather data. Yu (1985), in pointing out the numerous distortions that can be introduced by the manner in which data are processed, attempted to correct and avoid many of these problems. Wrolstad (1986) has analyzed AVO in the North Sea and tried to understand some of the subtle perturbations in the amplitudes by comparing high-resolution offset synthetics with the gathers in the vicinity of well control. Much of this earlier work has been directed toward correcting or displaying prestack amplitude data in order to observe meaningful behavior in the offset domain.

*Saudi Arabian Oil Company (Saudi Aramco), Dhahran, Saudi Arabia, Formerly Arabian American Oil Company (Aramco).

FACTORS AFFECTING AMPLITUDES

The problems in analyzing seismic amplitudes are well known to research and interpretation geophysicists. The principal factors which produce amplitude variations or errors include, but probably are not limited to, the following:

1. Partitioning: This factor includes reflection and transmission effects and offset-dependent behavior due to the physical properties of various media in which waves are transmitted or reflected. This is the feature to be exploited in AVO analysis.
2. Source and receiver characteristics: This factor includes those effects due to different source strengths and receiver gains, and attenuation due to source- and receiver-array geometries (non-normal incidence).
3. Instrumentation: Included are the output responses of the particular equipment used as well as cable noise and attenuation.
4. Processing: This is perhaps the single most important factor which can affect amplitudes. Just a few of the many processing functions which can seriously modify amplitudes are trace equalization and balancing, spherical divergence and other time- and/or offset-variant gain functions, bandpass and other noise filtering schemes, deconvolution, wavelet shaping, and multiple suppression procedures.
5. Attenuation and Absorption: These factors include geometrical spreading losses and specific dissipation or Q effects, as well as anelasticity, anisotropy, and other frequency-dependent phenomena.
6. Layer Tuning: The effects of layer thickness relative to wavelength are known to have considerable influence on the absolute recorded amplitudes, ranging from significant amplitude increases when the reflections from the top and bottom of the target layer constructively interfere (at about 1/4 wavelength) to complete collapse of the signal when the layers become very thin. Furthermore, due to hyperbolic convergence, the degree of amplitude interference can be offset dependent if very long offsets (large local incident angles) are used in the analysis.

7. Multiple and Mode-Conversion Interference: This factor is most insidious when unrecognized, but similar moveout multiples and mode conversions (*P-SV*) interfere with the primary signals of interest and distort the amplitude behavior versus offset. Also included here, but easily recognizable, are the large distortions of the signal caused by crossing multiples and conversions at specific offset ranges within a gather.

8. Shallow Velocity and Density Variations: If variations in the physical properties occur laterally in the layers above the target horizon over dimensions less than the extent of the shot or CDP gather, the effect on amplitudes at different offsets can be significant.

9. Near Surface Geology: This factor includes variability in the weathering layer, water table, presence of sand dunes, salt flats, etc.

10. Noise: Included here are random as well as temporally and spatially coherent ambient, cultural, and signal-generated noise fields. Obviously, if the signals cannot be seen above the noise, then AVO analysis is impossible.

Clearly, unless corrected for or otherwise eliminated, the presence of any or all of the above factors would invalidate the measurement of amplitudes. To correct for all of these effects would require a knowledge beyond that which we normally possess about a particular seismic acquisition system or geophysical prospect. Worse yet would be to assume we know how to correct but in fact do not; such practices permit an interpreter to "prove" preconceived notions or already-known facts about an area with the use of amplitudes. However, by using relative, instead of absolute, amplitudes, and by applying suitable analysis techniques, most of the error-causing factors can be either eliminated exactly or at least minimized. With such a procedure, subsequent interpretations of AVO changes may be made on a firmer basis.

The feasibility of this approach in detecting gas/oil contacts in the Arabian Gulf was demonstrated in Chiburis (1984); the technique was modified and extended to similarly detect oil/water contacts in an onshore area of eastern Saudi Arabia (Chiburis, 1987a). With improvements in the algorithms and in the development of interactive analysis schemes, the technique has been applied to a number of areas, both land and marine, in Saudi Arabia (Chiburis, 1987b); the target reservoirs of these areas are composed of sands as well as carbonates and variously contain fluids of gas, oil, or water (brine) at depths of 2100 m to 4100 m.

THEORY

The single hypothesis as earlier proposed (Chiburis, 1984) remains the same: given any porous lithologic unit, the behavior of *P*-wave reflection AVO will measurably vary, depending on whether the pore spaces are filled with water, gas, or oil. This demands, of course, that no, or few, other physical changes are occurring within the particular transmitting and reflecting system. Such AVO variations are most easily explained by modeling different Poisson's ratios of the bulk lithologic unit for the different fluid types.

Although few data are available on in situ changes in compressional and shear velocities and density (at seismic bandwidths) due to different contained fluids as a function of porosity, differential pressure, and temperature, laboratory experiments on these effects suggest that significant changes do occur depending on whether the fluid is gas, oil, or water (e. g., Wyllie et al., 1958; Geertsma and Smit, 1961; King, 1966; Nur and Simmons, 1969; Takeuchi and Simmons, 1973; Domenico, 1974; Elliot and Wiley, 1975; Gregory, 1976; Toksoz et al., 1976). These changes do not involve only one of the physical parameters of the rock unit, but are in combination and may actually modify the velocity in directions opposite to one another. In particular, for porous sandstone, compressional velocities decrease when the contained fluid is changed from brine to oil or gas, whereas the shear velocity either does not change or increases significantly (King, 1966). Also, Toksoz et al. (1976) have shown that pore-aspect ratios can have large effects on both compressional and shear velocities, and that compressional velocity is relatively more sensitive to fluid type than shear velocity.

When free gas is present in a porous medium, Poisson's ratio, σ, is low, usually between 0.1 and 0.2, whereas for fully brine-saturated porous rock σ is often greater than 0.35; for oil, σ is intermediate. It should be pointed out that σ is a function of the ratio of shear (β) and compressional (α) velocities denoted by $\gamma = \beta/\alpha$:

$$\sigma = (0.5 - \gamma^2)/(1 - \gamma^2). \quad (1)$$

Alternatively,

$$\gamma^2 = (0.5 - \sigma)/(1 - \sigma). \quad (2)$$

Using laboratory results from King (1966) for sandstone (25 percent porosity) at depths of about 2000 m containing different saturants, and a blocked sonic and density marine model obtained from well logs in the Arabian Gulf, offshore Saudi Arabia, the earth model given in Table 1 was used for calculation. Plane-wave amplitudes (via the Zoeppritz equations) from the three fluid models out to a maximum offset of 2000 m and normalized to zero offset are shown in Figure 1. The gas case shows an increase in AVO of 11 percent, the oil an increase of 4 percent, and the brine a decrease of 5 percent. The feature to be exploited here is the difference in the AVO behavior of prestack gathers analyzed along a seismic line when moving off a structure containing hydrocarbons; the point at which the AVO pattern changes (a 9 percent to 16 percent decrease for this example) suggests that a

Table 1. Sonic and density model, Arabian Gulf. Poisson's ratio of upper layers assumed to be 0.33, except for the water layer. Compressional velocity (V_p) and shear velocity (V_s) for target formation are from King (1966) for sandstones (25 percent porosity) at 2000 m depth containing dry gas, oil, and brine.

D_z, m	V_p, m/s	V_s, m/s	Density	Poisson's Ratio
50	1500	0	1.00	0.50
210	1830	915	2.20	0.33
260	2290	1145	2.25	0.33
195	3050	1525	2.27	0.33
180	3960	1980	2.50	0.33
325	3810	1905	2.15	0.33
300	4870	2435	2.40	0.33
90	3810	1905	2.35	0.33
75	2440	1220	2.35	0.33
120	4270	2135	2.60	0.33
140	3475	1732	2.20	0.33
90	3700	1850	2.40	0.33
100	3275	2100	2.10	0.15 Gas
100	3392	2025	2.10	0.22 Oil
100	3450	1950	2.10	0.27 Brine

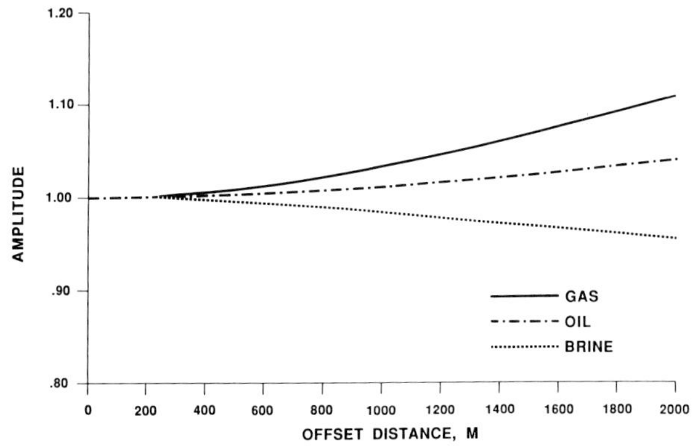

Fig. 1. Theoretical amplitude variation with offset (normalized to zero offset) for three multilayer models (see Table 1) with target layers containing gas, oil, or brine.

different fluid content may be responsible. The only questions are whether these levels of amplitude change can be observed on actual data.

AVO PROCEDURES

Processing

Prestack processing must be held to a minimum. Only those functions which are necessary to improve signal-to-noise ratios (S/N) of the appropriate signals should be used. A routine processing sequence might be the common-depth-point (CDP) sort, deconvolution, datum and residual statics, and normal moveout (NMO). (The only reason for applying statics and NMO is to reasonably flatten the events such that an interactive picker is easier, more efficient, and convenient to use.) Because amplitude ratios are used in the AVO procedures developed in this paper, there is no need to apply gain functions such as spherical divergence; interactive event pickers have gain applied for display purposes only. If necessary, a vertical stack (simple sum) of several adjacent gathers can be made, which usually produces a worthwhile improvement in S/N, especially in areas of low dip. In severe multiple areas, more complex demultiple procedures may have to be applied.

Event Correlation

Utilizing the experience of interpreters and geologists familiar with the prospect, the target and reference events are identified on the stacked sections with the aid of well log information and synthetics. Particular note is made of the expected polarities, target layer thicknesses, regions of potential multiples and mode conversions obscuring the signals, and regions of "fade" on the stacks. If sufficient well data are available, note is also made as to any possible changes in stratigraphy, lithology, porosity, and bed geometry of the layers involved (both target and reference). In most cases, it makes no difference whether the top or the bottom of the target formation is picked; the AVO effects remain nearly the same, at least in a general behavioral sense (this is easily demonstrated from offset-dependent synthetic studies using carefully blocked well logs, suitable wavelets, and interactive measurements).

Analysis Procedures

The current analysis procedures for exploiting the concepts of (AVO) with surface seismic data, as presented in Chiburis (1987a, b), are composed of the following steps:

1. Interactive measurements of peak amplitudes of target $T(x)$ and reference $W(x)$ events at offsets x, within a gather: This is an extremely important and necessary step; the difficulties in trying to automatically pick prestack peak amplitudes are numerous, principally due to errors in NMO, signal contamination, selection of proper time windows, etc. This step minimizes the effects of Factor 7, (Multiples and Mode Conversions) and possibly Factor 10, (Noise).

2. Normalization of target amplitudes to reference amplitudes on a trace basis using

$$R(x) = [T(x)/T_a]/[W(x)/W_a], \qquad (3)$$

where $R(x)$ is the amplitude ratio at offset x, and T_a and W_a are the average amplitudes of the target and reference events within the gather. This step eliminates exactly the effects of Factors 2 and 3 (Source and Receiver Characteristics and Instrumentation, excluding array response, which, however, is reduced with this step), and lessens the effects of Factors 4, (Processing), 5 (Attenuation and Absorption), 8 (Velocity and Density Variations), and 9 (Near-surface Geology).

With this step, the resultant set of amplitude ratios is bound on the low side by zero (when $T(x)$ is equal to zero) and on the high side by infinity (when $W(x)$ is equal to zero) yielding a one-sided, and unstable distribution. The use of least-squares procedures, which demand a normal error distribution, is inappropriate in such cases. However, each ratio in the set can be modified with the following step.

3. If $R(x) > 1$, the ratio is to be replaced as

$$R(x) = 2 - 1/R(x). \qquad (4)$$

With the application of this condition, the ratios are bound as $0 \leq R(x) \leq 2$. The resultant least-squares error distribution is now Gaussian and equal weight can be given to deletion of outliers whether on the high- or low-side of the least-squares curve fit.

4. Least-squares (parabolic) curve fit the form

$$F(x) = A + Bx^2 \qquad [1 \leq x \leq N] \qquad (5)$$

where N is the number of traces in the gather. With uniformly spaced traces, it is immaterial whether trace number or offset is used. The intent here is not to attempt an analytical representation or an approximation of the Zoeppritz equations, but only to determine amplitude trends from the data using limited offsets.

5. Optionally, delete extreme errors or weight the observations according to the degree of signal contamination. This step can help in reducing the effects of Factor 7 (Multiple and Mode Conversions).

6. Re-fit if deletions are made.

7. Form the AVO difference (relative to zero-offset) as

$$\begin{aligned} \text{AVO difference} &= [F(N) - F(O)]/F(O) \\ &= (A + BN^2 - A)/A \\ &= (B/A) * N^2. \end{aligned} \qquad (6)$$

In order to compare the degree of AVO difference obtained from different lines (which may have been shot with different N), all values are arbitrarily calculated as

$$\text{AVO difference} = (B/A) * 1000 \qquad (7)$$

This step in conjunction with Step 2, virtually eliminates the effects of Layer Tuning (Factor 6). Tuning determines the absolute reflection strength but has minimal effects on relative offset behavior, especially if limited ranges are used (less than about 2000 m).

The above procedures yield a single AVO difference for each gather along a line. The results are displayed on a line basis for the area under investigation.

Although the remainder of this paper will deal with AVO differences, as defined above, uses of the term "AVO" when results are given will imply "AVO differences".

An example of the flow for the above procedures is shown in Figure 2. The (a) plot shows the measured target $T(x)$ and reference $W(x)$ amplitudes versus trace number (offset), normalized to their respective amplitudes (T_a and W_a) within the gather (Step 1). The (b) plot shows the raw ratios $R(x)$ (Step 2). The (c) plot shows the conditioned ratios (Step 3) as a solid curve with no symbols on the conditioned-ratio curve (Step 5); and the final least-squares curve fit as the smooth curve with symbols (Step 6). The result for this particular gather, which is positioned above an oil-filled carbonate reservoir, is that the amplitude behavior of the target event relative to the reference event increases with offset, and the AVO is about 1.5 (we still do not place more than qualitative significance on the actual value of the AVO, only in its overall trend).

In this paper, the maximum offsets used for AVO analysis were restricted to 2000 m or less because of far-offset contamination caused by multiples and mode conversions. Such limited offsets are usually sufficient to determine AVO trends, but not so far that the analyses become unstable.

It should be noted that the above procedures exploit the relative aspects of AVO analysis:

1. Amplitude normalization on a trace basis.
2. Amplitude-ratio normalization to zero-offset.
3. Comparison of AVO differences along a line.

AVO RESULTS

Six areas have been studied thus far in Saudi Arabia: three offshore in the Arabian Gulf, and three onshore in the Eastern Province. Two of the offshore areas have gas targets (one of these areas was our first successful AVO application; Chiburis, 1984), and one has oil and an unknown fluid, all at depths of about 2000 m. Two of the onshore areas have oil targets at depths greater than 3000 m (one of these areas was our first success at detecting oil/water contacts with AVO; Chiburis, 1987a), and one has two separate gas targets at depths exceeding 4000 m. Each of these areas, and their results, are discussed in the following sections.

Land Area 1

The target in Land Area 1 (Chiburis, 1987a) is an oil-filled carbonate at a depth of about 3000 m. The relevant physical parameters are given in Table 2. The initial seismic data selected for the amplitude analysis are from an east-west line which crosses the surface-projected location of the oil/water contact once on the western edge of the field. The data for the analysis were CDP gathers spaced at 22.5 m along the line; eleven adjacent gathers were summed to improve S/N. Examples of the summed gathers are shown in Figure 3; the target event is at about 1700 ms and the reference event at about 2100 ms (times are relative to the surface). Because of far-offset noise, only the near 30 traces at 45 m spacing were used. Every fifth of the summed gathers (112.5 m apart) was analyzed for a total of 57. Starting over the top of the field with CDP 1, the AVO results from the first 24 gathers are shown in Figures 4a and 4b, in which the AVO values from the first 12 gathers are increasing, while those from gathers 13–24 are decreasing.

The AVO values are plotted (one AVO per gather) for the entire line in Figure 5. The (a) plot shows the actual values, the (b) plot shows a three-gather smooth, and the (c) plot shows the smoothed AVO in which the base line for plotting has been arbitrarily set to zero. The results for this line show distinct changes in the AVO behavior between gathers 13 and 14, a position which correlates within one CDP of the best estimate of the oil/water contact for this field.

It is rather interesting that this particular line was

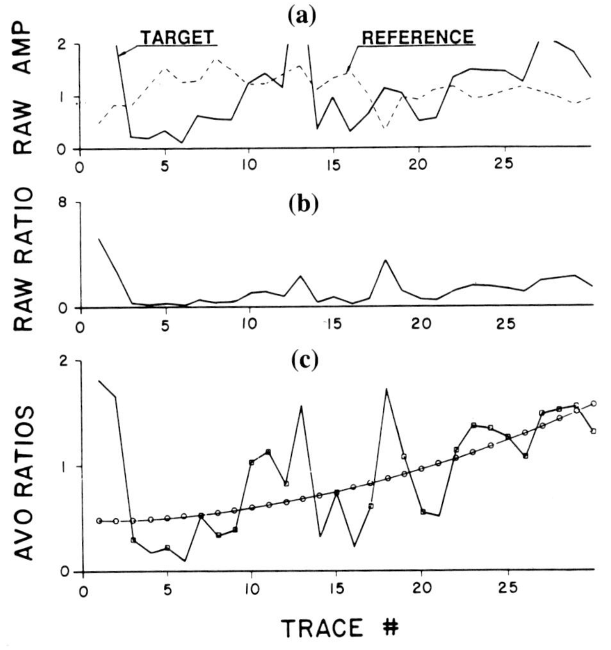

Fig. 2. AVO data for a single gather. (a) measured target (solid line) and reference (dashed line) amplitudes; (b) amplitude ratios; (c) conditioned ratios and least-squares fit.

Table 2. Physical properties of target layer when oil- and water-filled: compressional velocity, density, porosity, and thickness) and overlying layer (compressional velocity and density) for Land Area 1.

LAND AREA 1

OVERLYING LAYER
LIMESTONE
P = 6000 M/S
D = 2.7

3200 M

RESERVOIR
CALCARENITE

	OIL - FILLED	WATER - FILLED
P	4500 M/S	4900 M/S
D	2.51	2.57
ϕ	20 %	14 %
dz	50 M	30 M

LINES : 5
VSTACK : 11/5 - 11/10
NTR : 20 - 40
DX : 45 - 90 M

the first one analyzed for AVO with a view toward detecting oil/water contacts. Although the amplitudes were picked interactively, the analysis was done automatically by arbitrarily deleting outliers greater than one standard deviation from the least-squares fit of the conditioned ratios. Furthermore, the results are virtually the same even if the reference event is eliminated from the normalization procedure. However, few of the other lines analyzed for the eight areas demonstrate such a strong and robust AVO result, and the reference event cannot be eliminated from AVO analysis in general. It is not clear at this time why the AVO results for this particular line are so large and distinctly different for the oil and water cases, but the reasons may be due to better equipment and acquisition parameters and to simpler geology.

The production stack of this line (with times from a datum instead of the surface) is shown in Figure 6, with the position of the oil/water contact indicated at the top of the section and the target and reference events indicated at the sides. Based on the parameters given in Table 2, the top of the brine-filled reservoir provides an impedance contrast of about −0.13 and

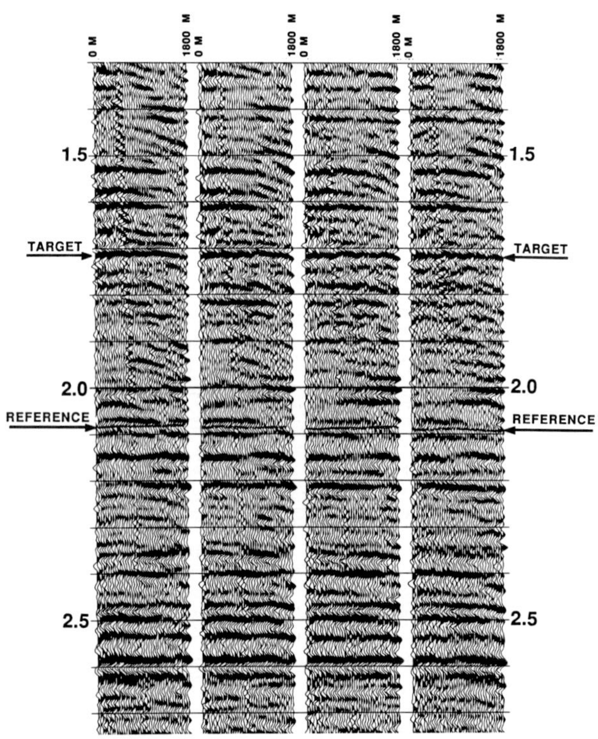

Fig. 3. Examples of CDP gathers used for amplitude measurements. Target (peaks) and reference (troughs) events are as indicated.

the oil-filled reservoir about −0.18. However, the stacked data in Figure 6 do not show the "bright spot" as would be predicted from a 50 percent impedance change, a result perhaps due to stack smearing.

It is possible that the abrupt AVO change observed in Figure 5 may be related, in part, to porosity differences as well as to changes in fluid type. Reservoir porosities over most of the structural highs in Saudi Arabia are generally larger than those on the flanks of the highs (note the porosities given in Table 2). Hence, the sharp AVO change may be due to both a change in fluid type as well as a change in porosity, both of which would tend to alter Poisson's ratio of the rock unit.

In all, five lines were analyzed for this area; the results for each line, displayed in the manner of Figure 5, are shown in Figure 7 superimposed on the unmigrated and uncorrected time structure of the field. The dashed line in this figure shows an earlier estimate made of the limits of the oil based on a best effort at an appropriate time/depth conversion. By using AVO information, the limits could undoubtedly be more closely estimated. Line A to the north, shows no positive AVO effect and indeed does not cross the oil/water contact (note the dry well position at about gather 18 along the line). Line D (5-fold vertical stack, every 10th gather analyzed) shows a gap in the positive AVO between gathers 16 and 20; the reason for the gap is not known, although this portion of the line might be analyzed at a closer spacing than every 10th gather with a possibility of smoothing out the trends.

It should be noted that the edges of the positive AVO regions (with an inferred mappable fluid contact) appear to span about 20 ms or 30 ms in the time structure map shown in Figure 7. Based only on AVO results, the limits of the oil can be determined without knowing the true relationship between time and depth.

The results for Land Area 1 are considered excellent and strongly suggest that AVO studies can closely delineate oil/water contacts for this moderately deep carbonate reservoir.

AVO Sensitivity

It is instructive at this point to view the sensitivity of the AVO differences that would result from a 45-layer land model, given in Table 3, in which the physical parameters of the target and overlying layers are varied. With the definition of AVO (Step 7 of the Procedures) and the selection of the target event (Layer 32 at 3000 m depth) and a reference event (Layer 45 at 4100 m depth), Poisson's ratio of both the

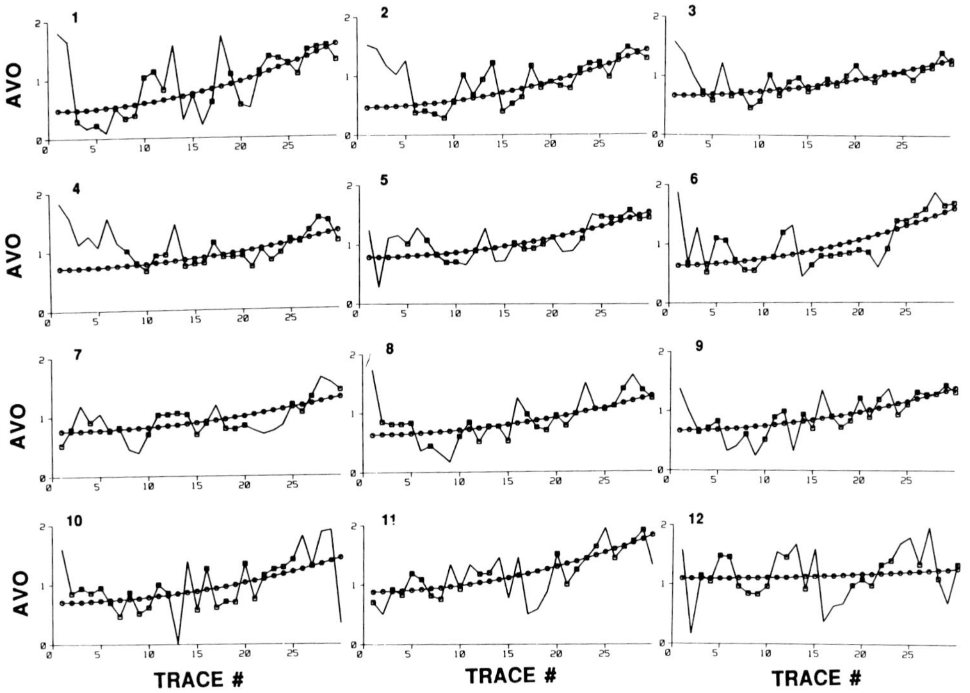

Fig. 4(a). Trace amplitude ratios and AVO curve fits for gathers 1–12, Line E, Land Area 1.

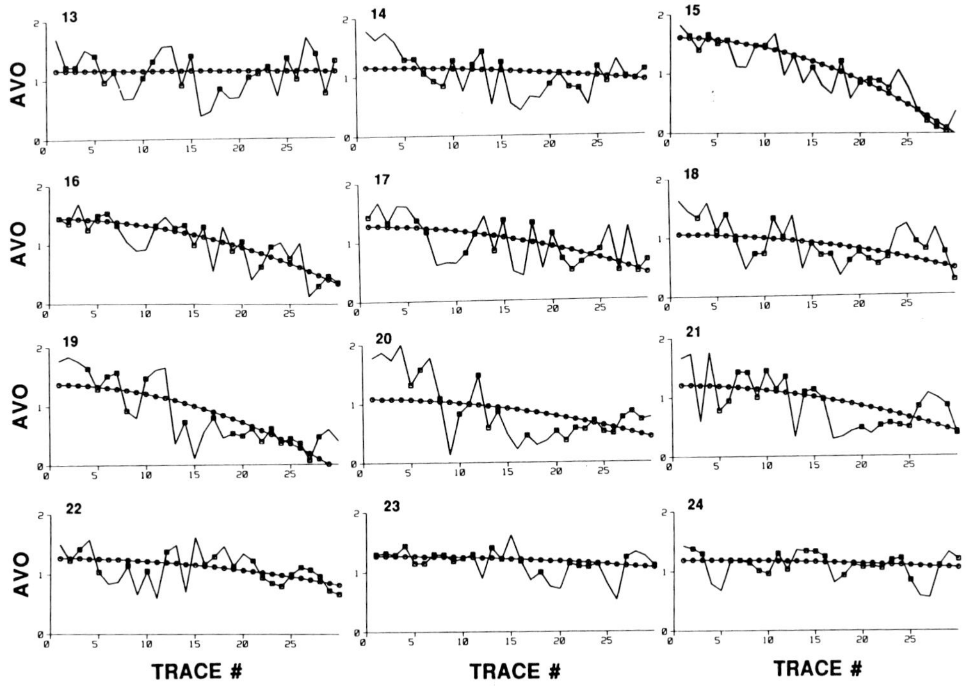

Fig. 4(b). Trace amplitude ratios and AVO curve fits for gathers 13–24, Line E, Land Area 1.

target and overlying layer are varied (via compressional velocity) continuously between 0.0 and 0.45. By plotting AVO calculated at an offset of 2000 m, the results are as shown in Figure 8, in which the (a) plot has a base AVO value of zero instead of −1 as in the (b) plot. A positive AVO in these plots indicates a target amplitude increase with offset relative to the reference event. The point to be demonstrated in Figure 8 is that, if the overlying Layer 31 has a constant Poisson's ratio, say 0.35, and if the target Layer 32 changes from a low Poisson's ratio, say 0.2 as might be for oil, to a higher one, say 0.35 as might be for water, when analyzed for differences along a particular seismic line, the AVO value would decrease from about +0.1 or +0.2 to about −0.05 or −0.1, a relative change of between 15 and 40 percent. Amplitude changes of this magnitude are apparently measurable on real data (as shown in Figure 5). Hence, for this model, if Poisson's ratio increases in a target formation, the AVO will decrease, all other parameters remaining the same. (Note: the region of AVO singularities seen in Figure 8, in which the values swing from −1 to +1 for small changes in Poisson's ratio, is due to those particular combinations of layer parameters which produce values of $F(O)$ near zero).

Figure 9 shows the results obtained by varying Poisson's ratio by changing the S-wave velocity, in which the same decreasing AVO occurs by increasing Poisson's ratio of the target. Because AVO has been calculated with only P-waves involved in the sequence and because only S-wave velocity has been varied, there is no region of singularities as in Figure 8.

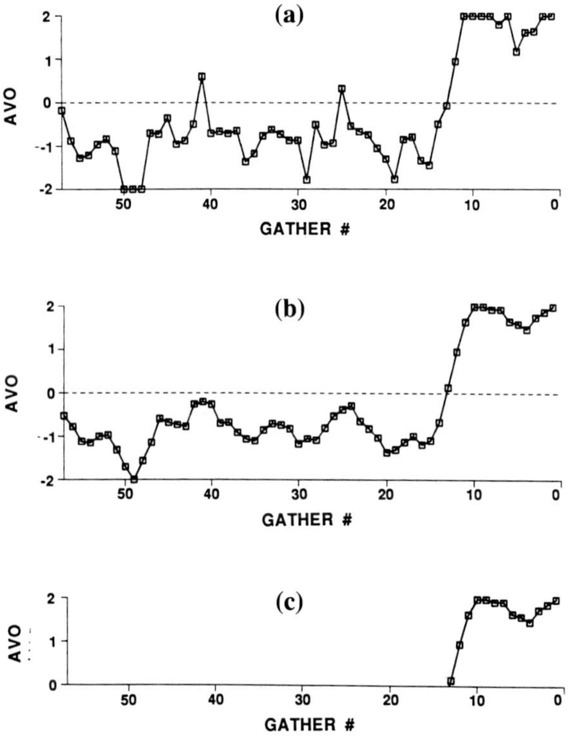

Fig. 5. AVO results for Line E, Land Area 1. (a) complete AVO results; (b) Smoothed AVO results; and (c) Results for AVO values greater than zero.

Table 3. Blocked land model from eastern Saudi Arabia. Poisson's ratios are assumed to be 0.33 throughout. Densities, where unknown or indeterminate, are assigned values of 2.400.

D_z, m	V_p, m/s	V_s, m/s	Density
80	716	358	2.300
60	2042	1021	2.330
86	2698	1484	2.400
13	3208	1604	2.400
17	3586	1793	2.400
12	2988	1494	2.400
15	2117	1058	2.400
6	2313	1657	2.400
58	2771	1385	2.400
8	3278	1639	2.400
5	1979	990	2.400
2	3586	1793	2.400
10	2651	1325	2.400
143	4689	2345	2.400
183	3586	1793	2.400
12	3810	1905	2.400
14	3387	1693	2.400
8	2032	1016	2.400
43	3810	1905	2.400
11	5861	2931	2.400
32	5878	2939	2.400
105	4064	2032	2.400
18	3048	1524	2.400
7	2438	1219	2.400
10	3278	1639	2.400
11	4763	2381	2.400
82	3810	1905	2.400
11	4689	2345	2.400
12	3810	1905	2.400
9	2771	1385	2.400
10	4046	2032	2.400
30	2291	1146	2.400 Target formation
177	4618	2309	2.400
93	4064	2032	2.400
32	3586	1793	2.400
18	2032	1016	2.400
26	4482	2241	2.400
37	2032	1016	2.400
30	4354	2177	2.400
29	3387	1693	2.450
61	2258	1129	2.300
18	2903	1451	2.400
81	4064	2032	2.450
55	3387	1693	2.350
100	2345	1172	2.300 Reference formation

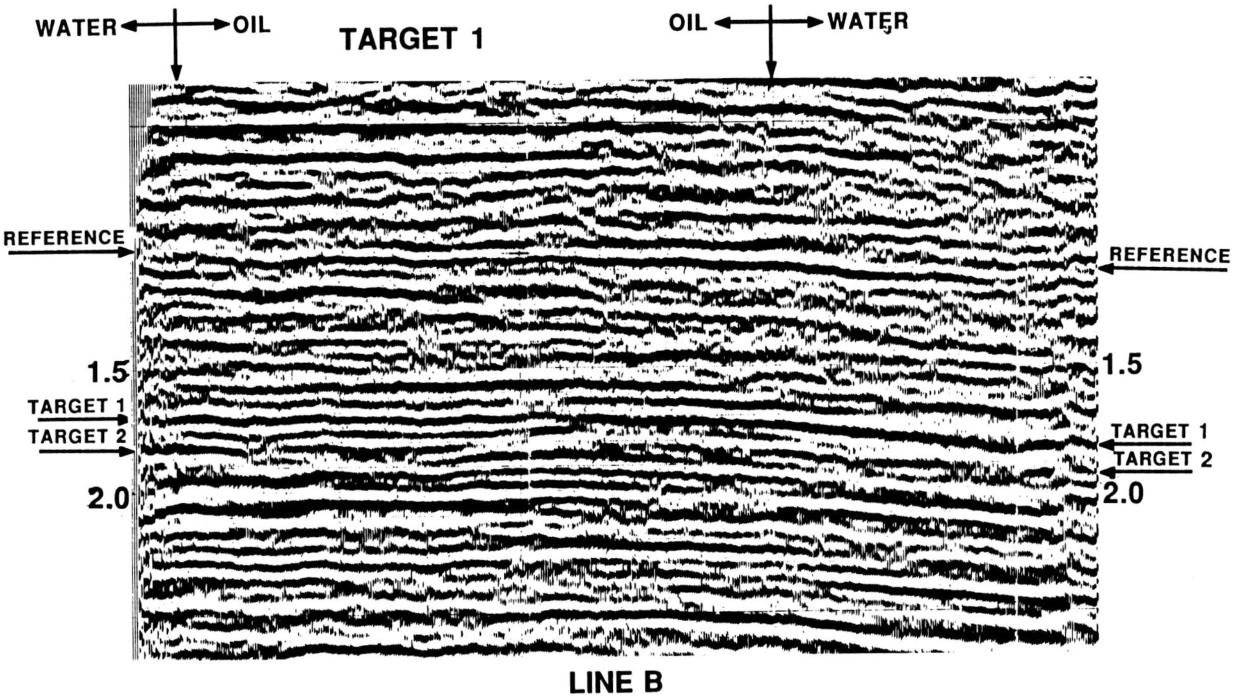

Fig. 12. Production stack of Line B, Land Area 2, showing references (troughs) and two targets (peaks) and presumed surface location of oil/water contacts for shallower target.

Table 4. Physical properties of target layer when oil- and water-filled: compressional velocity, density, porosity, and thickness and overlying layer (compressional velocity and density) for Land Area 2.

LAND AREA 2

OVERLYING LAYER
LIMESTONE
P = 6000 M/S
D = 2.7

3200 M

RESERVOIR
CALCARENITE

OIL - FILLED	WATER - FILLED
P = 4600 M/S	5000 M/S
D = 2.55	2.59
ϕ = 13 %	13 %
dz = 50 M	30 M

LINES : 5
VSTACK : 5/10
NTR : 20
DX : 90 M

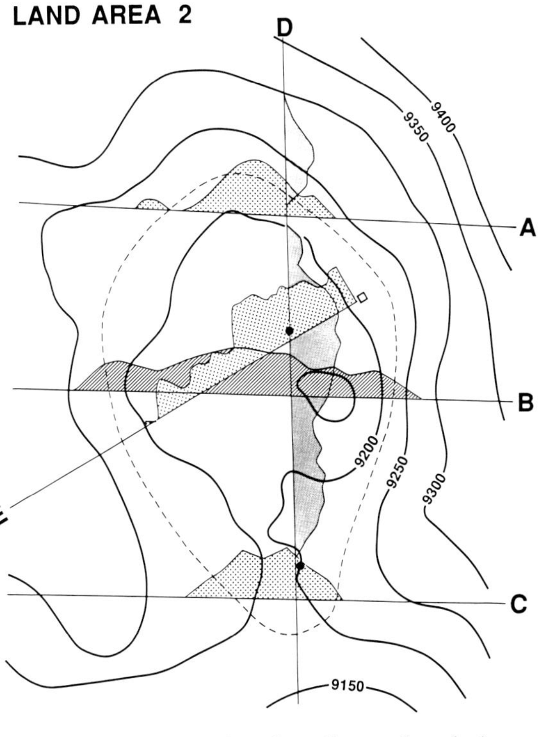

Fig. 13. AVO results for five lines, Land Area 2. Dashed line represents pre-AVO estimate of oil/water contact for shallower Target 1. Contours represent uncorrected depths (ft).

relative aspects of AVO are considered; for absolute amplitudes, corrected or otherwise, the effects could be substantial.

Land Area 3

Land Area 3 contains two spatially separated structural highs with the same target reservoir at about 4100 m depth (2800 ms), the deepest of all the areas studied. The target is fractured limestone containing dry gas. The relevant physical data are given in Table 5. The reference event is a shallower carbonate at about 2600 ms. The objectives in this area are to detect the gas/water contacts of the deep target reservoir for both structures.

Only two lines were analyzed, one of which crosses at least parts of both structures. The production stack of one of these lines is shown in Figure 15, which displays no bright spot. Because of very weak target signal strengths, 51-fold vertical stacks had to be made in order to obtain measurable amplitudes; although this large a fold produces about 600 m horizontal smearing at the level of the target, the AVO interpretation would not be affected because of the emphasis on relative spatial changes in AVO along the lines.

From the AVO results for this area, which are shown in Figure 16, it can be seen that both structures exhibit positive AVO. These results are confirmed from numerous well data throughout the entire area. The important conclusion to be made from this area is that AVO analyses need not be restricted to shallow targets; although angles of incidence at the level of the target as calculated from blocked sonic logs are only of the order of 16 degrees, the AVO results appear to be sensing the fluid contacts. However, because of poor signal quality, the results should be accepted only with fair confidence. More data would be needed to confirm the analysis.

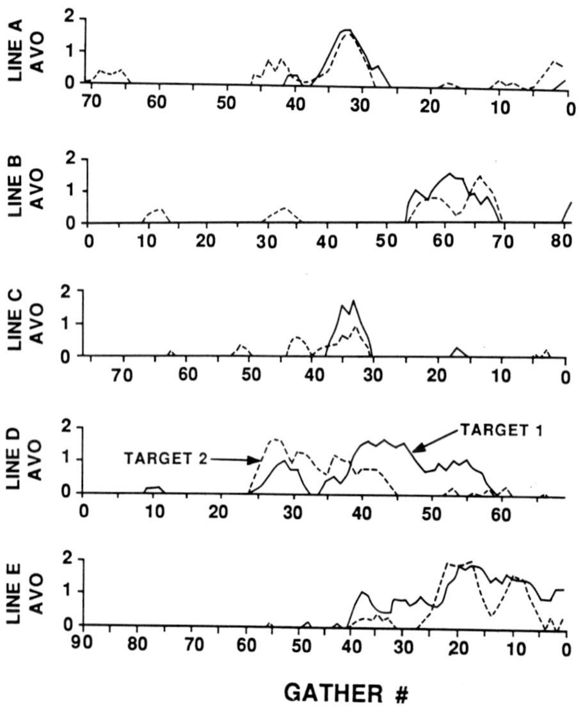

Fig. 14. Comparison of AVO results for five lines and for two reservoirs, Land Area 2.

Marine Area 2

The target in Marine Area 2 (Chiburis, 1987b) is the same oil carbonate reservoir as that in Land Area 1 and the same as the first target in Land Area 2. The depth of the reservoir in this area is about 2400 m. The relevant physical data are given in Table 6. The reference event is about 300 ms earlier than the target. Three lines were analyzed. The production stack from one of these lines is shown in Figure 17. Again, there is no evidence of the expected "bright spot", based on the calculated impedances from Table 6.

The area is characterized by two, possibly three, separate structures. The structure to the east (referred to as Structure I) is known to contain oil in the target reservoir; the structure to the west (Structure II and

Table 5. Physical properties of target layer when oil- and water-filled: compressional velocity, density, porosity, and thickness and overlying layer (compressional velocity and density) for Land Area 3.

LAND AREA 3

OVERLYING LAYER
ANHYDRITE
P = 6100 M/S
D = 2.8

4000 M

RESERVOIR
FRACTURED LIMESTONE

	DRY GAS-FILLED	WATER - FILLED
P	5000 M/S	5400 M/S
D	2.55	2.7
ϕ	15 %	12 %
dz	35 M	35 M

 # LINES : 2
VSTACK : 51/10
NTR : 30
DX : 48 M

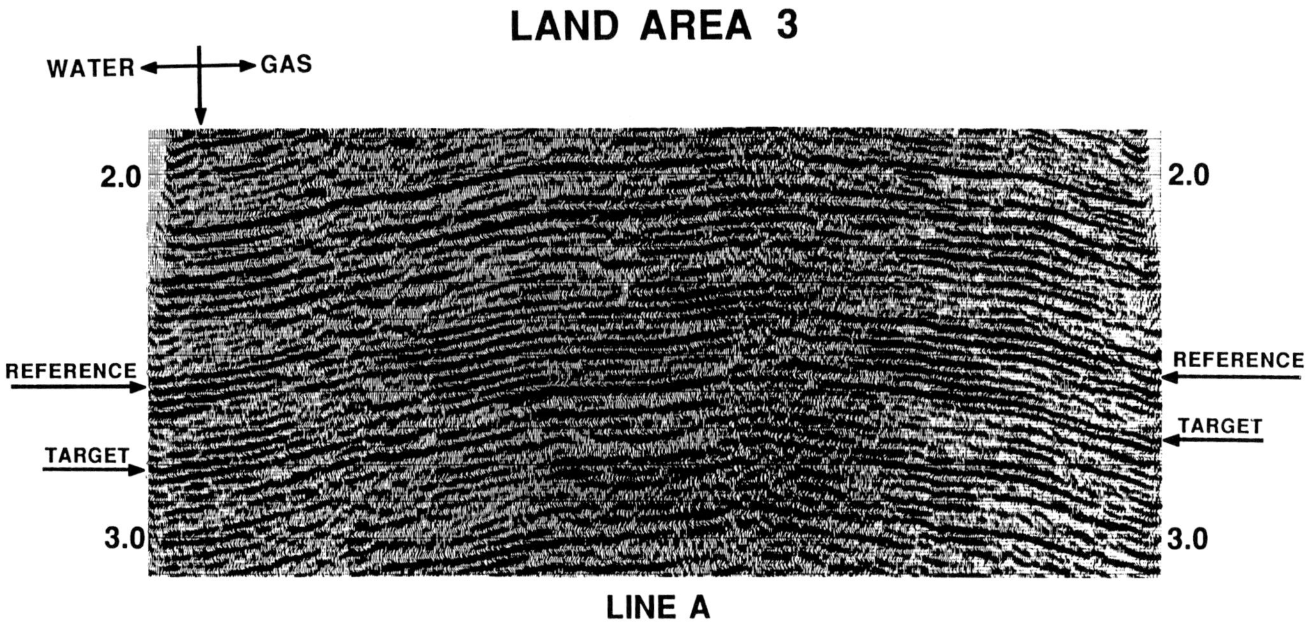

Fig. 15. Production stack of Line A, Land Area 3, showing target (peaks) and reference (troughs) events and presumed surface location of oil/water contact.

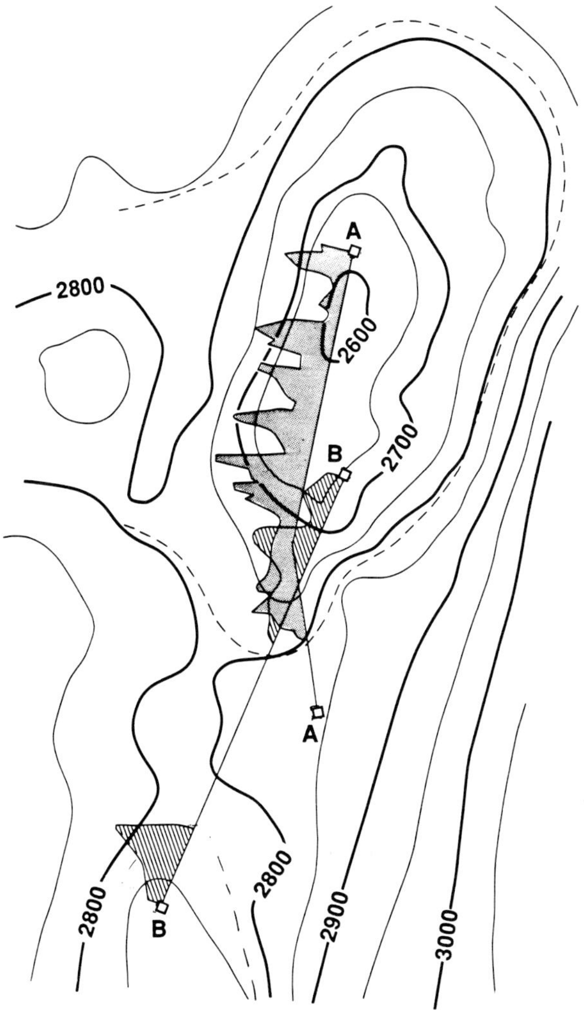

Fig. 16. AVO results for two lines, Land Area 3. Dashed lines represent pre-AVO estimate of gas/water contact for two structures. Contours are uncorrected times (ms).

Table 6. Physical properties of target layer when oil- and water-filled: compressional velocity, density, porosity, and thickness and overlying layer (compressional velocity and density) for Marine Area 2.

MARINE AREA 2

OVERLYING LAYER
LIMESTONE
P = 5800 M/S
D= 2.7

2400 M

RESERVOIR
CALCARENITE

OIL - FILLED	WATER - FILLED
P = 4600 M/S	4900 M/S
D = 2.45	2.5
ϕ = 17 %	14 %
dz = 30 M	30 M

LINES : 3
VSTACK : 11/10
NTR : 30
DX : 60 M

possibly Structure III to the South of Structure II) contains an unknown fluid because no wells have been drilled in the vicinity.

Of the three lines analyzed, one is oriented north-south across Structure I, one is north-south over Structures II and III, and one is east-west across Structures I and II.

The AVO results for three lines are shown in Figure 18. Structure I shows positive AVO from both lines; the oil/water contacts could almost certainly be closely delineated with more seismic coverage. However, the more interesting results are those from structures II and III, which contain unknown fluids; the east-west line shows strongly positive AVO across Structure II; the north-south line not only shows positive AVO across Structure II, but positive AVO into Structure III as well. The apparent gap in the positive AVO between Structures II and III was confirmed by analyzing every gather across the gap: the AVO remains negative. A compelling interpretation of these results is that the AVO behavior strongly implies the presence of oil in the target reservoir for both Structures II and III and further suggests that the two structures are disconnected in their accumulations. This interpretation will also have to await future drilling programs.

Marine Area 3

Marine Area 3 is the last area to be studied. The target is the same gas sand as Marine Area 1 at about the same depth of 2100 m. Unfortunately, the data from two lines were so poor that the target could not be picked with any degree of certainty. On almost every gather (more than 300 in all), the signals were either catastrophically broken up by multiples and/or mode conversions or they were incoherent virtually from trace to trace within the gathers. Although numerous processing schemes were attempted, the efforts were unsuccessful. The principle reasons appear to be due to the variable stratigraphic nature of the target (and associated stringers), the severe thinning and thickening throughout the area, and the overwhelming multiple contamination. This exercise points out the obvious: unless sufficient S/N ratios can be achieved for both target and reference events, an AVO analysis is likely to prove fruitless.

CONCLUSIONS

Six areas have been studied thus far for behavior of AVO in Saudi Arabia. The areas are located both offshore and onshore with target reservoirs composed

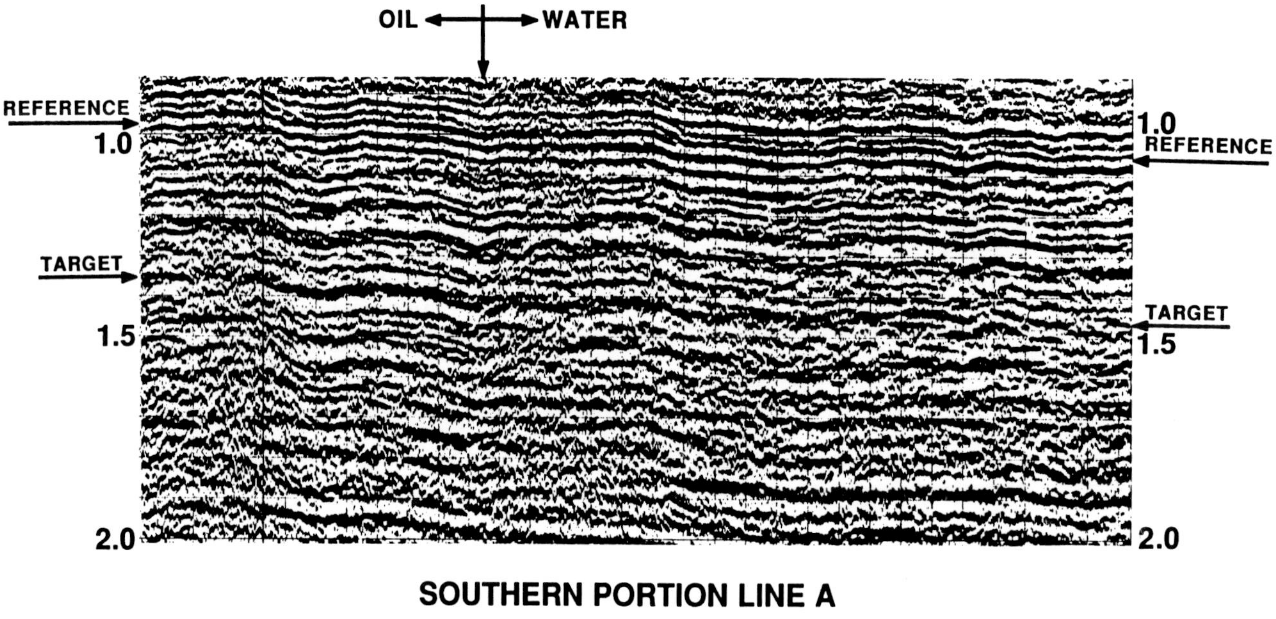

Fig. 17. Production stack of Line A (southern portion), Marine Area 2, showing target (peaks) and reference (troughs) events and presumed surface location of oil/water contact.

of sandstones and carbonates (calcarenite or fractured limestone) containing fluids of gas, oil, or water, and are at depths of between 2100 m and 4100 m. The reservoirs range in thickness from 30 m to 50 m with porosities from 12 to 20 percent. Generally, the API indices of the oils are in the range of 30–36, the gases are dry, and the brines are heavy.

Because of the various errors which may be present in recorded amplitudes, analysis procedures were developed to eliminate or minimize their effects. The key procedures are:

1. Normalization of target and reference amplitudes to their respective mean amplitudes within the gather.
2. Normalization of target amplitudes to reference amplitudes on a trace basis.
3. Deletion of outliers from least-squares parabolic curve fits of the AVO ratios.
4. Normalization of the final least-squares curve fits to the nearest offsets.
5. Comparison of AVO behavioral changes along a seismic line.

These procedures emphasize the spatial changes in the relative AVO for a prospect area.

Within the six areas studied, there are four gas targets, four oil, and one (or two) unknown. The AVO results were excellent for Land Area 1 (oil/water), good for Land Area 2 (oil/water, two targets), Marine

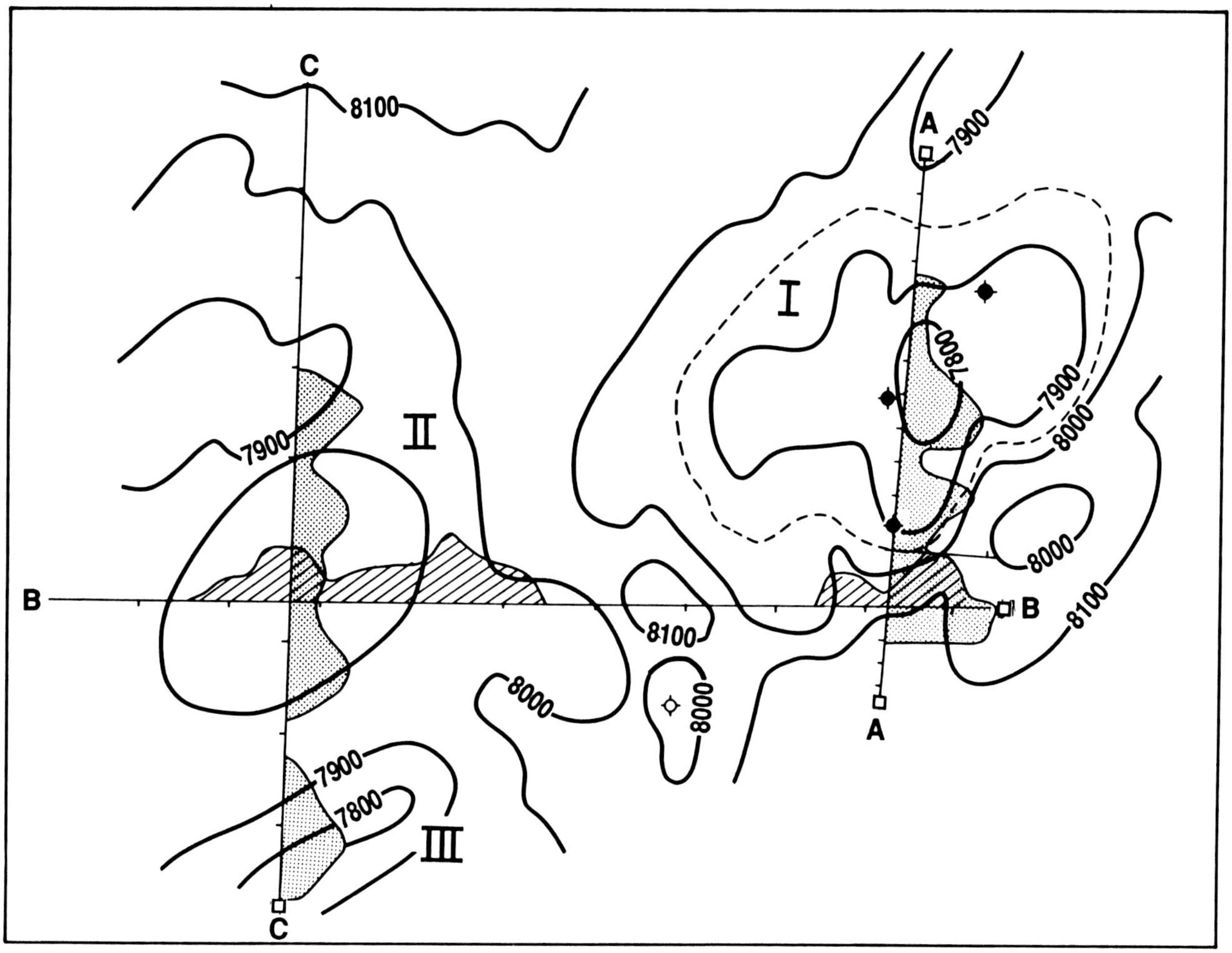

Fig. 18. AVO results for three lines, Marine Area 2. Dashed line represents pre-AVO estimate of oil/water contact for Structure I. Structures II and III contain unknown fluids. Contours are uncorrected depths (ft).

Area 1 (gas/oil), and Marine Area 2 (oil/water; two, possibly three, structures), and fair for Land Area 3 (gas/water; two structures).

Due to poor signal strengths, the AVO analysis for Marine Area 3 (gas/oil) was not successful. A trivial conclusion is that signals must be observable in order to analyze the AVO behavior in an area.

Because of the positive AVO results obtained for Structures II and III (with unknown fluids) in Marine Area 2, and because a nearby structure with the same target formation is known to contain oil, the implications are strong that these unknown fluids are also oil. This implication suggests that the AVO technique as developed has a high potential as an exploration, as well as an exploitation, tool.

One of the areas studied, Land Area 2, has two oil reservoirs separated in depth by about 150 m. Both targets yield positive AVO results over the oil, and the spatial limits of the deeper reservoir correlate with those of the shallower one, although the exact limits of both are unknown. These results suggest that transmission effects through overlying hydrocarbons are minimal and need not be of concern in a general analysis.

Because the AVO behavioral changes observed in the studied areas correlate with the known or estimated spatial positions of the fluid contacts (whether gas/oil, oil/water, or gas/water), we conclude that the interpretation of AVO changes being due to changes in fluid content is at least reasonable. The physical parameter believed to be most responsible is Poisson's ratio (changing via P- and/or S-wave velocity). However, the areas studied in Saudi Arabia often have higher porosities over the structural highs than on the flanks; that a difference in porosity may at least be a contributing factor to the observed AVO behavior cannot be discounted.

Although theoretical AVO changes are only expected to be of the order of 10–15 percent when going from hydrocarbons to water, the behavioral changes are apparently measurable (at least statistically from trends of curve fits) with actual gather data.

Finally, it is concluded that an analysis of AVO can, with sufficiently dense surface seismic coverage, closely locate the spatial positions of fluid contacts (whether gas/oil, oil/water, or gas/water) if suitable care is taken to condition the amplitude ratios using the procedures outlined herein. Moreover, the AVO technique can perhaps assist in delineation drilling programs to determine the optimum positions for outposting wells.

ACKNOWLEDGMENTS

The author expresses sincere thanks to the Saudi Arabian Oil Company (Saudi Aramco), formerly known as the Arabian American Oil Company (Aramco), for encouraging all aspects of the development of this work and to the Ministry of Petroleum and Mineral Resources of the Kingdom of Saudi Arabia for granting permission to publish this paper. Appreciation is also expressed to Professor Milo M. Backus, with whom helpful and provoking discussions were held, and to the reviewers who offered a number of constructive suggestions.

REFERENCES

Chiburis, E. F., 1984, Analysis of amplitude versus offset to detect gas/oil contacts in the Arabian Gulf: Presented at the 54th Ann. Internat. Mtg., Soc. Expl. Geophys.

Chiburis, E. F., 1987a, The analysis of amplitude versus offset to detect hydrocarbon contacts in Saudi Arabia: Presented at the 5th Middle East Oil Technical Conference, Soc. Petr. Eng., Manama, Bahrain.

Chiburis, E. F., 1987b, Studies of amplitude versus offset to detect hydrocarbon contacts in Saudi Arabia: Presented at the 57th Ann. Internat. Mtg., Soc. Expl. Geophys.

Domenico, S. N., 1974, Effect of water saturation on seismic reflectivity of sand reservoirs encased in shale: Geophysics, **39**, 759–769.

Elliott, S. E., and Wiley, B. F., 1975, Compressional velocities of partially saturated unconsolidated sands: Geophysics, **40**, 949–954.

Gassaway, G. S., and Richgels, H. J., 1983, SAMPLE: Seismic amplitude measurements for primary lithology estimation: Presented at the 53rd Ann. Internat. Mtg., Soc. Expl. Geophys.

Gassaway, G. S., 1984, Effects of shallow reflectors on amplitude versus offset (seismic lithology) analysis: Presented at the 54th Ann. Internat. Mtg., Soc. Expl. Geophys.

Geertsma, J., and Smit, D. C., 1961, Some aspects of elastic wave propagation in fluid-saturated porous solids: Geophysics, **26**, 169–181.

Gregory, A. R., 1976, Fluid saturation effects on dynamic elastic properties of sedimentary rocks: Geophysics, **41**, 895–921.

King, M. S., 1966, Wave velocities in rocks as a function of changes in overburden pressure and pore fluid saturants: Geophysics, **31**, 50–73.

Nur, A. M., and Simmons, G., 1969, The effect of saturation on velocity in low porosity rocks: Earth Plan. Sci. Letters, **7**, 183–193.

Onstott, G. E., Backus, M. M., Wilson, C. R., and Phillips, J. D., 1984, Color display of offset dependent reflectivity in seismic data: Presented at the 54th Ann. Internat. Mtg., Soc. Expl. Geophys.

Ostrander, W. J., 1984, Plane-wave reflection coefficients for gas sands at nonnormal angles of incidence: Geophysics, **49**, 1637–1648.

Takeuchi, S., and Simmons, G., 1973, Elasticity of water-saturated rocks as a function of temperature and pressure: J. Geophys. Res., **78**, 3310–3320.

Todd, C. P., and Backus, M. M., 1985, Offset-dependent

reflectivity in a structural context: Presented at the 55th Ann. Internat. Mtg., Soc. Expl. Geophys.

Toksoz, M. N., Cheng, C. H., and Timur, A., 1976, Velocities of seismic waves in porous rocks: Geophysics, **41**, 621–645.

Wrolstad, K. H., 1986, Offset-dependent amplitude analysis by viscoelastic lithological modeling: Presented at the 56th Internat. Mtg., Soc. Expl. Geophys.

Wyllie, M. R. J., Gregory, A. R., and Gardner, G. H. F., 1958, An experimental investigation of factors affecting elastic wave velocities in porous media: Geophysics, **23**, 459–493.

Yu, G., 1985, Offset amplitude variation and controlled amplitude processing: Presented at the 55th Ann. Internat. Mtg., Soc. Expl. Geophys.

A Model-Based Analysis of AVO in the Sacramento Valley

Ming-Ren Hong, John P. Castagna‡, and Charles J. Sicking§*

INTRODUCTION

The variation of reflection coefficients with source-to-receiver spacing in multioffset seismic data is known to contain information about lithology and pore fluid content of subsurface rocks (e.g., Ostrander, 1984). A key component of amplitude variation with offset (AVO) analysis is a methodology for amplitude calibration so that rock/fluid properties can be interpreted vertically and laterally away from a borehole with a high degree of confidence. Ostrander (1984) used onshore data from the Sacramento Valley to show that certain amplitude variations in the prestack seismic data were associated with gas-filled reservoirs in clastic sediments. We present a model-based methodology for AVO analysis and apply it to a land dataset from the Sacramento Valley.

SEISMIC AND WELL DATA

Figure 1 shows a true relative amplitude [reflection amplitude preservation (RAP)] seismic line exhibiting classic "bright spots" from the Sacramento Valley. The data were recorded using primacord and a split-spread geometry. The offset range is between 660 ft (201 m) and 5720 ft (1744 m) and the maximum fold is 24. There are multiple bright spots near common-depth-point (CDP) 91 between two-way times 1.3 and 1.6 s. Two other bright spots are evident at 2.0 s, near CDP 60 and CDP 100. Our objective is to analyze the AVO response for these bright spots and to relate this response to the presence of hydrocarbons.

Well #1 is located on the seismic line at CDP 91. The density, sonic, and computed-lithology logs are shown in Figure 2. Several zones between 4600 ft (1402 m) and 5800 ft (1769 m) show low density and high sonic transit times. The synthetic *S*-wave sonic log, shown in Figure 2, was derived from the conventional wireline logs using the techniques described in Castagna et al. (1993a and b) elsewhere in this volume. In this well, there is a 70-ft (21.3 m) thick gas sand at a depth of 4700 ft (1433 m) and two 20-ft (6.1-m) thick gas sands at a depth of 5100 ft (1555 m) and 5750 ft (1753 m). These sands are indicated by arrows in Figure 2. The bright spots in Figure 1, labeled 1, 2, and 3, correspond to these gas sands. The well did not penetrate the deeper bright spots at 2.0 s. In the following discussion our analysis indicates that these bright spots are also gas related.

*Saudi Aramco, formerly ARCO Oil and Gas Co.
‡ARCO Oil and Gas Co.
§ARCO Exploration and Production Technology; formerly ARCO Oil and Gas Co.

PROCESSING

A key requirement for successful AVO analysis is to process the data in such a way that preserves the offset dependent reflectivity. For the Sacramento Valley data, there was a variation in charge size and a variation in the weathering layer. In addition, source-generated noise causes a significant amount of amplitude variation from trace-to-trace within a CDP. The processing strategy used to address these issues is shown in Figure 3. The key ingredients to the strategy are attenuation compensation, surface consistent source and receiver deconvolution, wavelet shaping, and signal-to-noise ratio (S/N) enhancement by lateral summing of adjacent CDP gathers and partial stacking. At every stage of the processing, quality control displays were generated to ensure that the amplitude information was not corrupted.

The purpose of attenuation compensation is to make the data stationary in time and to allow the use of single window deconvolution filters. Thus, the final processed wavelet is relatively invariant with time and the AVO analysis is valid over a large time window.

Surface consistent deconvolution makes the data stationary in space by correcting for the source, receiver, and near surface variations. As a result, the final processed wavelet remains stable along a line. A final deconvolution step is required to correct the wavelet of the data to zero phase. The filters applied at this step should be designed so that the same filter is applied to all the traces in any given CDP. This

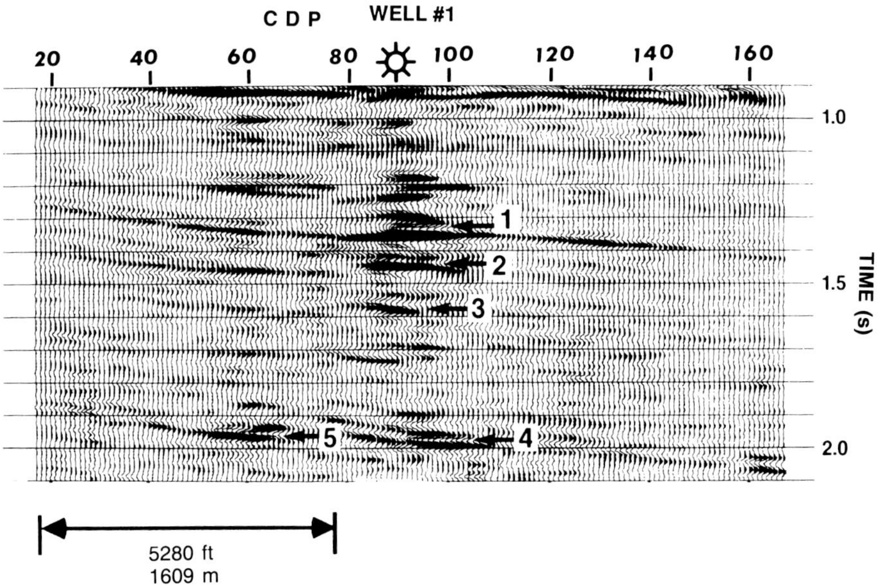

Fig. 1. Stacked section of Line A, Sacramento Valley. Bright spots of interest are numbered. See Figure 10 for a larger version of this line

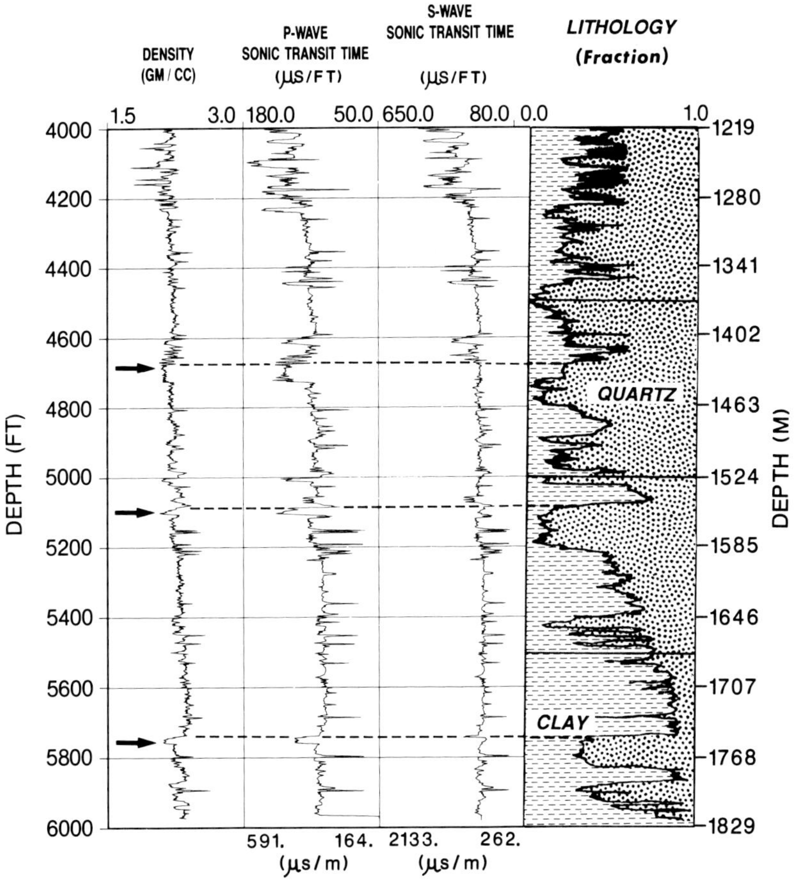

Fig. 2. Logs from Well #1, Sacramento Valley. The *S*-wave sonic transit-time log is estimated from the *P*-wave transit-time log and lithology. The zones of interest are indicated by arrows.

ensures that the AVO response is not compromised by a trace independent operation. After a careful velocity analysis and normal moveout (NMO) correction, the S/N of the AVO response is enhanced by summing adjacent traces within each CDP (partial stack).

Effects not treated by this processing strategy, such as overburden-reflectivity losses and thin-bed effects, are handled by the normalization technique used in Chiburis (1987) and by modeling (see following section).

COLOR DISPLAY

Todd (1986) described a method of displaying AVO information in color: the traces in a CDP gather are converted so that each trace represents a constant angle of incidence versus time. Each of these modified gathers is then reduced to three traces by stacking the low, middle, and high angles of incidence. For each gather, the three traces are displayed in different colors (near equals red, middle equals green, far equals blue) and are superimposed. Increasing AVO should show bluish tone using this color scheme.

The color display provides a means of scanning a large volume of seismic data for AVO anomalies and highlighting zones for further analysis. This kind of display is also good for identifying residual moveout caused by incorrect velocities which appears as a color fringing (Todd, 1986). Finally, because stacking is involved, the process gives some noise reduction. When a whole seismic section is plotted in this manner, the continuity and structure are clearly visible.

Figure 4 shows the superimposed color display for the Sacramento Valley data. The "hatched" zones represent negative amplitudes, whereas "nonhatched" zones denote positive amplitudes. A seven-intensity scale is used for each color. Reflections which are high amplitude at all offsets produce white events. The five bright spots shown in Figure 1 are clearly visible and better defined in Figure 4. However, they do not show the bluish tone, which would indicate the amplitude increase versus offset commonly expected for gas sands.

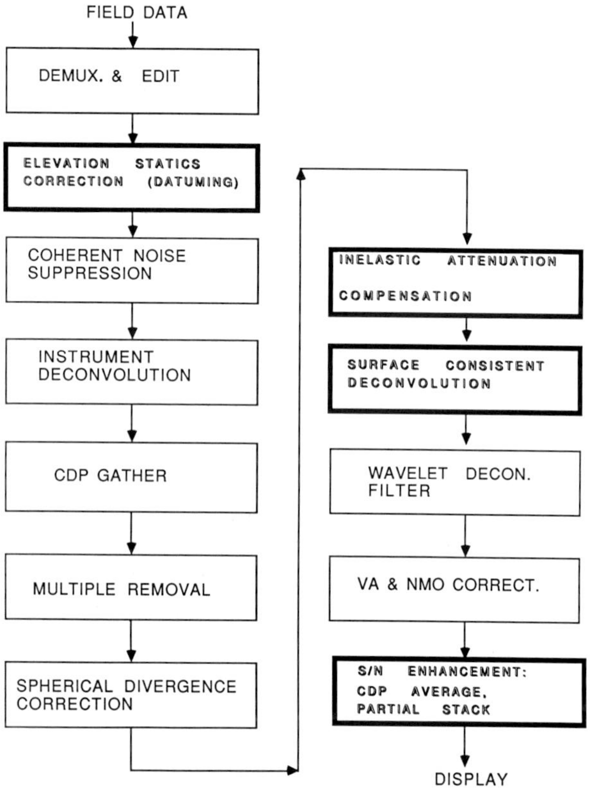

Fig. 3. Recommended seismic processing flow of land data for AVO analysis. Key steps of the flowchart are highlighted.

MODELING

The "bright spots" highlighted in Figures 1 and 4 are analyzed through a detailed comparison of the seismic data with synthetics derived from numerical modeling. Utilizing the well log data for input, two kinds of models were run; 1-D convolutional synthetic seismograms and full elastic wave-equation synthetic prestack gathers.

The convolutional synthetic seismogram and the stacked traces at the well location are shown in Figure 5a. After trying many wavelets, the best match was obtained using a 25 Hz Ricker wavelet. This match provided a starting model for additional work and indicated there were no severe problems with the well logs in the target zones. A convolutional S-wave synthetic (also with a 25 Hz Ricker) is shown in Figure 5b for comparison. Note that the gas sand reflections are not as "bright" on the S-wave synthetic. Further improvements to the synthetics could be achieved by careful editing of the logs and wavelet extraction.

Prestack synthetic gathers were generated using a full elastic wave equation modeling program (Young et al., 1983). The inputs to this program were obtained by blocking the density, P-wave sonic and simulated S-wave sonic logs (see Figure 6). The quality of the synthetics generated by the program is sensitive to the degree of log blocking. To study the impact of block-

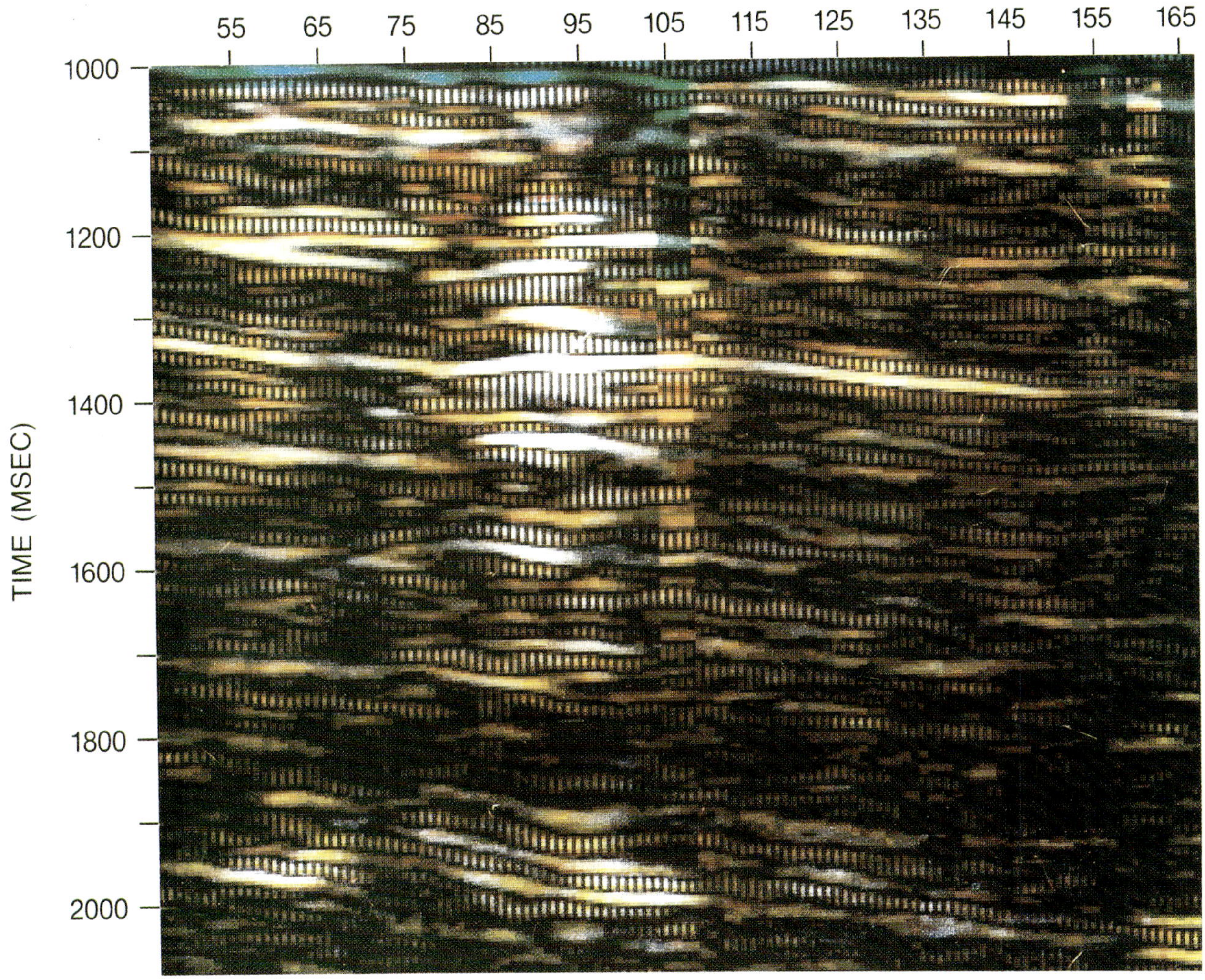

Fig. 4. Color overlay for line A, Sacramento Valley, California. This is produced by superimposing three partial stacks as described in text.

ing, the logs were blocked to produce average layer thicknesses of 5 ft (1.52 m), 10 ft (3.05 m), and 20 ft (6.1 m). This yielded three multilayered earth models with 227, 435, and 711 layers, respectively. The corresponding NMO corrected CDP gathers are shown in Figure 7. The AVO response is significantly different for the 227 layer model, but essentially identical for the 435 layer and the 711 layer models. Consequently, the 435 layer model was chosen for further analysis.

The elastic wave equation modeling program included P-wave to S-wave conversions at each interface, multiples, and the array response appropriate for the acquisition geometry. Linear arrays of 24 geophones spaced at 10 ft (3.05m) intervals were used; corresponding to a seismic trace spacing of 220 ft (67 m). Figure 8 compares synthetics computed using the 435 layer model with (8a) and without (8c) multiples and mode conversions, to the processed prestack seismic gather at the well locations. Note that the gas

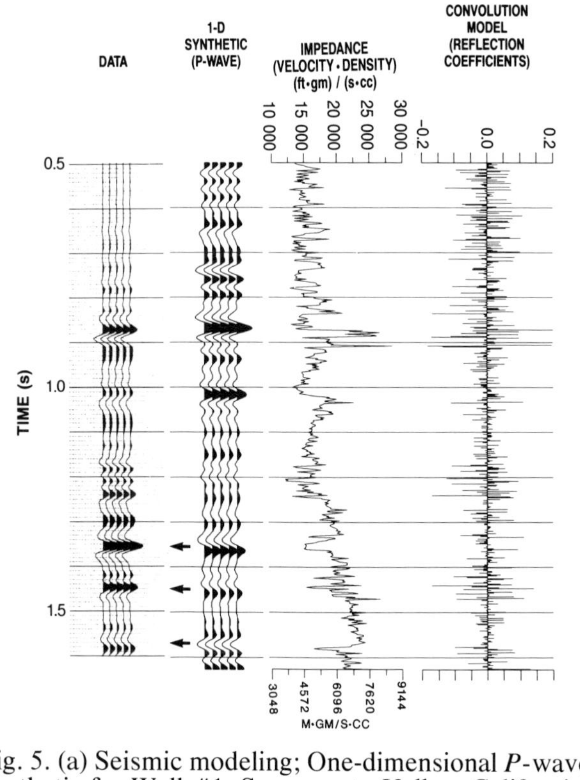

Fig. 5. (a) Seismic modeling; One-dimensional P-wave synthetic for Well #1, Sacramento Valley, California. The zones of interest are indicated by arrows.

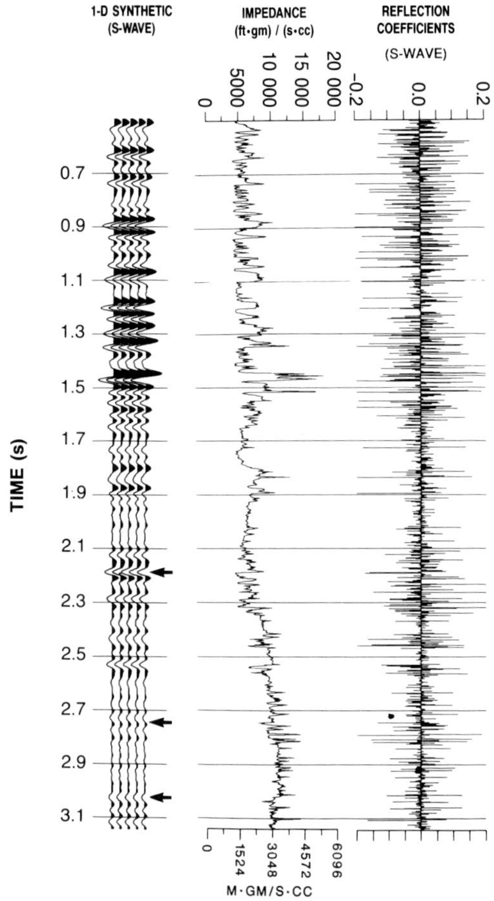

Fig. 5. (b) Convolution modeling; One-dimensional S-wave synthetic using synthetic S-wave impedance log.

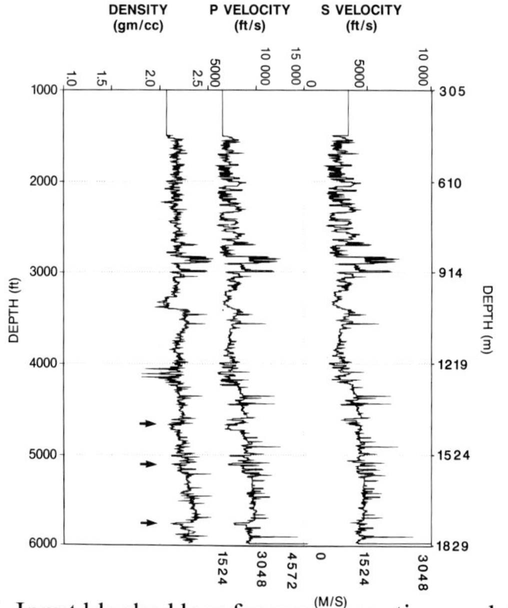

Fig. 6. Input blocked logs for wave-equation modeling. A total of 435 layers are generated in this example. The arrows indicate the zones of interest.

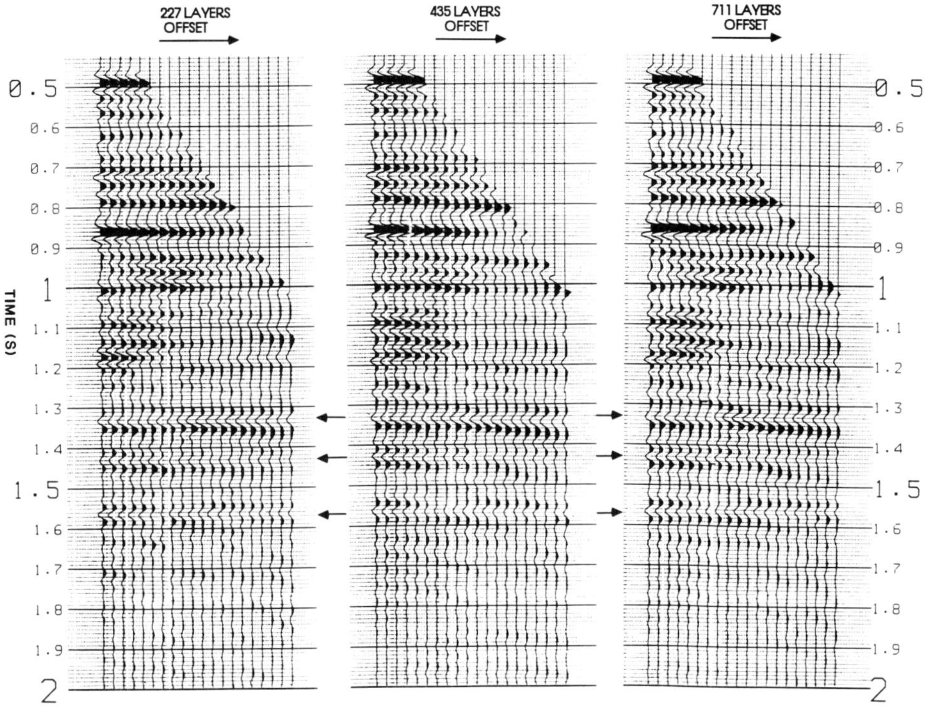

Fig. 7. A comparison of prestack synthetic seismograms for various multilayer earth models. In the final analysis, the 435 layer model was used for amplitude calibration.

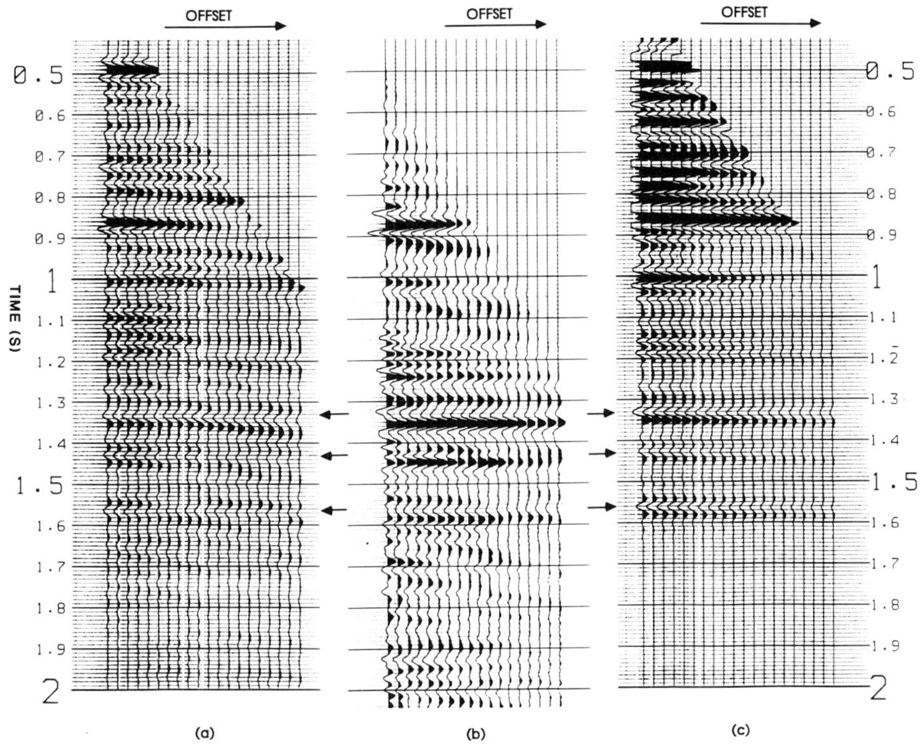

Fig. 8. Comparison of prestack data at wellsite (CDP 91) with synthetics. (a) Synthetic with mode conversions and multiples, (b) data at wellsite, and (c) synthetic with primaries only.

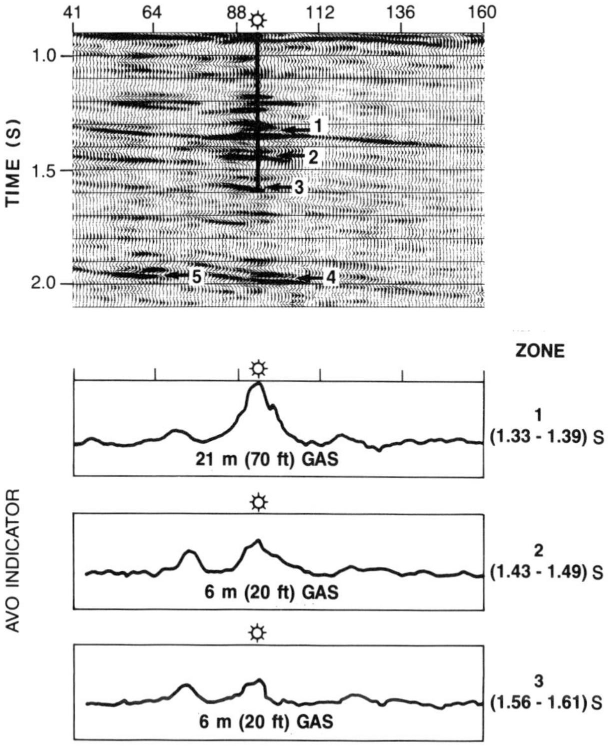

Fig. 9. Lateral AVO indicator (after Chiburis, 1987) for the three target events, normalized to the continuous event at .9 s.

sands (indicated by arrows) show decreasing AVO for both synthetic and real data. Compared to the event at about .9 s, however, the gas sand amplitude decreases less rapidly with offset. Thus, AVO measured relative to this reference event can be used to emphasize the gas sands.

CHIBURIS METHOD

Chiburis (1987) presented AVO studies from Saudi Arabia (onshore area, Arabian Gulf) where he suggested the use of a lateral AVO indicator whereby the target event is normalized on a trace-by-trace basis within a CDP gather to a reference event. In this process, the AVO slope (B) and intercept (A) terms are obtained by fitting the amplitudes to a curve of the form,

$$A + B \sin^2 \phi,$$

where ϕ is the local angle of incidence. The B/A response of the target event is then normalized by the B/A response of the reference event. We applied the Chiburis indicator to the Sacramento Valley data, choosing the reference event at about .9 s. The lateral AVO indicator is displayed in Figure 9 for the three

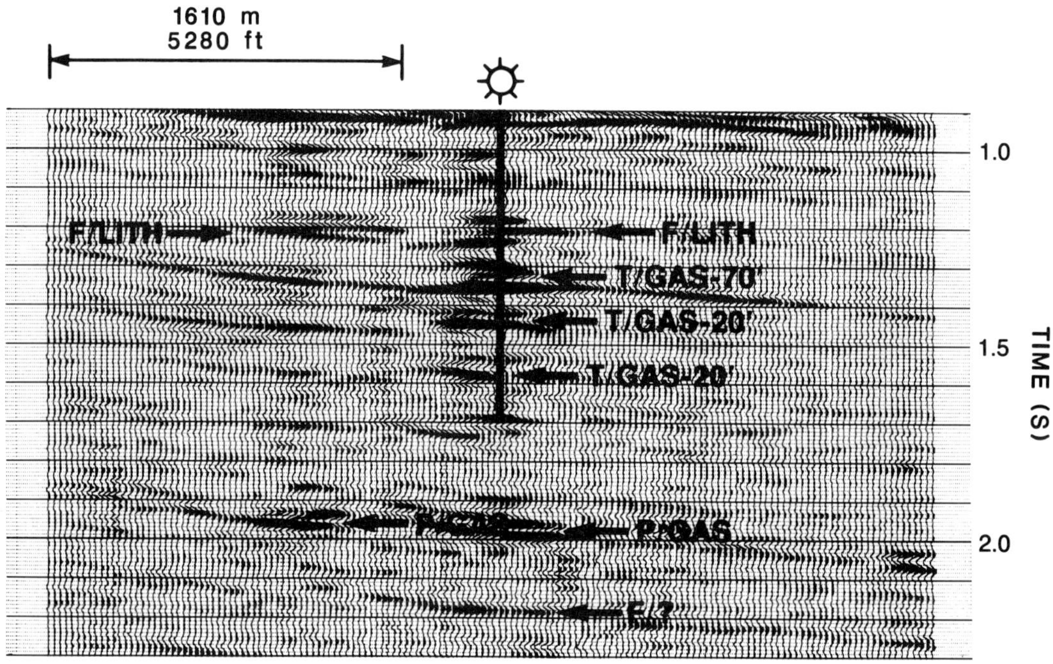

Fig. 10. The final interpretation of bright spots on Line A. T stands for true (hydrocarbon related) bright spots; P represents predicted hydrocarbon-related bright spot; F indicates false bright spots; LITH identifies a lithology related amplitude anomaly. Numbers are thicknesses of gas zones as identified from the well data.

gas sands; all three sands exhibit distinct AVO anomalies at the crest of the structure.

In addition to the three bright spots discussed, we examined five other events. Those labeled F/LITH in Figure 10 are interpreted to be false bright spots due to lithology changes unrelated to the presence of hydrocarbons, whereas those labeled P/GAS showed positive AVO anomalies and are predicted to contain hydrocarbons.

CONCLUSIONS

We have shown an example of the successful use of AVO analysis on land data. Our processing scheme included surface consistent deconvolution and partial stacking. Full elastic wave equation modeling was employed to model the prestack data. Gas-sand amplitude was found to decrease with offset, but at a slower rate than water-sand amplitude on the flank of the structure. The partial stack color display proved helpful in scanning a large amount of data for local AVO anomalies. The Chiburis lateral AVO indicator was then particularly helpful in delineating lateral variations of the AVO response of the target zone.

ACKNOWLEDGMENTS

We thank Dave Huggins and Steve Scott for supplying the data, Dan Miller for seismic processing, and P. C. Luh and C. Y. Young for the use of their algorithms and providing help with the AVO displays. Thanks to H. V. Tran for his help in implementation of the color display program. We thank ARCO Oil and Gas Company for its support and permission to publish this paper.

REFERENCES

Castagna, J. P., Batzle, M. L., and Kan, T. K., 1993(a), Rock physics: The link between rock properties and AVO response, *in* Castagna, J. P., and Backus, M. M., Eds., Offset dependent reflectivity: Society of Exploration Geophysicists, 135–171.

Castagna, J. P., Batzle, M., Tubman, K., Gaiser, J., and Burnett, M., 1993(b), Shear-wave velocity control, *in* Castagna, J. P., and Backus, M. M., Eds., Offset dependent reflectivity: Society of Exploration Geophysicists, 115–134.

Chiburis, E. F., 1987, Studies of amplitude-versus-offset in Saudi Arabia: Presented at the 57th Ann. Internat. Mtg., Soc. Expl. Geophys.

Ostrander, W. J., 1984, Plane wave reflection coefficients for gas sands at nonnormal angles of incidence: Geophysics, **49**, 1637–1648.

Todd, P., 1986, Isolation, display, and interpretation of offset dependent phenomena in seismic reflection data using offset to depth (ODR) tange partial stacking: M. A., Univ. Texas Austin.

Young, C. Y., Jarrard, R. D., and Kan, T. K., 1983, Prestack synthetic seismogram of elastic-layered earth: Presented at the 53rd Ann. Internat. Mtg., Soc. Expl. Geophys.

AVO Analysis in Low- and High-Porosity Gas Sand Reservoirs

*Jose Regueiro S.**

INTRODUCTION

The detection, identification, and quantification of thickness, porosity, and fluid content in reservoir rocks, using seismic reflection data, is a very important topic in the area of production geology and geophysics. Unfortunately, because the seismic response is a function of not one, but of all these parameters, the problem is not trivial. Even for those reservoirs where the fluid present is gas, and therefore in most instances, a relatively strong contrast between the elastic properties of these and those of the surrounding rocks is expected, the detection and quantification depend upon many other factors: degree of consolidation, ratio depth/thickness, lithologies involved, clay content, etc.

We describe, in their seismic perspective, two examples of gas reservoirs that could be considered extremes: (1) Los Lanudos gas field, which is characterized by shallow depths (less than 1200 m), high porosities (greater than 18 percent), sand thicknesses below tuning (less than 18 m), and where the reservoirs show as high-amplitude anomalies (bright spots) in the conventional seismic sections, and (2) the Macaira-Uveral gas field, which exhibits greater depths (over 2500 m), lower porosities (under 8 percent), gas thicknesses near or greater than the tuning thickness (greater than 25 m), and where the sandstone reservoirs also appear as high-amplitude anomalies on the stacked sections.

The integration of seismic data with direct modeling, both before and after stack, and with well information, allowed us to draw important conclusions regarding reservoir extensions, as well as changes in other properties (net gas thickness in one case, and net sand thickness/net porosity thickness for the other) which are important for advanced development of these fields.

*USB-FUNINDES, formerly with INTEVEP S. A.

SHALLOW HIGH-POROSITY RESERVOIRS

Los Lanudos gas field, located in the western part of Venezuela, produces from Eocene age sandstones of the upper Misoa formation. These sands have been interpreted as low-energy, shallow submarine bars, which were deposited in an estuarine environment. The main factor controlling gas accumulation is structure.

These reservoirs are shallow (under 1200 m) and exhibit porosities from about 18 to 25 percent. Seismically, the gas sands are characterized by high amplitude events (bright spots) on the conventional sections (Figure 1).

The analysis of well information (sonic-density logs) indicates that, while the water-saturated sands exhibit higher impedance than the shales, the gas sands are characterized by impedance values lower than the adjacent shales (negative reflection coefficient at the top of the sands), and also, that the contrast is larger at shallower depth and decreases as the sands get deeper (Figure 2). These curves show the possibility of having a zero impedance contrast, and thus the nondetection of gas sands at depths from 1000 to 1200 m by means of conventional seismic methods (Regueiro, 1985).

By looking at the seismic data before stack, as for example a distance-weighted common-midpoint (CMP) gather (Figure 3), we noticed a distinct change in the behavior of amplitude as a function of source-receiver distance (offsets from 100 to 1275 m) for the events representing the main lithologic types present in this area (coal beds, gas sands, water-saturated tight sands; Regueiro, 1986). This change by itself indicates the possibility of applying amplitude variation with offset (AVO) analysis for discriminating fluid/lithology types in this particular area, but also indicates that conventional stacking will average these AVO effects, which, for the case of the E reservoirs, will result in events with considerably higher amplitude than the real zero offset value (Figure 4). Therefore any attempt to correlate well information (impedance) with these conventional stacks is an impossible task, especially if

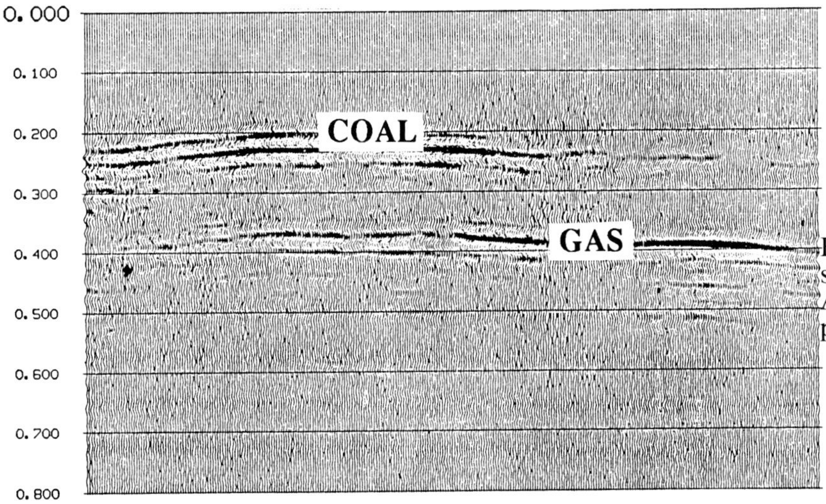

Fig. 1. Conventional stacked seismic section from Los Lanudos gas field. Amplitude is normalized to maximum per trace to emphasize anomalies.

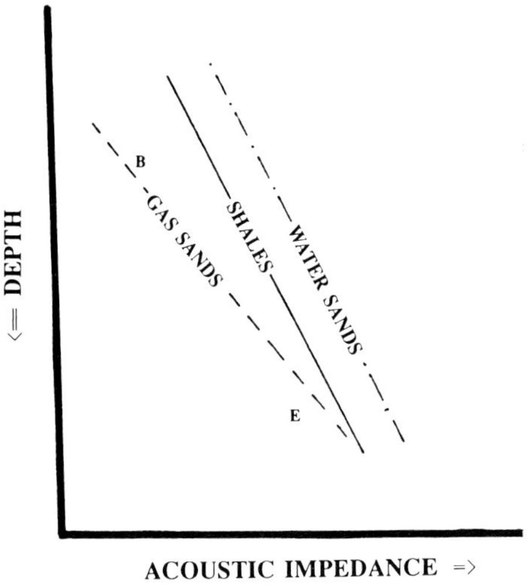

Fig. 2. Schematic representation of the observed trends of acoustic impedance as a function of depth, for shales, gas and water sandstones in Los Lanudos.

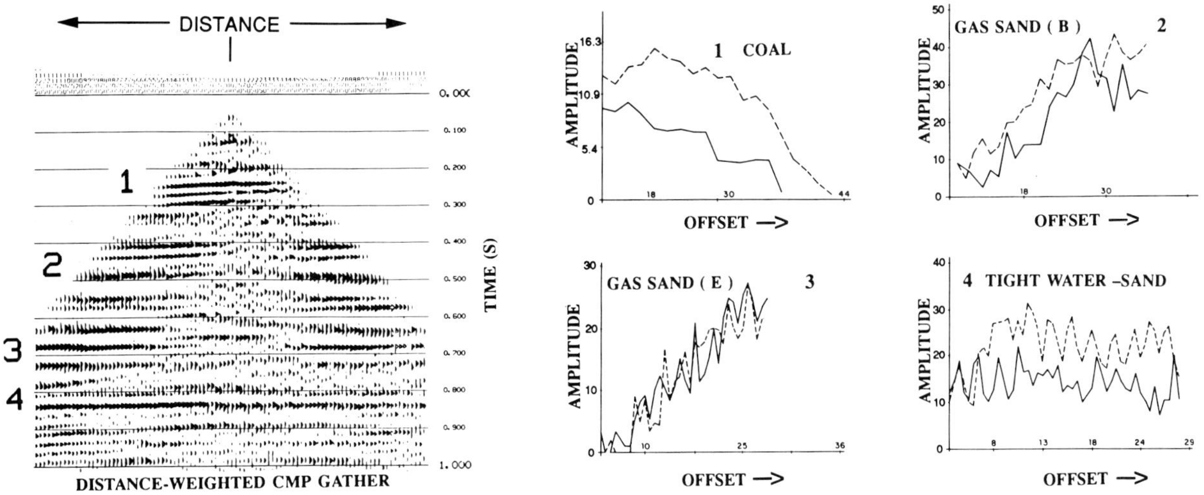

Fig. 3. Distance-weighted CMP gather and AVO curves showing trends for the main lithologies present in the study area (Los Lanudos). The solid and dashed AVO curves represent the amplitude variations as distance increases in both directions from the center ("split spread"). The horizontal axes show various angles of incidence in degrees, which increase with increasing offset.

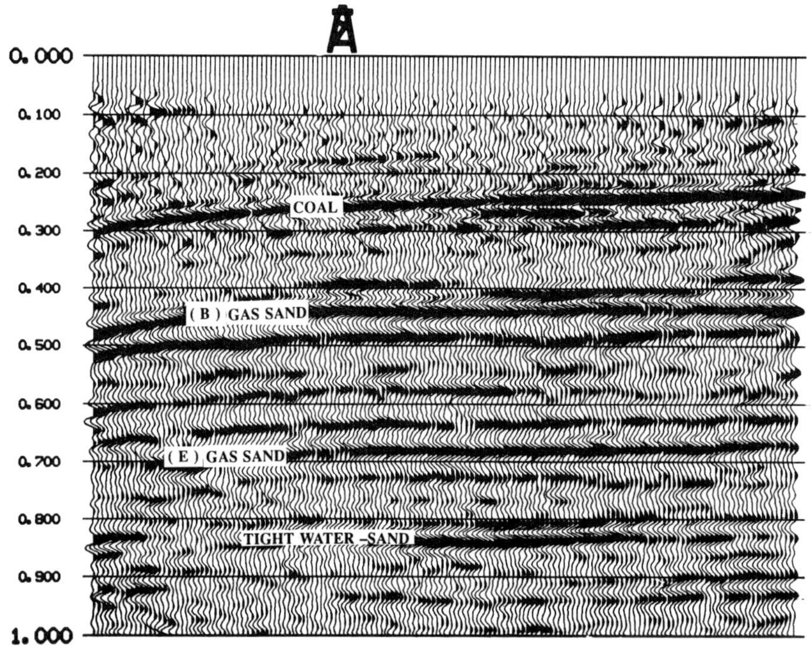

Fig. 4. Conventional stacked section showing events 1, 2, 3, and 4 of Figure 3, corresponding to coal, shallower gas sands (reservoirs B), deeper gas sands (reservoirs E), and the deeper water-saturated tight sand. This portion of the line is the same shown in Figures 7 and 8. The vertical axis is two-way traveltime in seconds.

a polarity reversal exists, as these data seem to show (event 3 on top of Figure 3), as the offset increases. For this particular case, the impedance contrast at zero offset for event 3 (E reservoir) is very small, as interpreted from well information, and therefore the contribution due to longer offsets dominates the conventional stacking procedure.

Figure 5 shows the results of AVO-modeling using Zoeppritz equations, of a reservoir with the characteristics of velocity and density of reservoir B for different values of Poisson's ratio. In Figure 6 we have drawn the observed amplitude values (normalized to the zero offset reflection coefficient) for two of the common depth points (CDPs) displayed. Notice that a good, qualitative relation between the observed and the calculated curves for low values of Poisson's ratio exists, thus indicating the possibility of using AVO for the detection and evaluation of these reservoirs. This use becomes important, primarily for the case of the deeper E reservoir, which may exhibit a zero impedance contrast with the surrounding shales. As an example, Figures 7 and 8 show the results of applying AVO analysis (Hilterman's linear approximation) to a portion of one line in this field (Donati and Regueiro, 1988). From this process we obtained two sections, namely a zero offset ("normal incidence") and a change in Poisson's ratio ("gradient") section. These sections have been plotted so that colors represent amplitude values of the indicated parameter, while the traces represent the values of the other term. Notice the high amplitude event corresponding to reservoir E on the change in Poisson's ratio

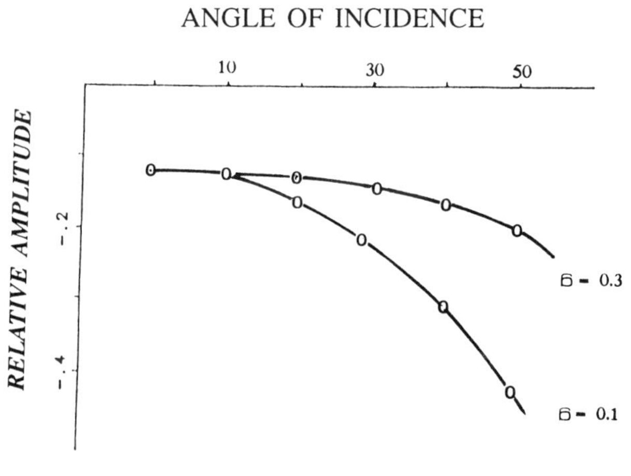

Fig. 5. Theoretical AVA curves for a model of the B reservoir (Los Lanudos) for different values of Poisson's ratio. The angle of incidence is given in degrees.

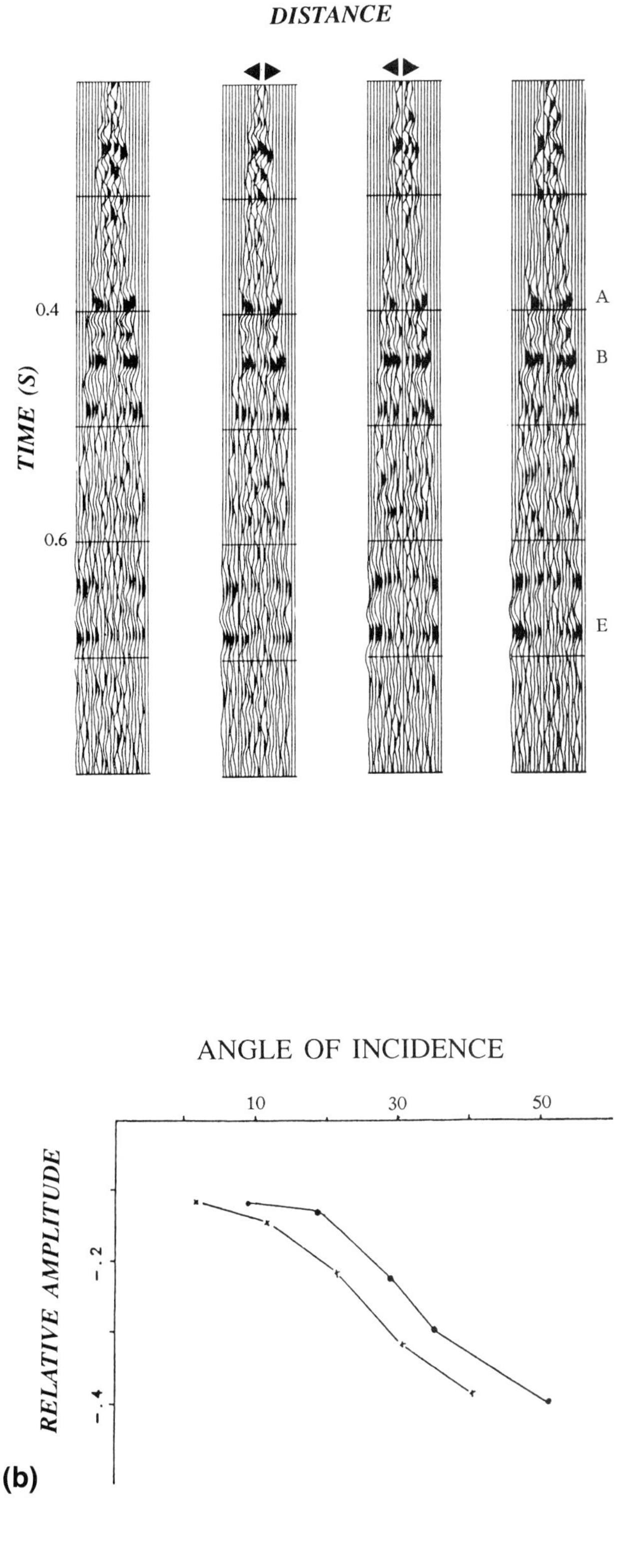

Fig. 6. (a) CMP gathers and (b) observed AVA curves, from Los Lanudos, showing an increase in amplitude (negative) as a function of incidence angle for event B. The angle of incidence is given in degrees.

section, and the lack of strength on the normal incidence (NI) section, which are in agreement with well data. A more quantitative treatment of this type of data, used in conjunction with well logs such as V_p/V_s, should produce important information regarding lateral changes in these elastic properties, which in turn, thru derived "empirical" relations, could yield interpretable changes in petrophysical parameters.

For the shallowest sandstones (*A*, *B*) maps of reservoir quality using conventional stacked seismic data and available well information were generated. For this case, the analysis of the theoretical behavior of seismic amplitude response as a function of porosity, thickness, and gas saturation indicated that an increase in any of these parameters produces an increase in the expected seismic amplitude. However, since sand thickness is very small, the separation of individual effects is difficult. By plotting the measured values of gas thickness (porosity times thickness times gas saturation), obtained from wells in this field, versus the observed calibrated seismic amplitude values, we noticed a direct relation, between these two parameters (Figure 9). Using this relation, maps were generated which showed reservoir quality in the form of gas thickness (Figure 10; Regueiro et al., 1986).

The lack of good impedance contrast for the deeper E reservoir, and because the data under analysis were obtained from two different seismic surveys, made obtaining detailed information on lateral changes or even on the limits of this reservoir from the conventional stacked data difficult. The most recent and largest survey used was intended for "high resolution" and thus was recorded using a very short station interval (15 m) which in turn gave very small offset (650 m) for the farther traces. AVO manipulations, such as gradient stacks and distance angle stacks, were found to be important tools to further study this reservoir (Regueiro, 1986).

DEEP LOW-POROSITY RESERVOIRS

Macaira-Uveral, a gas field located in the northern part of the Guarico state, central Venezuela, is characterized by large accumulations of gas in very thick sandstones (greater than 20 m), but with very low porosities (less than 8 percent) which makes production very poor in most of the wells. These reservoirs are deeper than the previous ones in Los Lanudos field, exhibiting an average depth of about 2700 m. In general, primary porosity has been lost due to diagenetic processes (Lambertini et al., 1987). These sands

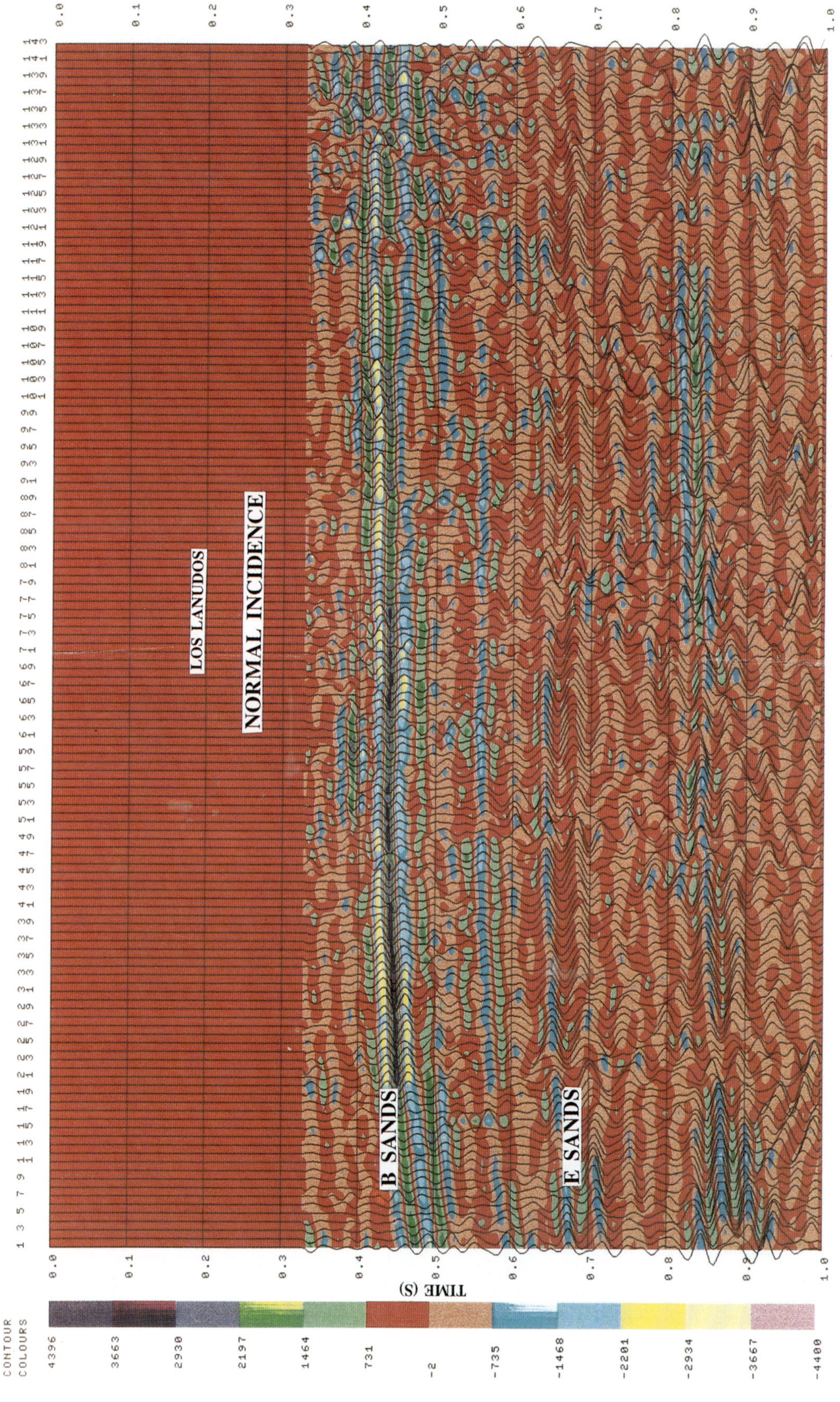

Fig. 7. "Normal Incidence" section, obtained from inversion of prestack amplitude data using Hilterman's approximation, from Los Lanudos. Wiggle traces represent the change in Poisson's ratio.

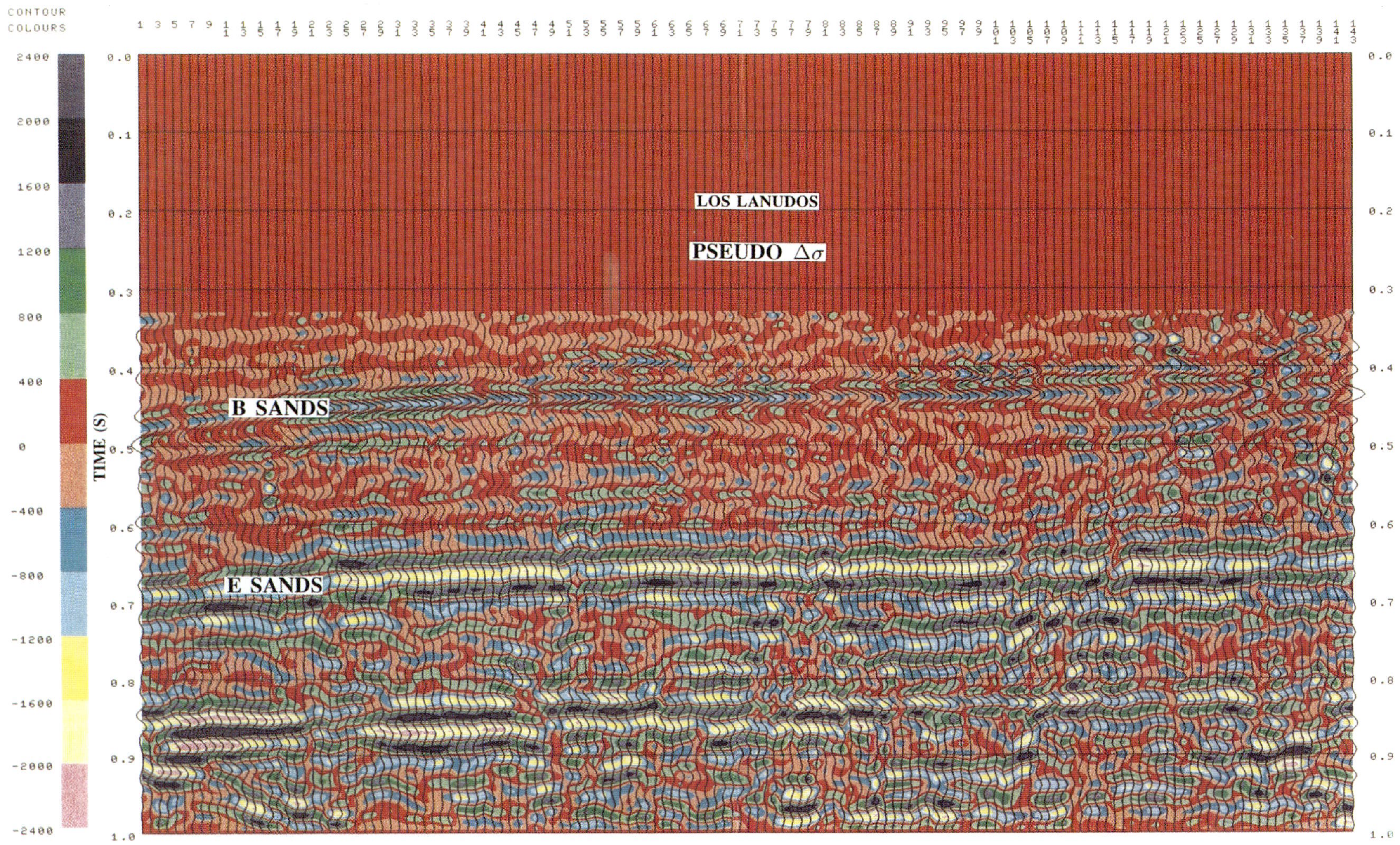

Fig. 8. "Pseudo-Delta-Poisson" section corresponding to the same data as in Figure 7. Wiggle traces represent the normal incidence values.

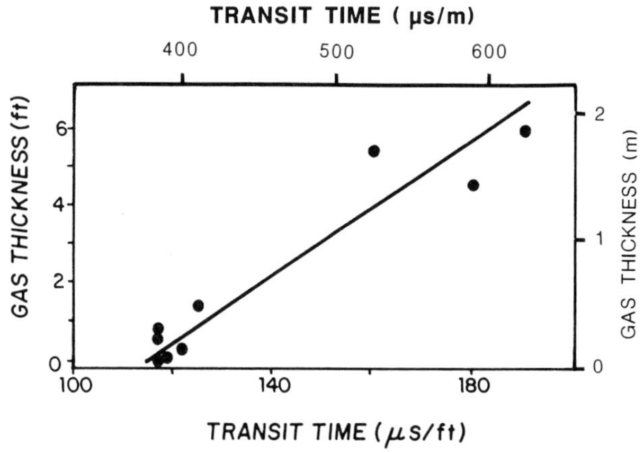

Fig. 9. Empirical relation between observed seismic pseudo-sonic values and the calculated gas thickness from well data for the B reservoir (Los Lanudos).

are of Oligocene age, and were deposited as offshore bars in a platform environment.

Seismically, the reservoirs appear as high amplitude events on the conventional stacked sections (Figure 11). Even though these anomalies resemble "bright spots", analysis of well data shows that gas sandstones exhibit higher velocities than their surrounding shales, thus indicating a positive reflection coefficient at the top of the sands (Figure 12).

AVO analysis shows (Figure 13) that there is very little change of amplitude as a function of angle for the available data (offsets from 300 to 2000 m), generally indicating a decrease in amplitude for greater angles of incidence. Theoretical modeling (Figure 14) also shows this tendency and agrees with the observed data but also indicated is the presence of a greater difference for angles between 20 and 30 degrees. Linear inversion of the prestack data did not show, as expected due to the very small changes in AVO response, any important interpretable result. Detailed interpretation of subtle changes from curve to curve, using well information, nonlinear inversion processes, and data recorded with longer offset (over 2000 m), may prove to be helpful in detecting "small" lateral changes in reservoir characteristics.

By studying the available well data and the theoretical influence of thickness and porosity, we concluded that the amplitude anomalies observed in the conventional stacked sections correspond to sandstones with thickness nearing the seismic tuning thickness (25–27 m). This seismic information was inverted and calibrated, and then used to generate maps of net-sand thickness. To do this, porosity changes were assumed not to influence the seismic response significantly. Well data showed that porosity in the area of study ranges between 2 and 6 percent, and that these changes are regional, due to diagenetic processes. The resulting maps (Figure 15) were useful, not only to delineate the different sandstone reservoirs, but also to infer the areas with better thickness development. Using the porosity trends interpreted with well data, maps were drawn which better represented trends in porosity-thickness (porosity times thickness) for each reservoir (Regueiro et al., 1986).

CONCLUSIONS AND DISCUSSION

The analysis of seismic data from two gas fields in Venezuela, both before and after stack, has shown:

(1) For the shallow, high porosity reservoirs,
   (a) Amplitude anomalies for the shallower A and B sandstones, in the conventional stacks, correspond to "true" bright spots, and the seismic response is highly influenced by changes in both porosity and thickness. For the deeper E reservoir, the impedance contrast is very small, thus making evaluation of this reservoir using conventionally stacked seismic data difficult.

REFERENCES

Donati, M. S., and Regueiro, J., 1988, Generation of normal incidence and Poisson's ratio contrast sections utilizing the phenomenon of amplitude as a function of angle of incidence: Proceedings of the IV Venezuelan Geophysical Congress, Caracas

Lambertini, R., Garcia, L. E., Klar, P., Gonzalez, C., and Aspiritxaga, I., 1987, Lithofacies, sedimentation, and diagenesis in the Tertiary of Northeast Guarico: Technical Report # 01763, 87, INTEVEP S. A.

Regueiro, J., 1985, Establishment of a methodology to delimit gas reservoirs in the Los Lanudos area: Technical Report # INT-01351, 85, INTEVEP S. A.

Regueiro, J., 1986, Amplitude versus offset: direct hydrocarbon detector?: Proceedings of the III Venezuelan Geophysical Congress, Caracas

Regeuiro, J., Leon, F., De Mena, J., and Casanova, M., 1986, Seismic stratigraphy in the Macaira-Uveral area: Technical Report # INT-EPCT-00005, 86, INTEVEP S. A.

Offset-Dependent Amplitude Analysis of Data from the Veslefrikk Field, Offshore Norway

*K. H. Wrolstad**

INTRODUCTION

The Veslefrikk Field is located on the Bergen High in the northern Viking Graben of the Norwegian North Sea (Figure 1). The target sands in this investigation are the fan delta facies of the Brent group (Oseberg Formation). They were deposited in mid-Jurassic time in a shallow water marine environment from sediments derived from erosion of basement rocks exposed by tectonic uplifts associated with rifting of the Atlantic. Rapid basin subsidence followed by deposition of marine shales occurred in the Cretaceous. These organic rich beds are generally the source of hydrocarbons that migrated into the Brent sands. The porosity of the Brent sands ranges from 10 to 25 percent depending usually on the extent of diagenetic calcite cementation that formed prior to basin subsidence. Seismic detection of either the extent of calcite cementation or the direct detection of hydrocarbons are important goals for the explorationist in this area. In this environment where the acoustic impedances of oil saturated target sands are commonly greater than that of surrounding shales, can AVO be a useful hydrocarbon indicator? Detailed elastic modeling was undertaken to answer this question.

The exploration seismologist has for many years correlated stacked traces with normal incidence synthetic traces generated from sonic and density logs. The stacked trace, however, has lost some information present in the prestack domain that may be related to lithology and fluid saturation. Gassaway and Richgels (1983), Ostrander (1984), and Yu (1985) have reported a correlation between increasing AVO and hydrocarbon saturation in sandstones. In complex lithologic sequences prestack data interpretation is often very difficult, however, and requires the aid of sophisticated modeling programs. The problem of how to construct elastic models [i.e., the specification of compressional velocity (V_p), vertical shear velocity (V_s), and density (ρ) from well log data], is addressed and insight is given on the effect of fine layering and interference on offset dependent amplitude variations in the prestack field data.

The method of approach was to compute the seismic response using predicted lithologic elastic parameters from well logs and to compare this with field data CMP gathers and offset-dependent amplitude attributes. Porosity, sand/clay fraction, and water/hydrocarbon saturation were derived from well log data using a log interpretation program. *S*-wave velocities were predicted from these data using the Gregory-Pickett-Biot method. The computed *S*-wave velocities along with the measured primary wave velocities and densities were specified for each interval of a constant Q multilayer model. Synthetic shot gathers were generated using viscoelastic modeling programs that include all converted phases and multiples. These synthetic seismograms were compared with field CMP gathers and were examined for offset-dependent amplitude anomalies characteristic of many hydrocarbon saturated sands.

THEORY

The averaging method discussed in Backus (1962) provides a solution to the problem of how to incorporate the rapid oscillations of log data into a model that honors the most important acoustic impedance changes. Backus showed that a stack of discrete, isotropic horizontal layers with total thickness, Δz, is equivalent to a homogeneous, transversely isotropic medium for seismic waves whose $\lambda \gg \Delta z$. The rapidly varying fine-layered medium is thus replaced by a "long-wave equivalent" transversely isotropic medium, whose density and vertical *P*- and *S*-wave velocities are averaged over layer thicknesses Δz, according to:

$$\bar{V}_P = \sqrt{\left\langle \frac{1}{\rho V_P^2} \right\rangle^{-1} \Big/ \langle \rho \rangle} \qquad (1)$$

*Unocal Science and Technology Division, Unocal Corporation, Brea, California, USA.

$$\bar{V}_S = \sqrt{\left\langle \frac{1}{\mu} \right\rangle^{-1} \Big/ \langle \rho \rangle} \qquad (2)$$

and $\rho = \langle \rho \rangle$, where μ is the rigidity, $\langle \ \rangle$ denote average and $^{-}$ denotes the "long-wave equivalent."

In our approach, the distance Δz is determined by ordering the acoustic impedance magnitudes in the log data. Then, the largest values are selected to limit the total number of layers in the model keeping Δz much less than the dominant wavelength in the data.

The S-wave velocities in equation (2) are computed by estimating the rigidity using the zero frequency limit of the Biot equation equivalent to Gassmann's equation (Gassmann, 1951).

$$\rho V_p^2 = K_b + 4\mu_b/3 + \frac{(1 - K_b/K_s)^2}{(1 - \phi - K_b/K_s)(1/K_s) + \phi/K_f}$$

where

μ_b = shear modulus of drained rock
K_s = Solid rock bulk modulus
K_f = fluid bulk modulus
ϕ = porosity

This simplified form of the Biot equation is useful at seismic frequencies. It is used to solve for the bulk modulus of the drained rock frame K_b, given an initial P-velocity, water saturation, porosity, an assumed dry rock Poisson's ratio σ, and densities and bulk moduli of the water, hydrocarbon and solid constituents of the rock. The drained frame modulus refers to rock that is not fluid saturated but still retains liquid molecules adhering to grain surfaces. Rigidity is then given in Geertsma, 1961 or Fung, 1965:

$$\mu_b = \frac{3}{4} K_b \left[\frac{3(1 - \sigma)}{1 + \sigma} - 1 \right],$$

and the shear velocity by, $V_s^2 = \mu_b/\rho$.

This procedure has been called the Gregory-Pickett solution of the Gassmann-Biot-Geertsma equation (Hilterman and Graul, 1984).

MODEL GENERATION

The use of the above method for clastic rocks having clay/sand mixtures is based upon two assumptions:

1. For unconsolidated sediments or sediments with a high clay fraction, the clay partially supports the quartz grains and, therefore, the clay and sand moduli and densities are averaged to give a composite solid matrix. Sediments belonged to this class if the clay/sand ratio was greater than 0.5.
2. For consolidated, well-cemented sandstones, having a low clay fraction, only quartz (or a mineral mixture) should compose the solid matrix, nor should the clay be averaged with the fluid constituents; it simply fills the pores and lowers effective porosity.

The fractions of sand and clay are then used to average the density and bulk modulus of the solid rock for class 1, whereas for class 2, the calcite and/or quartz minerals are used for the average and clay is excluded. Rocks belonging to class 2 will not necessarily be assigned the same properties, however, because the P-velocity, density, and porosity are varying from point-to-point in the log which affect the estimated shear velocity. The bulk moduli are computed from the average of the upper (Voigt) and lower bound (Reuss) methods (Hilterman and Graul, 1984).

The required input parameters for the Gregory-Pickett estimate of shear velocity were obtained from a well log interpretation program. The parameters output by this program that are used in the modeling, are the measured P-velocity from sonic, measured bulk density, water saturation, effective porosity, sand/clay fractions, calcite cement fraction, clay velocity, and clay density. In addition, lithologies other than clastics are detected such as coal, limestone, dolomite, tuff, chalk, salt, anhydrite, siderite, etc. If these lithologies are present, shear velocities are computed using fixed Poisson's ratios for each lithology obtained from laboratory data or the literature.

Limits are required on porosity and clay/sand ratios in the calculation of shear velocity in clastics. For example, the Gregory-Pickett method was used to estimate Vs only if the porosity was greater than 4 percent. The Castagna et al. (1985) mudrock $Vp - Vs$ relation was used unconditionally for clay-sand ratios greater than 20, and for ratios greater than 5, if the porosity was less than 0.1. The analyst must also decide on a representative density and bulk modulus for the hydrocarbons, brine, and accessory minerals that may be used in the estimate of shear velocity. The physical properties used in these calculations, including typical values for clay in the Brent interval, are listed in Table 1. The bulk moduli of clays were computed using,

$$K = \rho \left(V_p^2 - \frac{4}{3} V_s^2 \right),$$

from estimates of sonic traveltimes and densities obtained from the well log analysis. Shear velocities were obtained from the Castagna mudrock formula.

The model is thus generated in a semiautomatic fashion from the well log data once the appropriate decisions have been made regarding the test limits and properties of the fluids and minerals. The analyst must next choose the level of thin bed resolution required for the overburden and the target zone. The sonic log in the target zone was blocked by locating and ranking the 100 largest acoustic impedance changes. Averaging by the method of Backus was then done in between these selected layer tops. In this study, a low resolution overburden was chosen to match the major events in stacked seismic sections and velocity spectra at the well locations. The upper portion of the model was constructed using seismic data, and the lower portion including the target zone, having been derived from log data as described above, was merged with the upper portion at a marker horizon.

MODELING PROCEDURE

Synthetic seismograms were generated by the viscoelastic modeling program VESPA (Apsel, 1979), and by a proprietary UNOCAL ray tracing program. VESPA simulates the complete solution (all body waves, surface waves, converted phases and their multiples) without approximations or limiting assumptions for vertical seismic profiles (VSP) or surface seismic shot records. The source is a point explosion incident upon planar layers. Single receivers are located at each offset to approximate the shooting configuration in the marine cable. No receiver array attenuation is thus included; however, geometric spreading is taken into account. The ray tracing program was also used for mode selection tests.

Table 1. Elastic constants for synthetic modeling

Poisson's ratios for Average Water Saturated Rock	Well A (Oil)	Well B (Water)
Chert	.31	.31
Dolomite	.30	.30
Limestone	.311	.311
Coal	.40	.40
Bulk Moduli (GPa)		
Hydrocarbons	1.014 (API 40)	.021 (gas)
Water	2.39	2.39
Quartz grains	40.0	40.0
Calcite mineral	70.0	70.0
Dolomite mineral	80.0	80.0
Clay (Brent)	23.1	24.0
Density (gm/cc)		
Hydrocarbons	.825 (API 40)	.12 (gas)
Water	1.04	1.04
Quartz grains	2.65	2.65
Clay (Brent)	2.76	2.73

Average clay into rock frame if clay/sand ratio > .5.

RESULTS

Well A

Well A contained oil in the target sands of the Brent Group as indicated on the left-hand side of the interpreted well log data shown in Figure 1 by the low water saturation starting at 2590 ms two-way time (Base Cretaceous). Figure 1 also gives sand/shale/porosity fractions, residual and movable hydrocarbons and indices for calcite and organic matter fractions. Most of these quantities were used in model elastic parameter computations as described. Figure 2 depicts the resulting Backus averaged compressional and shear velocities and densities for the model from well A. The lithologic parameters used are listed in Table 1. The Q was held fixed at 100 for shear and 200 for compressional waves in all models. The water velocity used in this model was 4728 ft/s giving a water bottom two-way time depth of .244 s.

Figure 3 shows the comparison of NMO corrected field and synthetic prestack gathers at the well A location. The figure illustrates how elastic modeling can be used to help understand the AVO character of field data. Evaluation of the relative contributions of interference and offset dependent elastic reflection coefficients follow. The field data were common offset summed using three adjacent CMP gathers centered at the well. The synthetic data are plotted with reverse polarity to match the field data in which negative reflection coefficients have excursions to the right. The synthetic data was also phase rotated by 240 degrees to visually match the field data in the window of interest in order to compensate for unknown phase alterations acquired during propagation, acquisition, or processing. Predictive deconvolution was applied to the prestack field data followed by statistical wavelet shaping to zero phase. The filter length was 260 ms with a 160 ms prediction length. The deconvolution was not ideal and is responsible for some of the mismatch in Figure 3 due to the fact that multiple reverberation was suppressed somewhat by the predictive decon in the field data, but not in the model data. Offset dependent amplitude corrections for geometric spreading and receiver array attenuation have been applied to the field data and to the synthetic data depending on the type of program used. Although relative amplitudes do not match precisely, there is reasonable agreement in the event timing and offset dependent amplitude variation.

Figure 4 is a comparison of field and elastic synthetic stacked traces with the AVO information displayed in color by a program called Seismic Offset Dependent Amplitude Stack (SODAS). This program is similar to the method called "Colorstack" reported in Onstott (1986) in which amplitude relationships of NMO corrected near, mid, and far offset partial stack trace samples are assigned colors as shown by examples in the code on the right. High amplitudes in the far

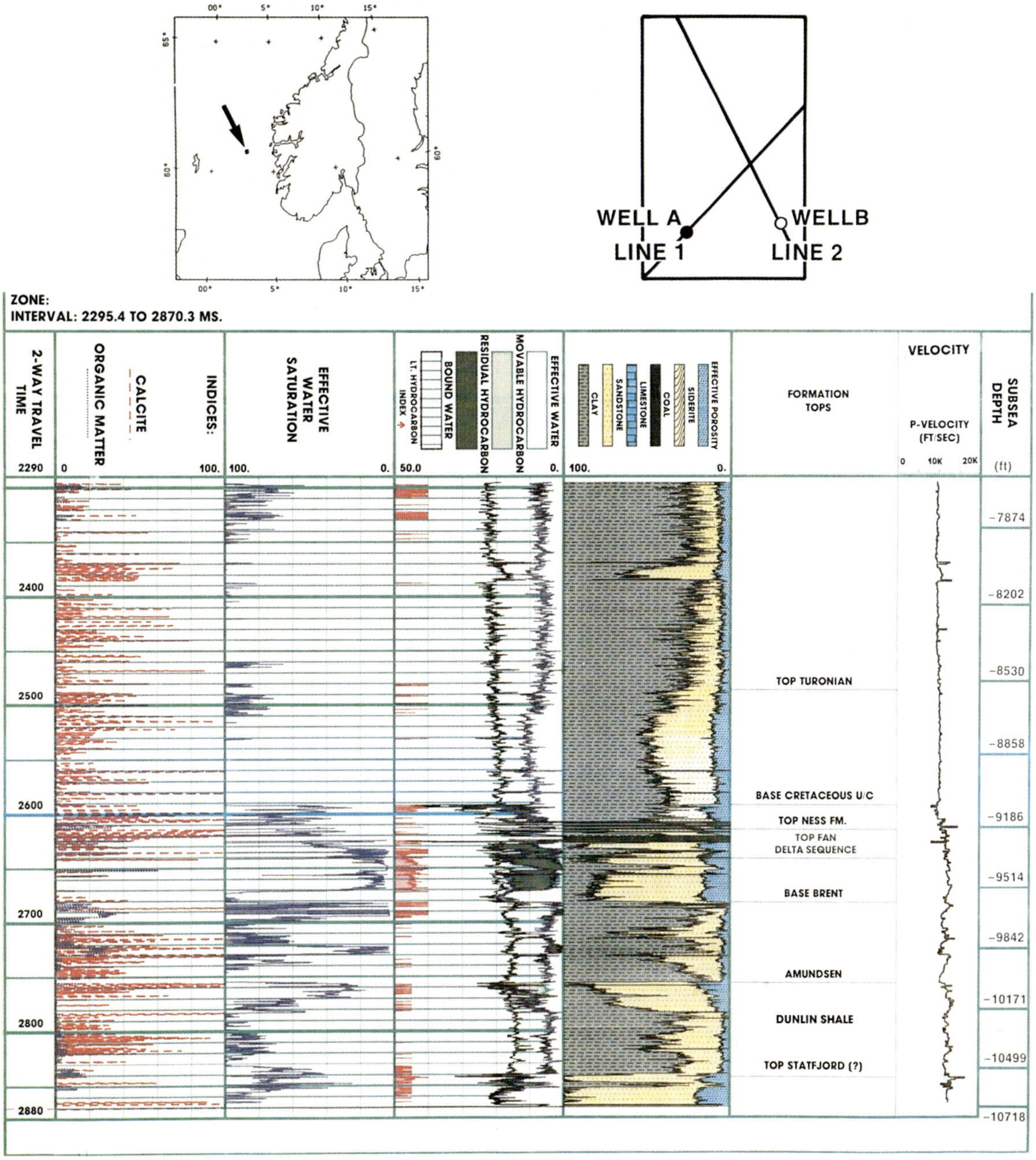

Fig. 1. Interpreted log data for Veslefrikk Well A.

offset traces will be assigned reddish colors, low amplitudes in the far offset traces will be assigned bluish colors, etc., and the overlay traces are the full stack. Color intensity is proportional to the stack trace amplitude. NMO residuals caused by stacking velocity errors or multiples will produce color fringing across wavelet cycles. This display can thus be used for section quality control as well as AVO interpretation.

Figure 5 is a comparison of NMO corrected elastic synthetic gathers with different mode selection options. These synthetics were computed by the Unocal ray tracing program. The same phase rotation of 240 degrees was applied to these data. These displays illustrate the relative contributions of different classes of events in the seismogram on prestack amplitude and phase.

The field data in Figure 4 shows increase in the absolute value of amplitude versus offset (hereafter referred to as "AVO increase") throughout the Brent. The synthetic data essentially agree most everywhere except for the prestack mismatch between 2.62 and 2.63 s (dark blue band). The explanation for this seismic response follows.

1. The AVO buildup (red) between 2.59 and 2.61 s [Event A in Figure 5(c)] is caused by the low

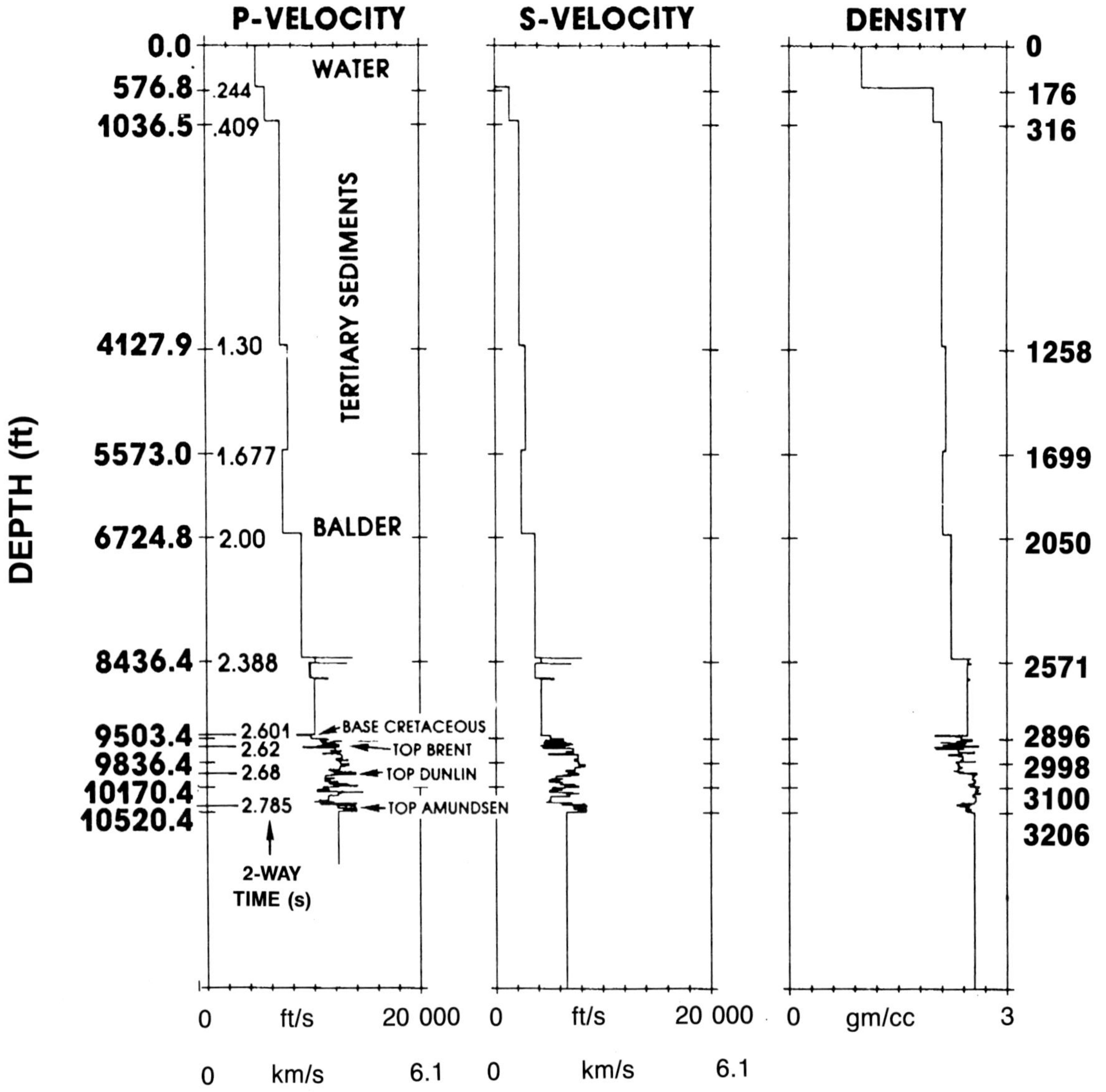

Fig. 2. Input elastic model for Veslefrikk Well A.

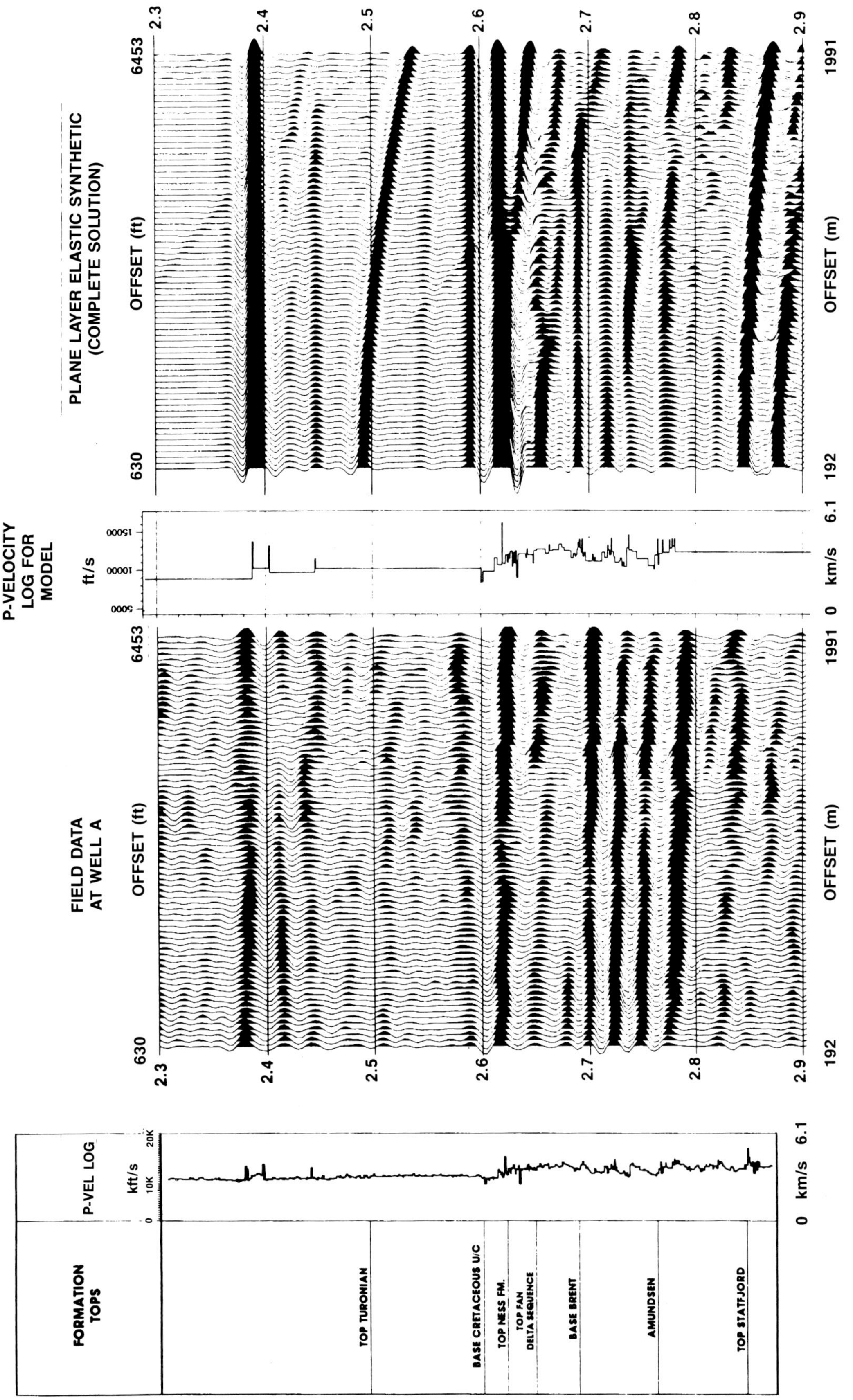

Fig. 3. Normal-moveout corrected field and elastic synthetic CMP gathers at Well A.

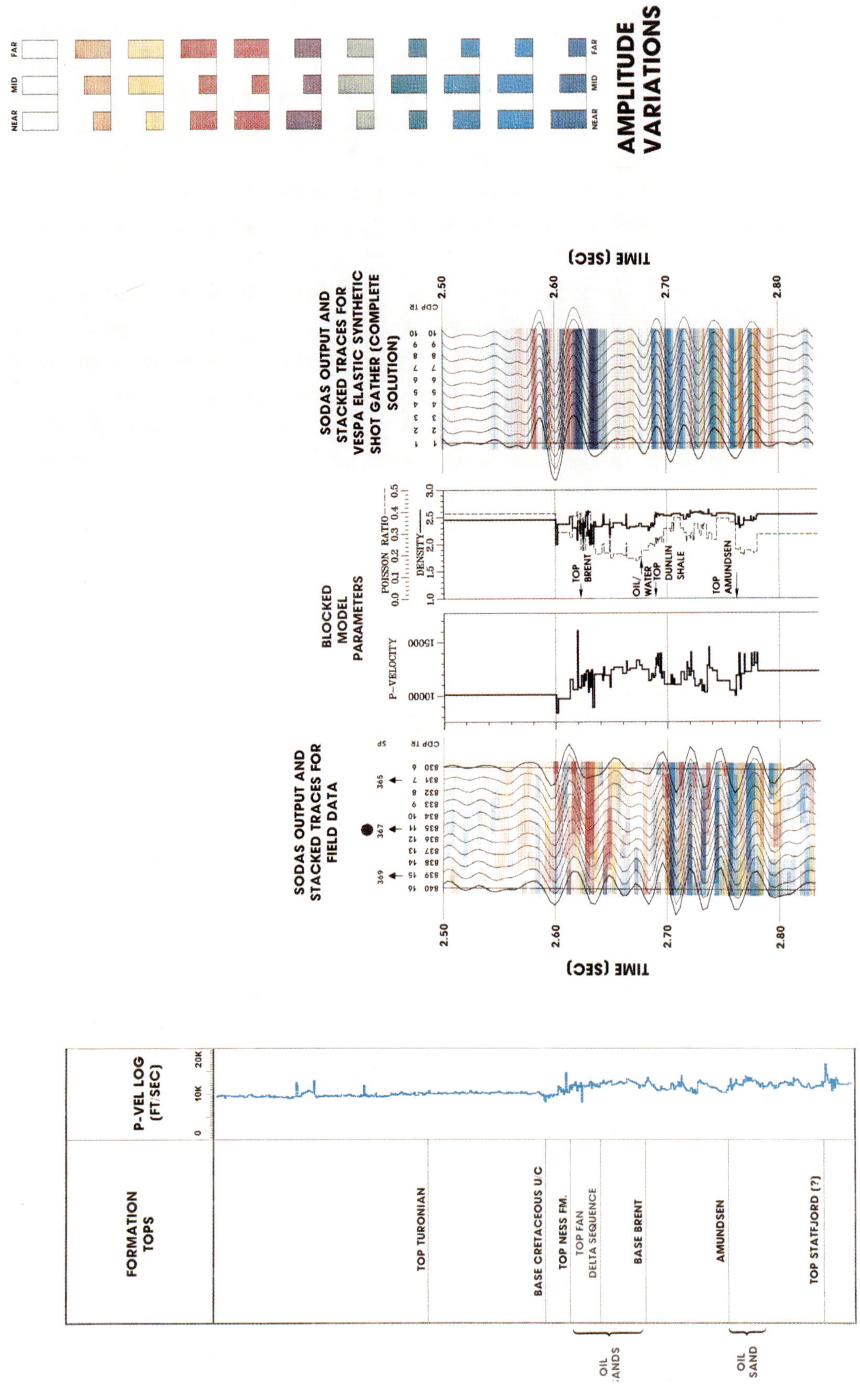

Fig. 4. SODAS displays of field data and VESPA elastic synthetic data at Well A.

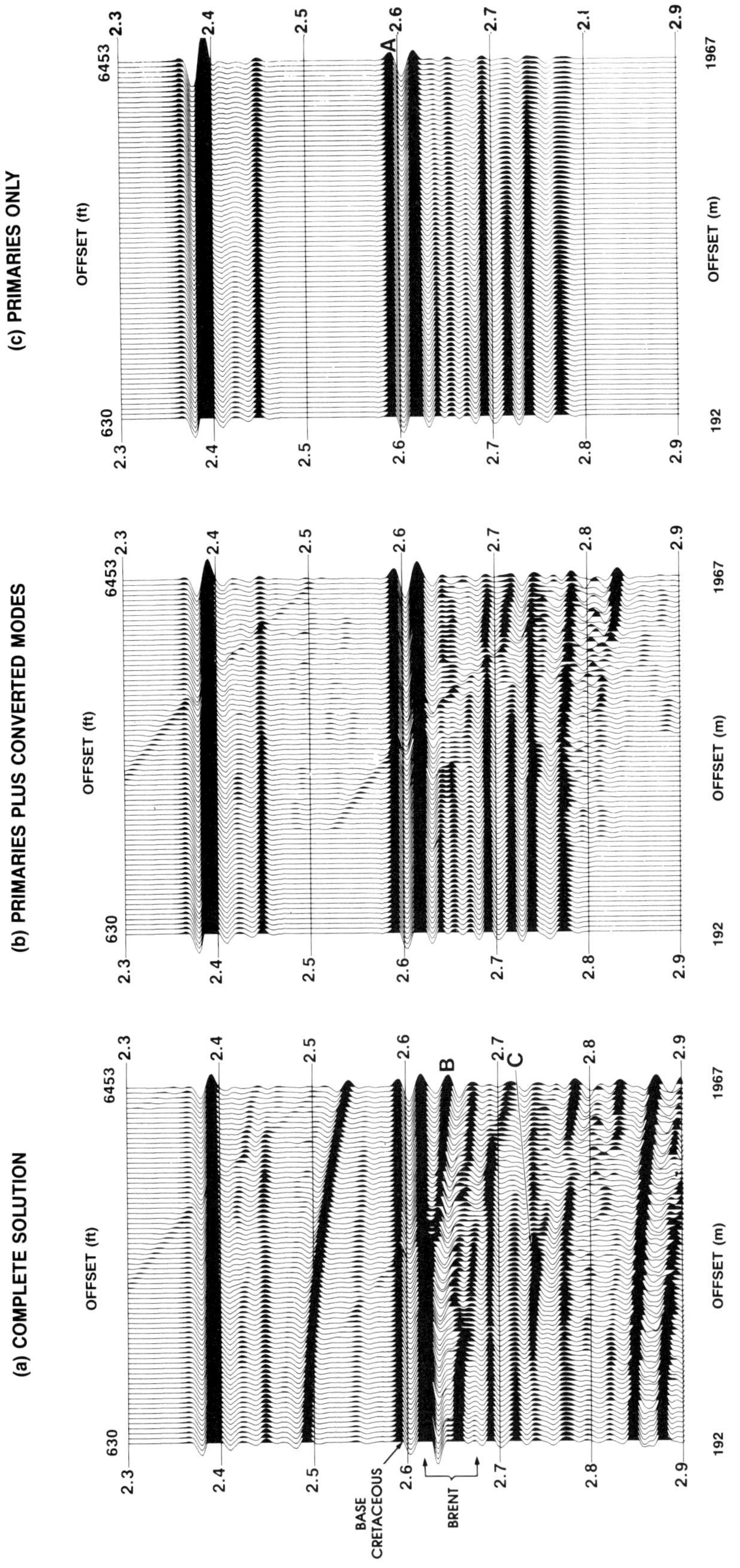

Fig. 5. Comparison of NMO corrected, ray-traced elastic synthetic gathers with different mode suppression options (240 degrees phase rotation) for Well A.

acoustic impedance, low Poisson's ratio, partial hydrocarbon saturated, sandy shales between the base Cretaceous and top Brent, including possibly a few layers in the uppermost Brent.

2. The far offset high amplitudes between 2.62 and 2.64 s in the field data identified as event B in the synthetics of Figure 5(a) are caused by two compressional water bottom pegleg events. One reflects from a Cretaceous sandstone at 2.39 s and the other from the Balder tuff marker at 2.0 s. The appearance of these events is seen in Figure 5(a) where compressional multiples are added to the elastic synthetic. The apparent AVO buildup is caused by residual NMO tuning. The divergence of the multiple from the primary at mid range offsets can also be seen in the field data as a doublet at 2.62 s. This effect influences the AVO response from 2.62 through 2.67 s, and therefore masks any direct detection of hydrocarbons in the mid Brent including the oil/water contact at 2.67 s. Thus, the AVO mismatch between the blue band in the synthetic data and the red band in the field data is due to differences in the amplitude and phase of the multiple. The multiple has been altered somewhat by predictive decon in the field data and may have been too strong in the synthetic data because of uncertainty in the water bottom reflection coefficient. Its presence and timing is still modeled accurately enough, however, to provide an explanation for the behavior of the field data at this location in time.
3. Between 2.7 and 2.8 s, cyclic velocity variations in the Dunlin Shale cause strong reflections that best match the synthetic data in Figure 5(b). Predictive deconvolution apparently attenuated the strong water bottom reverberation that reflects from the Balder Formation modeled in Figure 5(a) (Event C). The color fringing in Figure 4 is due to the fact that NMO was applied according to the correct primary velocity function, but converted modes produce some net residual NMO as shown in Figure 5(b). These are PSP conversions involving the Cretaceous sandstones at 2.39 s, the Brent sands, and impedance contrasts within the Dunlin Shale.

Figure 6 is a comparison between conventional normal incidence (primaries only) synthetic traces from the sonic and density log with the field and elastic synthetic stacked traces. The stacked elastic traces are a better match with the field data than the normal incidence synthetics. The shot gather synthetics were convolved with a 50 Hz Ricker in all cases. This seemed to give a good match after stack with a 30 Hz Ricker for the normal incidence synthetic because the stacking process acts as a low pass filter. As expected, the near offset traces with only compressional primaries [Figure 5(c)] match the normal incidence traces of Figure 6.

Well B

The Brent Group sands in Well B were water saturated as indicated by the well log analysis results in Figure 7, although there appears to be some partial hydrocarbon saturation in the Brent Group sands as well as in the shales between the base Cretaceous and the Brent. The partial saturation was assumed to be gas in the elastic modeling (Table 1).

The elastic model derived from the seismic and the log data is shown in Figure 8. The water depth in two-way time for this model is .27 s with a water velocity of 4796 ft/s. Figure 9 shows the resulting NMO corrected synthetic traces with the common offset summed field traces at the well site for comparison. Although the match is not exact, most events in the field data are represented in the elastic synthetic with the correct timing and approximate relative prestack amplitude variation. The SODAS synthetic and field AVO response is shown in Figure 10. The interpretation of these figures is as follows:

1. Between the base Cretaceous and top Brent there is corresponding AVO buildup in both field and elastic synthetic data. This is due to a combination of partial gas saturation in the effective porosity of "hot" organically rich shales and compressional multiple interference. For example, a low-velocity multiple at 2.55 s is a water bottom pegleg reflecting from the Balder tuff marker [Event A in Figure 11(a)]. The residual NMO causes apparent AVO buildup between 2.55 and 2.57 s. Event B in Figure 11(c) shows phase reversal at about 2/3 maximum offset which is a primary elastic effect. The red event at 2.64 in Figure 10 [Event C in Figure 11(c)] occurs at low Poisson's ratio sandy shales.
2. The AVO response at the top Brent at 2.71 s (Event D) is decreasing (blue) in both synthetic and field data due to high acoustic impedance sands with moderately low Poisson's ratios.
3. Within the Brent interval there is no AVO increase in the synthetic data and in the field data there is also no strong AVO increase except for

subdued pink and purple colors which indicate AVO decrease to midrange offsets, then AVO increase at far offsets. This may be attributed to multiple and noise interference with residual NMO manifested as color fringing across wavelet cycles. At 2.74 s Event E in Figure 11 is a low amplitude event that disappears at mid-range offsets in both synthetic and field data [Figures 9 and 11(a)]. This is caused by destructive interference from S-waves converted between reflectors in the Heather shales and the Brent [Figure 11(b)].

Figure 12 shows the correlation of normal incidence synthetic traces with the stacked field and elastic synthetic traces. Again the overall amplitude and phase match of the elastic synthetic with the field data is better than the normal incidence synthetic. An exception is the event at 2.68 s in the elastic synthetic

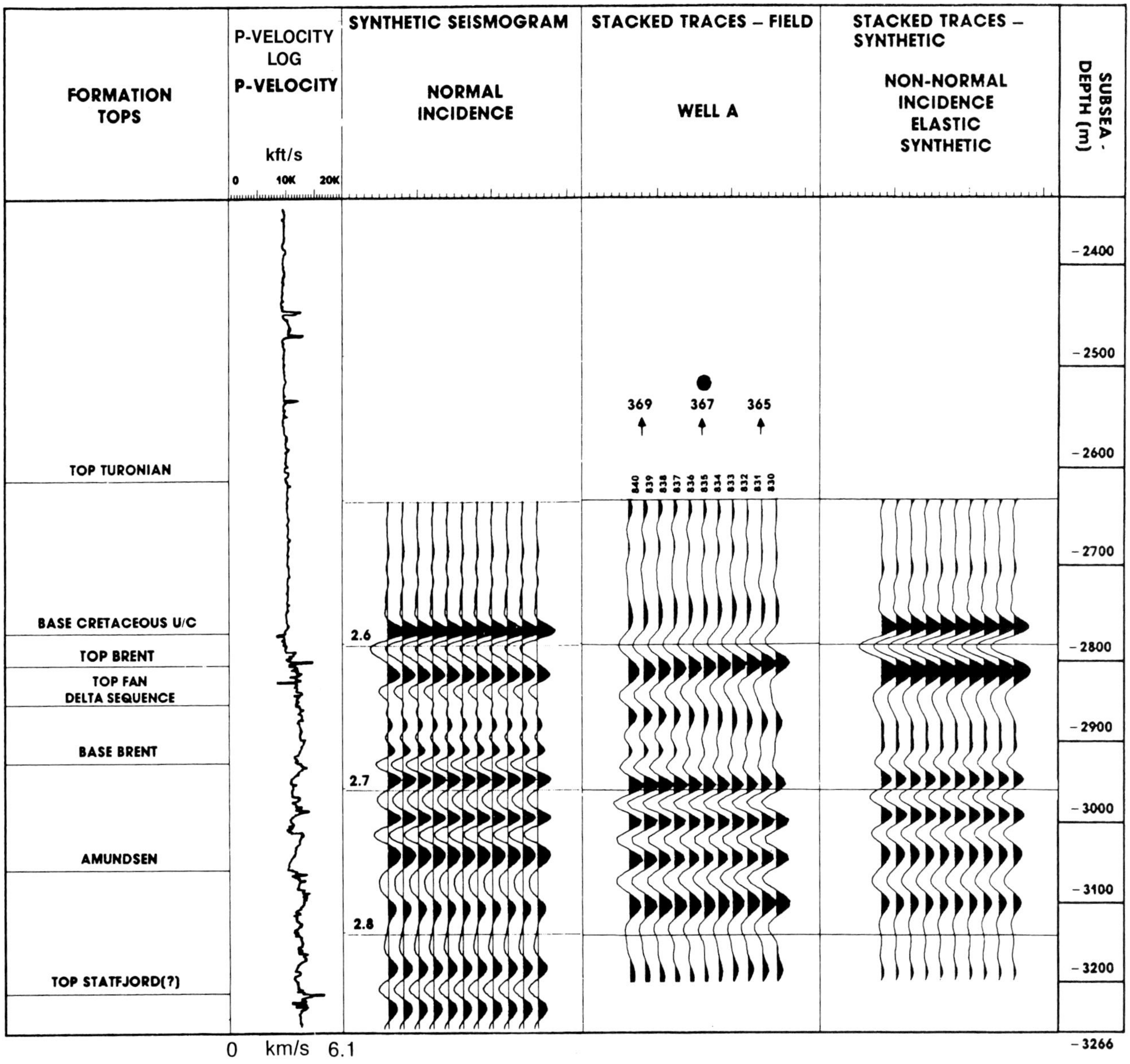

Fig. 6. Comparison of normal incidence primaries only synthetic traces with field and elastic synthetic stacked traces at Well A.

which is attributed to an abrupt step in the blocked acoustic impedance for the elastic model. Figure 12 also illustrates that the stacking process robustly attenuates events with residual NMO (compare Figure 11(c) to the stacked elastic synthetic traces in Figure 12).

CONCLUSIONS

This case study has demonstrated the usefulness of prestack elastic synthetic modeling for detailed lithologic interpretation. Figures 5 and 11 illustrate several

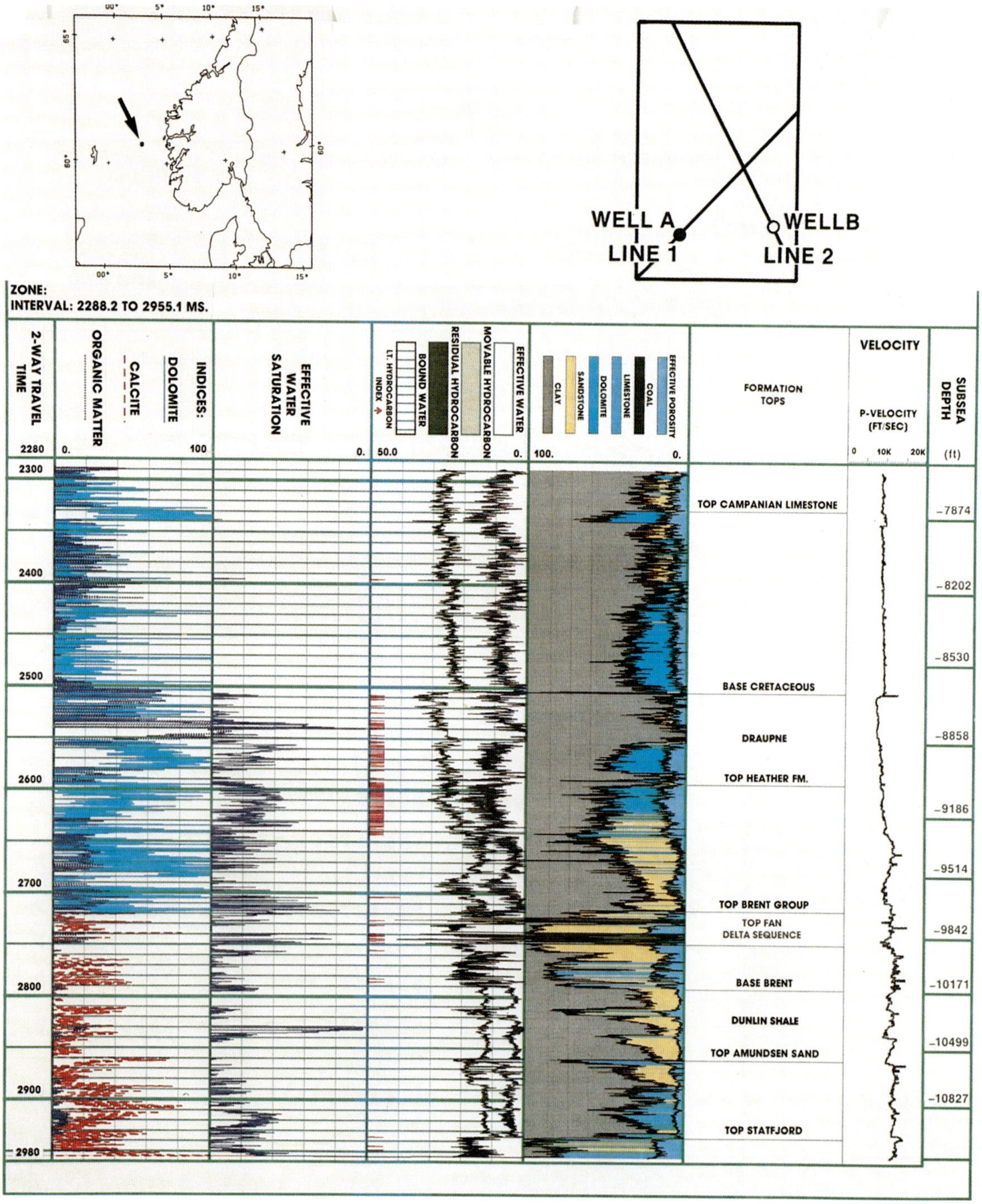

Fig. 7. Veslefrikk Well B interpreted log data.

important points in AVO analysis. The first is that interference patterns between primaries and multiples can dominate the prestack amplitude variations unless the primaries are isolated "bright spots." If we have done ideal processing to remove the multiples, there still may be a problem in interpreting AVO as a direct indicator of lithology because of locally converted S-wave events as is illustrated by Figures 5(b) and 5(c) or 11(b) and 11(c). It is therefore imperative to search for changes in the seismic waveform directly related to changes in lithology using borehole calibrated model based interpretation.

Processing has a strong effect on the seismic data. The synthetic and field data should be identically processed. Deterministic source signature deconvolution is recommended. If that is not done, as in this case, the synthetic wavelet phase should be matched as closely as possible to the field data in the target zone by phase rotation, but expect some degradation in interpretability. Nevertheless, as is typically the case, if processing steps for synthetic and field data are not identical, meaningful AVO interpretation can be made with the aid of elastic lithologic modeling.

Care must be exercised when interpreting amplitude buildup with offset as due to economic hydrocarbon saturation. AVO increase may occur if P-wave veloc-

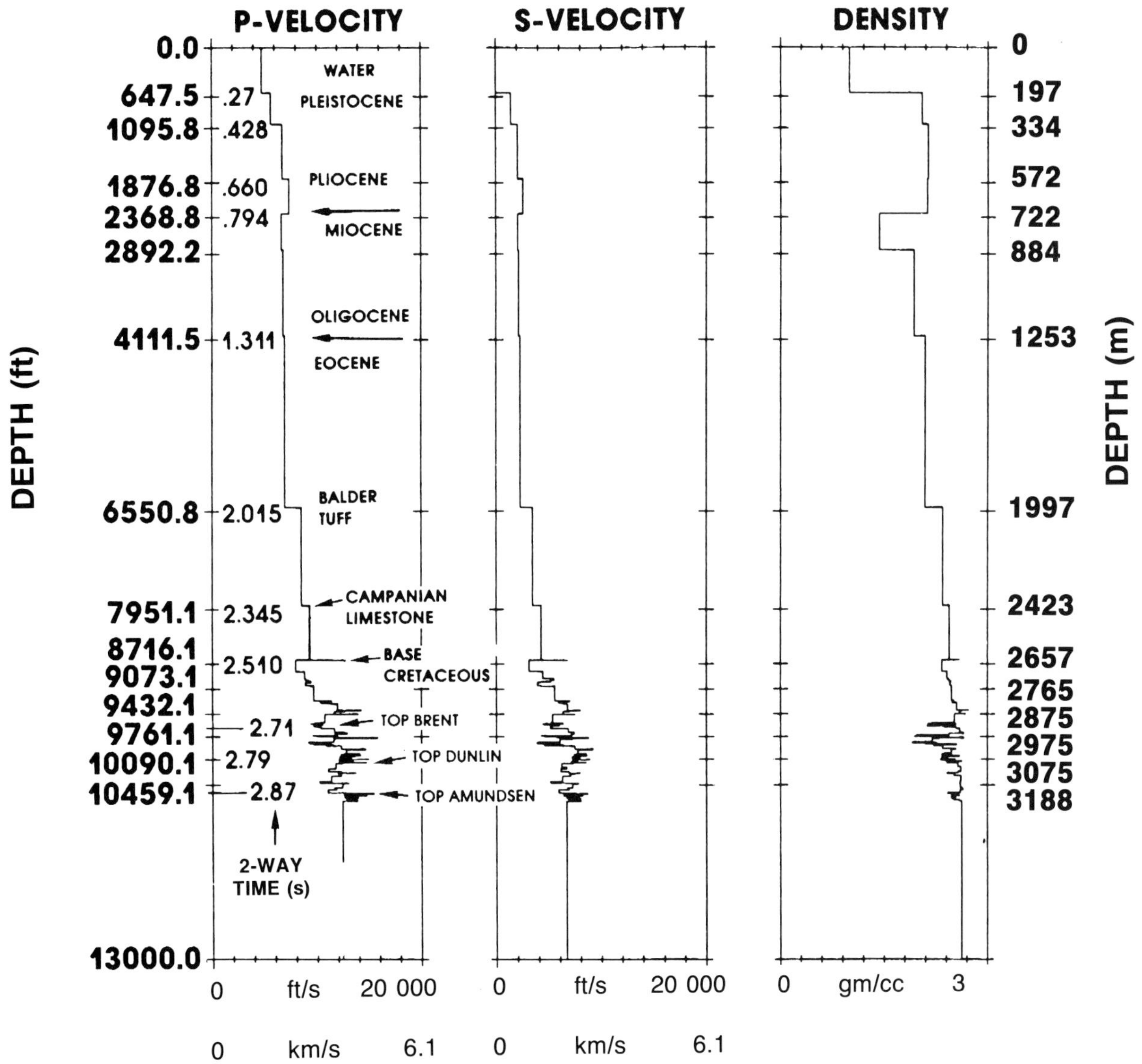

Fig. 8. Input elastic model for Veslefrikk Well B.

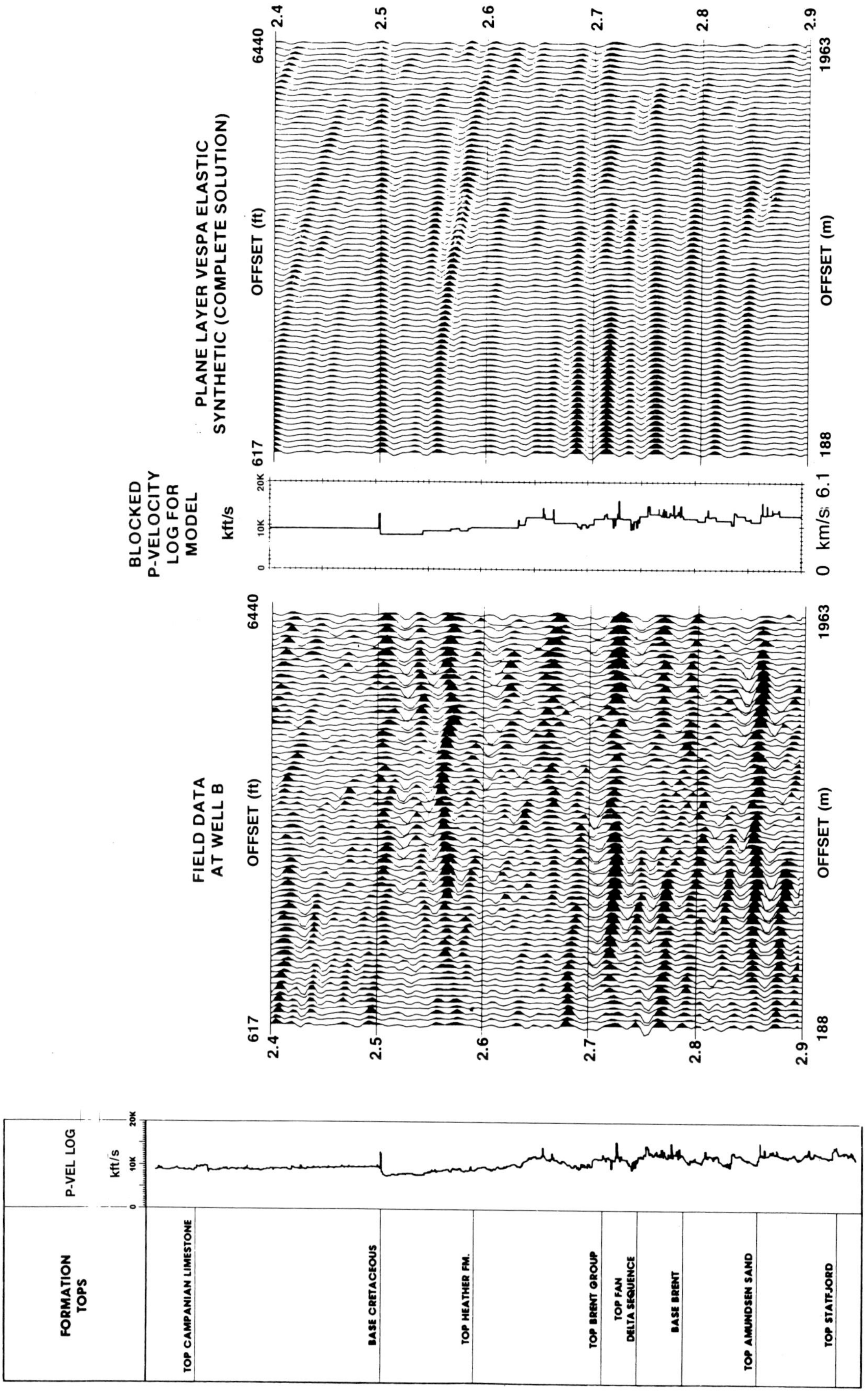

Fig. 9. Normal-moveout corrected field and VESPA elastic synthetic for Well B.

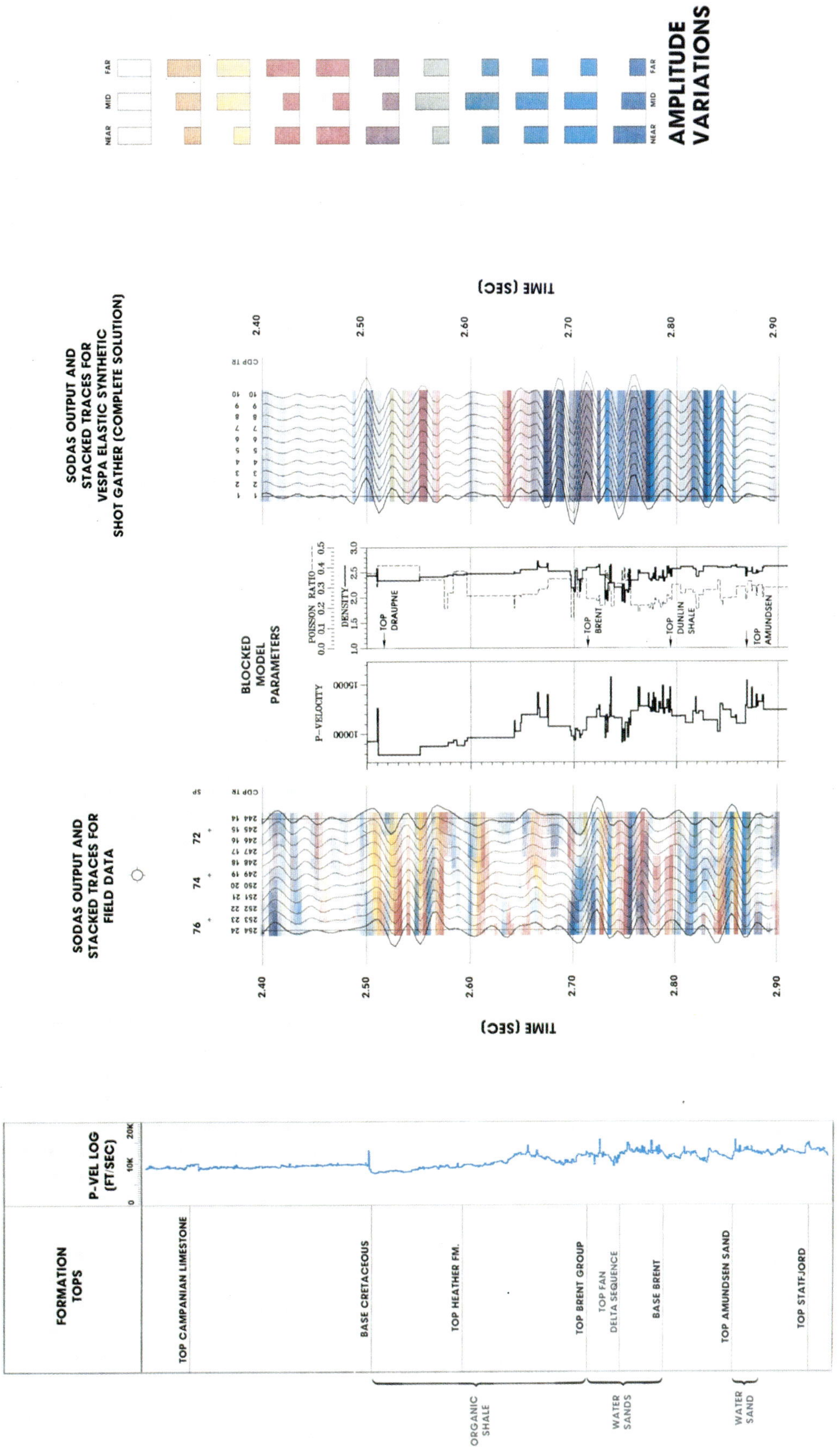

Fig. 10. SODAS displays of field data and VESPA elastic synthetic data at Well B.

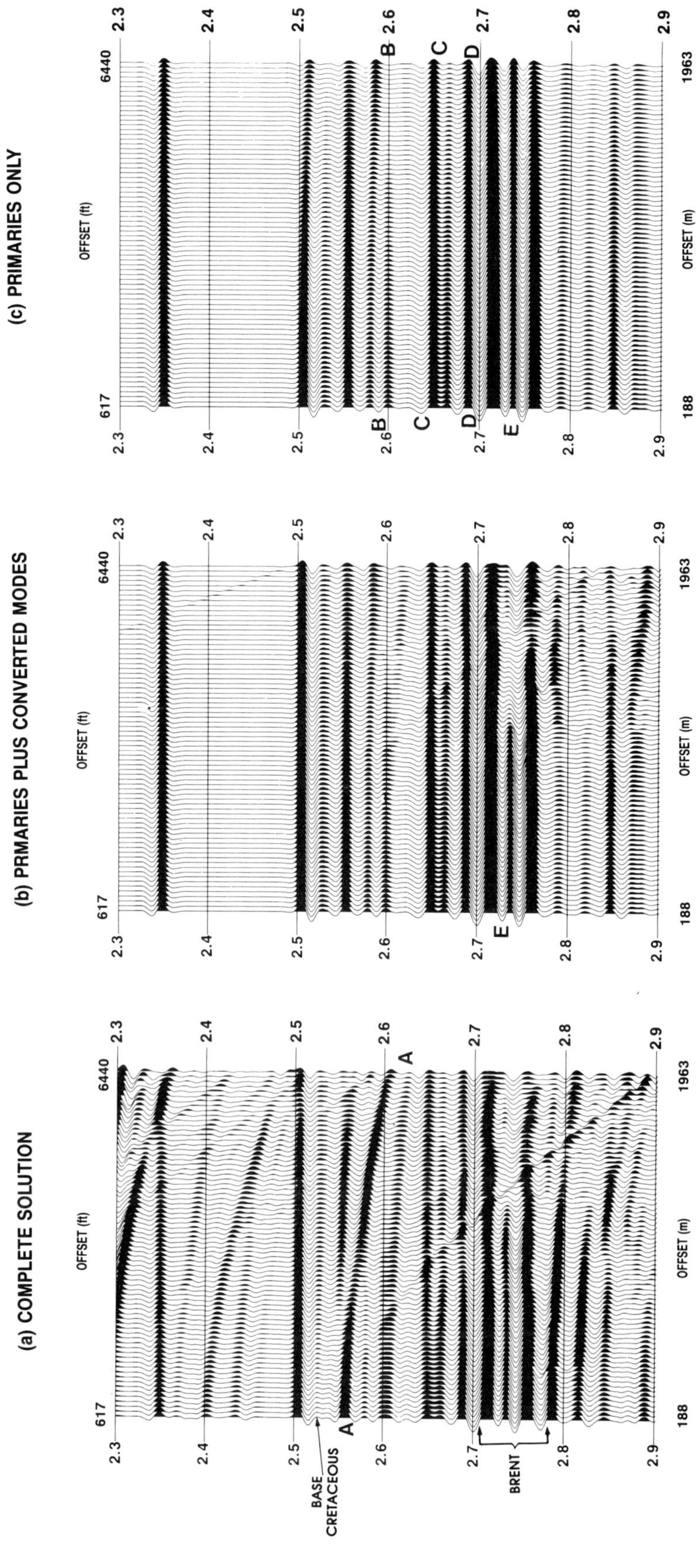

Fig. 11. Comparison of normal-moveout corrected, ray-traced elastic synthetic shot gathers with different mode selection options (240 degrees phase rotation) for Well B.

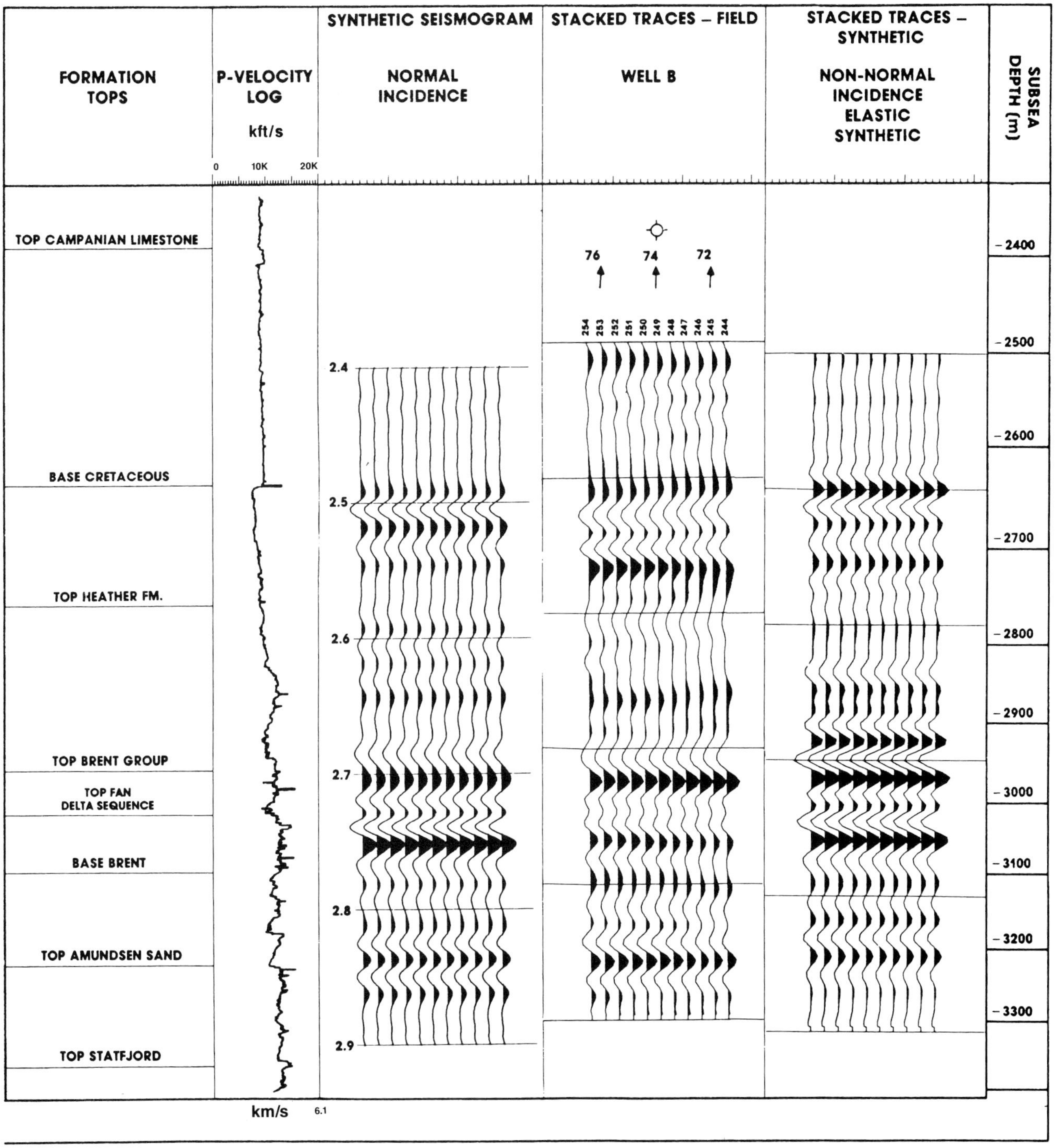

Fig. 12. Comparison of normal incidence primaries only synthetic traces with field and elastic synthetic stacked traces at Well B.

ity decreases due to hydrocarbon saturation in high porosity sands, but AVO increase may also occur in low velocity, partial gas saturated sandy shales surrounded by higher velocity wet sands. In the Brent interval, lithologies with high porosity and a high sand fraction generally have high velocity even with light hydrocarbon saturation. Thus, if velocity increases and Poisson's ratio decreases across a shale/sand boundary, amplitudes usually decrease with offset. In this case, it is more likely that anomalous amplitude buildup will occur at the oil/water interface within the sandstone formation, but this was difficult to interpret because elastic modeling showed that the AVO effect at the oil/water contact is very weak and difficult to observe in the presence of multiple interference.

The implications of the results of this study for the interpreter are that, in this geologic environment, formation tops should be carefully identified prior to AVO interpretation. For example, target formations should be mapped for structural closure and proximity to organic shale source rocks that AVO analysis may help to identify.

ACKNOWLEDGMENTS

The author acknowledges Terry O'Sullivan (Unocal) for a thorough well log interpretation, including the graphics plots; Marcia Spriggins (Unocal) for data plotting and program execution; Peter Western (Unocal Norge), Jan Grzywacz (Unocal Norge), Steve Schreiber (Unocal Norge), Milo Backus, and other reviewers for critically editing this manuscript and providing the data and much information regarding it and the geology. Thanks also to the operator, Statoil, and members of license group PL052 for permission to publish this paper.

REFERENCES

Apsel, R. J., 1979, Dynamic Green's functions for layered media and applications to boundary-value problems: Ph.D. Thesis, Univ. of California at San Diego.

Backus, G. E., 1962, Long-wave elastic anisotropy produced by horizontal layering: J. Geophys. Res., **67**, 4427.

Castagna, J. P., Batzle, M. L., and Eastwood, R. L., 1985, Relationships between compressional-wave and shear-wave velocities in clastic silicate rocks: Geophysics, **50**, 571–581.

Fung, Y. C., 1965, Foundations of solid mechanics: Prentice-Hall, Inc.

Gassaway, G. S., and H. J. Richgels, 1983, SAMPLE: Seismic amplitude measurements for primary estimation: Presented at the 53rd Ann. Internat. Mtg., Soc. Expl. Geophys.

Gassmann, F., 1951, Elastic waves through a packing of spheres: Geophysics, **16**, 673–685.

Geertsma, J., 1961, Velocity-log interpretation: The effect of rock bulk compressibility: Soc. of Petr. Eng. J. (AIME), **1**, 235.

Hilterman, F. J., and M. Graul, 1984, SEG Seismic Lithology Course Notes, (Unpublished) Soc. Expl. Geophys.

Onstott, G. E., 1984, Processing and display of offset dependent reflectivity in reflection seismograms, M.Sc. thesis, Univ. of Texas, Austin.

Ostrander, W. J., 1984, Plane wave reflection coefficients for gas sands at non-normal angles of incidence: Geophysics, **49**, 1637.

Yu, G., 1985, Offset-amplitude variation and controlled-amplitude processing: Geophysics, **50**, 2697–2708.

Interpretive Aspects of AVO—Application to Offshore Gulf Coast Bright-Spot Analysis

*Robert Estill and Keith Wrolstad**

INTRODUCTION

Amplitude variation with offset (AVO), as introduced in Ostrander (1984), is a useful technique for "bright spot" (amplitude anomaly) evaluation. The theoretical basis for AVO including correction factors and basic data processing steps is briefly discussed. AVO case histories are presented in which calculated elastic reflection coefficients using well logs are compared with corrected prestack field data amplitudes. Results show: (1) AVO correction factors are especially important for amplitude anomalies shallower than 5000 ft (1524 m) subsurface; (2) thin beds decrease reflectivity with offset; (3) density contrasts are difficult to separate from velocity contrasts; (4) a major problem for AVO interpretation is the false indication of producible hydrocarbons due to the presence of slight partial gas saturation; and (5) in many cases, detection of gas sands can be made by matching elastic synthetic and field AVO data.

THEORETICAL BASIS OF AVO

Bright spot evaluation using AVO is an application of the elastic wave equation prediction that reflection amplitudes will vary with angle of incidence or recording offset distance. AVO is directly related to acoustic impedance and Poisson's ratio contrasts which are in turn related to changes of the in situ fluid saturation.

The realization of potential petroleum exploration applications from elastic-wave reflection coefficient variations with angle of incidence, especially if there are contrasts in Poisson's ratio, started with Koefoed (1955), although Muskat and Meres (1940) previously tabulated solutions for plane-wave elastic reflection coefficients. The solutions given in Muskat and Meres (1940), and more recently in Tooley et al. (1965) and Shuey (1985) are essentially equivalent to those contained in the program used here for modeling. Recently, the interpretation of offset or angle of incidence dependent seismic reflection amplitude variations for the direct detection of hydrocarbons has been discussed in Backus (1982) and verified in field tests in Ostrander (1984).

In theory, for certain acoustic impedance contrasts, a gas sand can be distinguished from a brine-saturated sand or salt. In practice, a number of factors need to be considered in the interpretation of AVO results including reflector dip and depth, receiver array attenuation, inelastic attenuation, thin bed effects, and anisotropy.

DATA PROCESSING

The basic data processing sequence for AVO analysis in this study included:

(1) recover recording gain,
(2) apply spherical divergence correction,
(3) trace edit,
(4) sort and CDP gather,
(5) use signature deconvolution or mild predictive deconvolution if the far field signature is not available,
(6) try standard mute,
(7) apply migration before stack,
(8) use NMO by static shift to avoid NMO stretch (single events),
(9) apply common offset stack,
(10) trace balance with large (1 s) window that includes the zone of interest but excludes large amplitudes in the mute zone,
(11) correct for the known receiver array attenuation and correct for an approximate lower limit inelastic attenuation,
(12) scan bright spot events for prestack peak-trough amplitudes,
(13) reformat for display devices,
(14) display AVO of bright spots in NMO corrected gathers and plot,
(15) compare field data with theoretical multilayer elastic reflection coefficient model curves, and
(16) revise model and repeat Step 15 if necessary.

*UNOCAL Corporation

The migration before stack is important in areas of steep dip. As indicated in Resnick et al. (1987) analysis of AVO in areas of dip is problematic for several reasons. Since the reflection point for a given dipping event in a CMP gather varies with source to receiver offset (moving updip with increasing offset), the reflection angle depends on reflector dip. Also, NMO corrections depend on reflector dip, and mispositioned (i.e., unmigrated) events interfere with one another.

CORRECTION FACTORS FOR AVO ANALYSIS

A number of correction factors need to be considered in AVO processing and analysis including reflector dip, effect of depth, thin bed tuning, source and receiver array attenuation, inelastic attenuation and anisotropy. The discussion of differential correction for these factors from near to far offset follows.

The relationship between dip and angle of incidence is shown in Figures 1 and 2. Assuming a constant velocity medium, Figure 2 shows families of curves for the range of dips from zero to 90 degrees, where the ordinate $\Delta\Theta$, is the difference between the offset ray angle of incidence to a zero dip reflector and the angle of incidence to a dipping reflector. Note that the maximum error (or reflection point smearing) occurs at 45 degrees angle of incidence for small dips, and that the maximum error increases and moves to smaller angles of incidence for larger dips. This relationship can be used to estimate the effect of dip on reflection coefficients by noting the change in amplitude for a particular $\Delta\Theta$.

The effect of depth on the offset-dependent correction factors, array attenuation, geometric spreading, and inelastic attenuation are shown in Figures 3a–c. Notice that the difference between corrected and uncorrected reflection coefficient curves is less for successively deeper reflectors because the angle of incidence range decreases for a fixed-offset range. Anisotropy also affects the AVO of reflected *P*-wave events. For example, the effect of transverse isotropy on elastic *P*-wave reflection coefficients due to thin bedding in shales is illustrated in Figures 3d–f. The reflection coefficients are compared for an isotropic and transversely isotropic shale over isotropic gas and water sands. These results were computed from approximate relations given in Banik (1987), using the anisotropy parameter, δ, given in Thomsen (1986). Note that the AVO increase for anisotropic shale over an isotropic gas sand can be enhanced if the δ of shale is positive and can be diminished if δ is negative. Further work is necessary on the effect of general anisotropy on AVO.

Source and receiver array corrections are of the same order of magnitude as the spherical divergence correction for typical marine cables. The rate of atten-

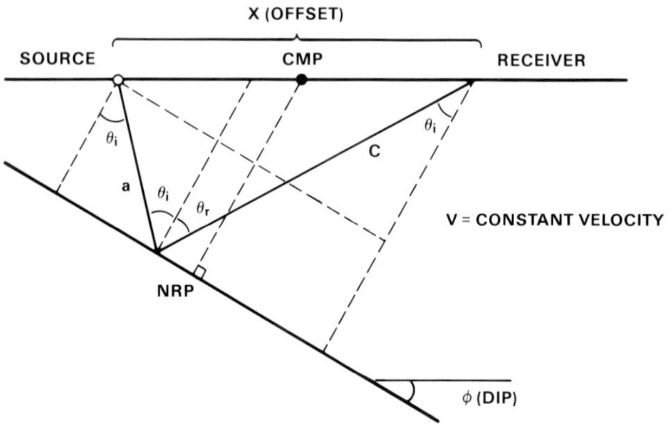

Fig. 1. Geometry for computing offset ray angle of incidence for dipping reflectors.

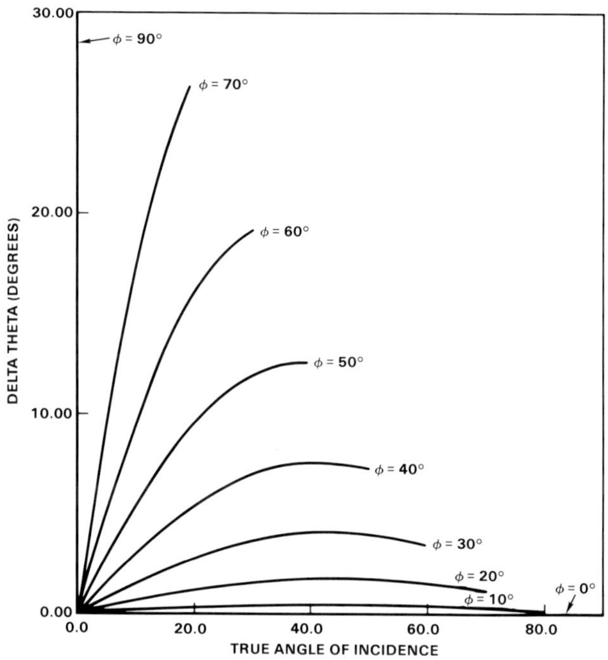

Fig. 2. The discrepancy ($\Delta\Theta$) between the true angle of incidence for an offset ray to a plane dipping reflector (dip = ϕ) and the angle of incidence to a zero dip reflector. This example assumes a constant velocity medium of 6992 ft/s (2131 m/s) and a fixed normal incidence 2-way traveltime of 2.12 s.

dipping about 20 degrees which causes the actual angle of incidence to be about 2 degrees less at the maximum offset than would be calculated if dip were neglected. In this case, the discrepancy at the maximum offset of 10 276 ft (3132 m) causes only about a 2.5 percent decrease in the combined magnitude of the inelastic and array attenuation corrections, thus the amplitude corrections for dips less than 20 degrees are negligible. Large dips, however, have significant deviations between the zero dip and dipping reflector angles of incidence as shown in Figure 2. Figure 9 is a color contour map of the reflection amplitudes for this event as a function of shotpoint and offset. The coverage is shown in Figure 6.

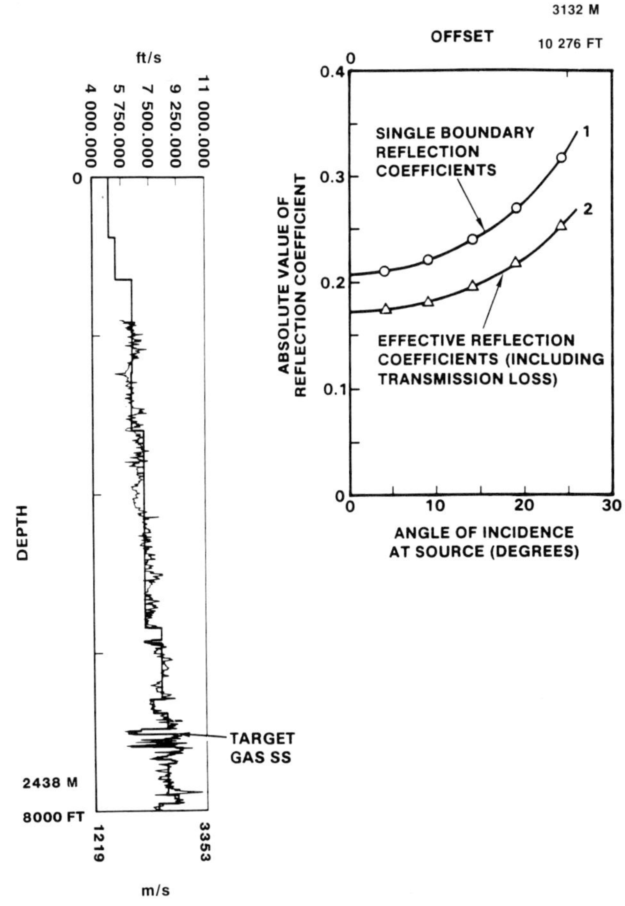

Fig. 7. Multilayer velocity model and theoretical reflection coefficients for top of gas sand reflector. Data are from Case 1. Model parameters are: shale P-velocity = 2580 m/s, shale Poisson's ratio = .401, shale density = 2.206 gm/cc, gas-sand P-velocity = 1839 m/s, gas-sand Poisson's ratio = .15, gas sand density = 2.027 gm/cc.

Case 2, from offshore Texas, had two gas-related bright spots overlying a water-sand reflector from a single well located at a tie point for several seismic lines. The two seismic sections considered here are line 1 (Figure 10) and line 2 (Figure 11), which show the spatial orientation of these features. These two lines intersect at the well but are not perpendicular. Figure 12 is a composite summarizing the predicted reflection coefficients for each target horizon. Figures 13 through 15 are the field data amplitudes from line 2 corrected for spherical divergence, receiver array attenuation (26 element tapered array), and inelastic attenuation (Q = 150). Also plotted are the effective reflection coefficients for the top of each target horizon in the multilayer velocity model shown in Figure 12. The gas-sand reflection amplitudes from line 2 for the anomaly at 1.52 s agree adequately with the effective reflection coefficients for the top of the sand (Figure 13 and Plot A in Figure 12). The resulting Poisson's ratio of .2 is reasonable for a gas-sand but suggests that it is not a clean sand as was verified by log analysis for this well.

The gas-sand reflection amplitudes for the reflector at 1.83 s also agrees well with theoretical effective reflection coefficients, and the Poisson's ratio of .10 suggests a cleaner sand than the one at 1.5 s. Note in Plot B of Figure 12 the large difference between the

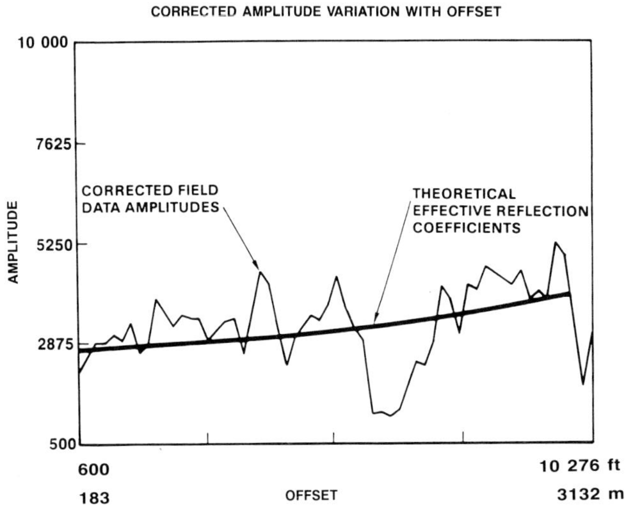

Fig. 8. Offset-dependent reflection amplitudes at the Case 1 well location with the multilayer model reflectivity function that includes transmission effects. The field data are peak-trough amplitudes with inverse corrections for spherical divergence, array attenuation, and inelastic attenuation (Q = 150).

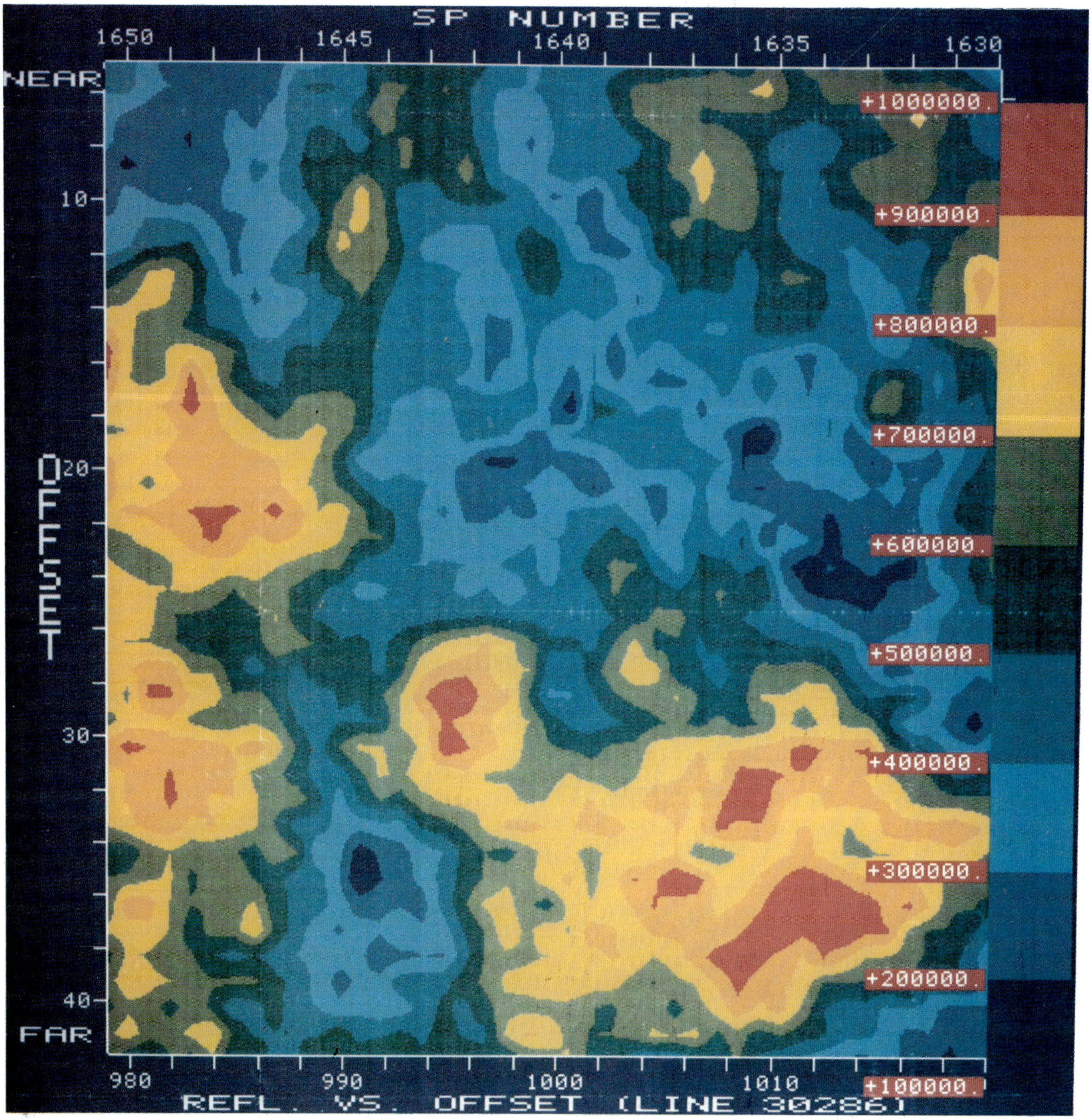

Fig. 9. Field data reflection amplitude as a function of offset at each CMP or shot point for the gas condensate reflector at 2.12 s for Case 1.

single boundary and the effective reflection coefficients. This difference is increased for both reflectors B and C due to the presence of overlying gas-sand layers. The elastic offset-dependent transmission coefficients of overlying gas sands tend to flatten the offset-dependent amplitudes of deeper reflectors. The water sand offset-dependent amplitudes for reflector C at 2.29 s decrease as is also predicted by the synthetic curves (Figure 15). This decrease of AVO is due to the fact that the acoustic impedance decreases whereas the Poisson's ratio increases across the reflecting boundary. Notice that the acoustic impedance and Poisson's ratio both decrease across the shale/gas-sand boundaries.

Figure 16 shows the lateral variability of the offset-dependent amplitudes for the gas reflector at 1.55 s from line 1. The amplitude response at the tie point matches reasonably well in AVO with that from Figure 13. This suggests that the observed variability in the offset-dependent amplitude response has a significant structural or stratigraphically controlled component, or the response is determined to a large degree by multiple or mode converted interference superimposed on the elastic reflection coefficients. Notice increasing AVO buildup at the well location. In some cases, hydrocarbon saturation is indicated only by relative lateral buildup rather than an absolute AVO increase.

Case 3 from offshore Texas shows a gas reservoir encountered at 5693 ft (1735 m) in a clastic section. A gas thickness of 35 ft (10.7 m) overlies 10 ft (3.05 m) of oil in a gross sand thickness of 75 ft (22.9 m). At the bottom of the reservoir are 7 ft (2.13 m) of water sand. The corresponding seismic event is the bright spot at 1.95 s in Figure 17. Figure 18 shows a log analysis using the digitized log data. The analog long-spaced sonic is also shown in the right hand track. Reflection coefficients were calculated using the 7-layer model shown in Figure 19 as derived from conventional velocity analysis. From the sonic, it is difficult to measure the gas-sand velocities which are apparently extremely low, with traveltimes slower than the mud

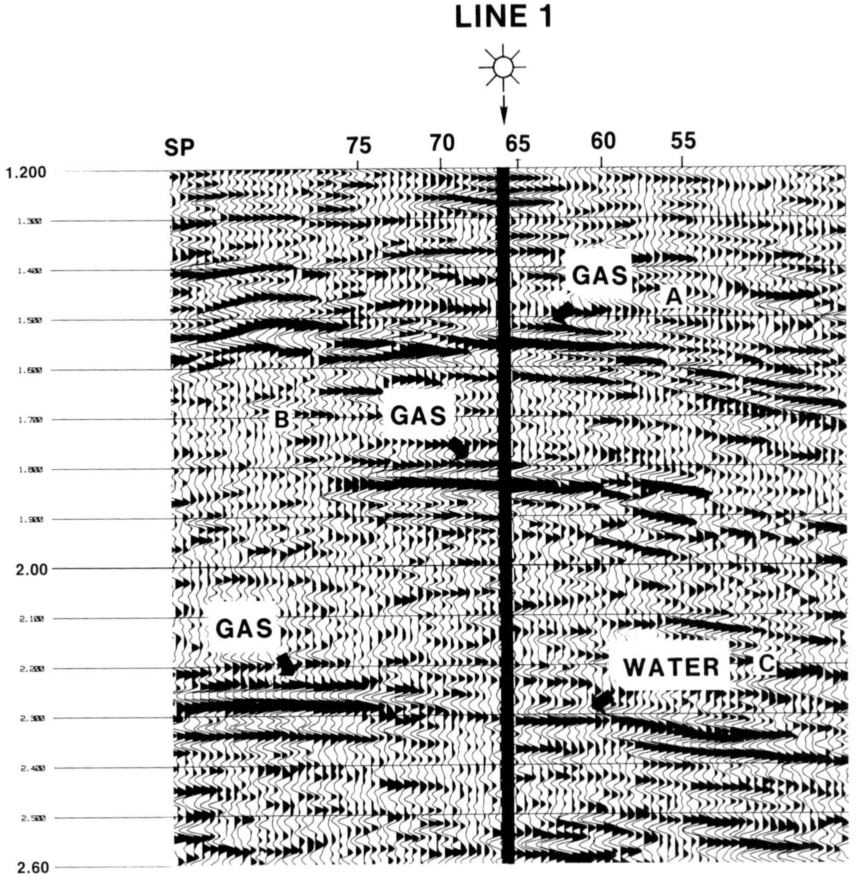

velocity, making the measurement highly unreliable. Therefore, the gas-sand compressional- and shear-velocities were estimated using the Gregory-Pickett method and Gassmann's equation as given in Castagna et al. (1985).

The starting point is a measurement of $V = 7092$ ft/s (2162 m/s) for 100 percent water saturation at a measured depth of 6155 ft (1876 m). Note that this is quite

Table 2. Model parameters for Figure 12.

	A	B	C
Shale V_p (m/s)	2285	2931	2839
Shale Poisson's Ratio	.426	.372	.372
Shale density (gm/cc)	2.152	2.345	2.353
Gas sand V_p (m/s)	1732	2277	
Gas sand Poisson's Ratio	.200	.100	
Gas sand density (gm/cc)	2.036	2.013	
Brine sand V_p (m/s)			2625
Brine sand Poisson's Ratio			.413
Brine sand density (gm/cc)			2.216

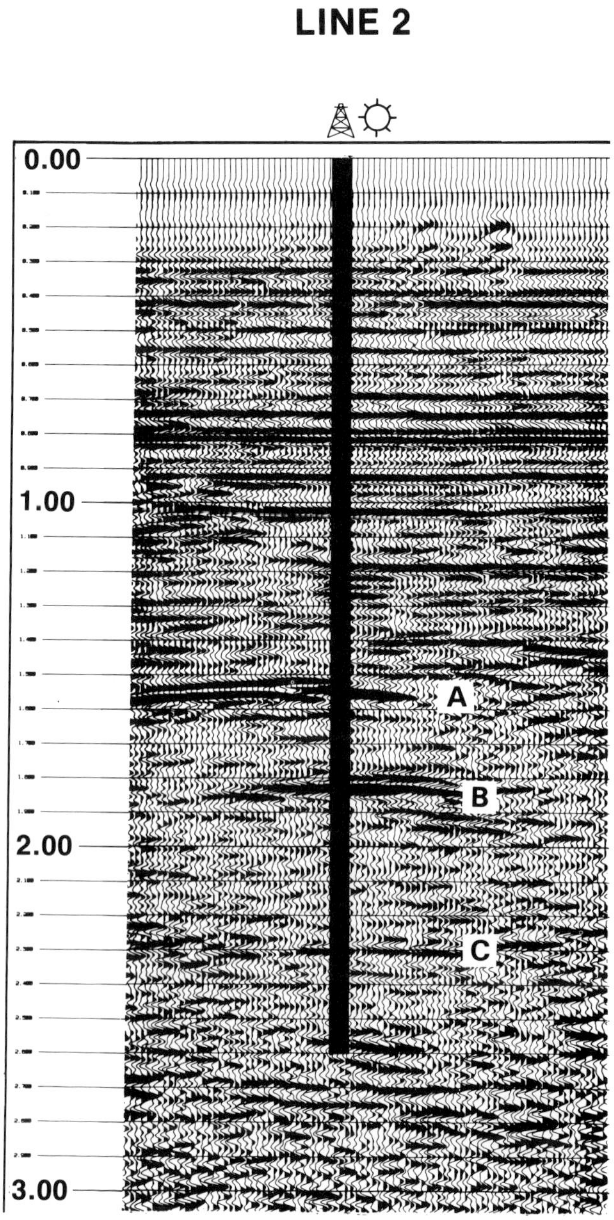

Fig. 11. Seismic section for Case 2 showing the same bright spot as in Figure 10.

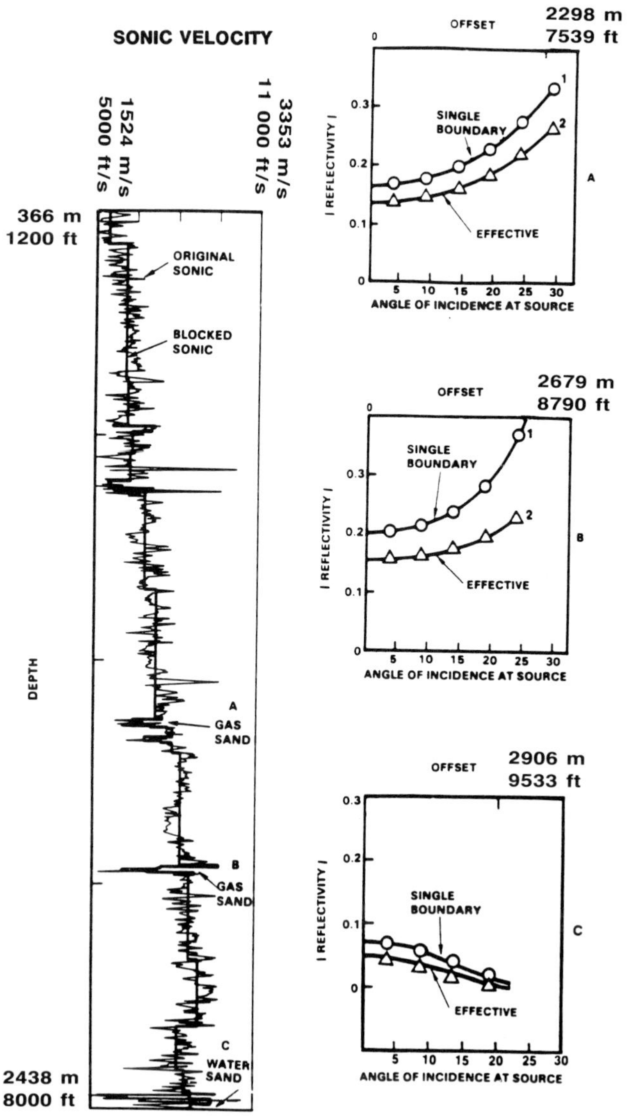

Fig. 12. Velocity model with single boundary (curves 1) and multilayer reflection coefficients (curves 2) for case 2 well. Model parameters are given in Table 2.

low for a vertical depth of 5600 ft (1707 m), and is responsible for an extremely low predicted compressional velocity of 4455 ft/s (1358 m/s) for 100 percent gas saturation using the Gregory-Pickett method. Using a hydrocarbon correction given in Timur et al. (1978), a conservative predicted gas-sand velocity is 5280 ft/s (1609 m/s). At this time, it is not known which method is most accurate, but in any case, the predicted gas-sand velocity is approximately less than or equal to the mud velocity. The resulting reflection coefficient curves for the top of the gas sand are shown in Figure 19. There should be good discrimination between water and gas sand in this case (curves 1, 2, or 3). The modeling predicts a larger increase in AVO than is observed (Figure 20). Possible explanations are multiples or converted phase interference, lateral variations in inelastic attenuation, or an insufficient number of layers in the model overburden.

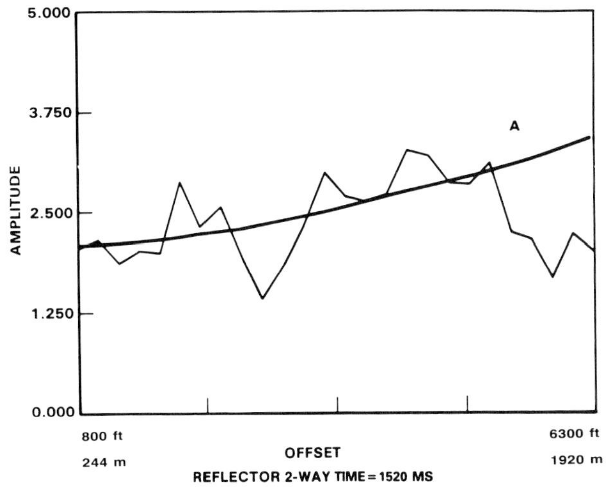

Fig. 13. Corrected AVO (line 2). Field data are for gas sand bright spot (Reflector A) at the Case 2 well location. Smooth curve is the theoretical effective reflectivity function. Reflector two-way time = 1.52 s.

Case 4 is an example of a confirmed salt bright spot offshore Texas. The salt was encountered at 14 500 ft (4419 m). The bright spot is shown in Figure 21 at 3.5 s of two-way traveltime. Figure 22 shows the multilayer velocity model from the well sonic and the single boundary and multilayer "effective" reflection coefficients for the top of the salt. The strong predicted decrease in AVO is verified by field data at this location shown in Figure 23. Figure 24 (see also Table 4) is another Gulf Coast example showing the dramatic difference in AVO from shale to salt, brine sand, and gas reflectors. The offset-dependent correction effects are negligible for deep bright spots such as these.

Case 5 is from offshore Texas and uses AVO attribute displays in conjunction with pattern recognition studies. Figure 25 is the standard seismic section with the well bore as indicated. Figure 26 is a Western Geophysical AVO attribute display (Ng, 1987) corresponding exactly in time and CMP range to Figure 25 with AVO buildup indicated in red. The AVO attribute displayed is the gradient defined by the least squares

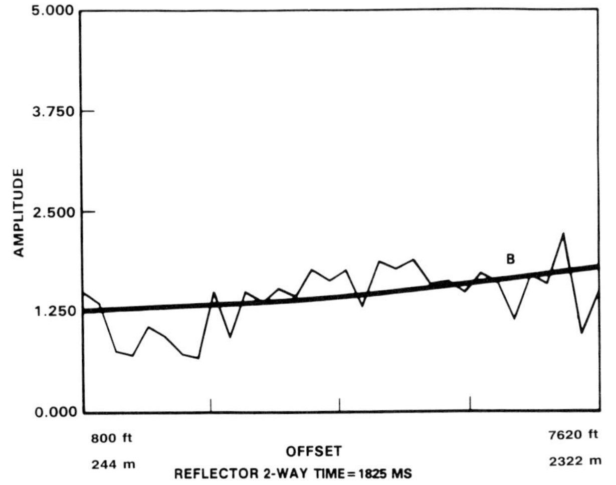

Fig. 14. Corrected AVO (line 2). Field data are for gas sand bright spot (Reflector B) at the Case 2 well location. Smooth curve is the theoretical effective reflectivity function. Reflector two-way time = 1.825 s.

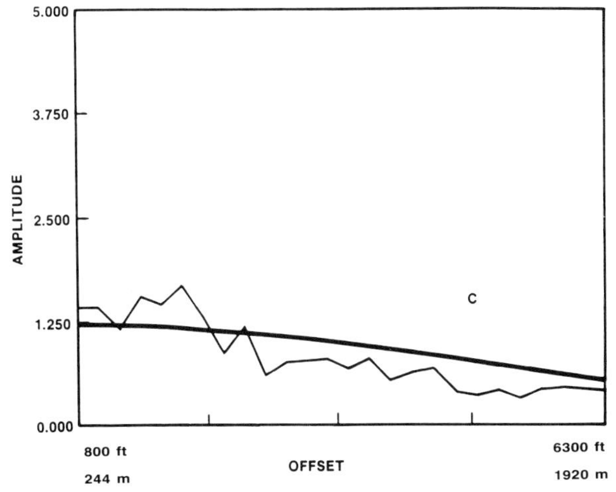

Fig. 15. Corrected AVO (line 2). Field data are for gas sand bright spot (Reflector C) at the Case 2 well location. Smooth curve is the theoretical effective reflectivity function. Reflector two-way time = 2.29 s.

slope of amplitude versus the squared sine of the computed angel of incidence. Note how the high-amplitude events on the standard variable area section of Figure 25 do not correlate directly with AVO buildup; yet the AVO attribute display indicates increase of reflectivity with offset on the seismic data at the pay zones in the No. 1 well location. The application of AVO analysis in conjunction with seismic pattern recognition studies is essential.

CONCLUSIONS AND DISCUSSION

The field data preparation and elastic model building procedures used in this study included event picking of NMO corrected data, sonic log blocking, and S-wave velocity calculations based on empirical published data. Hand editing was done on the picked events when necessary and target lithology properties were recalculated in some cases using the Gregory-Pickett method. The known offset-dependent corrections were applied including spherical divergence, receiver array attenuation, and inelastic attenuation. These corrections had moderate effect for reflectors at 1.5 s or less of two-way time. These correction factors are especially important for shallow reflectors to prevent AVO misinterpretation as discussed in Fulton and Darr (1984) and Gassaway (1985).

In most cases, direct detection of gas sands from the case histories could be made on the basis of increasing AVO, or in more difficult cases, relative lateral increases in AVO. Problems in detecting gas can occur if P-velocity and density are abnormally low as this will tend to flatten the amplitudes, and flattening can also occur due to offset-dependent transmission effects of overlying gas sands. Problems to be investigated further are the offset-dependent transmission effects of overlying cyclic thin beds, amplitude buildup due to anisotropy in shale, and absorption and dispersion effects as discussed in Hindlet (1986), Wright (1987), Martinez and McMechan (1987), Thomsen (1986), and Banik (1987), respectively.

A major problem for AVO interpretation is the false indication of producible hydrocarbons due to the presence of partial gas saturation. A few percent of gas by volume in the pore space will theoretically cause as much velocity drop as 100 percent gas saturation in the same rock. The parameter that may allow discrimination between partial and full gas saturation is the density. The density of partial gas-saturated sands has a linear trend with water saturation, whereas the velocity of partial gas-saturated sands is nonlinear and much lower than the 100 percent water saturated sands. A method to measure density accurately would

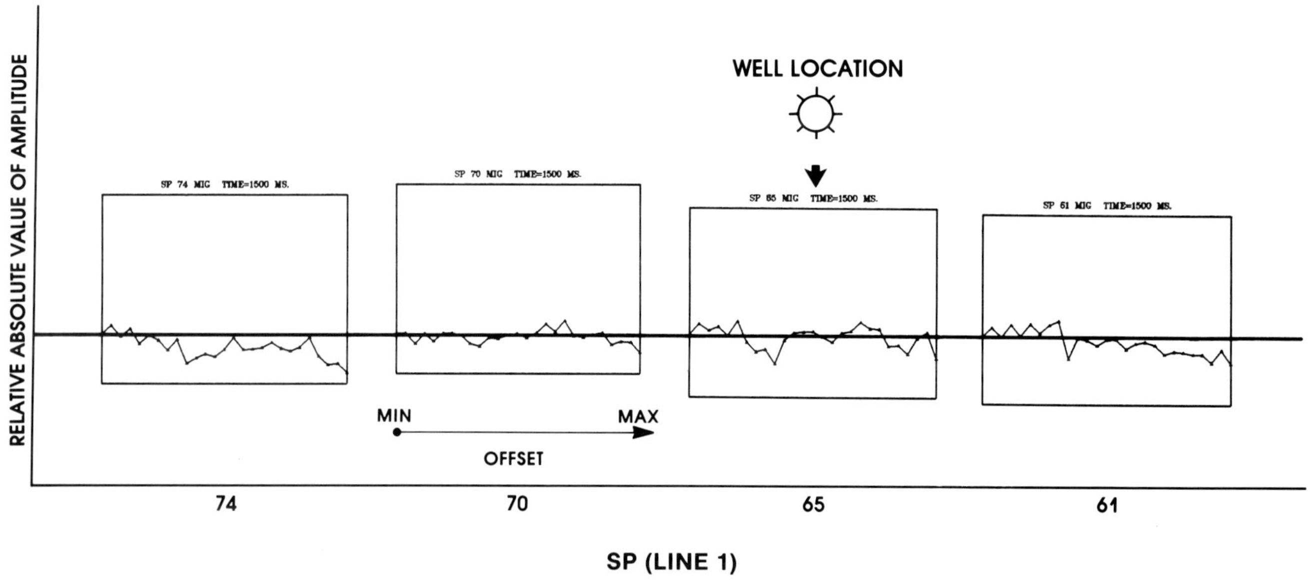

Fig. 16. Field data for Gulf Coast bright spot (Reflector A) for Case 2, after correction for spherical divergence, receiver array attenuation, and inelastic attenuation (Q = 150).

Table 3. Model parameters for Figure 19.

	Curve 1 Water Sand	Curve 2 Gas Sand Gassmann	Curve 3 Gas Sand Hydrocarbon Correction
Shale *P*-velocity (m/s)	1966	1966	1966
Shale Poisson's ratio	.462	.462	.462
Shale density (gm/cc)	2.061	2.061	2.061
Sand *P*-velocity (m/s)	2162	1358	1609
Sand Poisson's ratio	.417	.166	.2
Sand density (gm/cc)	2.204	1.984	1.984

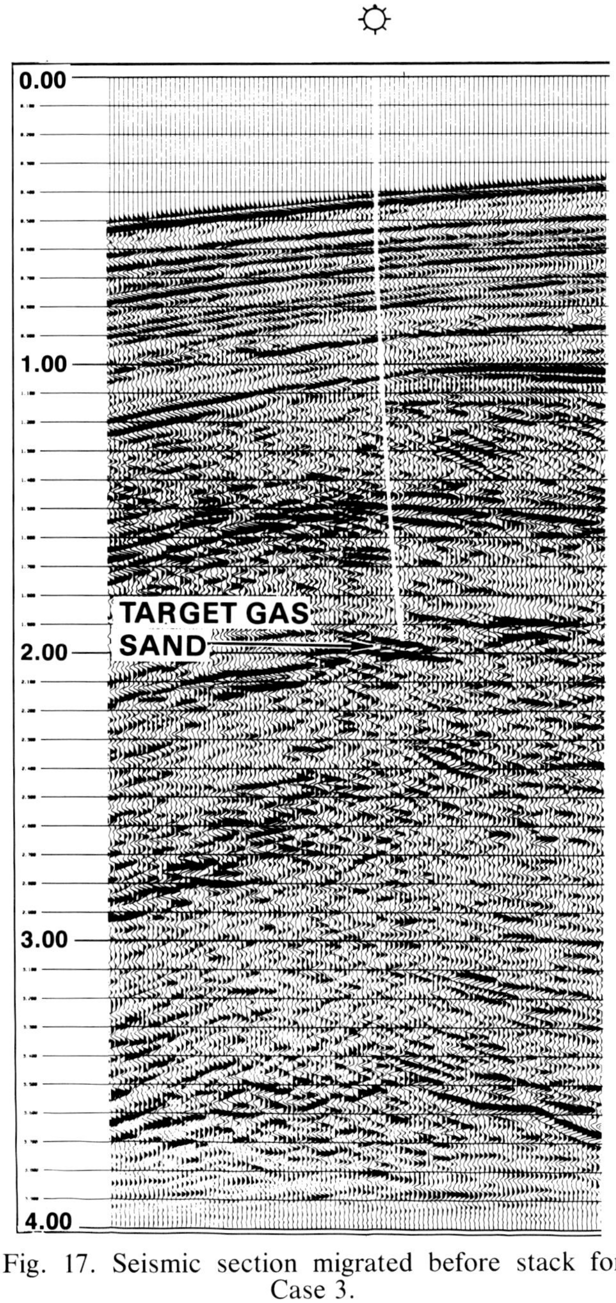

Fig. 17. Seismic section migrated before stack for Case 3.

be useful for this type of HCI interpretation. Heavier weight oil will also cause a flatter AVO response. Also, in some cases a shale/water sand reflector will show a mild increase in AVO, if, for example, a small velocity drop is accompanied by a small Poisson's ratio drop. The distinguishing factor in these cases is usually a larger normal incidence reflection coefficient for true gas sands. Caution is also advised for buildup caused by shale/water sand reflectors where the water sand *P*-velocity and Poisson's ratio is higher than that of the shale. Sandstone *P*-velocities higher than that for shale are common for older, more consolidated Gulf Coast sediments of Miocene Age or older (Engelmark et al., 1985). In this type of section, increasing AVO will more likely occur only for low specific gravity saturants. That is, the saturant density and matrix porosity must be such as to drop the acoustic impedance lower than that of the overlying shale.

The importance of density and *S*-wave velocities in the interpretation of AVO are often overlooked. Density and velocity contrasts are often difficult to separate in terms of acoustic impedance alone and thereby lead to ambiguity in AVO interpretation. Full wave sonic log measurements that provide in situ *S*-wave velocities and Poisson's ratios may yield more accurate elastic synthetics for comparison with field data. A careful consideration of both density and *S*-wave velocity will not only provide more physically realistic synthetic data but significantly reduce AVO interpretation uncertainties.

In spite of uncertainties, this study does reveal the advisability of checking AVO for oil or gas prospect evaluation, especially if there is the possibility of the presence of salt. Distinguishing salt from gas or light oil sand is easier than distinguishing gas sands from water sands or heavy oil sands. It is uncertain whether this method can presently distinguish unambiguously commercial gas sands from partial gas-saturated water sands.

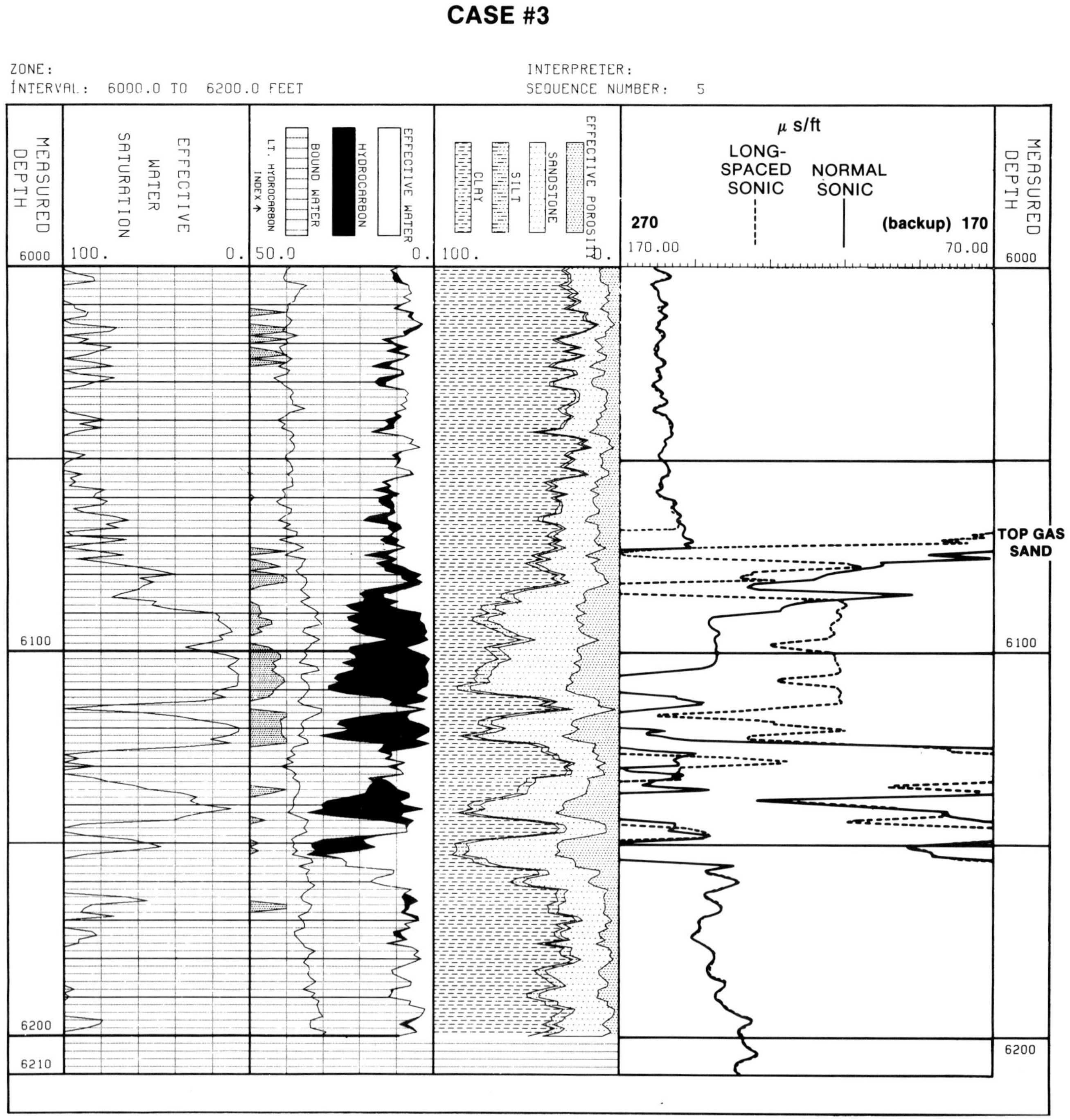

Fig. 18. Log analysis output for Case 3 showing long-spaced sonic, water saturation, sand fraction, and porosity.

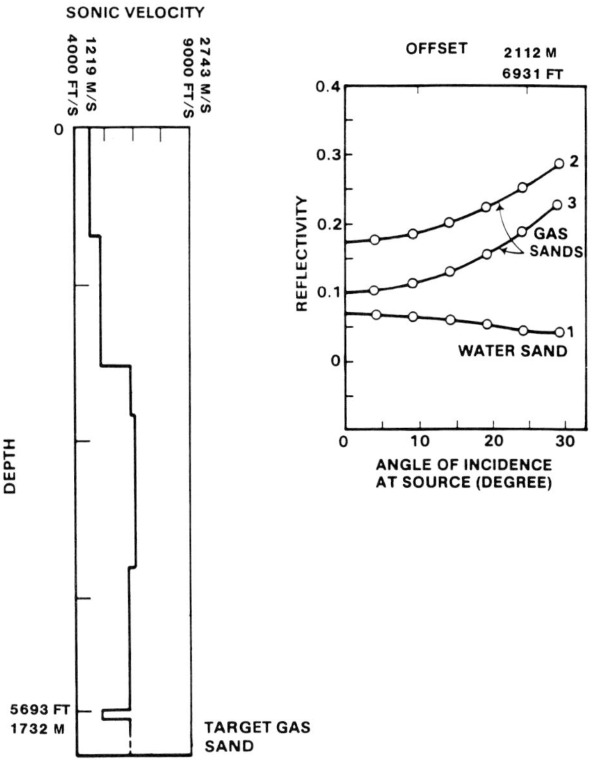

Fig. 19. Elastic reflection coefficients for multilayer model from Case 3. *P*-wave velocities and Poisson's ratios are calculated from the well log and the Gregory-Pickett method. Model parameters are given in Table 3. Sand porosity = 24%. Water sand is high impedance positive reflector, while gas sands are low impedance negative reflectors.

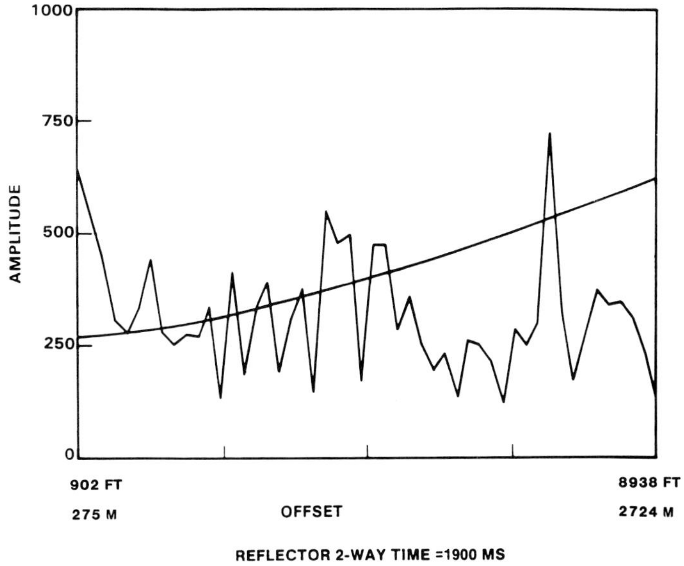

Fig. 20. Corrected offset-dependent reflection amplitudes from Case 3 at the well location. The smooth curve is the multilayer model effective reflection coefficients for the top of the gas sand (curve 3 in Figure 19).

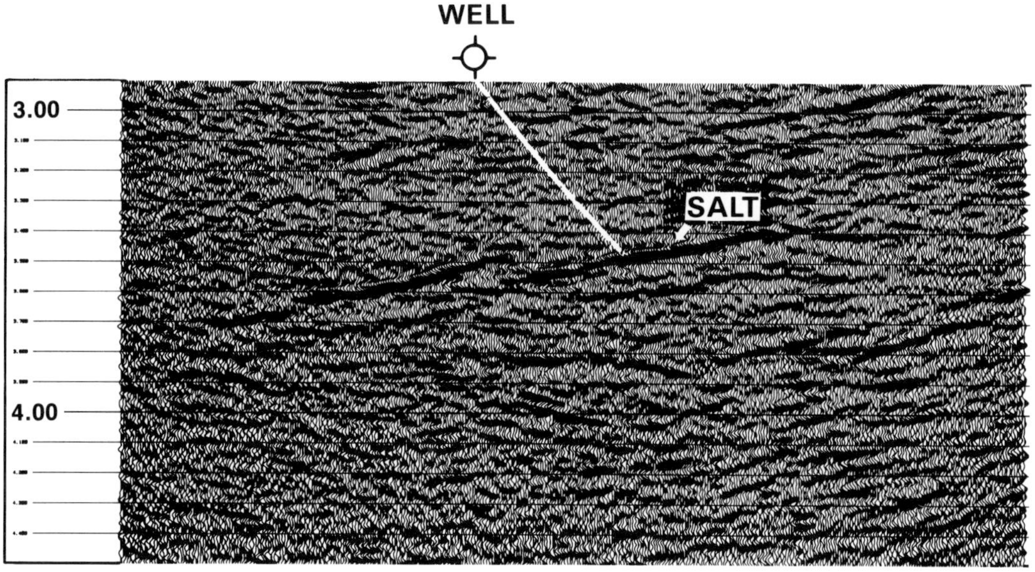

Fig. 21. Gulf Coast confirmed salt bright spot for Case 4.

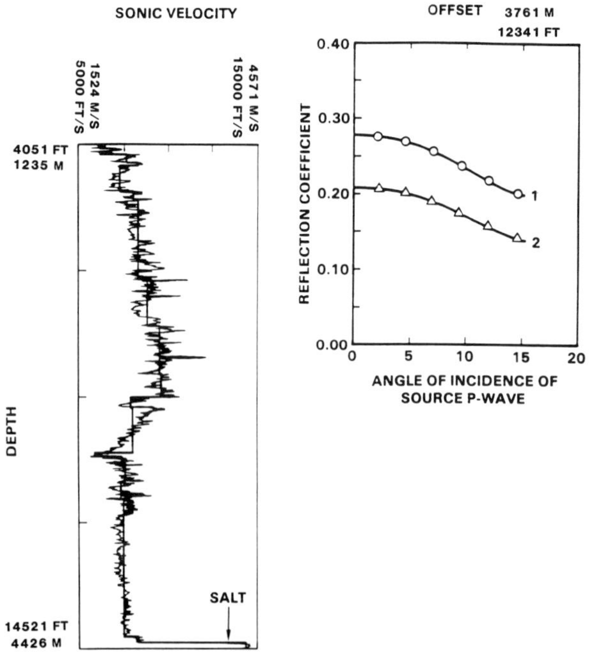

Fig. 22. Sonic log, multilayer model, and plane-wave elastic reflection coefficients for target horizon modeled as salt for Case 4. Model parameters are: shale *P*-velocity = 2474 m/s, shale Poisson's ratio = .41, shale density = 2.183 gm/cc, salt *P*-velocity = 4415 m/s, salt Poisson's ratio = .256, salt density = 2.164 gm/cc.

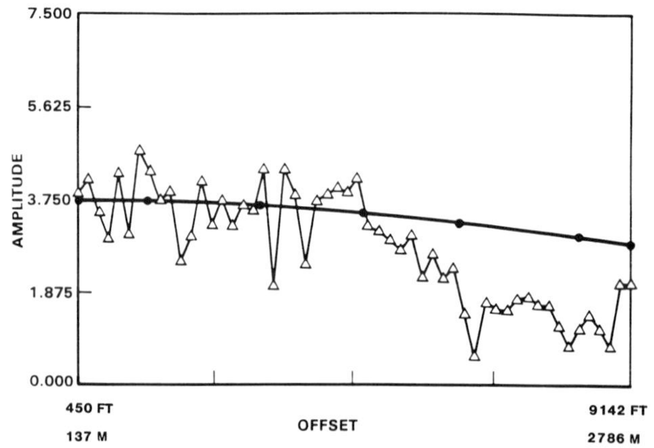

Fig. 23. Average AVO for Gulf Coast salt bright spot (Case 4) from field data at well location, corrected for spherical divergence, array attenuation, and inelastic attenuation ($Q = 200$). Superimposed is the theoretical multilayer effective reflection coefficients.

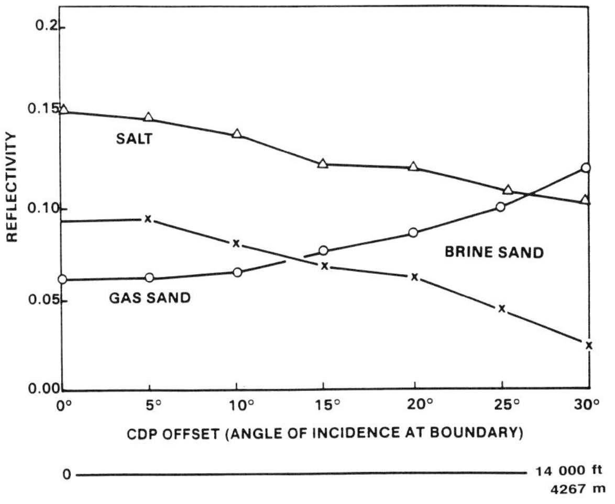

Fig. 24. Reflectivity versus offset for deep gas sand, brine sand, and salt. Model parameters are given in Table 4. Brine sand and salt are high impedance, positive reflectors, while gas gas sand is a negative low impedance unit.

Table 4. Model parameters for Figure 24.

Shale *P*-velocity (m/s)	2770
Shale Poisson's ratio	0.30
Shale density (gm/cc)	2.49
Gas-sand *P*-velocity (m/s)	2789
Gas-sand Poisson's ratio	0.06
Gas-sand density (gm/cc)	2.20
Brine-sand *P*-velocity (m/s)	3200
Brine-sand Poisson's ratio	0.24
Brine-sand density (gm/cc)	2.30
Salt *P*-velocity (m/s)	4267
Salt Poisson's ratio	0.25
Salt density (gm/cc)	2.20

ACKNOWLEDGMENTS

Many of the UNOCAL staff provided input for the AVO results reported here. In particular, Jack Schanck, Bob Myers, and Craig Cobb provided the primary data for this report including well and surface seismic data. Pat Hoffmann and Vance Lynch supported our efforts from the speculative introduction of AVO research to the routine use of AVO in exploration. Fred Aminzadeh (elastic modeling), and Bijan

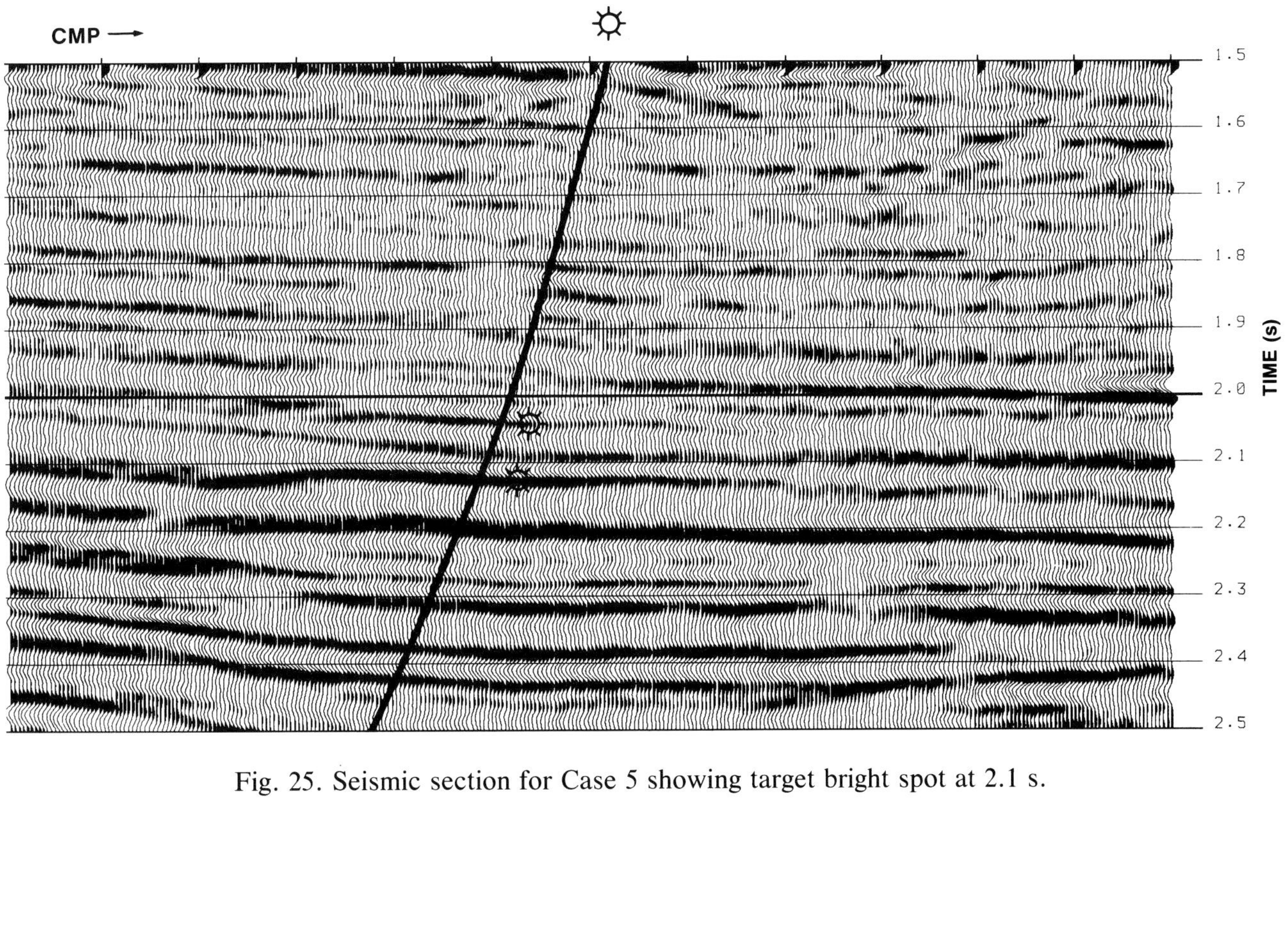

Fig. 25. Seismic section for Case 5 showing target bright spot at 2.1 s.

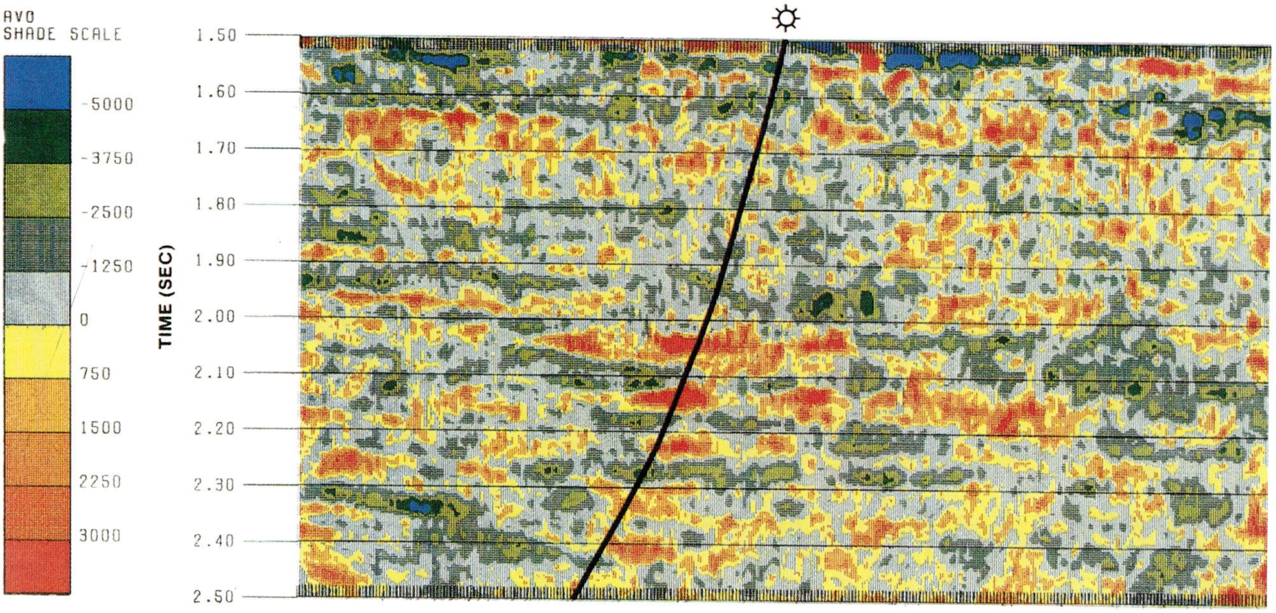

Fig. 26. AVO attribute section for Case 5.

Lashgari and Jerzy Trybek (pattern recognition) made substantial contributions to our data analysis techniques, and Marcia Spriggins processed and plotted much of the data shown in the figures.

REFERENCES

Backus, M. M., 1982, The reflection seismogram in a solid layered earth: Presented at the 52nd Ann. Internat. Mtg., Soc. Expl. Geophys.

Banik, N. C., 1987, An effective anisotropy parameter in transversely isotropic media: Geophysics, **52**, 1654–1664.

Castagna, J. P., Batzle, M. L., and Eastwood, R. L., 1985, Relationships between compressional- and shear-wave velocities in clastic silicate rocks: Geophysics, **50**, 571–581.

Engelmark, F., Hilterman, F., Liang, L. C., and Burns, C. S., 1985, Relating seismic lithology to Gulf Coast petrophysics: Presented at 55th Ann. Internat. Mtg., Soc. Expl. Geophys.

Fulton, T. K., and Darr, K. M., 1984, Offset panel: Geophysics, **49**, 1140–1152.

Gassaway, G., 1985, Effects of shallow reflectors on amplitude versus offset (seismic lithology) analysis, Offshore Technology Conference Report 4933.

Hamilton, E. L., 1980, Geoacoustic modeling of the sea floor: J. Acoust. Soc. Am., **68**, 1313–1340.

Hindlet, F., 1986, Amplitude decomposition of thin layer offset amplitude data: Annual Progress Review, Seismic Acoustics Laboratory, Univ. of Houston.

Koefoed, O., 1955, On the effect of Poisson's ratios of rock strata on the reflection coefficients of plane waves: Geophys. Prosp., **3**, 381–387.

Martinez, R., and McMechan, G., 1987, Analysis of absorption and dispersion effects in synthetic T-P seismograms: Geophysics, **52**, 1033–1047.

Muskat, M., and Meres, M. W., 1940, Reflection and transmission coefficients for plane waves in elastic media: Geophysics, **5**, 115–148.

Ng, P., 1987, Quantitative amplitude-versus-offset analysis: Western Geophysical Program—S6521.

Ostrander, 1984, Plane-wave reflection coefficients for gas sands at non-normal angles of incidence: Geophysics, **49**, 1637–1648.

Resnick, J. R., Ng, P., and Larner, K., 1987, Amplitude-versus-offset analysis in the presence of dip: Presented at 57th Ann. Internat. Mtg., Soc. Explor. Geophys.

Shuey, R. T., 1985, A simplification of the Zoeppritz equations: Geophysics, **50**, 609–614.

Thomsen, L., 1986, Weak elastic anisotropy: Geophysics, **51**, 1954–1966.

Timur, A., Alger, R. P., Kowalski, J. J., and Zemanek, J., 1978, Acoustic Logging: Soc. Prof. Well Log Analysts

Tooley, R. D., Spencer, T. W., and Sagoci, H. F., 1965, Reflection and transmission of plane compressional waves: Geophysics, **30**, 552–570.

Wright, J., 1986, Reflection coefficients at the pore-fluid contacts as a function of offset: Geophysics, **51**, 1858–1860.

Wright, J., 1987, The effects of transverse isotropy on reflection amplitude versus offset: Geophysics, **52**, 564–567.

Chapter V

Inversion

Inversion of *P*-Wave AVO

Emin Demirbag,* *Cahit Çoruh*,‡ *and John K. Costain*‡

INTRODUCTION

The estimation of *P*- and *S*-wave seismic velocities and densities of earth layers from the data sets recorded at the surface has long been one of the main research interests in reflection seismology. Although structural interpretations of the subsurface with sophisticated data processing and interpretation techniques have been facilitated, there remains a lot more to be done in obtaining quantitative information about the seismic parameters to lead to lithological interpretation that will open the ways to direct hydrocarbon exploration, groundwater exploration, and understanding the deep earth crust.

The reflection, transmission and mode conversion of the plane *P*-waves at a boundary as a function of incidence angle have been extensively investigated in the geophysical literature. The mathematical description of this physical process is given by Zoeppritz equations and can be found in Waters (1987) and Berkhout (1987). Curves and values of the *P*-wave reflection coefficients were tabulated for different models by several authors, i.e., Koefoed (1955, 1962) and Tooley et al. (1965). Various approximations to the Zoeppritz equations were developed in Bortfeld (1961), Aki and Richards (1980), Hilterman's unpublished course notes (1983), Shuey (1985), and Lörtzer et al. (1988) in order to make general conclusions about the importance of the seismic parameters in the reflection process, and to estimate the seismic parameters or their various combinations from observed amplitudes of reflection seismic data. The simplest approximation to the Zoeppritz equations can be found in Shuey (1985), Gelfand et al. (1986), and Lörtzer et al. (1988). This approximation is a trigonometric polynomial with two linear coefficients and is valid for small angles of incidence. This simple equation is used to obtain zero-offset seismic sections and gradient sections; however, it does not yield any quantitative information about the seismic parameters. Backus (1987) reviewed the status of amplitude-variation-with offset (AVO) analysis and underlined its importance in direct hydrocarbon detection, reservoir development, and production. His projections for future studies included practical AVO inversion methods. In the case of small contrast in the seismic parameters of upper and lower layers, Lörtzer et al. (1988) approached the inversion of AVO data by a linear least-squares minimization procedure using approximate Zoeppritz equations that are linear in the seismic parameter contrast. De Haas and Berkhout (1988, 1989) discussed the information content of different modes of reflection data and noted the improvement in the quality of inversion results because of the simultaneous use of different modes. They described a method based on convolutional reflection modeling for the inversion of multilayer AVO data.

The inversion method presented here estimates the *P*- and *S*-wave seismic velocities and the density ratio of two elastic media separated by a horizontal boundary. The method has for its basis the inversion of the plane wave Zoeppritz equations by the generalized linear inversion (GLI) method, and operates on *P*-wave reflection AVO data. The Zoeppritz equations (Waters, 1987) were converted into the time-offset domain (Demirbag and Çoruh, 1988) using the common-mid-point (CMP) geometry, Snell's law, and zero-offset two-way traveltime. The inversion operates on the AVO data corrected for the amplitude variations due to wave propagation and data acquisition although the corrections can be considered a part of the forward modeling in the inversion. The estimation of seismic parameters is controlled by physical constraints, and the results are further enhanced by the bootstrapping method to obtain the most likely solutions and their confidence limits.

To justify the proposed method, the linearity and uniqueness of the inversion were investigated by using

*Istanbul Technical University, Faculty of Mines, Dept. of Geophysical Engineering, Maslak 80626 Istanbul, Turkey.
‡Virginia Polytechnic Institute and State University, Department of Geological Sciences, 4044 Derring Hall, Blacksburg, VA 24061.

residual function maps (RFM) obtained by changing the values of two parameters while keeping the others at their true values. It is observed that the RFMs display a minimum around the true values of parameters except for the *S*-wave velocity pair. The minimum around the true values is better defined when the contrast between the seismic parameters of the media and offset-to-depth ratio (O/D) are increased. The inversion method was tested using model relative AVO data with and without noise. In a noise free model, the model parameters are exactly recovered when the O/D is about 2 or larger, depending on the contrast between the seismic parameters of the media. The results for smaller O/D (O/D $\simeq$ 1) diverge from the true values, especially for *S*-wave velocities, and indicate the importance of the O/D in the AVO data inversion because as the O/D increases, the relative AVO curves become more distinguishable. Tests with noisy AVO data indicate that noise is the most disturbing factor influencing the inversion method. In the case of noisy model AVO data, the parameters are not recovered correctly because of the statistical nature of the inversion formulation; however, the true parameters fall into the confidence limits of the estimated parameters if tight constraints (a range of acceptable solutions around the true parameters, i.e., ± 20 percent of the true parameters) are imposed on the solutions and the signal-to-noise ratio (S/N) is high. In addition to the O/D and noise, the inversion method is also sensitive to the auxiliary parameters, such as rms velocity and zero-offset two-way traveltime, that are used in the corrections and converted Zoeppritz equations. This inversion method is limited to isolated single reflection events on the CMP gathers. Furthermore, the densities of the media appear in the ratio of ρ_2/ρ_1 in the Zoeppritz equations; therefore, only the ratio of densities can be estimated.

ZOEPPRITZ EQUATIONS IN THE TIME-OFFSET DOMAIN

Forward modeling is used to determine the response of a physical system using mathematical relations. The variables in the relations can be classified as: (1) controlled variables to be adjusted by the user; (2) dependent variables to be observed as the response of the physical system; (3) the model parameters that summarize the system in terms of physical quantities. The modeled physical process in this study is the process of energy partition at a boundary that separates two elastic layers. The Zoeppritz equations and their approximations are the mathematical relations that predict the partition of energy between reflected and transmitted waves. The controlled variable for the conventional form is the angle of incidence of the arriving *P*-wave upon the boundary, the dependent variables are the reflection and refraction amplitudes of the resulting *P*- and *S*-waves, and the model parameters are the *P*- and *S*-wave velocities and densities of the layers.

Inverse modeling, on the other hand, operates on the response of a physical system to estimate the model parameters. However, it requires the knowledge of its counterpart forward operator to relate the observations and controlled variables to the model parameters. Therefore, the inversion of AVO data using the method presented requires that the Zoeppritz equations in the time-offset domain serve as the forward operator.

Conversion of the Zoeppritz equations into the time-offset domain

In reflection seismology, a CMP gather is the most suitable form of data from which to observe the variation of reflection amplitude with offset. The recording geometry for a CMP gather is well known, and the depth of reflectors can be identified by zero-offset two-way traveltime and average velocities. To use CMP gathers as input data in the inversion, the classical form of the Zoeppritz equations (Waters, 1987)

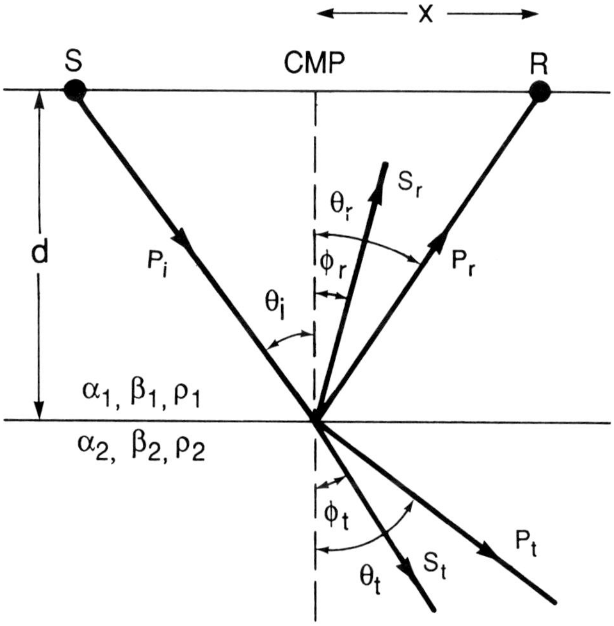

Fig. 1. CMP geometry and energy partition processes.

were converted into the time-offset domain. As a result of this conversion, the controlled variables in the forward model became half the source-receiver offset and the zero-offset two-way traveltime.

The model considered is a two-layer elastic medium with a single horizontal interface. Each layer is homogeneous and isotropic. In Figure 1 α_1, β_1, and ρ_1 represent the P- and S-wave velocity, and density for the upper layer, respectively; and α_2, β_2, and ρ_2 represent the same parameters for the lower layer. Assuming a plane P-wave to be incident on the boundary in the upper layer, the Zoeppritz equations predict the four types of waves that are generated at the boundary by energy partitioning process; reflected and refracted P- and S-waves. θ_i is the P-wave incidence angle, θ_r and ϕ_r are the P- and S-wave reflection angles, θ_t and ϕ_t are the P- and S-wave transmission angles, respectively. From Figure 1, the simple time-offset relation for P-waves is

$$t_x = \left(t_0^2 + \frac{4x^2}{\alpha_1^2} \right)^{1/2}$$

where t_0 is the zero-offset two-way travel time defined by $t_0 = 2d/\alpha_1$, and d and x are the depth of the reflector and half the source-receiver offset, respectively. In addition to the time-offset equation, Snell's law

$$\frac{\sin\theta_i}{\alpha_1} = \frac{\sin\theta_r}{\alpha_1} = \frac{\sin\theta_t}{\alpha_2} = \frac{\sin\phi_r}{\beta_1} = \frac{\sin\phi_t}{\beta_2}$$

and the ratio of densities $\Delta\rho = \rho_2/\rho_1$ are used to convert the Zoeppritz equations into the time-offset domain. The derivation of the trigonometric relations for the two-layer model is given in Appendix A. The Zoeppritz equations in the time-offset domain have the following form:

$$\begin{bmatrix} a_{11} & a_{12} & a_{13} & a_{14} \\ a_{21} & a_{22} & a_{23} & a_{24} \\ a_{31} & a_{32} & a_{33} & a_{34} \\ a_{41} & a_{42} & a_{43} & a_{44} \end{bmatrix} \begin{bmatrix} R_p \\ R_s \\ T_p \\ T_s \end{bmatrix} = \begin{bmatrix} b_{11} \\ b_{21} \\ b_{31} \\ b_{41} \end{bmatrix} \tag{1}$$

where R_p, R_s, T_p and T_s are the amplitudes of the reflected P- and S-waves and transmitted P- and S-waves, respectively; for a unit amplitude incident P-wave. Note that R_s, T_p, and T_s all refer to the same incident angle as that for the R_p reflection at offset X. The entries of the matrices are combinations of the controlled and known variables (x and t_0) and model parameters (α_1, β_1, α_2, β_2, and Δ_ρ). The elements of the matrices are

$$a_{11} = \frac{2x}{\alpha_1 t_x}, \qquad a_{12} = (1 - a_{22}^2)^{1/2},$$

$$a_{13} = -a_{11}\frac{\alpha_2}{\alpha_1}, \qquad a_{14} = (1 - a_{24}^2)^{1/2},$$

$$a_{21} = -\frac{t_0}{t_x}, \qquad a_{22} = a_{11}\frac{\beta_1}{\alpha_1},$$

$$a_{23} = -(1 - a_{13}^2)^{1/2}, \qquad a_{24} = -a_{11}\frac{\beta_2}{\alpha_1},$$

$$a_{31} = -2a_{11}a_{21}, \qquad a_{32} = a_{41}\frac{\alpha_1}{\beta_1},$$

$$a_{33} = -a_{11}a_{23}\frac{2\Delta\rho\beta_2^2}{\beta_1^2},$$

$$a_{34} = -\frac{\Delta\rho\alpha_1\beta_2}{\beta_1^2}(1 - 2a_{24}^2),$$

$$a_{41} = 1 - 2a_{22}^2, \qquad a_{42} = -a_{12}a_{22}\frac{2\beta_1}{\alpha_1},$$

$$a_{43} = -\frac{\Delta\rho\alpha_2}{\alpha_1}(1 - 2a_{24}^2),$$

$$a_{44} = a_{14}a_{24}\frac{2\beta_2\Delta\rho}{\alpha_1},$$

$$b_{11} = -a_{11}, \qquad b_{21} = a_{21},$$

$$b_{31} = a_{31}, \quad \text{and} \quad b_{41} = -a_{41}.$$

A multilayer model has also been developed by using the rms velocity α_{rms}, average velocity α_{ave}, and average depth d_{ave} for the conversion of the Zoeppritz equations in Appendix B. Note that for the two-layer model $\alpha_{rms} = \alpha_{ave} = \alpha_1$; therefore, the entries in Appendix B reduce to the entries given above. The use of this multilayer formulation is an area of continuing research.

Amplitude Measurements of Reflected Wavelets

Determination of the offset dependent reflection amplitudes is one of the most critical stages. Amplitude measurements of the reflected wavelets from trace to trace on a CMP gather may be obtained in a variety of ways. Among these are the rms amplitude of the reflected wavelet, peak amplitude of the reflected wavelet, peak-to-trough amplitude of the reflected wavelet, and rms amplitude of the crosscorrelation function of the reflected wavelet with a reference wavelet. The main task for these techniques is to measure the AVO reflection amplitude with high S/N by reducing the effect of noise. For this inversion the rms amplitudes of the reflected wavelets are measured by the Euclidean norm (Stoer and Bulirsch, 1980); if $u(t, x)$ is the reflected wavelet with length T, then its mean energy at time t_x is defined as

$$E(t_x, x) = \frac{1}{T}\int_{t_x - T/2}^{t_x + T/2} |u(t, x)|^2\, dt \tag{2}$$

where $u(t, x) = w(t)*R(t_x, x)$ is represented by the convolution process which can be reduced to multiplication for a single interface model, $w(t)$ is the downward traveling wavelet, and $R(t_x, x)$ is the reflection coefficient at time t_x and offset x. Then, the rms amplitude of the reflected wavelet is

$$\begin{aligned}\mathrm{amp}_u\ (t_x, x) &= [E(t_x, x)]^{1/2}\\ &= \left(\frac{1}{T}\int_{t_x - T/2}^{t_x + T/2} |u(t, x)|^2\, dt\right)^{1/2}, \end{aligned}\tag{3}$$

and

$$\begin{aligned}&\mathrm{amp}_u\ (t_x, x)\\ &= \left(|R(t_x, x)|^2\,\frac{1}{T}\int_{t_x - T/2}^{t_x + T/2} |w(t)|^2\, dt\right)^{1/2}. \end{aligned}\tag{4}$$

Assuming the shape of the downward traveling wavelet does not change with offset (attenuation free medium), its rms amplitude is defined by using the integral term in the last equation as follows:

$$\mathrm{amp}_w = \left(\frac{1}{T}\int_{t_x - T/2}^{t_x + T/2} |w(t)|^2\, dt\right)^{1/2}. \tag{5}$$

Equation (4) then simplifies to

$$\mathrm{amp}_u\ (t_x, x) = |R(t_x, x)|\mathrm{amp}_w. \tag{6}$$

Because the downward traveling wavelet and the absolute value of the zero-offset reflection amplitude are not available, the normalized or relative AVO can be defined as

$$|R_n(t_x, x)| = \frac{\mathrm{amp}_u\ (t_x, x)}{\mathrm{amp}_u\ (t_x, 0)} = \frac{|R(t_x, x)|}{|R(t_x, 0)|} \tag{7}$$

where $\mathrm{amp}_u(t_x, 0)$ can be obtained by the extrapolation of the measured AVO curve to zero-offset. Using the sign of the maximum value of $u(t, x)$, the polarity of the measured reflection can be recovered by

$$R_n(t_x, x) = \mathrm{sign}\ [\max\ |u(t, x)|]_{t \in T}|R_n(t_x, x)|. \tag{8}$$

Because the zero-offset amplitude $\mathrm{amp}_u(t_x, 0)$ is not always available, a curve fitting scheme is developed using the three-term linear approximate Zoeppritz equation (Lörtzer et al., 1988) for extrapolating the zero-offset reflection amplitude. The details of derivations are given in Appendix C.

GENERALIZED LINEAR INVERSION OF AVO DATA

Generalized linear inversion (GLI) has been extensively used in the solution of inverse geophysical problems. Aspects of the GLI method relevant to this study are discussed in detail in Jackson (1972, 1979), Lines and Treitel (1984), Tarantola (1987), and Menke (1989). The FORTRAN codes of the method can be found in Press et al. (1988) and Van Heeswijk and Christopher (1988). Recently, the inversion of AVO data on the basis of the GLI method has been formulated by several authors, i.e., Macdonald et al. (1987), Russell (1988), De Haas and Berkhout (1988, 1989), and Demirbag and Çoruh (1988, 1989).

Parameterization, Linearization, and Uniqueness of the Inverse Problem

The Zoeppritz equations represent nonlinear functions; therefore, the linearity and uniqueness of solutions of an inverse problem depend upon selection of the model parameter vector as well as suitability of the nonlinear function for linearization. In the time-offset domain, the controlled variables are half the source to receiver offset and zero-offset two-way travel time. The model parameters are *P*- and *S*-wave velocities and density of the lower and upper layers. The inverse problem is to find the model parameters given the source-receiver offsets, zero-offset two-way travel time, and the AVO data. Because densities of the layers appear in the form of ratios in the Zoeppritz equations, the following parameter vector was chosen;

$$\mathbf{P} = (\alpha_1, \beta_1, \alpha_2, \beta_2, \Delta\rho). \tag{9}$$

The linearity and uniqueness of the inversion of the Zoeppritz equations by the GLI method for the above model parameter vector was investigated using Residual Function Maps (RFMs) following Macdonald et al. (1987) and Menke (1989). The residual error function is

$$E(x, y) = \sum [f - \bar{f}(x, y)]^2 \quad (10)$$

where f is the forward model response for the chosen values of the model parameters and $\bar{f}$ is the response of the same function to different values of the chosen parameter pair (x and y) while the other model parameters are kept at their true values. Note that $E(x, y)$ is zero for the true values of parameters. RFMs are used in the investigation of linearization of nonlinear functions in a large neighborhood of the chosen model parameter values. Because the inversion by a linear function results in RFMs where the contours are elliptically closed around the true parameters, the RFMs are used to determine if the nonlinear function is eligible for linearization. If the RFMs show closed contours with one minimum implying a unique solution, then the nonlinear function can be used to formulate the GLI scheme by two-term Taylor series expansion.

The linearization and uniqueness of the inversion of the Zoeppritz equations were investigated for many two-layer models. An example is given here for the model parameters shown in Table 1. S-wave velocities were chosen to be half of the P-wave velocities, and the densities were assigned using Gardner's relation (Gardner et al., 1974). The boundary between the layers is located at a depth of 2000 m. This gives 1.33 s zero-offset two-way travel time, and the two-way traveltimes at other offsets are calculated from the time-offset relation for a two-layer model. The maximum source to receiver offset is 4000 m with an offset increment of 100 m, hence, the angle of incidence spans 45 degrees from zero to maximum offset. To generate a RFM, first the model response was calculated for a set of source to receiver offsets using the Zoeppritz equations and equation (8). This response corresponds to the function f in equation (10). The parameter vector has 5 elements, therefore, there are 10 combinations of model parameter pairs. For each of these combinations, the parameters in the chosen pair were assigned new values around a large neighborhood of their true values while the remaining three parameters were fixed at their true values. For each new value of the chosen parameter pair, the model response was calculated to obtain the function $\bar{f}$ in equation (10) again using equation (8). Having obtained the functions f and $\bar{f}$ in this way, the residual function was calculated and contoured as a RFM. The results of the RFMs are shown in Figures 2a through 2j. Note that the contours close smoothly around the true values of the model parameters that are marked by the crossing vertical and horizontal lines in the RFMs where the contour values change from 0.1 to 1.0 by increments of 0.1. Because of the scaling of reflection amplitudes, the contour values are dimensionless. In the case of the S-wave velocity pair in Figure 2j, a long valley is observed instead of a well defined minimum around the true values. However, the RFMs of S-wave velocities with the other model parameters show well defined minima at the true values. An important suggestion from these observations is that the S-wave velocities can be recovered from the inversion of the Zoeppritz equations. In general, the RFMs display well defined closing contours around the true values of the model parameters when there exists a critical angle ($\alpha_2 > \alpha_1$) because the model response changes rapidly as the angle of incidence gets closer to the critical angle. This rapid change causes function $\bar{f}$ to differ significantly from function f in equation (10). As a result, a well defined minimum is observed at the true values of the considered parameter pair. For model responses where there is no critical angle ($\alpha_2 < \alpha_1$), the functions show little change by offset; therefore, RFMs exhibit minima which are less well defined. However, a smaller contouring interval can be chosen to map closing contours around the true values of parameters and to investigate the uniqueness of the solution. The shape of the contours also indicates trade off between the parameters. For example, in Figure 2j the contours indicate that the upper layer S-wave velocity is more likely to have larger errors than the lower layer S-wave velocity. Inspection of the RFMs for many different noise free models indicated that the Zoeppritz equations can be inverted by the generalized linear inversion to obtain the model parameters defined in the model parameter vector **P**.

Table 1. Model parameters for RFMs. Velocities are in units of m/s. Densities are in units of g/cm^3

Layer	α	β	ρ
Upper	3000	1500	2.294
Lower	4000	2000	2.465

Formulation of the Generalized Linear Inversion

Inversion schemes require the knowledge of a forward operator that relates the model parameters to the model response through an equation or set of equations that are usually nonlinear in model parameters

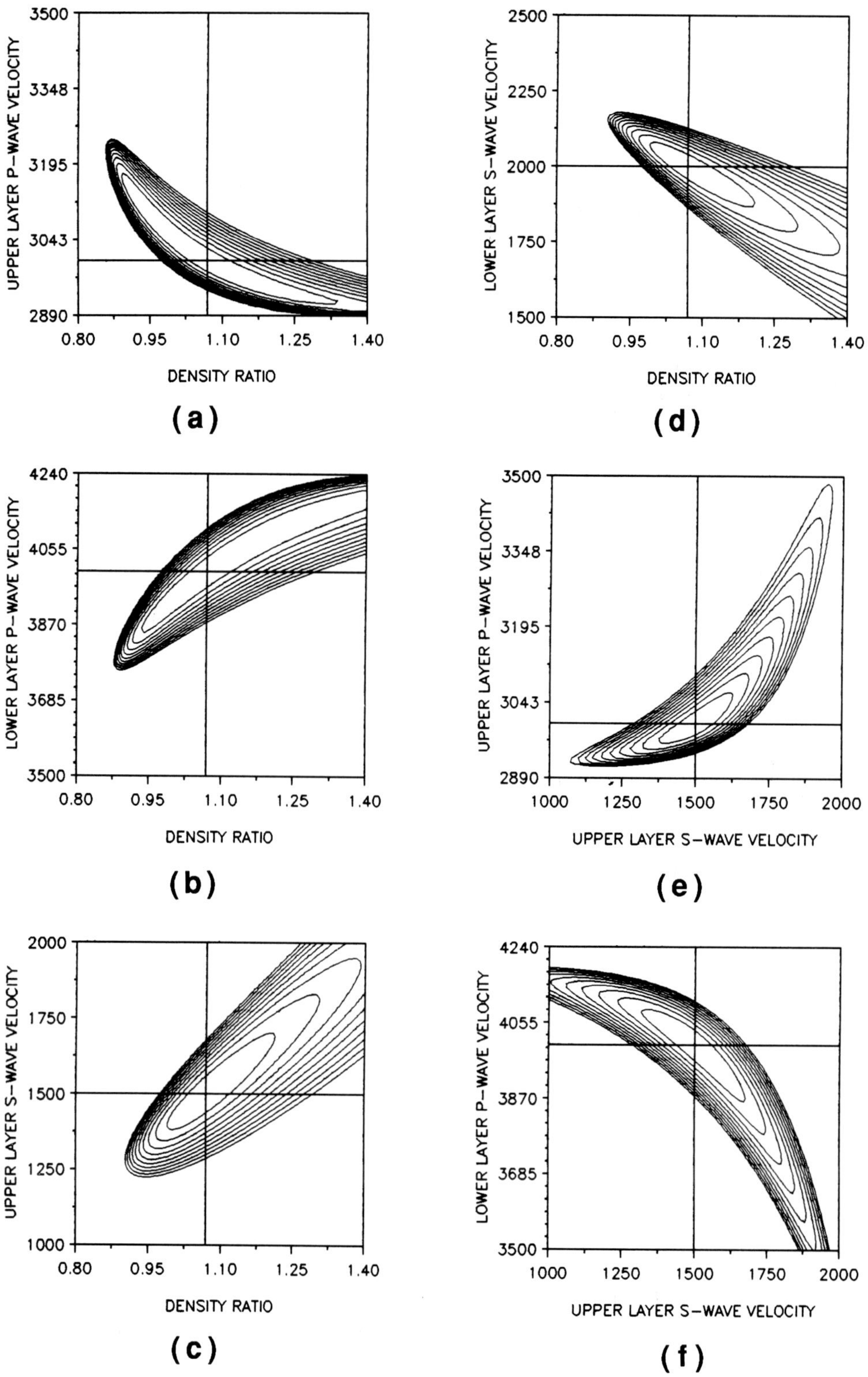

Fig. 2. Residual Function Maps (RFM). The exact parameters are marked by crossing hair.

and controlled variables. It is important that the forward operator simulates the corresponding modeled physical process, and it is parsimonious in the number of model parameters.

If f represents the forward operator as a function of model parameter vector $\mathbf{P}_M$ and controlled variable vector $\mathbf{X}$, and y represents the observed data as the outcome of the corresponding physical process for the chosen controlled variables, then the inverse problem is to determine the model parameter vector $\mathbf{P}_M$ given the observed data y and controlled variable vector $\mathbf{X}$. The forward operator f can be expanded into a truncated Taylor series around $\mathbf{P} = \mathbf{P}_0$ where $\mathbf{P}_0$ is called the initial parameter vector (Lines and Treitel, 1984):

$$f(\mathbf{P}, \mathbf{X}) = f(\mathbf{P}_0, \mathbf{X}) + (\mathbf{P} - \mathbf{P}_0) \left. \frac{\partial f(\mathbf{P}, \mathbf{X})}{\partial \mathbf{P}} \right|_{\mathbf{P}_0, \mathbf{X}} . \tag{11}$$

In matrix notation this equation can be given by

$$\underset{\sim}{\mathbf{F}} = \underset{\sim}{\mathbf{F}}_0 + \underset{\sim}{\mathbf{J}}\delta \tag{12}$$

where $\underset{\sim}{\mathbf{F}}$ is a column matrix and represents the linearized general model response, $\underset{\sim}{\mathbf{F}}_0$ is the column matrix and represents the model response to the initial parameter vector $\mathbf{P}_0$, $\underset{\sim}{\mathbf{J}}$ is the Jacobian matrix of the partial derivatives of the forward operator evaluated at the initial parameter vector and controlled variables at

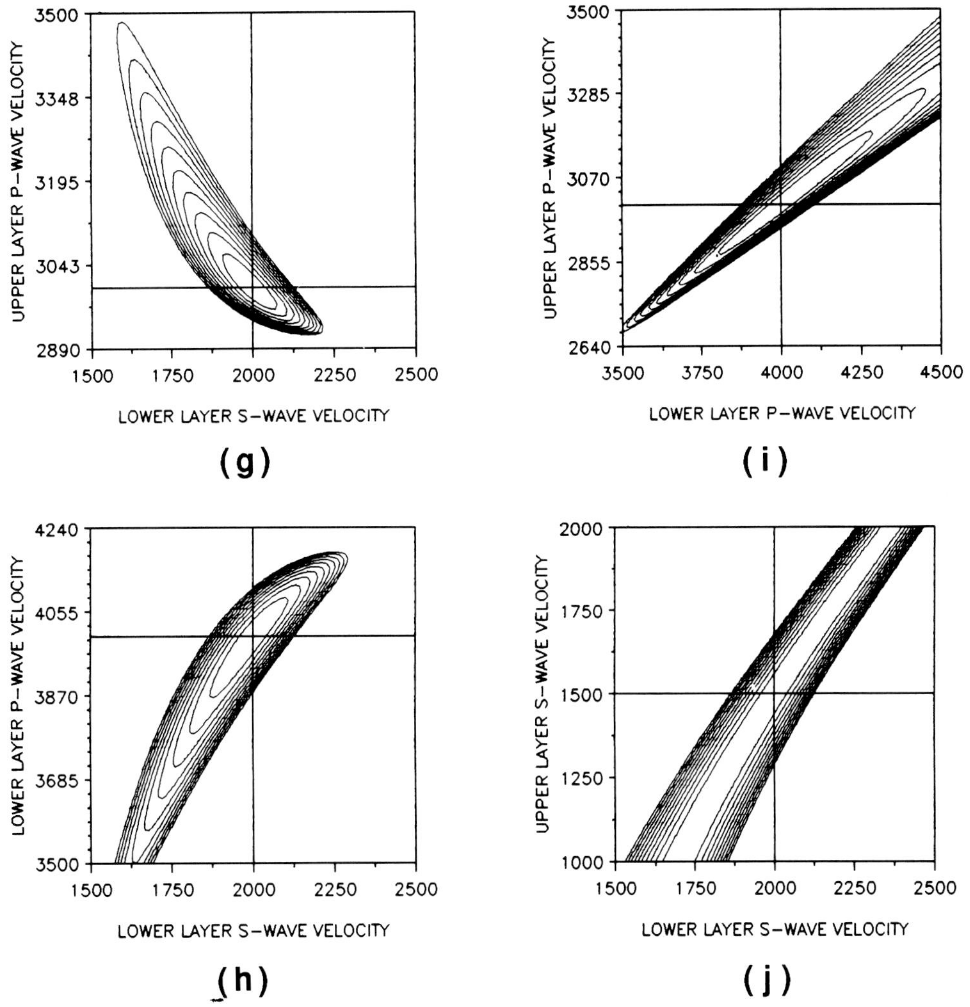

Fig. 2. continued

which the data were observed, and $\delta = (\mathbf{P} - \mathbf{P}_0)$ is the parameter change column matrix. The other step in the GLI scheme is to minimize the sum of squares of the differences between the model response and observed data. If E is the error function which represents the difference between the model response and the observed values, then

$$E = \underset{\sim}{\mathbf{Y}} - \underset{\sim}{\mathbf{F}} \text{ or } E = \underset{\sim}{\mathbf{Y}} - (\underset{\sim}{\mathbf{F}}_0 + \underset{\sim}{\mathbf{J}}\delta) \text{ or}$$

$$E + \underset{\sim}{\mathbf{J}}\delta = \underset{\sim}{\mathbf{Y}} - \underset{\sim}{\mathbf{F}}_0 \text{ or } E = \mathbf{D} - \underset{\sim}{\mathbf{J}}\boldsymbol{\delta}$$

where $\underset{\sim}{\mathbf{Y}}$ is the column matrix containing the observed data and $\mathbf{D} = \underset{\sim}{\mathbf{Y}} - \underset{\sim}{\mathbf{F}}_0$ is the discrepancy vector, or the difference between the observed data and the initial model response.

In the simplest least-squares solution, the cumulative squared error function $G(\delta) = E^T E$ is minimized with respect to the parameter change vector δ; however, this simple minimization procedure often runs into difficulties due to the singularity of the matrix $\underset{\sim}{\mathbf{J}}^T\underset{\sim}{\mathbf{J}}$. Singularity or near singularity of the matrix $\underset{\sim}{\mathbf{J}}^T\underset{\sim}{\mathbf{J}}$ causes the elements of the matrix δ to grow without bounds. Therefore, the procedure may diverge from the correct solution. To avoid this difficulty, a damped least square minimization method, the Levenberg-Marquardt method (Lines and Treitel, 1984), is used. In this scheme the solution is obtained in such a way that a cost function is minimized with respect to δ subject to the constraint $\delta^T\delta < \delta_0^2$ where δ_0 is a finite quantity. The cost function is defined by

$$G(\delta, \eta) = E^T E + \eta(\delta^T\delta - \delta_0^2) \tag{13}$$

where η is a Lagrange multiplier. The derivative of the cost function with respect to δ is set to zero for minimization

$$\frac{\partial G(\eta, \delta)}{\partial \delta} = -\underset{\sim}{\mathbf{J}}^T\mathbf{D} + \delta^T\underset{\sim}{\mathbf{J}}^T\underset{\sim}{\mathbf{J}} + \eta\delta^T = 0.$$

Finally, the solution for δ is given by

$$\delta = (\underset{\sim}{\mathbf{J}}^T\underset{\sim}{\mathbf{J}} + \eta\underset{\sim}{\mathbf{I}})^{-1}\underset{\sim}{\mathbf{J}}^T\mathbf{D} \tag{14}$$

where I is the identity matrix. This equation is used to solve for the parameter change vector from which

$$(\mathbf{P} - \mathbf{P}_0) = \delta \text{ or } \mathbf{P} = \delta + \mathbf{P}_0$$

is obtained. The model response is calculated by using the parameter vector $\mathbf{P}$, controlled variable vector $\mathbf{X}$ and the forward operator. If the discrepancy between the model response and the observed data is acceptable, the parameter vector $\mathbf{P}$ is declared to be $\mathbf{P}_M$, the result of the inversion procedure. Otherwise, $\mathbf{P}$ is taken as the initial model parameter vector $\mathbf{P}_0$, and the minimization procedure is repeated until an acceptable model parameter vector is found when a predefined drop in the error levels of two successive iterations is obtained. Adaptation of the GLI method to the inversion of AVO data is straightforward. The forward model is the Zoeppritz equations converted into the time-offset domain by using the time-offset relation and Snell's law. The controlled variables correspond to half the source to receiver offset and zero-offset two-way travel time. The model parameter vector consists of the *P*- and *S*-wave velocities of the lower and upper media and the density ratio of the media.

Physical Constraints on the Solutions

The inversion of AVO data by the GLI method for seismic parameters is formulated in a statistical sense, and it is affected by noise in the data. Therefore, the solutions from inversion might be quite different from the true values or they might even make no physical sense if some physical constraints are not imposed on the solutions. Three physical constraints can be imposed on the solutions.

Positive value solutions.—Solutions for *P*- and *S*-wave velocities and the density ratio cannot be zero or negative because they don't correspond to physical qualities. Therefore, any zero or negative value solution is rejected.

***P*-wave velocity and density.**—To include all possible relations between the *P*-wave velocity and density, the zero-offset reflection amplitude is defined by

$$R(0) = \frac{\alpha_2\Delta\rho - \alpha_1}{\alpha_2\Delta\rho + \alpha_1}$$

Therefore, if the polarity of the reflection at zero-offset is assumed to be positive, then $\alpha_2\Delta\rho > \alpha_1$ otherwise $\alpha_2\Delta\rho < \alpha_1$ must be satisfied.

Poisson's ratio constraint.—The range of Poisson's ratio for natural materials, such as rocks, is assumed to be [0.05, 0.45]. Therefore, the solutions for *P*- and *S*-wave velocities are expected to give a Poisson's ratio that falls into this range.

Notice that the last two constraints are related to all of the five parameters to be inverted. A solution, when found, is checked against these constraints. If the constraints are satisfied, then the solution is accepted; otherwise the solution is rejected, and the inversion is reinitialized.

The Most Likely Solutions and Confidence Limits by Bootstrapping

The GLI method is combined with the bootstrapping method to estimate the most likely model parameters and their confidence limits. As discussed in the previous sections, the GLI method uses an underlying mathematical model that represents the corresponding physical process and the observed data to arrive at a solution through a minimization and constraint satisfaction procedure. After the application of corrections for amplitude disturbing factors such as spherical divergence and vertical geophone response, the observed data can be viewed in two parts: (1) the AVO signature related to the seismic parameters; (2) the part related to noise. Here "noise" refers to the unorganized noise during the data acquisition and errors in the measurements of the reflection amplitudes. The data are assumed to be free from the organized noise, i.e., multiples and surface waves. The effect of noise on the inversion by the GLI method is unpredictable; therefore, the resulting solutions for the seismic parameters can be considered to be random variables somewhere around the true seismic parameters. For a given AVO data set, there exists only one true set of underlying seismic parameters that is unknown to us. If the data acquisition experiment were repeated many times with the same recording configuration, the noise on the acquired data would have been different. Then, the inversion of each of these data sets would have given a slightly different seismic parameter solution, each of which is a random variable around the true solution. Therefore, distribution functions of solutions could have a shape influenced by the distribution function of noise and its variance, both of which are unknown. But the most likely solutions could be assigned as some statistical measures of the distribution of solutions such as the mean or mode value.

In practice, it is too expensive and time consuming to repeat the same data acquisition procedure a large number of times. Therefore, we take advantage of the high computational power of computers to simulate the data acquisition procedure with different noise series added at each trial. Bootstrapping, introduced in Efron (1979, 1982) and Efron and Gong (1983), is a computational method to obtain statistical measures in a nonparametric statistical sense. What makes the bootstrapping method a nonparametric method is that it assumes neither the distribution type of noise nor its statistical measures, such as mean value and variance, nor any specific relation tying the modeled physical process to any statistical measures. The bootstrapping method is tied to the inversion of AVO data by coupling it with the GLI method using the following steps. Let R and $\bar{R}$ represent the observed and model AVO data, respectively, and ε represents the noise with unknown distribution function and variance. Then

$$R = f(t_0, x, \mathbf{P}) + \varepsilon \qquad (15)$$

and

$$\bar{R} = f(t_0, x, \mathbf{P}) \qquad (16)$$

can be written for the observed and model AVO data in the time-offset domain. Note that $\bar{R}$ is defined to be noise free forward model response calculated from the Zoeppritz equations in the time-offset domain. This model response is to be fit to the data R to get an estimation of the seismic parameter vector $\mathbf{P}$. An estimation of the seismic parameter vector $\mathbf{P}$ is obtained by $\mathbf{P}^*$ such that

$$\sum (R - \bar{R})^2$$

is a minimum, and the imposed constraints are satisfied.

The summation is taken over the number of data points available. The next step is to estimate the noise from the data. Let R^* represent the model response to the seismic parameter vector $\mathbf{P}^*$, then,

$$R^* = f(t_0, x, \mathbf{P}^*) \text{ and } \varepsilon^* = R - R^*. \qquad (17)$$

Notice that $\sum (\varepsilon^*)^2 \leq \sum (\varepsilon)^2$ because of the least-squares minimization. ε^* is useful in simulating the noise in the data. Having estimated the noise, the observed AVO data can be simulated many times as if each simulation corresponded to one repetition of the data acquisition in a real experiment. This is achieved by, first, assigning the probability of $1/n$ to each estimated noise value in ε^* where n is the number of data points, i.e., the number of traces in a CMP gather, and then, resampling as many as n times from the estimated noise ε^* by replacement at a time (Efron, 1982). The simulated data R_i^* for the ith experiment can be written as

$$R_i^* = f(t_0, x, \mathbf{P}^*) + \varepsilon_i^* \qquad (18)$$

where ε_i^* is the noise associated with the ith experiment. Notice that the simulation is achieved without any assumptions about the distribution type and variance of ε because the distribution type and variance of the resampled noise are automatically determined from the real data by curve fitting. Having simulated the data in this way, the inversion is applied to obtain a new estimation of the seismic parameters for the ith experiment by $\mathbf{P}_i^*$ such that

$$\sum (R_i^* - \bar{R})^2$$

is a minimum, and the imposed constraints are satisfied.

Repeating the noise resampling and inversion procedure for a large number of times, a collection of estimated seismic parameter vectors $\mathbf{P}_i^*$ is obtained. The projection of all of the vectors $\mathbf{P}_i^*$ onto their five dimensions, namely upper and lower layer P- and S-wave velocities and density ratio, gives the distribution of solutions for each seismic parameter. From these distributions the most likely solutions are determined as the mode values along with their 90 percent confidence intervals.

EXAMPLES OF APPLICATIONS WITH MODEL AVO DATA

Examples from the application of inversion to model data follow. The mainstream of the inversion procedure is highlighted. Limitations of the inversion method and the required form of the AVO data are discussed.

AVO Data Inversion Procedure

The inversion procedure of the AVO data is shown in Figure 3. Before the application of inversion, the AVO data are corrected for the disturbing effects of wave propagation and data acquisition. The corrected AVO data, CMP geometry, zero-offset two-way traveltime, and rms velocity are submitted to the GLI scheme to obtain an initial solution vector satisfying the imposed physical constraints. This initial solution vector provides a basis to calculate a model AVO response which is used in the estimation of noise on the AVO data by equation (17). Then a new noise series is resampled from the estimated noise and added onto the model AVO response to simulate the AVO data. The simulated AVO data are again inverted to find a solution vector that satisfies the imposed physical constraints. This simulation-inversion loop is repeated as indicated in Figure 3 by a backward branch. When a predetermined number of solution vectors are obtained (in this study various values were used, i.e., 100 to 1000), a distribution function of solutions is generated for each seismic parameter. Finally, the mode values of the distribution functions of solutions are declared as the most likely seismic parameters.

Inversion of Model AVO Data and Discussions

The first example is the inversion of a two-layer model AVO data without noise. The parameters of the two-layer model are given in Table 1. The maximum offset was chosen to be 4000 m long and the boundary was located at 2000 m depth giving a O/D of 2. The maximum angle of incidence is 45 degrees, which is less than the critical angle of 48 degrees for the model; therefore, the model AVO data fall in the pre-critical

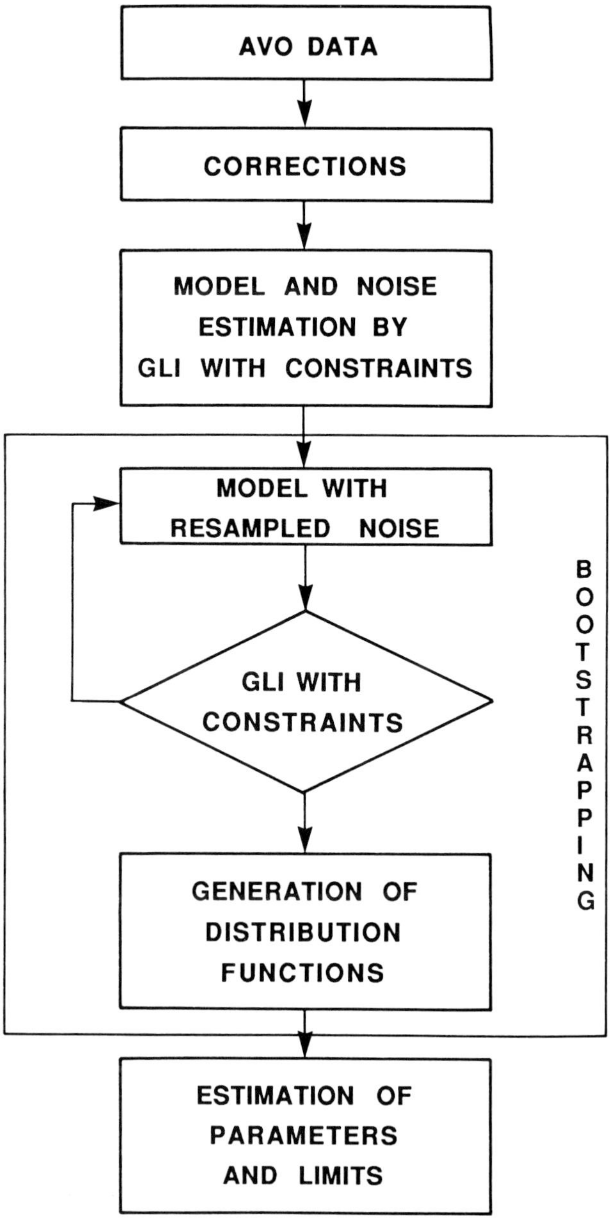

Fig. 3. AVO data inversion procedure.

region of offsets. First, the offset dependent reflection coefficients were calculated using the Zoeppritz equations in the time-offset domain, including the effects of spherical divergence and vertical geophone response for single source-receiver pairs. These coefficients were used to obtain the model CMP gather using a 25 Hz Ricker wavelet. This model CMP gather was used in the inversion procedure as follows. The amplitudes of the reflections were measured using equation (8) in a 150 ms window centered at the peak of the reflection. The model relative-AVO data are shown in Figure 4. Notice that the reflection amplitudes are normalized to the zero-offset reflection amplitude. This simulates normalized observed data because the reflection amplitudes cannot be observed in absolute terms. The zero-offset relative amplitude was also included in the inversion for the examples given here although it is not necessarily needed in the inversion. The relative AVO data along with the zero-offset two-way traveltime, source-receiver geometry, and upper layer P-wave rms velocity were used as input to the inversion procedure. The parameters in Table 1 were exactly recovered by the inversion in the absence of noise. The small differences are due to round-off error. After the model parameters are recovered, the zero-offset reflection amplitude is calculated to scale the input model relative-AVO data. The calculated AVO curve from the recovered model parameters and the scaled model AVO data by the estimated zero-offset reflection amplitude are shown in Figure 4. Notice the exact match between the input and output data sets.

The effect of the O/D on the inversion was tested using the same model AVO data with different O/Ds. As the O/D is decreased, the resolving power of the inversion method becomes less and results in an incorrect estimation of the model parameters. The inversion results for O/Ds between 0.5 and 2.0 are

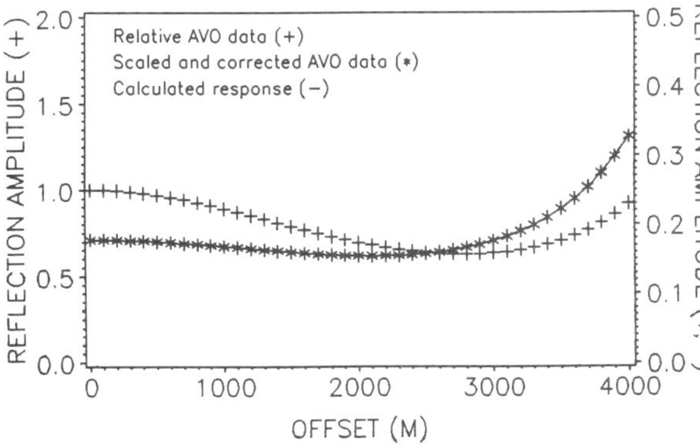

Fig. 4. Inversion of the noise free model AVO data.

Table 2. O/D test results.

O/D ratio	α_1	β_1	α_2	β_2	$\Delta\rho$
2.0	3000	1501	4000	2001	1.07
1.75	3000	1499	4000	2000	1.07
1.5	3001	1513	4003	2010	1.07
1.25	2985	1339	3968	1885	1.07
1.0	2947	1019	3887	1668	1.07
0.75	3033	2023	4078	2400	1.08
0.5	3408	1581	3938	1876	0.93

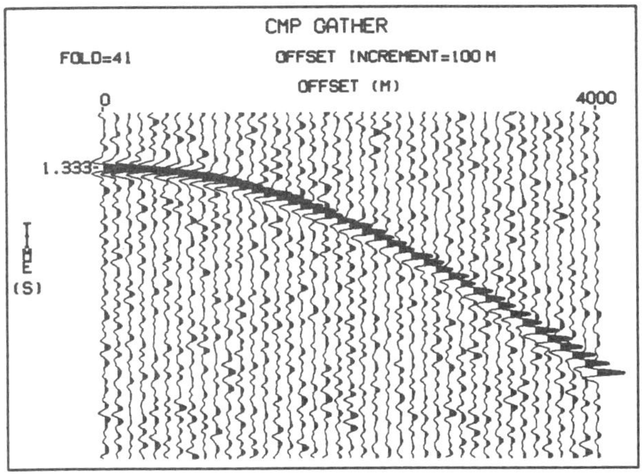

(a)

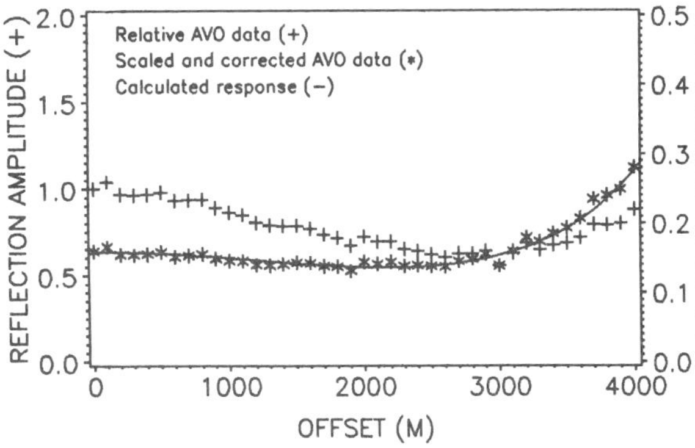

(b)

Fig. 5. Inversion of the model AVO data with noise. The model CMP gather with 20 dB S/N with respect to the average reflection amplitude is shown in the upper diagram (5a). The calculated model response (marked by the continuous curve) from the estimated parameters fits well the scaled noisy model data marked by '*' (5b).

shown in Table 2. Notice that when O/D $\geq$ 1.5, the model parameters were recovered well in opposition to the results of O/D < 1.5. It is also seen from Table 2 that for the smaller O/Ds, *S*-wave velocities were estimated with larger errors than the other parameters. This is a direct consequence of the fact that the inversion method is formulated only on the basis of *P*-wave AVO data. Additional results of other models suggest that an O/D of about 2 is necessary to obtain satisfactory results from the inversion of the *P*-wave reflection data, depending on the contrasts between the parameters of layers.

Model AVO data with noise were generated following similar steps and the same parameters. Traces of Gaussian noise and the noise free model CMP gather were summed to obtain a noisy CMP gather, and the reflection amplitude on each trace was subsequently calculated using equation (8). The CMP gather with noise and related model relative-AVO data are shown in Figures 5a and 5b, respectively. In spite of the noise, the trend in the AVO data is easily noticeable due to high S/N. The S/N for this example is 20 dB relative to the average reflection amplitude in the CMP gather. Even though the reflection events aligned on a move-out hyperbola in a CMP gather can be recognized at much lower levels of the S/N due to their continuity, the AVO data at these lower S/N levels are so deteriorated by noise that the general trend in the reflection curve is lost. Therefore, a high S/N was chosen in the example to observe the general trend in the noisy model AVO data. The distribution functions of solutions from the inversion are seen in Figure 6, and the results are summarized in Table 3. The true parameters are marked by a solid triangle on the horizontal axis in the distribution functions of solutions. The stars in the middle and on the edges mark the estimated values of parameters and 90 percent confidence limits, respectively. The true parameters were not exactly recovered by the inversion due to noise. However, because of tight physical constraints (a range of acceptable solutions around the true parameters, i.e., $\pm$20 percent of the true parameters) on the solutions, the true parameters are still within the confidence limits for this example, except in the case of lower layer *S*-wave velocity. The experiments showed that the inversion with no constraints and small O/Ds gives unsatisfactory results when noise is introduced. The model response for the estimated parameters and the scaled model AVO data are displayed in Figure 5b. Notice that the calculated AVO curve from the estimated parameters fits the model AVO data well although the estimated parameters are not the true ones.

CONCLUSIONS

The inversion method presented uses the plane wave Zoeppritz equations in the time-offset domain and is capable of estimating the *P*- and *S*-wave velocities and the density ratio from the observed *P*-wave AVO data when S/N and O/D are high.

Input data to the inversion scheme are in the form of reflection amplitudes which are measured on seismic traces by the square root of the Euclidean norm within a time window covering the reflected wavelet. Interference of overlapping reflected wavelets from two or more consecutive reflectors deteriorates the amplitude information of individual reflections. Hence, it is required that the reflections under consideration are free from the effect of any major interference.

The necessary corrections such as spherical divergence and vertical geophone response are time and space dependent and require some auxiliary information such as rms velocity function. If the corrections are applied using inaccurate auxiliary information, the observed AVO data set is either over or under corrected; therefore, the parameters are estimated erroneously.

The feasibility of this inversion method was checked by the RFMs which show minima around the true parameters except in the case of the *S*-wave velocity pair; however, the RFMs of the *S*-wave velocities with other parameters show closed contours around the true parameters, implying the uniqueness of the inverse problem in the absence of noise.

The formulation of the inversion method for AVO data and the algorithm developed are based on pre-critical reflection data. Therefore, the maximum offset should be long enough to observe a significant change in the AVO data while it falls into the pre-critical offset range. Model tests indicated that an O/D of about 2 is necessary to obtain satisfactory results, depending on the contrast in the seismic parameters of layers. AVO data with a small O/D (O/D < 1) give unsatisfactory results because of the uncertainty in the definition of AVO curve.

Besides the small O/D, noise is the most disturbing factor in the inversion of AVO data. When S/N and O/D are small, the noise obscures the change in the amplitude with offset; as a result, parameter estimation by least-squares minimization suffers, and the parameters are estimated with large errors.

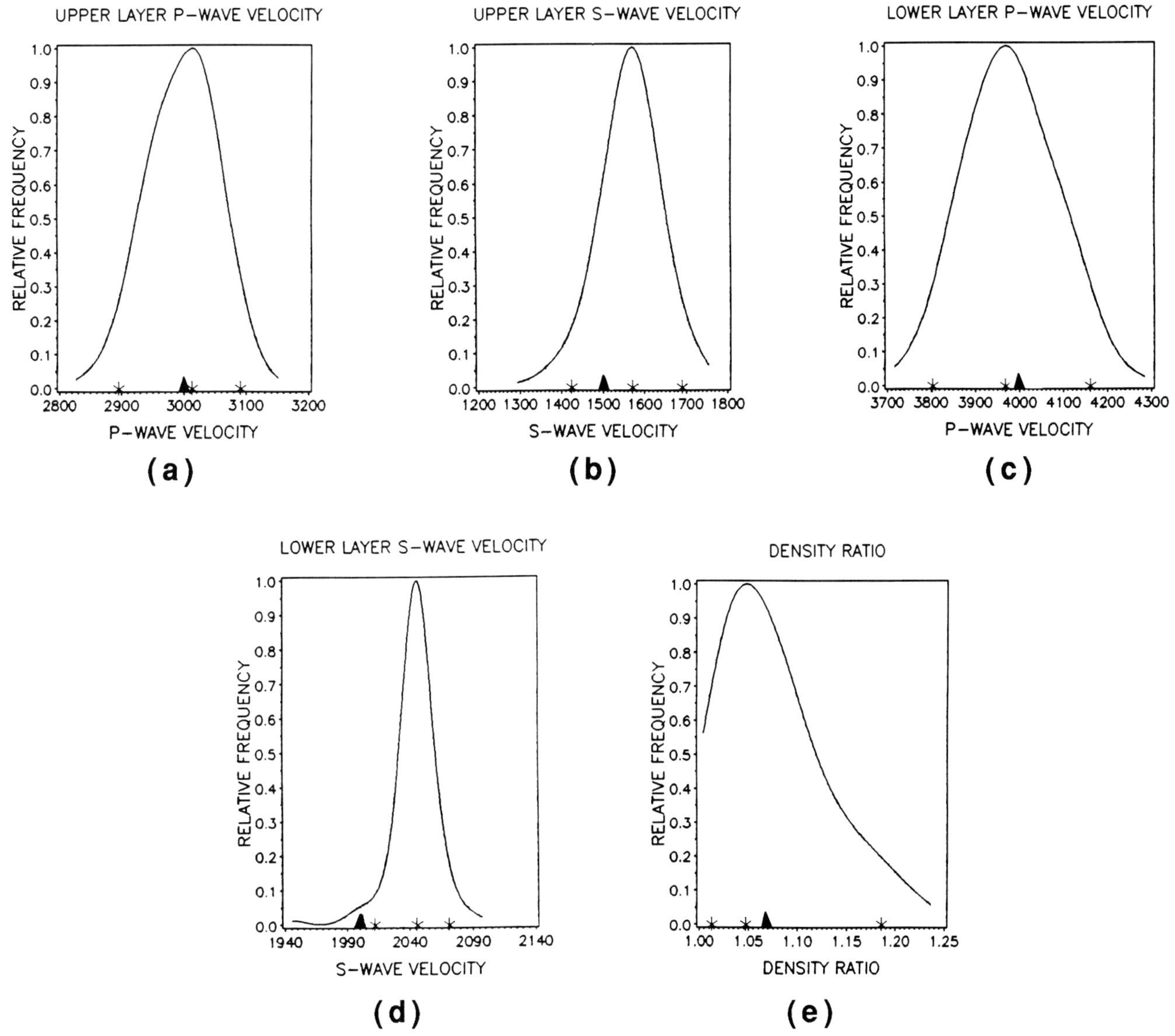

Fig. 6. Distribution functions of solutions. The most likely parameters are assigned as the peak values (modes) of the distribution functions. The solid triangle indicates the location of the actual parameters. The star at the peak value marks the most likely parameters and the stars on the edges define 90 percent confidence limits for the most likely parameters. The calculated AVO curve using the most likely parameters is shown in Figure 5b.

Table 3. Inversion results for model AVO data with noise.

Parameter	Lower Limit	Estimated	Exact	Upper Limit
α_1	2895	3013	3000	3091
β_1	1424	1571	1500	1690
α_2	3804	3967	4000	4161
β_2	2012	2046	2000	2071
$\Delta\rho$	1.01	1.05	1.07	1.19

ACKNOWLEDGMENTS

We thank Dr. Panickos Palettas of the Department of Statistics at VPI & SU for his contributions to the statistical concerns. We also thank Susan Sharpless Hubbard for her excellent review of the manuscript.

APPENDIX A: CONVERSION OF TRIGONOMETRIC FUNCTIONS INTO THE TIME-OFFSET DOMAIN

The Zoeppritz equations were converted into the time-offset domain to be used with CMP gather. The conversion is achieved by replacing the trigonometric functions with their corresponding equalities given in terms of the *P*- and *S*-wave velocity of lower and upper layers, two-way traveltime, and CMP geometry. The derivation of relations follows a stepwise substitution procedure of trigonometric functions into each other. Referring to Figure 1 and using Snell's law and general trigonometric relations, the following list can be used with the Zoeppritz equations:

$$\sin\theta_r = \frac{2x}{\alpha_1 t_x}, \qquad \cos\theta_r = \frac{2d}{\alpha_1 t_x} = \frac{t_0}{t_x},$$

$$\sin\theta_t = \frac{\alpha_2}{\alpha_1}\sin\theta_r = \frac{2x\alpha_2}{\alpha_1^2 t_x},$$

$$\cos\theta_t = \left(1 - \frac{4x^2\alpha_2^2}{\alpha_1^4 t_x^2}\right)^{1/2},$$

$$\sin\phi_r = \frac{\beta_1}{\alpha_1}\sin\theta_r = \frac{2x\beta_1}{\alpha_1^2 t_x},$$

$$\cos\phi_r = \left(1 - \frac{4x^2\beta_1^2}{\alpha_1^4 t_x^2}\right)^{1/2},$$

$$\sin\phi_t = \frac{\beta_2}{\alpha_1}\sin\theta_r = \frac{2x\beta_2}{\alpha_1^2 t_x},$$

$$\cos\phi_t = \left(1 - \frac{4x^2\beta_2^2}{\alpha_1^4 t_x^2}\right)^{1/2},$$

$$\sin 2\theta_r = \frac{4xt_0}{\alpha_1 t_x^2}, \qquad \cos 2\phi_r = 1 - \frac{8x^2\beta_1^2}{\alpha_1^4 t_x^2},$$

$$\cos 2\phi_t = 1 - \frac{8x^2\beta_2^2}{\alpha_1^4 t_x^2},$$

$$\sin 2\theta_t = \frac{4x\alpha_2}{\alpha_1^2 t_x}\left(1 - \frac{4x^2\alpha_2^2}{\alpha_1^4 t_x^2}\right)^{1/2},$$

$$\sin 2\phi_r = \frac{4x\beta_1}{\alpha_1^2 t_x}\left(1 - \frac{4x^2\beta_1^2}{\alpha_1^4 t_x^2}\right)^{1/2}, \quad \text{and}$$

$$\sin 2\phi_t = \frac{4x\beta_2}{\alpha_1^2 t_x}\left(1 - \frac{4x^2\beta_2^2}{\alpha_1^4 t_x^2}\right)^{1/2}.$$

These relations were used to convert the Zoeppritz equations into the time-offset domain by direct substitution.

APPENDIX B: THE ZOEPPRITZ EQUATIONS IN THE TIME-OFFSET DOMAIN FOR A MULTILAYER MODEL

The Zoeppritz equations in the time-offset domain are also developed for a multilayer model. The model is a stack of horizontal, linear elastic, homogeneous, and isotropic layers. In this case the rms and average velocities appear in the equations. Let d_{ave}, α_{rms}, and α_{ave} be the average depth, rms velocity, and average velocity to the considered boundary, respectively. The other variables and seismic parameters are the same as those of the two-layer model defined in the paper. Using the CMP geometry, Snell's law as well as the following relations for time-distance and average depth-velocity

$$t_x = \left(t_0^2 + \frac{4x^2}{\alpha_{rms}^2}\right)^{1/2} \quad \text{and} \quad d_{ave} = \frac{t_0\alpha_{ave}}{2},$$

the Zoeppritz equations were converted into the time-offset domain for the multilayer model. In this case, the entries of equation (1) are given as follows:

$$a_{11} = \frac{2x}{\alpha_{rms} t_x}, \qquad a_{12} = \left(1 - \frac{4x^2\beta_1^2}{A^2}\right)^{1/2},$$

$$a_{13} = -\frac{2x\alpha_2}{A}, \qquad a_{14} = \left(1 - \frac{4x^2\beta_2^2}{A^2}\right)^{1/2},$$

$$a_{21} = -\frac{\alpha_{ave}}{\alpha_{rms}}\frac{t_0}{t_x}, \qquad a_{22} = \frac{2x\beta_1}{A},$$

$$a_{23} = -\left(1 - \frac{4x^2\alpha_2^2}{A^2}\right)^{1/2}, \qquad a_{24} = -\frac{2x\beta_2}{A},$$

$$a_{31} = \frac{4x\alpha_{ave}t_0}{\alpha_{rms}^2 t_x^2}, \qquad a_{32} = \frac{\alpha_1}{\beta_1}\left(1 - \frac{8x^2\beta_1^2}{A^2}\right),$$

$$a_{33} = \frac{4x\Delta\rho\beta_2^2}{\alpha_{rms}\beta_1^2 t_x}\left(1 - \frac{4x^2\alpha_2^2}{A^2}\right)^{1/2},$$

$$a_{34} = -\frac{\Delta\rho\beta_2\alpha_1}{\beta_1^2}\left(1 - \frac{8x^2\beta_2^2}{A^2}\right),$$

$$a_{41} = 1 - \frac{8x^2\beta_1^2}{A^2},$$

$$a_{42} = -\frac{4x\beta_1^2}{\alpha_1^2\alpha_{\mathrm{rms}}t_x}\left(1 - \frac{4x^2\beta_1^2}{A^2}\right)^{1/2},$$

$$a_{43} = -\frac{\alpha_2\Delta\rho}{\alpha_1}\left(1 - \frac{8x^2\beta_2^2}{A^2}\right),$$

$$a_{44} = -\frac{4x\beta_2^2\Delta\rho}{\alpha_1^2\alpha_{\mathrm{rms}}t_x}\left(1 - \frac{4x^2\beta_2^2}{A^2}\right)^{1/2},$$

$$b_{11} = -a_{11}, \qquad b_{21} = a_{21},$$

$$b_{31} = a_{31}, \quad \text{and} \quad b_{41} = -a_{41}.$$

where $A = \alpha_1\alpha_{\mathrm{rms}}t_x$. When $\alpha_{\mathrm{rms}} = \alpha_{\mathrm{ave}} = \alpha_1$, these entries are reduced to those of the two-layer model.

APPENDIX C: ESTIMATION OF ZERO OFFSET REFLECTION AMPLITUDE

After the corrections, AVO data are scaled by zero-offset reflection amplitude to relate them to the model response. The zero-offset reflection amplitude is not available from the field data and needs to be estimated by the extrapolation of observed reflection amplitudes to zero-offset. Because the explicit values of the seismic parameters of the layers are not known prior to the inversion, a trigonometric polynomial approximation to the Zoeppritz equations is used, and the zero-offset reflection amplitude is estimated herein using the linear least-squares procedure.

Let $R(\theta)$ and $\bar{R}(\theta)$ represent the model and observed reflection amplitudes, respectively. Beginning with the following approximation equation given in Lörtzer (1988),

$$R(\theta) = a + b\sin^2\theta + c\tan^2\theta$$

the derivatives of the sum of the squares of the differences between the model and observed data with respect to coefficients are set to zero,

$$E = \sum[\bar{R}(\theta) - R(\theta)]^2$$
$$= \sum[\bar{R}(\theta) - (a + b\sin^2\theta + c\tan^2\theta)]^2$$

$$\frac{\partial E}{\partial a} = \frac{\partial E}{\partial b} = \frac{\partial E}{\partial c} = 0.$$

This leads to a system of linear equations which can be formulated into a matrix form,

$$\begin{bmatrix} n & \sum\sin^2\theta & \sum\tan^2\theta \\ \sum\sin^2\theta & \sum\sin^4\theta & \sum\tan^2\theta\sin^2\theta \\ \sum\tan^2\theta & \sum\tan^2\theta\sin^2\theta & \sum\tan^4\theta \end{bmatrix}\begin{bmatrix} a \\ b \\ c \end{bmatrix} = \begin{bmatrix} \sum\bar{R}(\theta) \\ \sum\bar{R}(\theta)\sin^2\theta \\ \sum\bar{R}(\theta)\tan^2\theta \end{bmatrix}.$$

The trigonometric functions are replaced by the following relations,

$$\sin\theta = \frac{2x}{\alpha t_x} \quad \text{and} \quad \tan\theta = \frac{2x}{\alpha t_0}$$

where α is the upper layer P-wave velocity for a two layer model and rms velocity for a multilayer model. However, the correctness of the value of α is not significant because the extrapolation is based on the best fit approximate Zoeppritz curve to the AVO data. Therefore, a conventional estimation of α is good enough for the extrapolation purpose. The matrix equation in the time-offset domain is expressed as

$$\begin{bmatrix} n & \sum\left(\frac{2x}{\alpha t_x}\right)^2 & \sum\left(\frac{2x}{\alpha t_0}\right)^2 \\ \sum\left(\frac{2x}{\alpha t_x}\right)^2 & \sum\left(\frac{2x}{\alpha t_x}\right)^4 & \sum\left(\frac{2x}{\alpha t_0}\right)^2\left(\frac{2x}{\alpha t_x}\right)^2 \\ \sum\left(\frac{2x}{\alpha t_0}\right)^2 & \sum\left(\frac{2x}{\alpha t_0}\right)^2\left(\frac{2x}{\alpha t_x}\right)^2 & \sum\left(\frac{2x}{\alpha t_0}\right)^4 \end{bmatrix}$$
$$\times\begin{bmatrix} a \\ b \\ c \end{bmatrix} = \begin{bmatrix} \sum\bar{R}(x, t_0) \\ \sum\bar{R}(x, t_0)\left(\frac{2x}{\alpha t_x}\right)^2 \\ \sum\bar{R}(x, t_0)\left(\frac{2x}{\alpha t_0}\right)^2 \end{bmatrix}.$$

Coefficients a, b, and c can be determined from this equation. Note that when $x = 0$ $(\theta = 0)$, $R(0) = a$ is obtained which corresponds to the zero-offset reflection amplitude. Having obtained $R(0) = a$, the observed AVO data are scaled with respect to coefficient *a* so that the zero-offset reflection amplitude becomes unity, and the amplitudes at other offsets are adjusted accordingly.

REFERENCES

Aki, K., and Richards, P. G., 1980, Quantitative seismology: Theory and methods: W. H. Freeman and Co.

Backus, M. M., 1987, Amplitude versus offset: A review: 57th Ann. Internat. Mtg., Soc. Expl. Geophys., Expanded Abstracts, 359–364.

Bortfeld, R., 1961, Approximations to the reflection and transmission coefficients of plane longitudinal and transverse waves: Geophys. Pros., **9**, 485–502.

Berkhout, A. J., 1987, Applied seismic wave theory: Elsevier Science Publ. Co., Inc.

De Haas, J. C., and Berkhout, A. J., 1988, On the information content of P-P, P-SV, SV-SV, and SV-P reflections: 58th Ann. Internat. Mtg., Soc. Expl. Geophys., Expanded Abstracts, 1190–1194.

De Haas, J. C., and Berkhout, A. J., 1989, Practical approach to nonlinear inversion of amplitude versus offset information: 59th Ann. Internat. Mtg., Soc. Expl. Geophys., Expanded Abstracts, 27–36.

Demirbag, E., and Çoruh, C., 1988, Inversion of Zoeppritz equations and their approximations: 58th Ann. Internat. Mtg., Soc. Expl. Geophys., Expanded Abstracts, 1199–1203.

Demirbag, E., and Çoruh, C., 1989, Inversion of multilayer amplitude-versus-offset data: 59th Ann. Internat. Mtg., Soc. Expl. Geophys., Expanded Abstracts, 709–712.

Efron, B., 1979, Computers and the theory of statistics: Thinking the unthinkable: Society for Industrial and Applied Mathematics, **21**, 460–480.

Efron, B., 1982, The jackknife, the bootstrap, and other resampling plans: Regional Conference Series in Applied Mathematics, Society for Industrial and Applied Mathematics, **38**, 839–842.

Efron, B., and Gong, G., 1983, A leisurely look at the bootstrap, the jackknife, and cross-validation: American Statistician, **37**, 36–48.

Gardner, G. H. F., Gardner, L. W., and Gregory, A. R., 1974, Formation velocity and density: The diagnostic basis for stratigraphic traps: Geophysics, **39**, 770–780.

Gelfand, V., Ng, P., Nguyen, H., and Larner, K., 1986, Seismic lithology modeling of amplitude versus offset data: 56th Ann. Internat. Mtg., Soc. Expl. Geophys., Expanded Abstracts, 334–337.

Jackson, D. D., 1972, Interpretation of inaccurate, insufficient, and inconsistent data: Geophys. J. Roy. Astr. Soc., **28**, 97–109.

Jackson, D. D., 1979, The use of a priori data to resolve non-uniqueness in linear inversion: Geophys. J. Roy. Astr. Soc., **57**, 137–157.

Koefoed, O., 1955, On the effect of Poisson's ratios of rock strata on the reflection coefficients of plane waves: Geophys. Prosp., **3**, 381–387.

Koefoed, O., 1962, Reflection and transmission coefficients for plane longitudinal incident waves: Geophys. Prosp., **10**, 304–351.

Lines, L. R., and Treitel, S., 1984, Tutorial: a review of least-squares inversion and its application to geophysical problems: Geophys. Prosp., **32**, 159–186.

Lörtzer, G. J. M., De Haas, J. C., and Berkhout, A. J., 1988, Evaluation of weighted stacking techniques for AVO inversion: 58th Ann. Internat. Mtg., Soc. Expl. Geophys., Expanded Abstracts, 1204–1208.

Macdonald, C., Davis, P. M., and Jackson, D. D., 1987, Inversion of reflection traveltimes and amplitudes: Geophysics, **52**, 606–617.

Menke, W., 1989, Geophysical data analysis: Discrete inverse theory: Academic Press, Inc.

O'Doherty, R. F., and Anstey, N. A., 1971, Reflections on amplitudes: Geophys. Prosp., **19**, 430–458.

Press, W. H., Flannery, B. P., Teukolsky, S. A., and Wetterling, W. T., 1988, Numerical recipes: The art of scientific computing: Cambridge Univ. Press.

Russell, B. H., 1988, Introduction to seismic inversion methods: SEG Continuing Education Course Note Series, Vol. 2, Soc. Expl. Geophys.

Shuey, R. T., 1985, A simplification of the Zoeppritz equations: Geophysics, **50**, 609–614.

Stoer, J., and Bulirsch, R., 1980, Introduction to numerical analysis: Springer-Verlag.

Tarantola, A., 1987, Inverse problem theory: methods for data fitting and model parameter estimation: Elsevier Science Publ. Co.

Tooley, R. D., Spencer, T. W., and Sagico, H. F., 1965, Reflection and transmission of plane compressional waves: Geophysics, **30**, 552–570.

Van Heeswijk, M., and Christopher, G. F., 1988, Iterative method and FORTRAN code for nonlinear curve fitting: Computers and Geosciences, **14**, 489–503.

Waters, K. H., 1987, Reflection seismology: A tool for energy resource exploration: John Wiley and Sons, Inc.

Elastic Inversion of Gulf of Mexico Data

James J. Carazzone and Leonard J. Srnka‡*

INTRODUCTION

An elastic inversion method based upon a wave equation for layered media has been tested on model data and seismic data from the Gulf of Mexico. Inverted P-wave velocity and density compare favorably with well-log measurements at two locations, and show extremely small depth misties. Inverted S-wave velocity was also computed, but no nearby S-wave velocity logs were available for comparison. Model results indicate that S-wave velocity information contained within the passband of the seismic wavelet is accurate to about 10 percent.

Reconstruction of subsurface elastic parameters from hydrophone data faces many difficulties, among them sensitivity to moveout-velocity errors (Spratt, 1987). Application of three-dimensional (3-D) radial Radon-transform techniques to the stratified situation permits utilization of the maximum range of recorded reflection angles. Use of this maximum data aperture helps to stabilize the separation of elastic parameters, and has benefits for the determination of source directivity and moveout velocity. Traveltime inversion in the Radon-transform domain can be carried out by picking only the normal-incidence events.

We apply nonlinear least-squares optimization to a line average of the water-bottom reflection to determine the source wavelet as a function of frequency and direction, and to update estimates of the water-bottom properties. An approximate elastic layer-stripping scheme is used to isolate P-wave reflectivity from the source-deconvolved data. Robust fits over a small traveltime window (10 to 20 ms) to the linearized elastic reflection curve (corrected for a stated model of P-wave attenuation) then produce estimates of the *log-steps* in P-wave velocity, S-wave velocity, and density. Integration of the log-step estimates and conversion to depth then produces *pseudo-log* predictions which can be compared with well measurements.

*Exxon Production Research Company
‡Exxon Exploration Company

RADON TRANSFORM

Although there are extra costs encountered in obtaining good Radon transforms, accurate plane-wave decomposition of well-sampled marine data is essential to our inversion approach and can be obtained in two steps. The first step is the computation of data-truncated transforms,

$$F_0(\xi,\,\omega) \equiv 2\pi(-i\omega)\int_0^\infty dR\;RJ_0 \times (\xi\omega R)f(R,\,\omega)\Omega(\xi,\,R), \qquad (1)$$

where ξ is the slowness or ray parameter, ω is the angular frequency, R is radial distance, i is the unit imaginary number, $J_0(x)$ is the ordinary Bessel function of zero order, and $f(R,\,\omega)$ is the Fourier transform of the data. The positive taper function $\Omega(\xi,\,R)$ is a linear filter which has value unity over data-supported ranges of ξ and R, and falls smoothly toward zero outside these ranges. Beam steering and trace interpolation are used to further increase the accuracy of the integration. The second step involves amplitude correction of the truncated transforms for the antenna pattern of the receiver arrays. Antenna compensation is used to correct the transforms at each $\xi - \omega$ point for wave number values up to a limit set by the group interval.

BACKGROUND *P*-WAVE VELOCITY

For a plane layered earth, the hyperbolic reflection patterns in the X-T domain are transformed to curves which appear as nested ellipses characterized by the interval velocity structure. The two-way traveltime for a P-wave reflection at depth z, starting from depth z_0, is

$$\tau(z,\,z_0,\,\xi) \equiv 2\int_{z_0}^{z}\frac{d\zeta}{V(\zeta)}\sqrt{1-\xi^2V^2(\zeta)}. \qquad (2)$$

A background P-wave velocity model is created by locating the best-fit, normal-incidence traveltimes taken from a CMP stacked section. Maximization of stack power along a number of trial moveout curves yields good results, particularly when combined with median filtering over adjacent locations. Low-frequency density and S-wave velocity can be created from this background P-wave velocity by applying regional empirical relations.

SOURCE DETERMINATION

Source directivity and frequency content are estimated from the data in the Radon-transform domain thereby avoiding complications due to changes in scale between the offset domain and the Radon-transform domain. A marine line will usually contain several hundred or more midpoint gathers spaced over several kilometers. A line average of the water-bottom reflection events produces a robust estimate of the source wavelet if water bottom properties remain relatively invariant along the line.

Let $\mathbf{G}(z, z', \xi, \omega | z_{fs} = 0)$ be the $\xi - \omega$ domain (displacement) elastodynamic Green's function for a water layer of thickness h over an elastic layered half space (z_{fs} stands for the free surface location). We assume that the air-water interface at $z_{fs} = 0$ approximates a pressure release surface. Wave equation analysis then shows that measurements of hydrostatic pressure at depth z due to a source of pressure at depth z' determines only the vertical-to-vertical component of the velocity $i\omega\mathbf{G}$ at $z = z' = 0$. The strong attenuation of horizontally moving energy by the arrays dictates that we consider only quantities with direct arrivals (denoted by $\mathbf{G}^{(\mathrm{dir})}$) removed. In situations where the water layer is thick enough to permit a clear separation of direct and reflected arrivals, the direct arrival can be removed by a time domain editing step. Thus along the elliptic curve defining the water-bottom reflection from a *simple water bottom*, the source- and receiver-ghost corrected data approximately should fit the coefficient of reflection for P-waves (R_{pp}) at a fluid-solid interface:

$$\begin{aligned} i\omega[&G_{33}(0, 0, \xi, \omega | z_{fs} = 0) \\ &- G_{33}^{(\mathrm{dir})}(0, 0, \xi, \omega | z_{fs} = 0)]S_{\mathrm{acc}}(\xi, \omega) \\ &= \frac{V_w^2}{i\omega} \frac{M'(d_r, d_s, \xi, \omega)}{f_g(d_r, \xi, \omega) f_g(d_s, \xi, \omega)} \\ &\approx \frac{-2}{Z_p^{(w)}(\xi)} R_{pp}(V_w, \rho_w, V_p, V_s, \rho, \xi) \\ &\times S_{\mathrm{acc}}(\xi, \omega) e^{\frac{2i\omega h}{V_w}\sqrt{1 - V_w^2 \xi^2}} \end{aligned} \tag{3}$$

where V_w is water velocity; ρ_w is water density; V_p, V_s, and ρ are the P-wave velocity, S-wave velocity, and density of the water bottom, and S_{acc} is the accelerative source function. We denote the vertical-to-vertical motion with the subscript 33. The factor $Z_p^{(w)}(\xi)$ is the effective P-wave impedance of the water layer. The ghosting factor f_g for tow depth d is

$$f_g(d, \xi, \omega) \equiv \frac{\sin\left[(\omega d / V_w)\sqrt{1 - V_w^2 \xi^2}\right]}{\omega \sqrt{1 - V_w^2 \xi^2}}. \tag{4}$$

The quantity M' is the amplitude signal in $\xi - \omega$ domain with the direct arrival removed. The accelerative source function S_{acc} is expressed as a series of even-order Legendre polynomials in the ratio of ξ to the maximum Radon parameter value $\xi_{\max}$:

$$S_{\mathrm{acc}}(\xi, \omega) = \sum_{\ell = 0}^{m} a_{2\ell}(\omega) P_{2\ell}(\xi / \xi_{\max}). \tag{5}$$

where the complex coefficients $a_{2\ell}$ depend upon frequency. This source model is motivated by the assumption that the source array will generate a nearly isotropic far-field signature at the lower frequencies and a (possibly) complex distribution of outgoing energy at the higher frequencies. Solution of a non-linear, least-squares optimization problem for the coefficients and the parameters V_p, V_s, and ρ yield source directivity, wavelet shape, and water-bottom properties. These quantities are determined within the framework of an iterative numerical process begun at an initial guess for the elastic properties of the water bottom. Ambiguities exist between the changes in water-bottom impedance and overall source scale, and between elastic water-bottom properties and source directivity. Non-linear model aspects associated with full trace multiple deconvolution and parameter reconstruction can be used to guide selection of the initial elastic water-bottom model.

DECONVOLUTION AND LAYER STRIPPING

The estimated source function is replaced by a suitable minimum-phase wavelet over the passband of the measured data. The resulting minimum-phase deconvolved data are then mixed with an elastic wave equation computation. We use a process designed to merge the data smoothly with a low-frequency Green's function computed from the background model constructed using the traveltime analysis. The result approximates an elastic impulse response function. When the initial source scale is correctly set, and care is taken to filter out departures from the minimum-phase condition (cf. Carazzone and Srnka, 1987), we have found layer stripping deconvolution to be a stable and rapid process for extracting primary P-wave reflectivity from the impulse response. Our procedure is similar in spirit to that employed in Singh et al. (1989) but differs slightly in detail. A medium which primarily supports the propagation of P-waves such as the Gulf of Mexico subbottom can be viewed as weakly elastic. An effective acoustic-like wave equation can easily be derived to describe the effects of S-wave generation on P-wave reflections. To lowest order in shear-velocity contrast the affect of the S-wave generation modifies the effective (depth and slowness dependent) P-wave impedance by a multiplicative factor $C(z, \xi)$ given by

$$C(z, \xi) \equiv (1 - 2V_s^2(z)\xi^2)^2 + 4\xi^2[V_s^3(z)/V_p(z)] \times \sqrt{1 - V_s^2(z)\xi^2}\,\sqrt{1 - V_p^2(z)\xi^2}.$$

The acoustic layer stripping process (Bube and Burridge, 1983) inverts for impedance as a function of slowness-dependent traveltime. The layer stripping process accounts for elastic effects which influence the data only through a modification of the behavior of P-wave reflectivity as a function of slowness. No explicit correction for elastic effects is needed during the layer stripping process. Other elastic wave phenomena such as shear- and converted-wave arrivals are left unaccounted for by this approach. A full treatment of all elastic effects in three-component elastic layer stripping is given in Carazzone, 1986.

PARAMETER EXTRACTION

At each point in P-wave normal-incidence time, the reflectivity represents the influence of local variations in the elastic parameters together with accumulating changes in frequency content resulting from loss of energy to P-wave attenuation. To lowest order in δV_p, δV_s, and $\delta\rho$ and leading order in $1/\omega$, the reflected P-wave contribution to the elastic vertical velocity response measured at a pressure release surface at depth h (i.e., the bottom of the water layer) is (Sacks and Symes, 1987)

$$\begin{aligned} i\omega[G_{33}(h, h, \xi, \omega|z_{fs} = h) &- G_{33}^{(\mathrm{dir})}(h, h, \xi, \omega|z_{fs} = h)]S_{\mathrm{acc}}(\xi, \omega) \\ &= i\omega S_{\mathrm{acc}}(\xi, \omega)K_p^2(h, \xi)\int_h^\infty \frac{dz'}{V_p(z')} \\ &\times \sqrt{1 - V_p^2(z')\xi^2}\, e^{i\omega\tau(z', h, \xi)} \\ &\times \exp\left[\frac{-|\omega|}{2}\int_h^{z'} \frac{d\zeta}{V_p(\zeta)}\frac{1}{\sqrt{1 - \xi^2 V_p^2(\zeta)}}\frac{1}{|Q_\sigma(\zeta, \omega)|}\right]\delta R_{pp}(z', \xi) \end{aligned} \tag{6}$$

where $K_p^2(h, \xi)$ is an elastic P-wave admittance factor which reduces to $4/Z_p(\xi)$ for an acoustic medium. The factor $\delta R_{pp}(z', \xi)$ is the linearized P-wave to P-wave reflection coefficient at depth z'.

$$\delta R_{pp}(z', \xi) = \frac{1}{2}\left[(1 - 4V_s^2(z')\xi^2)\frac{\delta\rho(z')}{\rho(z')} + \frac{1}{1 - V_p^2(z')\xi^2}\frac{\delta V_p(z')}{V_p(z')} - 8\xi^2 V_s(z')\frac{\delta V_s(z')}{V_s(z')}\right]. \tag{7}$$

The quality factor Q_σ is the ratio of the magnitude of the imaginary part of $\sigma = \lambda + 2\mu$ to its real part. The moduli are required to have nonpositive imaginary parts to force reflected energy to decay with ray-path length. A S-wave quality factor Q_μ can be defined similarly. Both may be frequency and depth dependent. The linearized reflection coefficient of equation (7) will reduce to a linearization of equation (3) in the acoustic limit. Likewise, the Bube-Burridge reflectivity defined as the fractional change in *effective* P-wave impedance, $C(z, \xi)Z_p(z, \xi)$, leads to equation (7) in the limit of small variations in the medium parameters. The elastic layer-stripping process corrects for the K_p^2 factor, while time-to-depth conversion for each plane wave corrects for the phase factor. Compensation for the P-wave attenuation factor is done in the frequency domain using a supplied attenuation model.

These corrections produce a *linear*, least-squares, optimization problem for the log-steps in the elastic parameters:

$$\min_{\delta V_p/V_p,\ \delta V_s/V_s,\ \delta\rho/\rho} \sum_\xi \mu^2(\xi) \times [\delta R_{pp}(z, \xi; \delta V_p/V_p, \delta V_s/V_s, \delta\rho/\rho) - d^{\mathrm{corrected}}(z, \xi)]^2.$$

The quantity $d^{\mathrm{corrected}}(z, \xi)$ is the reflectivity data compensated for attenuation. The factor $\mu^2(\xi)$ represents an appropriate (normalized) confidence level. The minimization process is carried out over the values of the log-steps within a small window centered at depth z. We employ a packaged procedure LSEI by Hanson and Haskell (1978 and 1979) which allows the minimization process to be carried out subject to equality and inequality constraints. A number of different constraints can be used for stabilizing the inversion process. In the following discussion of results we adjusted the bounds on the three log-steps using the integrated low-angle absolute reflectivity at each depth level. This constraint helps to stabilize the inversion process but can lead to errors as discussed in the model test section below. Once computed, the log-steps are integrated to obtain estimates of the *P*- and *S*-velocity and density profiles. Moving-window rms statistics are used to force agreement between trends in the predicted *pseudo-logs* and the background model.

MODEL TEST

Synthetic time-offset domain traces were computed for a plane-layer elastic model by means of the reflectivity method (cf. Rendleman and Levin, 1990). The data collection aspects of the model are typical of marine surveys recently conducted in the Gulf of Mexico. A common-midpoint gather consisting of 180 traces was computed at a time sample rate of 0.002 s with a uniform offset spacing of 15 m (49.2 ft). The minimum shot-to-receiver offset was 100 m (328 ft) and the maximum offset was 2785 m (9137 ft). Synthetic receiver arrays of 7 uniformly spaced and uniformly weighted hydrophones distributed over 15 m (49.2 ft) were included in the computation. The source was modeled as an isotropic radiator of hydrostatic pressure located at a depth of 5 m (16.4 ft). The receivers were located at a constant depth of 4 m (13.1 ft). Source and receiver ghosting were included in the model with a surface reflection coefficient of −0.85. The source wavelet approximated a bandlimited Ricker wavelet with peak energy at 35 Hz and with minimal energy below 12 Hz or above 90 Hz. Because of the relatively large high frequency content of this particular wavelet, spatial undersampling was found to be worse than in actual field measurements.

The geologic model was based upon composite well and seismic measurements in the Mississippi Canyon for prospects displaying oil and gas related amplitude anomalies. The model contained 22 elastic layers each having *P*-wave and *S*-wave attenuation. The specific subsurface model was formulated and generated independently by L. J. Srnka. The input model is given in Table 1. The inversions were performed by J. J. Carazzone in a blind test without knowledge of the input model in order to provide unbiased results.

The computed data with amplitude adjustments to compensate for spherical divergence, display a strong amplitude anomaly between 2.6 and 2.7 (Figure 1). The resulting plane-wave decomposition, Figure 2, was subsequently found to match the prediction from a plane-wave ($\xi - \omega$) elastic wave equation model to within a 10 percent error over the precritical range. The plane-wave decomposition displays a number of artifacts such as linear noises related to missing data at near and far offsets. Its general quality is slightly lower than the quality of actual field data. This is due to the large degree of spatial undersampling relative to the frequency content of this data.

Source estimation was carried out assuming an isotropic source. An approximately correct normal-incidence reflection coefficient at the water bottom (+0.23) was supplied to the inversion process to fix

Table 1. Model for Inversion Test. Thickness in feet, velocity in feet per second, and density in specific density units.

Layer	Thickness	Density	V_s	V_p	Q_μ	Q_σ
1	800	1.00	10	5000	1000	10000
2	300	1.50	2120	5300	50	150
3	500	1.77	2350	5700	50	150
4	880	1.95	2650	6200	50	150
5	1500	2.05	3100	6800	50	150
6	800	2.15	3750	7450	50	150
7	1000	2.25	4350	8200	50	150
8	700	2.18	4100	8000	50	150
9	800	2.21	4600	8500	50	150
10	680	2.25	4860	9000	50	150
11	90	2.38	4860	8500	50	150
12	1200	2.25	4860	9200	50	150
13	250	2.29	5280	9900	50	150
14	70	2.05	5820	9300	50	150
15	190	2.28	5100	9700	50	150
16	120	2.25	5265	9850	50	150
17	60	2.27	5025	9770	50	150
18	40	2.29	5525	9950	50	150
19	80	2.31	5000	10750	50	150
20	210	2.37	5400	11500	50	150
21	1430	2.42	6050	12300	50	150
22	∞	2.48	6700	13500	50	150

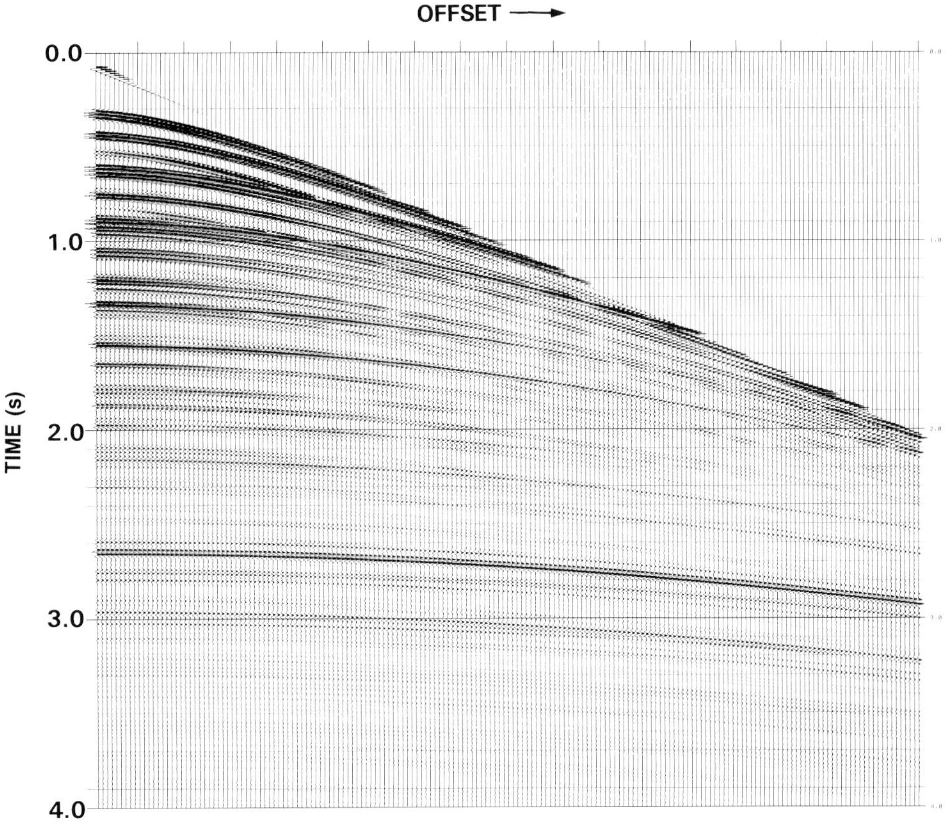

Fig. 1. Model data test offset gather.

scales. The least-squares source fitting process was subsequently found to reproduce the known model wavelet to within less than a 1 percent error.

Interval velocity estimation used traveltime picks taken from a conventional stack. Stacking velocities were supplied by the modeling program and (unfortunately) not independently interpreted from the data. Since the interval velocity analysis step used the supplied stacking velocities merely to window the 13 picked primary events, the inverted results do not rely upon perfectly accurate stacking velocities. The 13 primary reflections selected are listed in Table 2 together with their estimated and actual traveltimes. Picking errors were small, absolute error averaging a mere 0.11 percent.

INVERSION RESULTS FOR MODEL DATA

The accuracy of the interval velocity and depth inversion derived from the traveltime-ellipse fitting process is summarized in Tables 3, 4, and 5. Time-to-depth conversion errors were extremely small. The largest depth error of 1.11 percent (layer 3) was generated by an event picking error. The average absolute depth error was 0.42 percent. The average absolute thickness error was 1.48 percent. Interval velocity errors were less than 3.75 percent and absolute error averaged 1.35 percent.

Reflectivity inversion based upon the elastic layer stripping process correctly reproduced many of the

Table 2. Error in traveltime picks. Used for traveltime inversion. Time in seconds.

Traveltime Pick #	Estimated Time	Actual Time	% Error
1	0.3200	0.3200	0.00
2	0.4320	0.4332	−0.28
3	0.6040	0.6086	−0.76
4	0.8920	0.8925	−0.06
5	1.3340	1.3337	0.02
6	1.5480	1.5485	−0.03
7	1.7920	1.7924	−0.02
8	1.9660	1.9674	−0.07
9	2.1560	2.1556	0.02
10	2.3280	2.3279	0.00
11	2.5880	2.5888	−0.03
12	2.7180	2.7179	0.01
13	3.0200	3.0221	−0.07

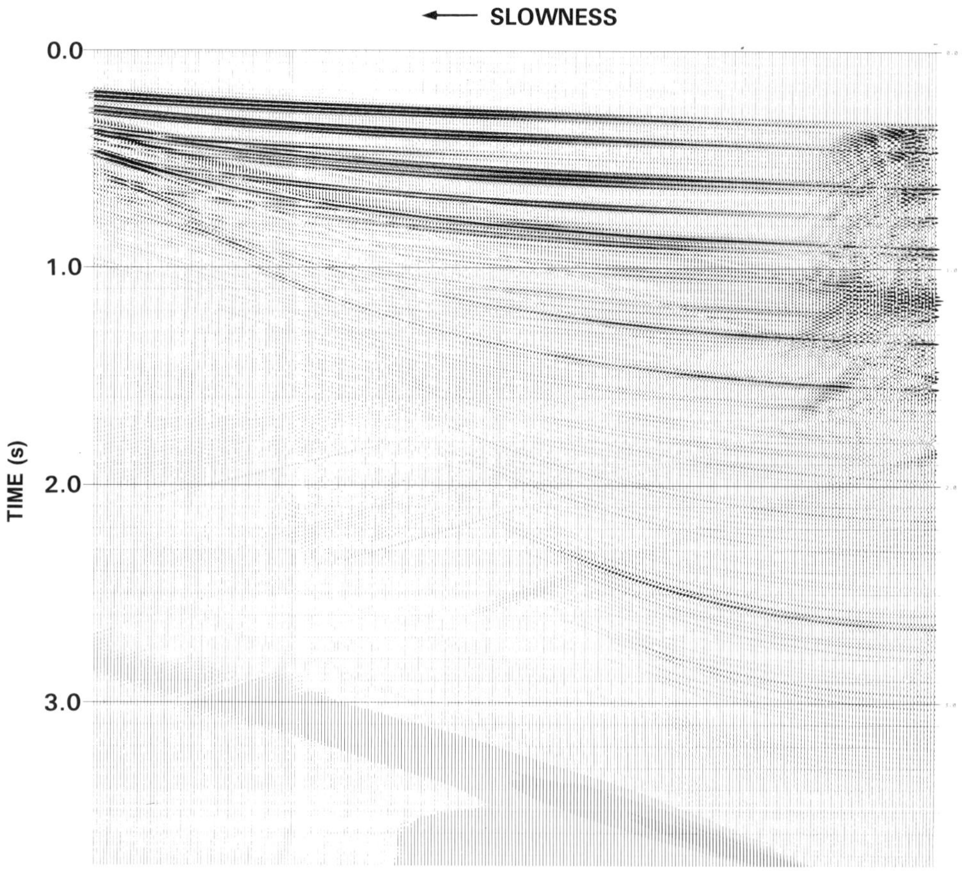

Fig. 2. Model data test P-tau gather.

essential aspects of the actual model on distance scales probed by the wavelet. A five-point time window (10 ms) was used in the inversion process. Our results are shown in Figure 3 to Figure 8. where we have displayed comparisons between the inverted parameters and the input model. We provide plots of the basic velocity and density parameters. Figures 3, 4, and 5, as well as plots of the parameter combinations ρV_p, ρV_s and V_p/V_s, Figures 6, 7, and 8.

A qualitative comparison of the inverted results near

Table 3. Macro-interval thickness errors from traveltime inversion. Thickness in feet.

Pick #	Estimated Thickness	Model Thickness	% Error
1	800.0	800.0	0.00
2	293.8	300.0	−2.07
3	488.4	500.0	−2.32
4	889.5	880.0	1.08
5	1504.0	1500.0	0.27
6	791.6	800.0	−1.05
7	1021.1	1000.0	2.11
8	725.9	700.0	3.70
9	787.2	800.0	−1.60
10	788.8	770.0	2.44
11	1200.4	1200.0	0.03
12	632.8	630.0	0.44
13	1859.5	1820.0	2.17

Table 4. Depth errors from traveltime inversion. Depth in feet.

Pick #	Estimated Depth	Model Depth	% Error
1	800	800	0.00
2	1094	1100	−0.56
3	1582	1600	−1.11
4	2472	2480	−0.33
5	3976	3980	−0.11
6	4767	4780	−0.27
7	5788	5780	0.15
8	6514	6480	0.53
9	7302	7280	0.30
10	8090	8050	0.50
11	9291	9250	0.44
12	9924	9880	0.44
13	11783	11700	0.71

the simulated gas sand at 2.6 s shows that all three elastic parameters correctly display the model behavior of a high-impedance rigid seal overlying a thin low-impedance reservoir layer. A quantitative comparison of the basic velocity and density parameters from the surface down to 3.0 s gives the following reflectivity inversion results:

Parameter	% rms Error	% Average Error
V_p	5.7	−3.0
ρ	4.6	+1.5
V_s	13.6	−9.6

Table 5. Interval P-wave velocity errors from traveltime inversion. Velocities in feet per second.

Pick #	Estimated V_p	Model (rms) V_p	% Error
1	5000	5000	0.00
2	5200	5300	−1.89
3	5712	5700	0.21
4	6156	6200	−0.71
5	6821	6800	0.31
6	7398	7450	−0.70
7	8370	8200	2.07
8	8297	8000	3.71
9	8329	8500	−2.01
10	9173	8940	2.61
11	9199	9200	−0.01
12	9809	9762	0.48
13	12318	11984	2.79

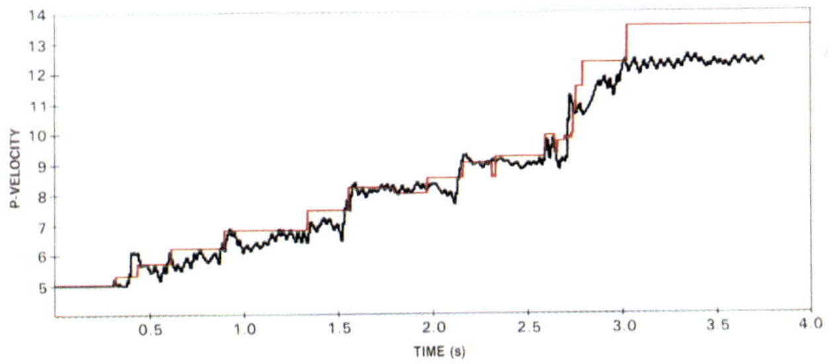

Fig. 3. Model inverted *P*-wave velocity with comparison.

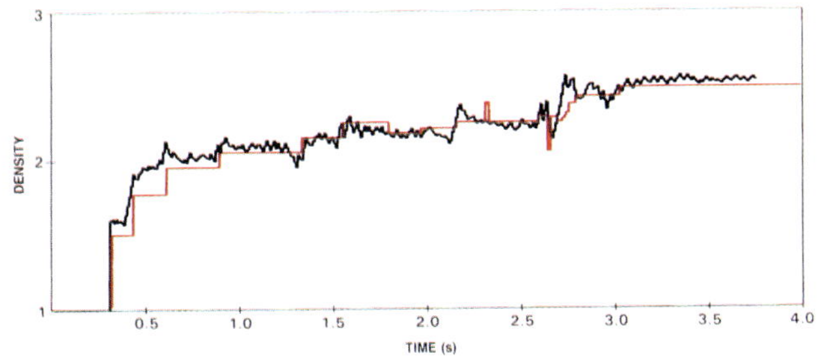

Fig. 4. Model inverted density with comparison.

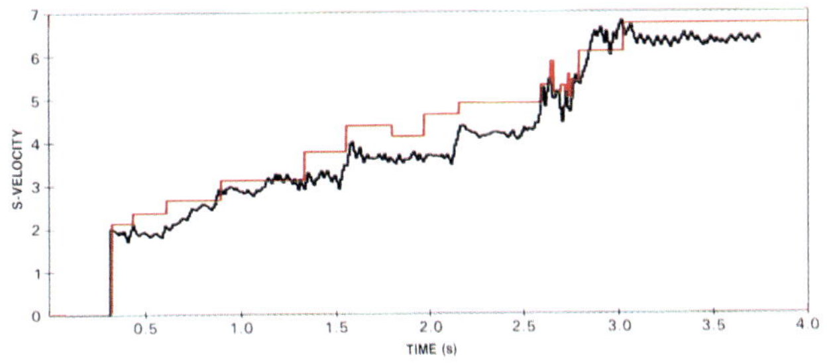

Fig. 5. Model inverted *S*-wave velocity with comparison.

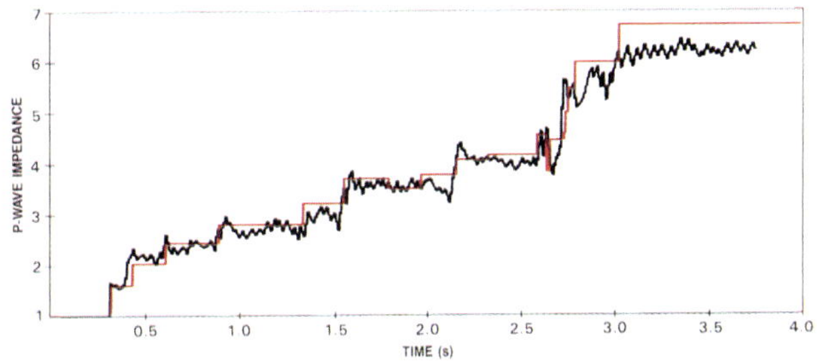

Fig. 6. Model inverted ρV_p with comparison.

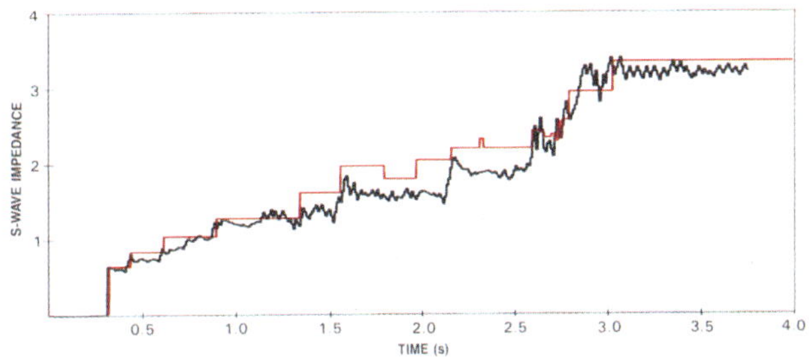

Fig. 7. Model inverted ρV_s with comparison.

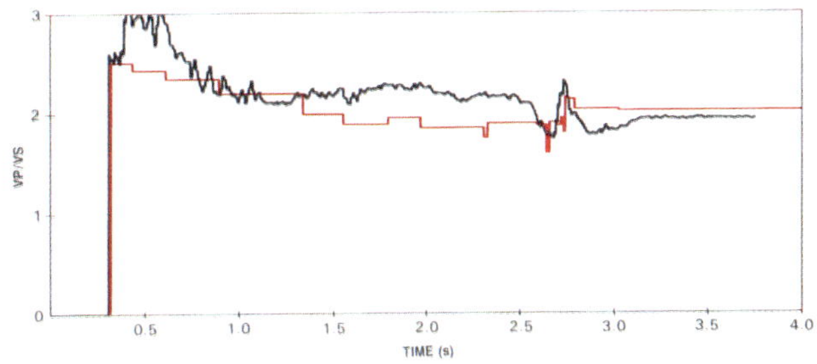

Fig. 8. Model inverted V_p/V_s with comparison.

A negative average error implies a systematically low inverted result. Reflectivity inversion has provided wavelet-scale resolution and material property information while decreasing the accuracy of the low frequency (traveltime related) *P*-wave velocity information. Inspection of the results indicates that nearly all discrepancies between the inverted parameters and the input model are due to bandwidth limitations (with one exception located near 2.3 s). The background estimate for the shear velocity, for example, was systematically low throughout the middle of the inverted model, accounting for most of its increased error relative to the other two parameters. The high quality of the inverted density results seen in the above error estimates is consistent with numerous inversion tests we have made using this process. This is partially due to the success of our empirical relationship between density and low-frequency interval *P*-wave velocity.

An exception, previously noted, to the general explanation of inversion errors as due to bandwidth limitations is the failure of this inversion experiment to detect a narrow layer near 2.3 s which has exactly cancelling changes in density and *P*-wave velocity, giving rise to a reflection event from a zero acoustic impedance contrast at normal incidence. We believe this error is caused by a constraint used in the linear least-squares process which limited the fractional elastic parameter changes based upon an estimate of absolute low-angle reflectivity. No effort has been made in the results reported here to adjust the inversion process using knowledge of the model.

FIELD DATA TEST

Our method was tested on a marine survey from the Gulf of Mexico. A portion of a near-trace CMP stack of this data, shown in Figure 9, is an area of the Gulf which has a strong gas-sand DHI near 2.3 s. The seismic data consisted of 511 shot gathers (shot interval 22.6 m) collected with a 301-channel hydrophone streamer having a total length of 4500 m (14765 ft). Each hydrophone group contained 17 equally weighted phones. No low-cut filters were used; a high-cut filter was applied at 110 Hz. The data sampling rate was 0.002 s; total record length was 5 s. Data

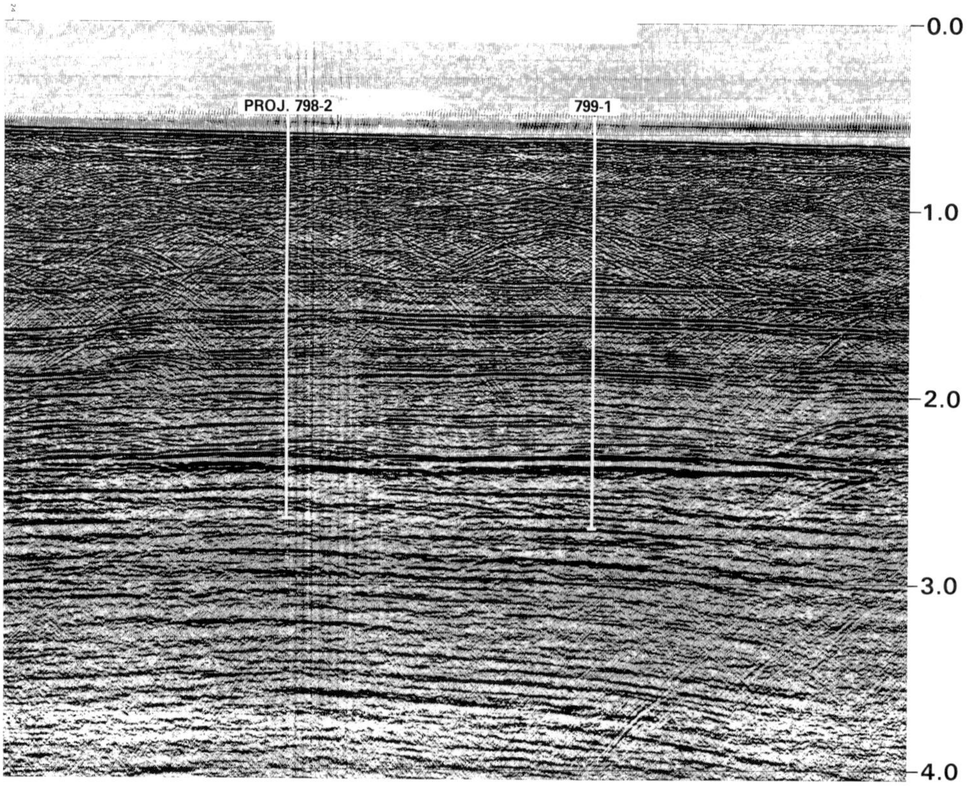

Fig. 9. Marine data common midpoint stack.

processing was done in true-amplitude format with no gain adjustments. The shot data contained mild low-frequency noises. These problems were removed by shot-order deconvolution.

The cylindrical Radon transform was applied to merged midpoint sorts of the deconvolved data after interpolation to one-half the original hydrophone group interval. Three adjacent midpoint gathers were merged. Antenna compensation was used to obtain a true plane-wave decomposition. A typical plane-wave gather is shown in Figure 10. The source estimation process was carried out using a line average of the water-bottom event. The source directivity was found to be fairly complex for these data, requiring 24 terms in the Legendre series to obtain a final-fit error of less than 5 percent. The reflectivity spectrum was assumed to range from 8 to 85 Hz. A five time-sample-point window was used in the parameter extraction step.

FIELD DATA INVERSION RESULTS

Figures 11, 12, and 13 show the inverted elastic-parameter curves for the velocities and density with comparison to nearby well measurements of P-velocity and density. Figures 14, 15, and 16 show the inverted ρV_p, ρV_s and V_p/V_s curves with comparison to the well-log values of ρV_p. No attempt has been made to correct or filter the well data except for the application of small check-shot corrections. The inverted results have been shifted downward by less than 0.004 s to tie the DHI located near 2.3 s (this corresponds to a depth error of less than 1 percent).

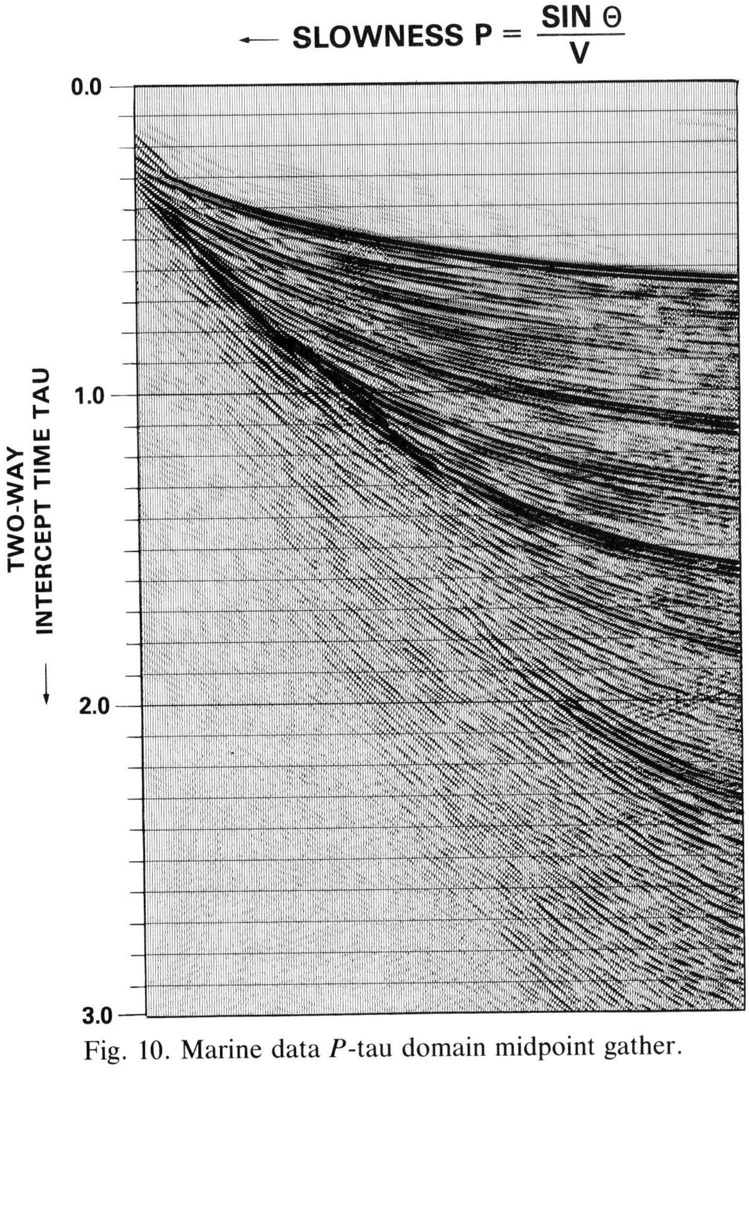

Fig. 10. Marine data P-tau domain midpoint gather.

Although not perfect, many important features of the well measurements have been properly predicted by the inversion process. The strong decreases in P-velocity and density within the gas sand are correctly represented. The most serious error from an interpreter's point of view is the confusion between a density increase and a P-velocity increase just above the gas sand. The inversion process becomes progressively more inaccurate beneath the top of the gas sand, perhaps due to low signal penetration. Much of the error in inverted density beneath the gas sand could be eliminated by adjusting the basement value of density in the background model.

CONCLUSIONS

We have illustrated the success of a wave equation based method for recovering the elastic parameters of layered models and (nearly) layered geologies. Our results suggest that recovery of P-wave velocity and density information contained within the passband of the seismic wavelet is possible in some

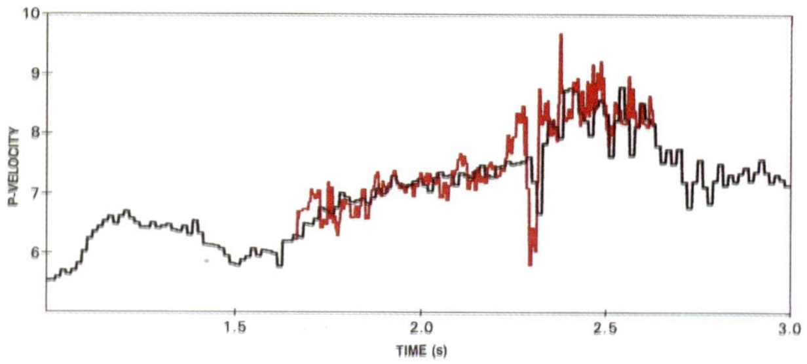

Fig. 11. Marine data P-wave velocity inversion with sonic log comparison.

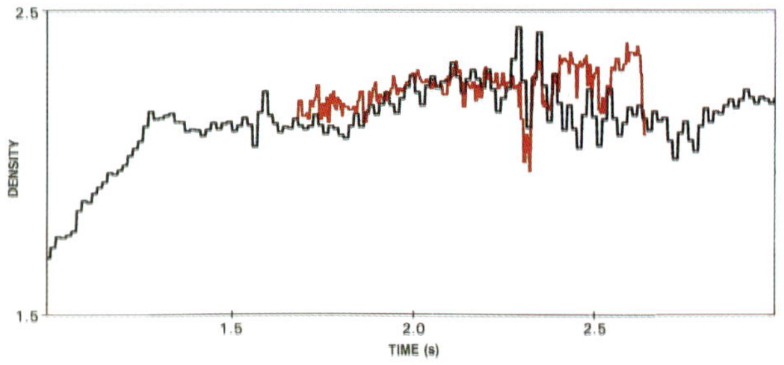

Fig. 12. Marine data density inversion with density log comparison.

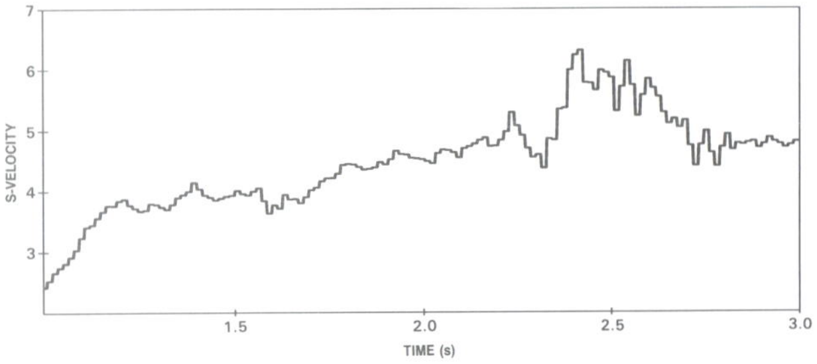

Fig. 13. Marine data S-wave velocity inversion.

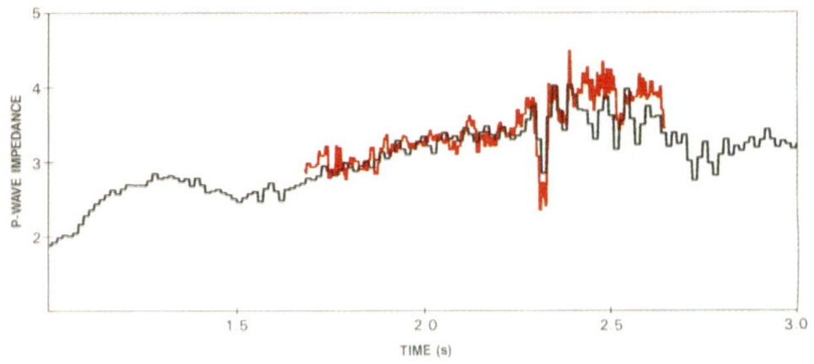

Fig. 14. Marine data inverted ρV_p with well comparison.

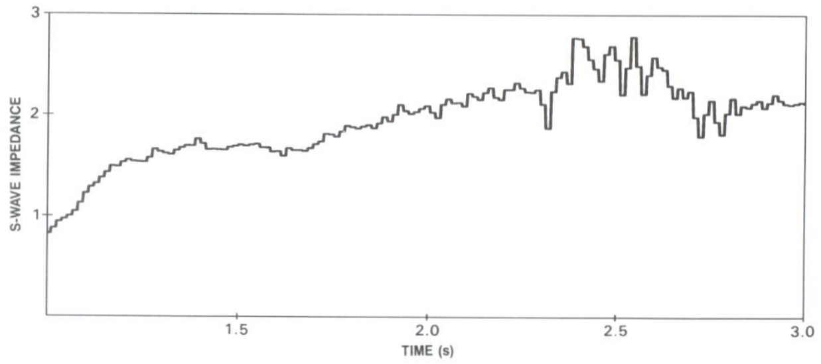

Fig. 15. Marine data inverted ρV_s.

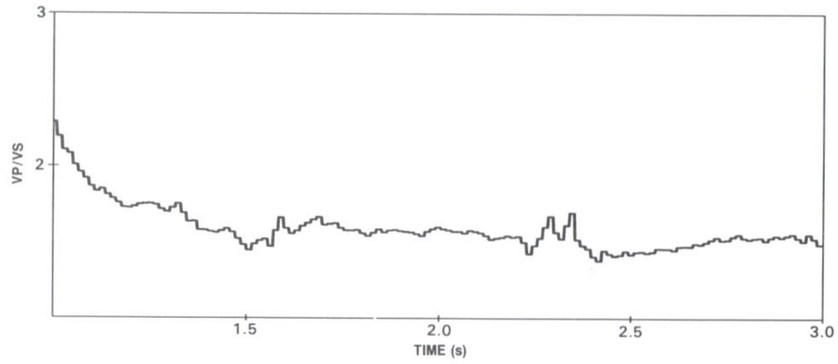

Fig. 16. Marine data inverted V_p/V_s.

situations. Accuracy of inverted S-wave velocity in real data situations remains to be confirmed with well based measurements. Our model study suggests that most inversion errors can be attributed to frequency bandwidth limitations. This study shows that (reflectivity) inverted P-wave velocity and density can be accurate to about 5 percent; inverted S-wave velocity can be accurate to about 15 percent; while depth conversion is accurate to about 1 percent. Interval velocity information obtained from purely traveltime inversion only is accurate to about 2 percent.

REFERENCES

Bube, K. P., and Burridge, R., 1983, The one-dimensional inverse problem of reflection seismology: SIAM Rev., **25**, 497–559.

Carazzone, J. J., 1986, Inversion of P-SV seismic data: Geophysics, **51**, 1056–1068.

Carazzone, J. J., and Srnka, L. J., 1987, Riccati impedance inversion of marine seismic data: Expanded Abstracts, 57th Ann. Internat. Mtg., Soc. Expl. Geophysicists, 824–828.

Hanson, R. J., and Haskell, K. H., 1978, Sandia Laboratories Tech. Repts.: Sand77-0552.

Hanson, R. J., and Haskell, K. H., 1979, Sandia Laboratories Tech. Repts.: Sand78-1290.

Rendleman, C. A., and Levin, F. K., 1990, Seismic exploration on a floating ice sheet: Geophysics, **55**, 402–409.

Sacks, P., and Symes, W., 1987, Recovery of the elastic parameters of a layered half-space: Geophys. J. Roy. Astr. Soc., **88**, 593–620.

Singh, S. C., West, G. F., Bregman, N. D., and Chapman, C. H., 1989, Full waveform inversion of reflection data: J. Geophys. Res., **94**, 1777–1794.

Spratt, S., 1987, Effect of normal moveout errors on amplitude versus offset-derived shear reflectivity: 57th Ann. Internat. Mtg., Soc. Expl. Geophys., Expanded Abstracts, 634–637.

Chapter VI

Multicomponent AVO

Linearized AVO Inversion of Multicomponent Seismic Data

G. J. M. Lörtzer and A. J. Berkhout‡*

INTRODUCTION

Exploration and exploitation of hydrocarbon reserves is becoming increasingly complex. To be successful, an increasing amount of information is required. In recent years much attention has been devoted to the information present in reflection amplitude variation with offset (AVO) (Balogh et al., 1986; Gelfand et al., 1986; Spratt, 1987; Chiburis, 1987; Ball, 1987; Lörtzer and Berkhout, 1988, 1989; Haas and Berkhout, 1988, 1989; Rijssen and Herman, 1988; SEG workshop, 1988). In principle, the offset variations of the *P-P* reflection amplitudes contain information on all the elastic parameter contrasts (*P*-wave velocity, *S*-wave velocity, and density), while the zero-offset reflection amplitude contains only information on the acoustic impedance.

All AVO inversion methods that derive estimates for the elastic parameters of the subsurface, can be subdivided according to two criteria (Figure 1):

(1) The type of forward model used, and
(2) The incorporation of wavelet interference.

With the first criterion, the true, nonlinear Zoeppritz equation can be used to relate the elastic properties (contrasts) to the variations in the plane wave *P-P* reflection amplitude with offset (or angle of incidence). On the other hand, a linear approximation of that equation may be used. The second criterion concerns the wavelet interference effects such as "tuning." These effects can be directly incorporated in the inversion, left in the results to be removed by subsequent processing, or neglected altogether (i.e., working with "picked" amplitudes).

In this study we restrict ourselves to methods that use a linear approximation of the Zoeppritz equation as forward model for the *P-P* AVO, and leave the wavelet interference in the inversion results. The proposed methods are put in a general theoretical framework in the discussion of existing methods. Next, the general framework is extended to incorporate statistical noise and multi-component measurements. One advantage of the statistical approach is the possibility to calculate accuracy measures for the results, which is illustrated in the discussion of result analysis. Subsequently, the problems associated with the practical implementation are addressed and the whole approach is illustrated with an example using simulated data. Finally, some remarks and a critical review of the results obtained are given in the conclusions.

EXISTING METHODS

In all existing methods for linearized AVO inversion (Smith and Gidlow, 1986, 1987; Gelfand et al., 1986; Balogh et al., 1986; Spratt, 1987) the nonlinear Zoeppritz equation for the elastic angle dependent plane-wave *P-P* reflection coefficient $d(\phi)$ is linearized. Following Aki and Richards (1980 p. 153) $d(\phi)$ is written as a weighted sum of the *relative contrasts* in the elastic parameters *P*-wave velocity V, *S*-wave velocity W, and density D

$$d(\phi) = g_1(\phi)\Delta V/V + g_2(\phi)\Delta W/W + g_3(\phi)\Delta D/D, \tag{1}$$

where ϕ equals the angle of the incident *P*-wave, ΔV equals contrast in the *P*-wave velocity over an interface, and V equals the average of the *P*-wave velocity over the interface. Similar definitions apply for ΔW, W, ΔD, and D. In the original formulation of Aki and Richards (1980) the functions $g_1(\phi)$, $g_2(\phi)$, and $g_3(\phi)$ read:

$$g_1(\phi) = \frac{1}{2} + \frac{1}{2}\tan^2(\phi), \tag{2}$$

$$g_2(\phi) = -4\frac{W^2}{V^2}\sin^2(\phi), \text{ and} \tag{3}$$

*Shell International Petroleum, MAT. B. V., P.O. Box 162, The Hague, NL 2501 AN, Netherlands.
‡Delft University, Physics Dept., P.O. Box 5046, Delft, GA 2600, Netherlands.

$$g_3(\phi) = \frac{1}{2} - 2\frac{W^2}{V^2}\sin^2(\phi). \quad (4)$$

The method described in Balogh et al. (1986) has the source receiver offset h as an independent variable, instead of the incident angle ϕ. This variation has been accomplished by substituting an approximate ϕ-h relation for plane horizontal layers in equation (1). Then the macro P-wave velocity model of the overlying layers enters the expression explicitly. However, it should be realized that any method implicitly needs a macro P-wave velocity model in a prior step to convert offset to angle of incidence. Note that the functions $g_2(\phi)$ and $g_3(\phi)$ contain a factor W/V. So, it is necessary to make some assumption about this factor. Similar to the macro P-wave velocity model, a macro W/V model is used since the detailed model is not available in practical situations.

Suppose that the reflections are known for a limited number of angles (N). We can then define the data vector **d** that contains these reflection coefficients

$$\mathbf{d} = [d(\phi_1), d(\phi_2), \ldots, d(\phi_N)]^T. \quad (5)$$

Defining the forward matrix $\mathbf{G}$ as $G_{ij} = g_i(d_j)$ and the parameter vector $\mathbf{x} = (\Delta V/V, \Delta W/W, \Delta D/D)^T$, we have

$$\mathbf{d} = \mathbf{G}\mathbf{x}. \quad (6)$$

The least-squares inverse of $\mathbf{G}$, yielding the estimates for the parameter **x**, can then be found analytically [which is one of the advantages of the linearization (e.g., Menke, 1984, p. 41 and Tarantola, 1987, p. 58)]:

$$\hat{\mathbf{x}} = (\mathbf{G}^T\mathbf{G})^{-1}\mathbf{G}^T\mathbf{d}, \quad (7)$$

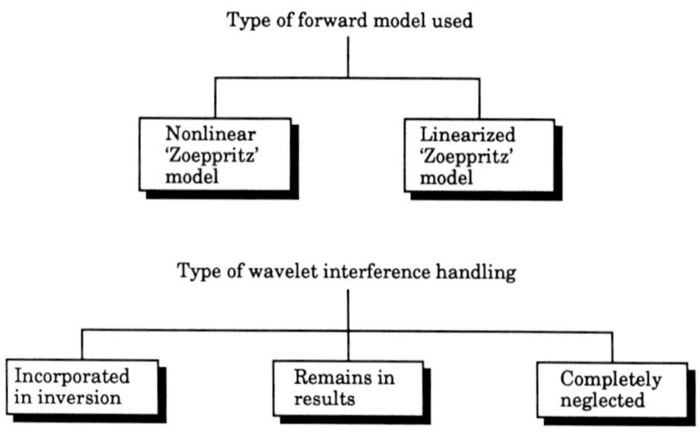

Fig. 1. Subdivision criteria for AVO inversion methods.

which minimizes the squared data mismatch

$$F(\hat{\mathbf{x}}) = \min_{\mathbf{x}} \{[\mathbf{d} - \mathbf{G}\mathbf{x}]^T[\mathbf{d} - \mathbf{G}\mathbf{x}]\}. \quad (8)$$

Equation (7) may be written as

$$\hat{\mathbf{x}} = \mathbf{H}\mathbf{d}. \quad (9)$$

Equation (9) shows that the estimates for the parameters (i.e., the estimated relative contrasts) are found as the dot product of the corresponding row of least-squares inverse matrix $\mathbf{H}$ and the data vector **d**. This is nothing more than the weighted adding or *weighted stacking* of the amplitudes in **d** with the elements of the row of $\mathbf{H}$ as weights. The term "weighted stacking" was first introduced in Smith and Gidlow (1986, 1987).

Since the estimates are found as a linear combination of the data, we can take any linear combination of the original parameters (i.e., $\Delta V/V$, $\Delta W/W$, and $\Delta D/D$) as replacement. This replacement represents the same information, but in a way that may allow easier interpretation. Smith and Gidlow (1986, 1987) suggest the following linear combinations to yield the relative contrasts in

- Acoustic impedance $\Delta Z/Z = \Delta V/V + \Delta D/D$, (10a)
- Shear impedance $\Delta Y/Y = \Delta W/W + \Delta D/D$, (10b)
- Poisson ratio (pseudo) $\Delta Q/Q = \Delta V/V - \Delta W/W$, (10c)
- Fluid factor $\Delta F = \Delta V/V - k[W/V]\Delta W/W$, (10d)

where k is a constant.

The relative (pseudo) Poisson ratio contrast ($\Delta Q/Q$) is claimed to be a lithology indicator. The last, the fluid factor (ΔF), is claimed to be a hydrocarbon indicator. It shows deviations from an (empirical) relation between P-wave and S-wave velocities called the mudrock-line, see Castagna et al. (1985 p. 572), valid for *water bearing* silici-clastic deposits

$$V = V_0 + kW. \quad (11)$$

Deviations from this relation are associated with gas.

Unfortunately, the matrix $\mathbf{G}^T\mathbf{G}$ in equation (7) will be close to singular, meaning that the calculation of its inverse is very unstable and sensitive to noise or errors in the elements of $\mathbf{G}^T\mathbf{G}$. The degree of singularity or *conditioning* of the matrix $\mathbf{G}^T\mathbf{G}$ can be shown with the singular value decomposition of the forward matrix $\mathbf{G}$ (e.g., Menke, 1984, p. 119). Combinations of parameters that correspond to small singular values are called weak, or ill-resolved, and cannot be estimated accurately (i.e., they are too sensitive to the noise on the data). It appears that even for a considerable range of angles of incidence there is at least one (very) weak parameter combination of the form $\mathbf{x} = (1, \alpha, -1)^T$, with α being a model dependent constant. If we display this in 3-D parameter space, with the three parameters set up in a right handed coordinate system (Figure 2), the weak parameter combination yields a line L of almost equally possible solutions in parameter space. Models with parameter combinations varying along this weak direction L ($\Delta V/V + \alpha\Delta W/W - \Delta D/D =$ const.) do not show significant differences in the simulated seismic data generated for those models. However, models with parameter combinations varying according to $\Delta V/V + \Delta D/D =$ const., in the plane perpendicular to this weak direction do show significant differences in the simulated seismic data, i.e., they are well resolved. This implies that the relative acoustic impedance contrast as defined by equation (10*a*) is well resolved and can be estimated correctly, corresponding with our expectations.

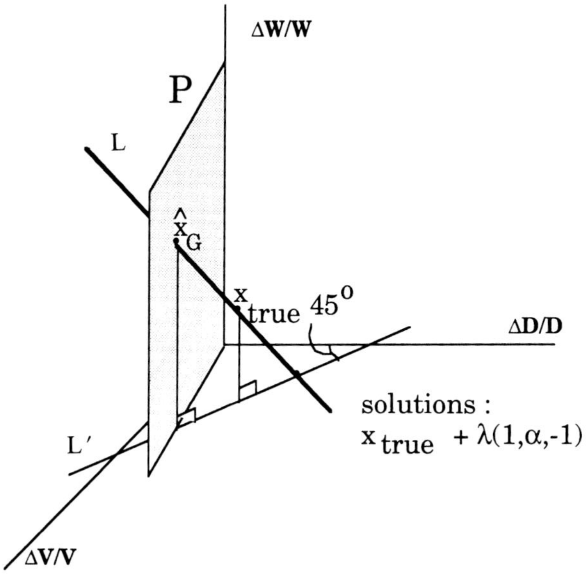

Fig. 2. Graphic interpretation of the ill-resolved direction (line L) and the solutions satisfying the Gardner relation (plane *P*). The solutions of the unstabilized inversion problem will all lie on line L, which is defined by the true model vector $\mathbf{x}_{\text{true}}$ and the weak direction $(1, \alpha, -1)^T$ with α being a model-dependent constant. The position of the solution on L depends on the noise on the seismic data. The line L′ is the projection of L on the $\Delta V/V$-$\Delta D/D$ plane, making a 45° angle with the $\Delta V/V$ and the $\Delta D/D$ axes. The parameter combinations satisfying the Gardner relation lie in the plane *P*.

The methods listed differ in the way they cope with this instability problem. Gelfand et al. (1986) propose to take $W/V = 1/2$ (i.e., a Poisson model), while Smith and Gidlow (1987, p. 998) use an empirical relation between the *P*-wave velocity and density, which reads for the relative contrasts

$$\Delta D/D = \frac{1}{4}\,\Delta V/V. \tag{12}$$

The so-called Gardner relation (Gardner et al., 1974) in terms of the relative contrasts is substituted into the original equation (1). This substitution has an adequate stabilization effect following directly from Figure 2: The parameter combinations that satisfy equation (12) lie in a plane P, which has a considerable angle with the line L. This type of stabilization is to be preferred over a purely mathematical one, like diagonal stabilization, since it is geologically more meaningful.

STATISTICAL APPROACH

In all existing methods for linearized AVO inversion the noise on the data and the scatter around the stabilizing empirical relations are neglected. Thus, the first improvement to the existing methods is to incorporate this kind of uncertainty. We use the Bayesian inversion approach (Jackson and Matsu'ura, 1985; Tarantola, 1987; Duijndam, 1988 a, b), which can elegantly handle deterministic forward models and noise on the measurements (*seismic* information), together with a sensible incorporation of any available additional (*prior*) information with a corresponding degree of confidence (scatter). The empirical relations are considered as additional information. Both types of information (seismic and additional) are described in terms of *probability density functions* (pdf's), where the statistical properties of the seismic observations (i.e., the offset/angle dependent reflection amplitudes) and the scatter around the empirical relation (generally from regression analysis of cross-plots) is modeled as an additional "noise" term $\mathbf{n}$, meaning that equation (6) changes to

$$\mathbf{d}_p = \underset{\sim}{\mathbf{G}}_p \mathbf{x} + \mathbf{n}_p, \tag{13}$$

where the subscript p refers to the primary *P-P* reflectivity information. Symmetrically with equation (13), we write for all empirical relations

$$\mathbf{d}_e = \underset{\sim}{\mathbf{G}}_e \mathbf{x} + \mathbf{n}_e, \tag{14}$$

where the subscript e refers to the empirical information. Note that the noise term $\mathbf{n}_e$ describes the scatter around the empirical relations. In order to keep the favorable properties of the linearized inversion, only linearized empirical relations between the parameters are considered. Hence, $\underset{\sim}{G}_e$ is typically determined by linear regression analysis of crossplots. Examples of such empirical relations are the Gardner relation, equation (12), between the relative P-wave velocity and density contrasts and the mudrock line relation, equation (11), between relative P-wave velocity and S-wave velocity contrasts

$$\Delta V/V = k[W/V]\Delta W/W. \tag{15}$$

Note that interfaces satisfying equation (15) will show a fluid factor ΔF equal to zero.

In the remainder it will be assumed that the noise on the seismic amplitudes and the scatter around the empirical relations have Gaussian distributions with zero mean and covariance matrices $\underset{\sim}{C}_p$ and $\underset{\sim}{C}_e$ respectively. Note that the elements on the main diagonal of $\underset{\sim}{C}_p$ (the so called variances) refer to the absolute noise content in the data. The noise is generally computed from an estimate of the signal-to-noise ratio and the total energy in the data. The off-diagonal elements refer to the correlation between the different noise samples (correlated noise). Note also that $\underset{\sim}{C}_e$ follows directly from the aforementioned regression analysis (scatter). Under the Gaussian assumption it can be shown that again a (statistically weighted) least-squares solution is obtained (Tarantola, 1987 p. 188; Duijndam, 1988a p. 883)

$$\hat{\mathbf{x}} = (\underset{\sim}{G}_t^{*T}\underset{\sim}{G}_t^{*})^{-1}\underset{\sim}{G}_t^{*T}\mathbf{d}_t^{*}, \tag{16}$$

where $\underset{\sim}{G}_t^* = \underset{\sim}{C}_t^{-1/2}\underset{\sim}{G}_t$ is the statistically weighted forward matrix and $\mathbf{d}_t^* = \underset{\sim}{C}_t^{-1/2}\mathbf{d}_t$ is the statistically weighted data vector. In equation (16) the subscript t refers to the *total* amount of information and the forward matrix $\underset{\sim}{G}_t$, the covariance matrix $\underset{\sim}{C}_t$, and the data vector $\mathbf{d}_t$ are partitioned as follows:

$$\underset{\sim}{G}_t = \begin{bmatrix} \underset{\sim}{G}_p \\ \underset{\sim}{G}_e \end{bmatrix}, \quad \underset{\sim}{C}_t = \begin{bmatrix} \underset{\sim}{C}_p & \underset{\sim}{O} \\ \underset{\sim}{O} & \underset{\sim}{C}_e \end{bmatrix}, \quad \mathbf{d}_t = \begin{bmatrix} \mathbf{d}_p \\ \mathbf{d}_e \end{bmatrix}, \tag{17}$$

where $\underset{\sim}{O}$ is a null matrix with the proper dimensions. Using equation (17) in equation (16) yields

$$\hat{\mathbf{x}} = (\underset{\sim}{G}_p^{*T}\underset{\sim}{G}_p^{*} + \underset{\sim}{G}_e^{*T}\underset{\sim}{G}_e^{*})^{-1} \begin{bmatrix} \underset{\sim}{G}_p^{*} \\ \underset{\sim}{G}_e^{*} \end{bmatrix}^T \begin{bmatrix} \mathbf{d}_p^{*} \\ \mathbf{d}_e^{*} \end{bmatrix}, \tag{18}$$

which minimizes the squared total mismatch

$$F(\hat{\mathbf{x}}) = \min_{\mathbf{x}} [(\mathbf{d}_p^* - \underset{\sim}{G}_p^*\mathbf{x})^T(\mathbf{d}_p^* - \underset{\sim}{G}_p^*\mathbf{x}) + (\mathbf{d}_e^* - \underset{\sim}{G}_e^*\mathbf{x})^T(\mathbf{d}_e^* - \underset{\sim}{G}_e^*\mathbf{x})]. \tag{19}$$

For other distributions of the noise and the scatter, other types of solutions are found (Duijndam, 1988a, p. 884; Crase et al., 1990, p. 529), e.g., in case of Laplace (Double exponential) distributions, the optimum solution in the Bayesian sense is found as a least-absolute value (L_1-norm) solution.

The solution of the inverse problem is determined by the relative sensitivity of $\underset{\sim}{G}_p^*$ and $\underset{\sim}{G}_e^*$. Parameters that suffer from an ill-conditioned matrix $\underset{\sim}{G}_p^{*T}\underset{\sim}{G}_p^{*} = \underset{\sim}{G}_p^T\underset{\sim}{C}_p^{-1}\underset{\sim}{G}_p$ due to an insensitive forward model $\underset{\sim}{G}_p$ and/or a high noise level $\underset{\sim}{C}_p$, have their solution primarily determined by the empirical information. However, the solution is not entirely forced to satisfy the empirical relations as with hard substitution. The answer is allowed to deviate from them if the data contain strong enough information in favor of this deviation. This type of *soft* stabilization, where the answer is determined by the quality of the data (noise level), the theoretical information (forward model), and the additional information (empirical relations) is very elegant and more realistic.

Note that in practical situations additional information is always available and should always be used, even if only wide ranges can be specified. The inclusion assures the most accurate solution, i.e., the solution with the smallest standard deviation.

Later we will propose a measure to determine to what degree the solution depends on the additional information. In those situations where this measure indicates a large dependence, the involved seismic data has little to add to what we knew already. The use of any alternative linear inversion method does not change this fundamental fact.

MULTICOMPONENT EXTENSION

Until now only P-P reflection data were considered. However, the statistically weighted stacking formalism developed in the previous section allows the incorporation of multicomponent reflection data very easily. In principle the multicomponent data includes not only S-S reflection data but also P-S and S-P converted reflection data. Even if the inversion is to be applied to CMP gather data, the inclusion of converted waves suffers from a severe reflection point smear, so they are not considered in the remainder of this paper and we will concentrate on the effect of additional S-S reflection data. Rijssen and Herman (1988) and Haas

Table 1. Coefficients in linearized expressions for angle dependent *P-P* and *S-S* reflection amplitudes.

i	Rel. Contr.	$g_i(\varphi)$ $(P\text{-}P)$	$g_i(\psi)$ $(S\text{-}S)$
1	$\Delta V/V$	$1/2 + 1/2 \tan^2\varphi$	0
2	$\Delta W/W$	$-4 \frac{W^2}{V^2} \sin^2\varphi$	$-1/2 - 1/2 \tan^2\psi + 4 \sin^2\psi$
3	$\Delta D/D$	$1/2 - 2 \frac{W^2}{V^2} \sin^2\varphi$	$-1/2 + 2 \sin^2\psi$

and Berkhout (1989) already showed that for the case of elastic inversion using the nonlinear forward model, the combination of *P-P* and *S-S* reflection data contains the major part of the information on all the elastic parameters.

For the *S-S* reflection amplitude, that is the *SV-SV* reflection amplitude, as a function of the incident angle ψ, an expression similar to equation (1) can be derived (Aki and Richards, 1980 p. 154). Only the expressions for the functions $g_i(\psi)$ are different (see Table 1 for an overview).

Note the dissymmetry in the coefficients for $\Delta V/V$, which is equal to zero for the *S-S* reflection amplitudes, meaning that the linearized expressions for *S-S* reflection data contain absolutely no information on the relative *P*-wave velocity contrast.

For the solution of the inverse problem an expression similar to equation (15) is obtained. However, the *total* amount of information consists of three terms: *P-P* reflection data, *S-S* reflection data, and empirical information. Both the *P-P* and *S-S* reflection data are seismic information. Accordingly, the forward matrix $\underset{\sim}{\mathbf{G}}_t$, the covariance matrix $\underset{\sim}{\mathbf{C}}_t$, and the data vector $\mathbf{d}_t$ are partioned as follows:

$$\underset{\sim}{\mathbf{G}}_t = \begin{bmatrix} \underset{\sim}{\mathbf{G}}_p \\ \underset{\sim}{\mathbf{G}}_s \\ \underset{\sim}{\mathbf{G}}_e \end{bmatrix}, \quad \underset{\sim}{\mathbf{C}}_t = \begin{bmatrix} \underset{\sim}{\mathbf{C}}_p & \underset{\sim}{\mathbf{O}} & \underset{\sim}{\mathbf{O}} \\ \underset{\sim}{\mathbf{O}} & \underset{\sim}{\mathbf{C}}_s & \underset{\sim}{\mathbf{O}} \\ \underset{\sim}{\mathbf{O}} & \underset{\sim}{\mathbf{O}} & \underset{\sim}{\mathbf{C}}_e \end{bmatrix}, \quad \mathbf{d}_t = \begin{bmatrix} \mathbf{d}_p \\ \mathbf{d}_s \\ \mathbf{d}_e \end{bmatrix}. \tag{20}$$

The *P-P* reflection data, the *S-S* reflection data, and the empirical information are assumed to be uncorrelated. Of course $\underset{\sim}{\mathbf{C}}_p$, $\underset{\sim}{\mathbf{C}}_s$ and/or $\underset{\sim}{\mathbf{C}}_e$ may contain non-zero, off-diagonal terms, to represent also correlated noise (multiples, converted waves, etc.) and correlated scatter. The full expression for the solution reads similar to equation (19)

$$\hat{\mathbf{x}} = (\underset{\sim}{\mathbf{G}}_p^{*\,T}\underset{\sim}{\mathbf{G}}_p^* + \underset{\sim}{\mathbf{G}}_s^{*\,T}\underset{\sim}{\mathbf{G}}_s^* + \underset{\sim}{\mathbf{G}}_e^{*\,T}\underset{\sim}{\mathbf{G}}_e^*)^{-1} \begin{bmatrix} \underset{\sim}{\mathbf{G}}_p^* \\ \underset{\sim}{\mathbf{G}}_s^* \\ \underset{\sim}{\mathbf{G}}_e^* \end{bmatrix}^T \begin{bmatrix} \mathbf{d}_p^* \\ \mathbf{d}_s^* \\ \mathbf{d}_e^* \end{bmatrix}. \tag{21}$$

Of course the inversion can be done without the *P-P* reflection data using only the *S-S* reflection data. The resulting equations are identical to equations (17) and (19) with the subscript p replaced by s.

RESULT ANALYSIS

The seismic community is becoming less willing to accept only an estimate for the parameters in any inversion problem. In addition to the parameter estimates, the solution should contain at least the following two measures to quantify the quality of the inversion results (result analysis):

(1) Accuracy measure, and
(2) Resolution measure.

The accuracy measure shows what uncertainty can be expected for the parameter estimates while the resolution measure shows to what degree the solution is determined by the seismic information and to what degree by the empirical (a priori) information.

In Bayesian inversion the result analysis is an integral part of this approach. In general, the standard deviation in the solution is used for the accuracy measure of the parameter estimates. The standard deviation [σ] is defined as the square root of the diagonal elements of the covariance matrix, $\sigma_i = \sqrt{C_{ii}(\hat{x})}$. An important advantage of having a linear forward model is that this covariance matrix can be calculated in a closed form (Tarantola, 1987, p. 196; Duijndam, 1988b, p. 908)

$$\underset{\sim}{\mathbf{C}}_{\hat{x}} = (\underset{\sim}{\mathbf{G}}_a^{*\,T}\underset{\sim}{\mathbf{G}}_a^* + \underset{\sim}{\mathbf{G}}_e^{*\,T}\underset{\sim}{\mathbf{G}}_e^*)^{-1}, \tag{22}$$

where the subscript a denotes the seismic (AVO) information consisting of either *P-P* reflection data or *S-S* reflection data alone or in combination. Expression (22) which is independent of the actual seismic measurements, depends only on their statistical properties, and can therefore be determined prior to data acquisition to determine the optimum recording parameters (spread length, group interval, etc.)

For an objective resolution measure of the solution, we propose to use

$$K_i = \frac{\sigma_i^+}{\sigma_i^-} = \frac{\sqrt{(\mathbf{G}_a^{*\,T}\mathbf{G}_a^{*} + \mathbf{G}_e^{*\,T}\mathbf{G}_e^{*})_{ii}^{-1}}}{\sqrt{(\mathbf{G}_a^{*\,T}\mathbf{G}_a^{*})_{ii}^{-1}}}, \quad (23)$$

being the ratio of the standard deviation *with* empirical information σ_i^+ over the standard deviation *without* the empirical information σ_i^-. K_i is bounded to the interval [0, 1]. The two extremes correspond with "totally determined by the *empirical* input" and "totally determined by the seismic input" respectively, with the turning point at $K_i = 1/2$. Note that increasing both $\mathbf{C}_a$ and $\mathbf{C}_e$ with a factor b^2 does not alter the value of K_i, but increases the standard deviation of the parameter estimates with a factor b. In principle we would like an inversion solution with (very) low uncertainties (i.e., small standard deviation σ_i^+) and primarily determined by the seismic input (i.e., large resolution ratio K_i) to prevent potential bias from incorrect empirical information.

As said, even though the accuracy and resolution measures depend on $\mathbf{G}_a$, $\mathbf{C}_a$, $\mathbf{G}_e$, and $\mathbf{C}_e$ they are independent of the actual seismic input (seismic and empirical) and can be calculated before the actual acquisition. To determine the required minimum angle range, Figure 3 is shown as an example for P-wave velocity input. Based on a maximum requested uncertainty and a minimum requested resolution ratio, the minimum angle range can be determined. A different approach is shown in Figure 4, where uncertainty and

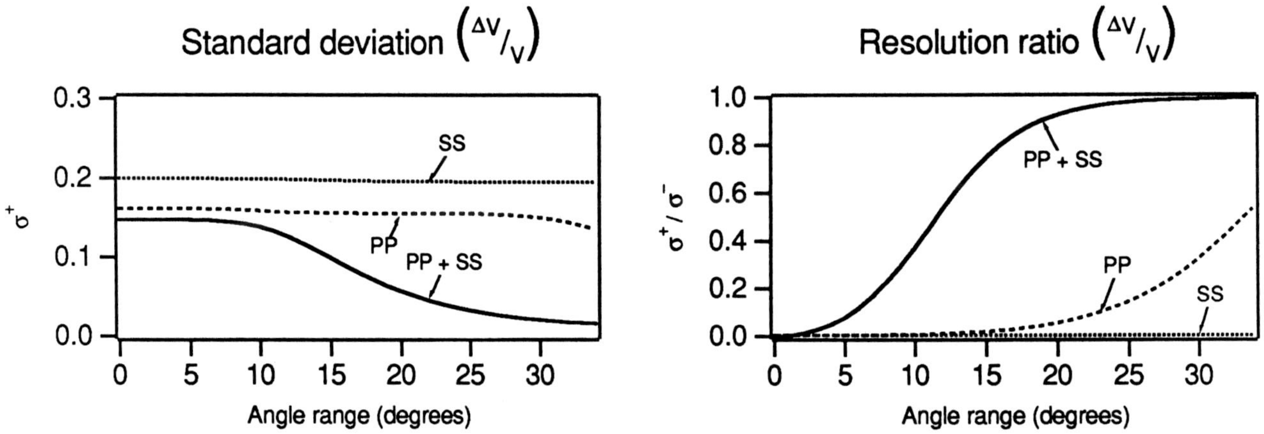

Fig. 3. Standard deviation and resolution ratio as a function of the range of angle of incidence from 0° to maximum angle in 1° increments ($\mathbf{C}_p = 0.01\mathbf{I}_p$, $\mathbf{C}_s = 0.02\mathbf{I}_s$ and $\mathbf{C}_e = 0.20\mathbf{I}_e$, where $\mathbf{I}$ is the unity matrix of the correct dimension).

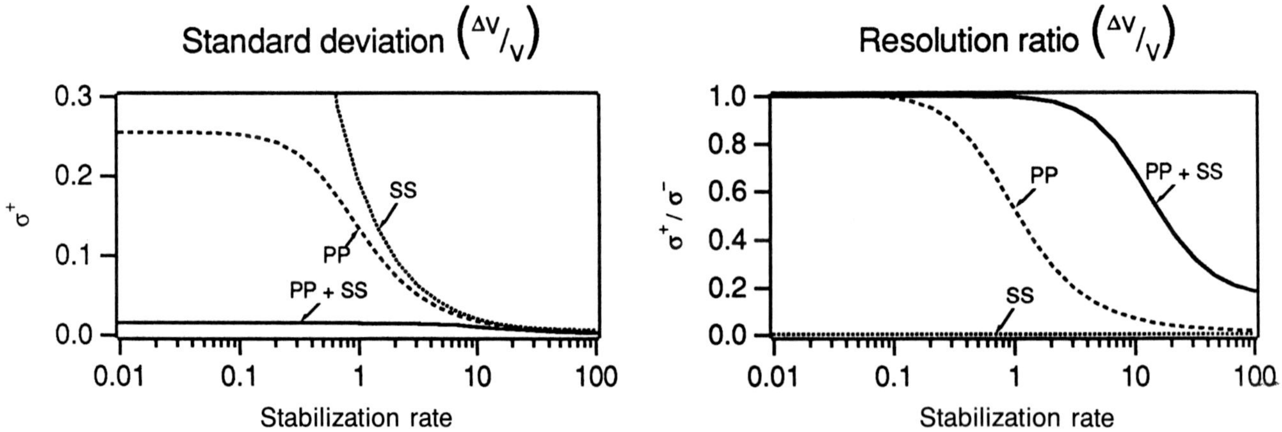

Fig. 4. Standard deviation and resolution ratio as a function of the stabilization rate. Same covariance matrices as in Figure 3 and angles of incidence range from 0° to 25° with 1° increment. The stabilization rate creates an additional stabilization: A rate of α results in a decrease of the standard deviations with a factor $1/\alpha$.

resolution ratio are displayed as a function of the stabilization rate, which acts as an additional scaling factor on $\underset{\sim}{C}_e$ (the covariance matrix of the empirical information): At a stabilization rate of 1, the original $\underset{\sim}{C}_e$ is used, while for a stabilization rate of 100, $1/100\underset{\sim}{C}_e$ is used. Thus a stabilization rate of 100 decreases the standard deviation (assumed scatter) by a factor of 100 and increases the importance (the weights in equation 18) of the empirical information. Figure 4 can be used to determine the maximum achievable resolution K_i for a certain maximum allowed uncertainty level σ. With respect to a certain range of angles of incidence, number of traces, noise level on the seismic data, and scatter around the empirical relation, the maximum allowed uncertainty level corresponds in Figure 4a with an additional stabilization rate for the empirical information. For this stabilization rate, Figure 4b gives the corresponding resolution. If the resolution is not satisfactory, it may be improved by (in decreasing order of importance):

(1) Increasing the offset (range); this may result in a conflict with the linearity assumption in the forward model;
(2) Decreasing the noise on the seismic input (by better processing); or
(3) Increasing the subsurface coverage within the same offset/angle range.

PRACTICAL ASPECTS

The weighted stacking for the relative contrasts in the elastic parameters is not to be done on the AVO of a single interface but on a whole seismic data set, e.g., in approximately 1-D media a CMP-gather and in structurally complex media a CDP-gather (Berkhout and Wapenaar, 1990). If the data in the CMP-gather have been NMO corrected, this implies we have for every zero offset time a reflectivity vector as defined by equation (5). Inversion, i.e., applying the weighted stacking procedure, yields estimates of the relative parameter contrasts for every zero offset time.

If we assume that

(1) the seismic wavelet does not vary with offset,
(2) the NMO corrected CMP-gather can be modeled by convolution of the wavelet and *aligned* spiky reflectivity traces, and
(3) the weights for the weighted stacking vary little over the effective length of the wavelet,

then it can be shown that the linearly estimated relative parameter contrasts will be a convolution of the estimated spiky contrasts and the wavelet. So, for example, the estimated $\Delta W/W$ contrast will be a band limited version of the actual (spiky) $\Delta W/W$ contrast. This is important when interpreting results from thin layers with interfering reflections. The spiky relative contrasts can be retrieved by applying a "poststack" stratigraphic inversion procedure (Duijndam, 1987, 1988a, b; Gelfand and Larner, 1984; Brac et al., 1988) to the weighted stacking results. Or in other words, a weighted stack result is the preferred input to poststack stratigraphic inversion.

For sake of completeness, the following list gives some of the most important factors affecting the inversion results, some of which may have a more significant effect than others:

(1) Errors in the offset to angle conversion (accuracy of the macro model);
(2) Unwanted amplitude effects (propagation effects, source/receiver directivity);
(3) Approximation of the Zoeppritz model (linearity assumption);
(4) NMO correction errors (stretch, erroneous velocities);
(5) Random noise, statics; and
(6) Correlated noise (multiple reflections, converted waves etc.).

Obviously, the effects of these factors (especially 1, 2, 4, 5, and 6) should be minimized as much as possible prior to the inversion by appropriate preprocessing (see, Berkhout and Wapenaar, 1990).

Table 2 lists the required additional processing to arrive at aligned, angle-dependent, reflection amplitudes for a CMP or CDP-gather.

EXAMPLE

We now show the statistically weighted stacking results in an experiment on simulated multicomponent seismic data, allowing us to compare the inversion results with the true model. Figure 5 shows schematically this true model, which consist of a sandstone layer encased in identical shales. The fluid-fill in the sandstone changes from brine to gas going from left to right and is simulated by a linearly decreasing P-wave velocity V while the S-wave velocity W and density D remain almost the same (Figure 6). The shooting geometry is shown in Figure 7, with offsets ranging from 0 to 2400 m, resulting in a maximum angle of incidence of approximately 35°. At each of the 6 indicated positions (Figure 5) a simulated NMO corrected CMP has been generated, using the following scheme, where a perfect S-wave two-way traveltime

(TWT) conversion to P-wave TWT is assumed, using the following forward scheme.

(1) Offset to angle conversion using raytracing.
(2) Angle dependent reflectivity calculation using a nonlinear forward model (i.e., the Zoeppritz equations for incident P-waves). Note that no NMO stretch is applied.
(3) Band limitation by convolution with a 20Hz zero phase Ricker wavelet.
(4) Addition of Gaussian distributed uncorrelated random noise in the seismic band (5-100 Hz). The noise samples for the P-P reflectivity section have a standard deviation [σ] of 0.01 (S/N ratio = 10dB) and those for the S-S reflectivity section have a σ of 0.02 (S/N ratio = 12dB).

In the statistically (soft) stabilized inversion both the Gardner relation equation (12) and the mudrock relation equation (15) will be used as empirical information. In case of the deterministically (hard) stabilized inversion only the Gardner relation will be used, as hard substitution of equation (15) will yield a fluid factor, equation (10d) equal to zero. Table 3 lists the actual values for the coefficients of the empirical relations, as used in the inversion example. The Gardner coefficients are correct for the brine filled part of the model. However, for the gas filled part they are not correct, since in general the presence of gas is still to be proven. For the mudrock line a combination of V_o and k [see equation (11)] is taken such that the correct W/V ratio is obtained, but with a high value of k. This will cause a large contribution of the $\Delta W/W$ estimate in the fluid factor [equation 10(d)].

As only two empirical relations are used in the example of the statistically weighted inversion, the

Table 2. Additional processing for AVO recovery.

CMP oriented approach
- Filtering (multiple attenuation and ground roll removal)
- Correction for source-receiver directivity
- Spherical divergence correction
- NMO correction
- Offset to angle conversion (based on macro model).

CDP oriented approach
- Decomposition into P- and S-waves (for multicomponent data),
- Surface related multiple elimination,
- Prestack migration (based on macro model).

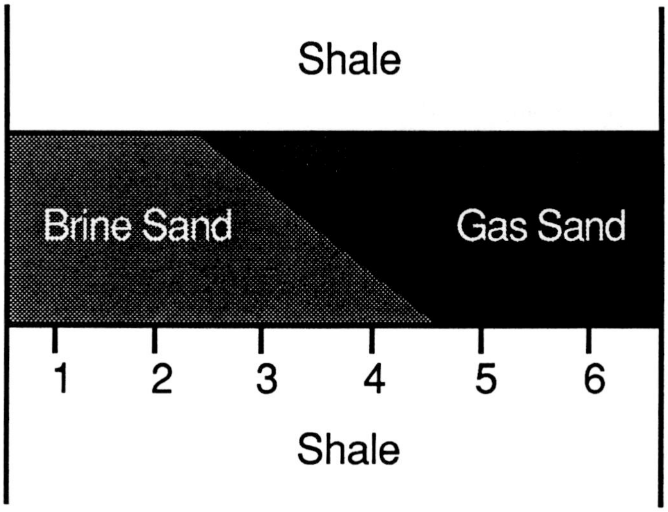

Fig. 5. Model for experiment on simulated data. The boundary between brine and gas sand is only drawn to show the gradual change in pore fluid fill. At locations 1 to 6 simulated P-P reflectivity and S-S reflectivity sections are calculated using the correct, nonlinear equations.

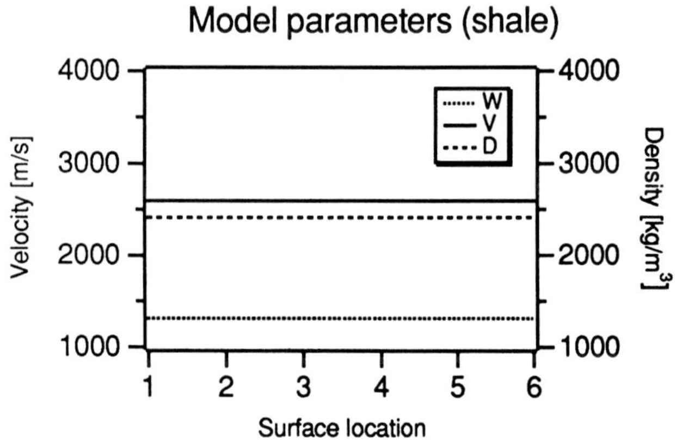

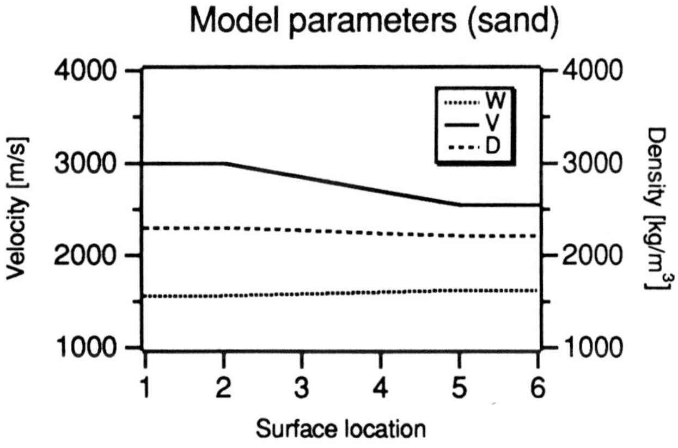

Fig. 6. Variation of elastic parameters with lateral position.

empirical covariance matrix is singular. To calculate its inverse a very small diagonal stabilization is used to prevent numerical problems. If three or more independent empirical relations are incorporated in the inversion, this stabilization for the calculation of the standard deviation due to the empirical information alone is not necessary. Similarly, the denominator of equation (23) is singular when only *S-S* reflection data are used. To prevent numerical problems again a very small diagonal stabilization was used. If *P-P* reflection data or a combination of *P-P* and *S-S* reflection data are used, this stabilization is not necessary.

Figure 8 shows the *P-P* sections at the four loca-

Table 3. Actual values for the coefficients in the empirical relations, as used in the inversion example.

Empirical coefficient	Value
Gardner (Equation 12)	−0.33
Mudrock (Equation 15), *k*	3.0
Mudrock (Equation 15), *W*/*V*	0.42

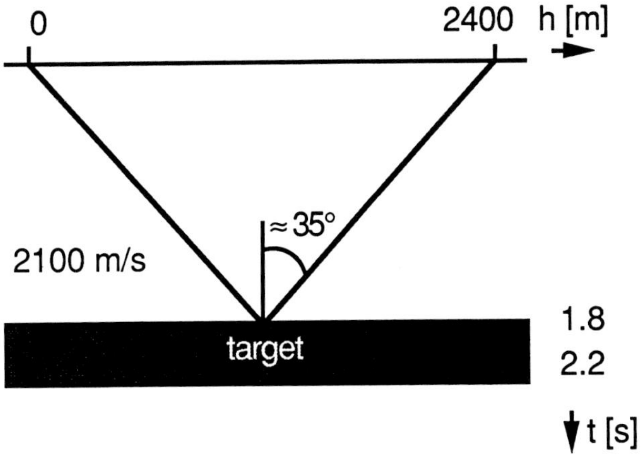

Fig. 7. Shooting geometry for simulated data. The layer is 52 ms thick, while the target considered for the inversion starts at 1.8 s. and is 400 ms thick. Maximum angle of incidence is ±35.

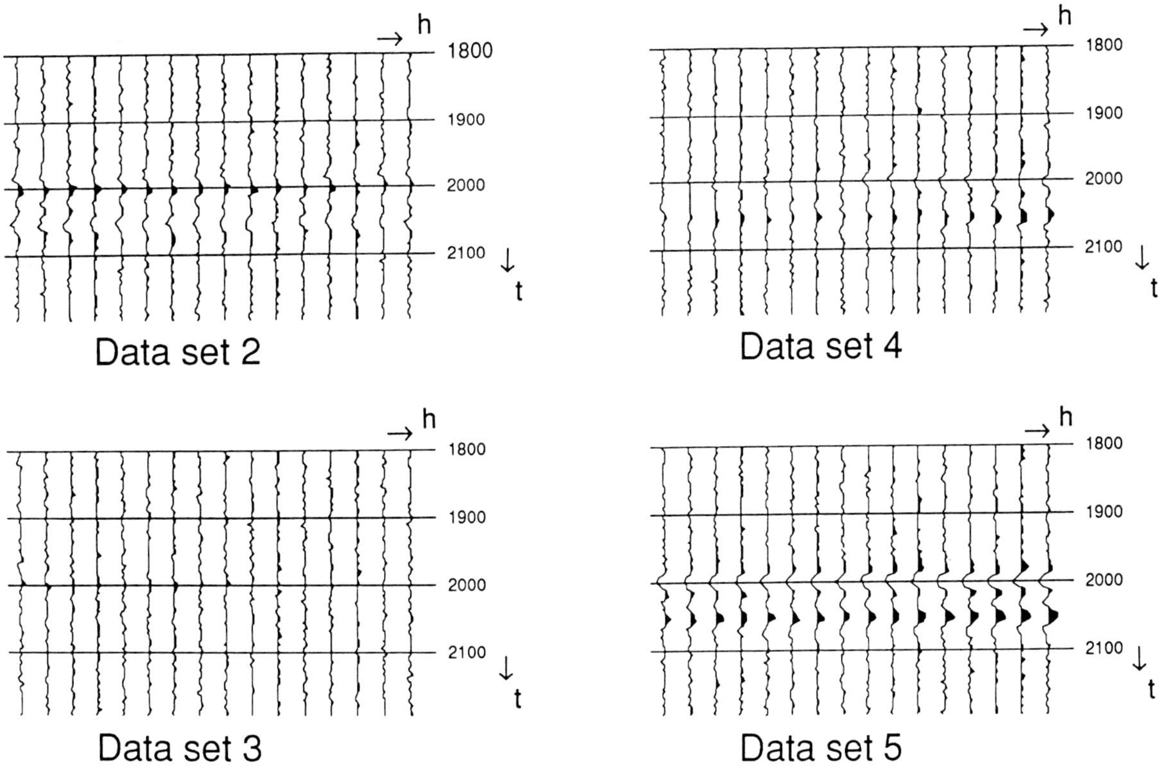

Fig. 8. Noisy sections of *P-P* reflectivity at locations 2 thru 5. The offset *h* increases from left to right. The section at location 1 is similar to that of location 2, they only differ in noise realization. A similar statement can be made for the sections at locations 5 and 6. Note the dramatic increase of the reflectivity with offset in the gas-bearing part, which is absent in the brine-filled part.

tions 2-5. The data at location 1 only differ in noise realization with those at location 2, and similarly for locations 5 and 6. Figure 9 shows the *S-S* sections at the corresponding four locations. In the inversion all 31 channels of each *P-P* section and/or *S-S* section will be used. Close examination of the NMO-corrected CMP gathers shows that in the gas part the *P-P* section exhibits a significantly increasing AVO, and the *S-S* sections are approximately the same, but contain a polarity flip at approximately half the maximum offset. Therefore, significant errors will be introduced in the interpretation of conventionally stacked data, as is shown in Figures 10 and 11.

For the acquisition geometry shown in Figure 7 and the statistical parameters listed in Table 4, the following results are discussed:

Zero-offset reflectivity: Figures 12 and 13
Fluid factor contrast: Figures 14 to 17.

All figures are plotted with the same scaling factor for all output traces (no AGC). As follows from the coefficients in Table 1, the zero-offset reflectivity is a scaled version of the relative impedance contrasts. The zero-offset *P* reflectivity is a scaled version of the relative acoustic impedance contrast $\Delta Z/Z$ and, likewise, the zero-offset *S* reflectivity is a scaled version of the relative shear impedance contrast $\Delta Y/Y$. Figures 12 and 13 show the estimated *P* and *S* reflectivities as obtained from the *P-P* and *S-S* sections, respectively. Comparison with the results of Figures 10b and 11b, shows dramatic improvement in the amplitudes of the zero-offset reflectivity estimates obtained with statistically weighted stacking over those obtained with conventional stacking in the case of significant AVO variations.

For the fluid factor contrast (as a gas indicator) we will first show in Figure 14 the results obtained with deterministically weighted stacking of *P-P* data only. The effect of a highly ill-resolved relative density contrast is removed by hard substitution of the Gardner relation. The relative *S*-wave contrast is not stabilized at all. Hence, its sensitivity with respect to the noise in the data is not taken into account. In this

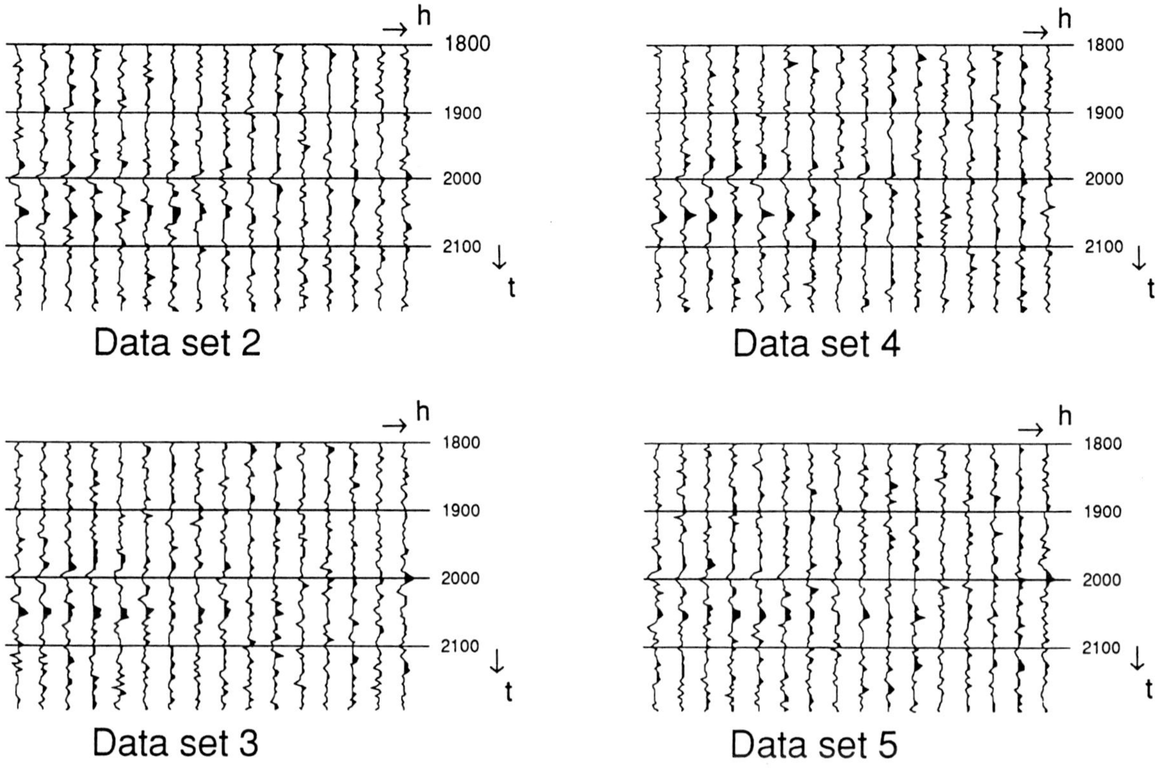

Fig. 9. Noisy sections of *S-S* reflectivity at locations 2 thru 5. As in Figure 8, the sections at locations 1 and 2 differ only in the noise realization and similarly for locations 5 and 6. Note that the sections show very similar behavior with offset for all locations. All sections show a polarity flip at about half offset which will have a devastating effect on the conventionally stacked results.

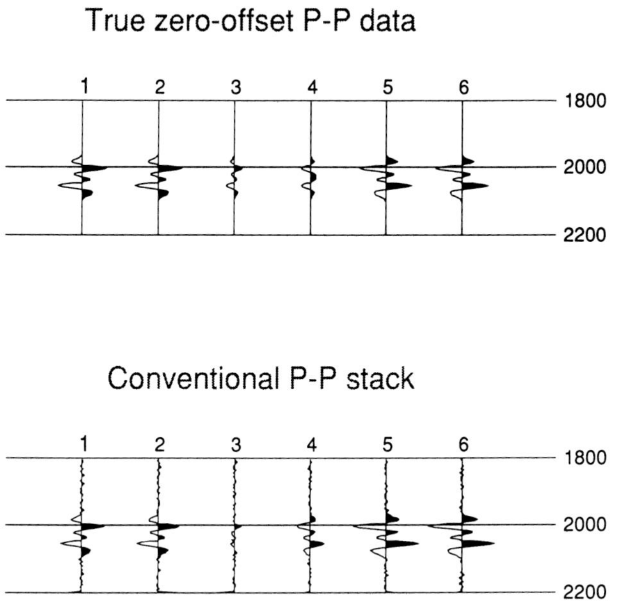

Fig. 10. True zero offset *P-P* reflectivity and the result obtained by simple (conventional) stacking. In the gas-bearing part the amplitudes are much too large, due to the increasing AVO.

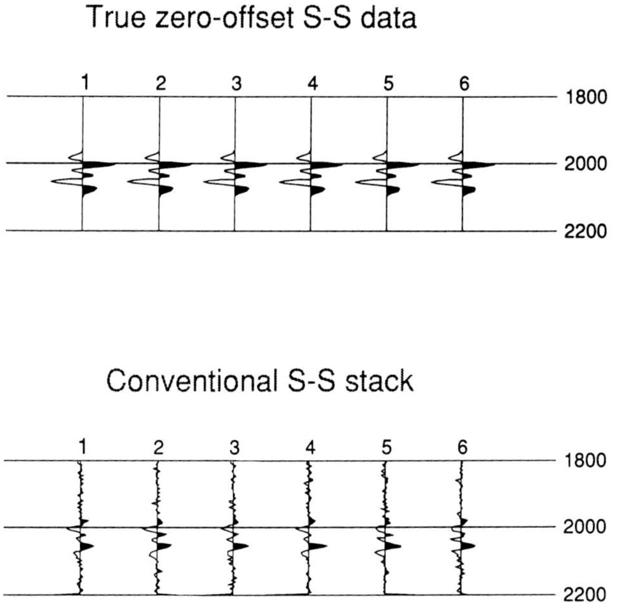

Fig. 11. True zero-offset *S-S* reflectivity and the result obtained by simple (conventional) stacking. As expected, the polarity flip results in stacked amplitudes that are too small.

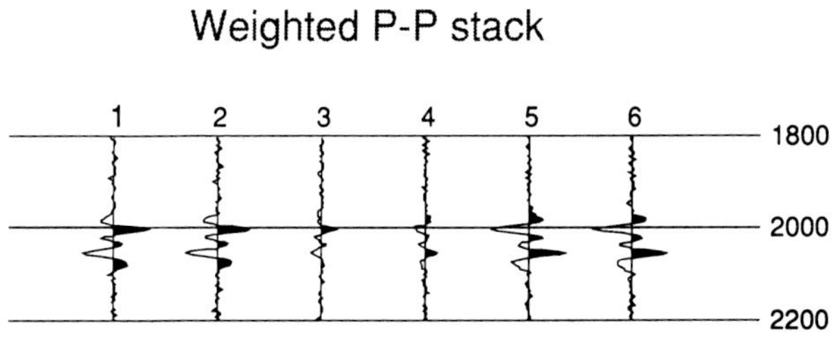

Fig. 12. Zero offset *P-P* reflectivity as calculated from the statistically weighted stacking results for the relative acoustic impedance contrasts from *P-P* data only. The amplitudes are much better than those obtained with conventional stacking.

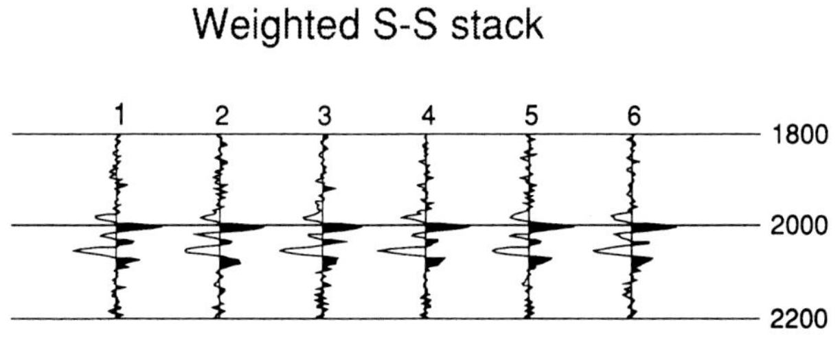

Fig. 13. Zero offset *S-S* reflectivity as calculated from the statistically weighted stacking results for the relative shear impedance contrasts from *S-S* data only. The amplitudes are much better than those obtained with conventional stacking.

Table 4. Statistical parameters (standard deviation) used in the example. The covariance matrices are assumed to be diagonal. The standard deviation of the seismic data corresponds with the values used in the noise generation. A rather large scatter around the empirical relations is assumed to allow for an unbiased (high K_i) solution.

Input type	σ
P-P data	0.01
S-S data	0.02
Gardner	0.20
Mudrock	0.20

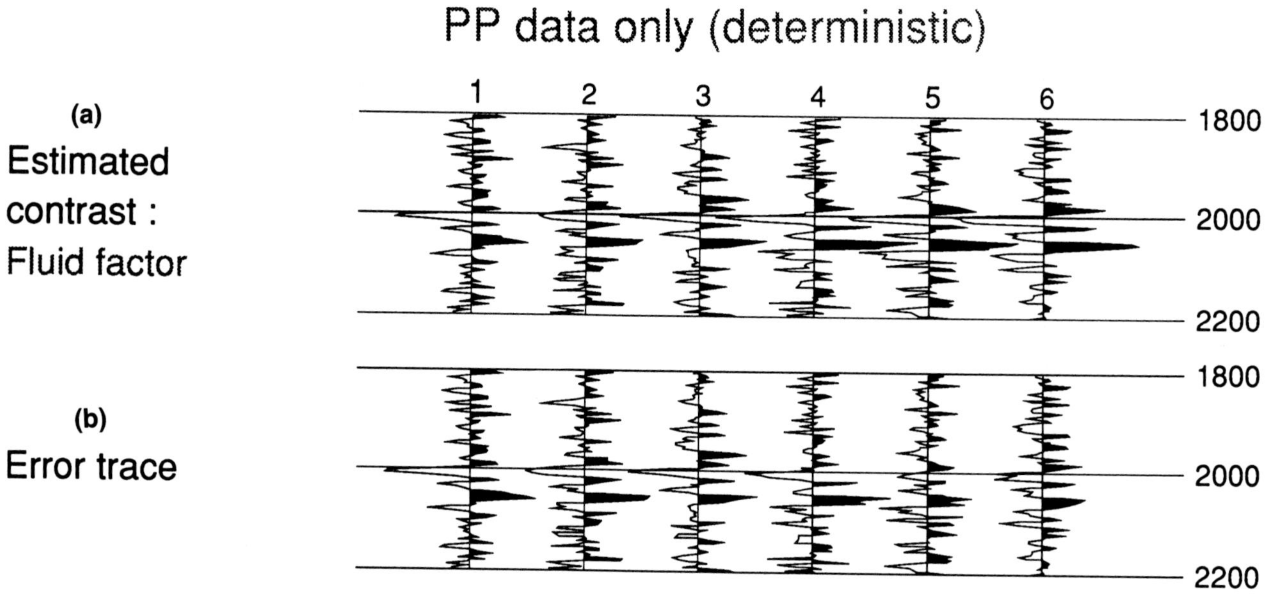

Fig. 14. Fluid factor results obtained with *deterministically* weighted stacking of *P-P* data only. (a) The estimates do show a significant negative over positive contrast in the gas-bearing part of the section. However, in deterministically weighted stacking the relative *S*-wave velocity contrast is not stabilized at all and its sensitivity with respect to the noise in the seismic input is not taken into account. (b) The actual errors in the estimated contrasts obtained by subtraction of the true bandlimited contrasts from the estimates.

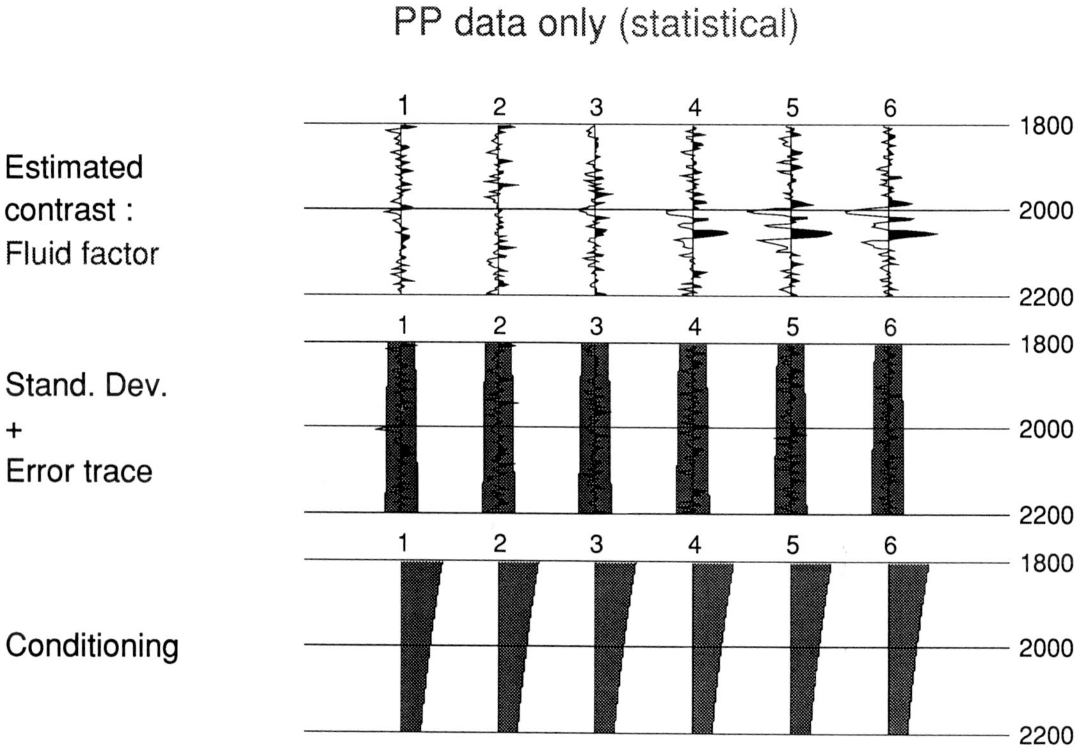

Fig. 15. Fluid factor results obtained with *statistically* weighted stacking of *P-P* data only. (a) The estimates are much better than the estimates of Figure 14a. (b) The actual error in the estimated contrasts, together with the standard deviation. (c) The conditioning of the estimated contrasts, which are approximately 1/2. So the results are determined by the seismic input for 50 percent. (The displayed range between two traces runs from 0 to 1, going from left to right).

experiment a high value of k is used in equation (9d), resulting in a large contribution of the (relatively) noisy S-wave contrasts to the fluid factor, which explains the noisy fluid factor results. Note that hard stabilization of the relative S-wave velocity contrasts using the mudrock relation, would yield a fluid factor result equal to zero (see also Figure 17). Because we have the true model available, we can calculate the true contrasts and the corresponding errors in the estimate, but in reality this, of course, is not possible.

If we look at the results in Figure 15, obtained with statistically weighted stacking of P-P data only, we see that the actual estimates are less noisy, than those of the previous figure (obtained with the deterministically weighted stacking). In the statistically weighted stacking, the sensitivity of the relative S-wave contrast with respect to noise in the seismic input is taken into account and a stabilized S-wave contrast estimate is used in the calculation of the fluid factor.

Moreover, much more information on the estimates is available. First, as shown in Figure 15, the standard deviations of the estimates are available as an accuracy measure. Due to a slowly decreasing angle range with depth, the standard deviation slowly increases with increasing depth. The error in the estimates is on the average well below standard deviation. So in practice, where we can not determine the *actual* errors in the estimates, the one standard deviation limit may serve as an estimator for the errors one can *expect* with respect to the anticipated noise levels on the seismic data and the scatter around the empirical relations. This is the first major advantage of the statistical approach over the deterministic one. The second advantage is the resolution ratio, or conditioning, curves as shown in Figure 15. These curves indicate whether the estimates are primarily determined by the empirical input or by the seismic input. Since in this case the conditioning is about one-half, the fluid factor contrasts are still determined by the seismic input for approximately 50 percent. The decreasing angle range causes the conditioning to become less for deeper parts of the target.

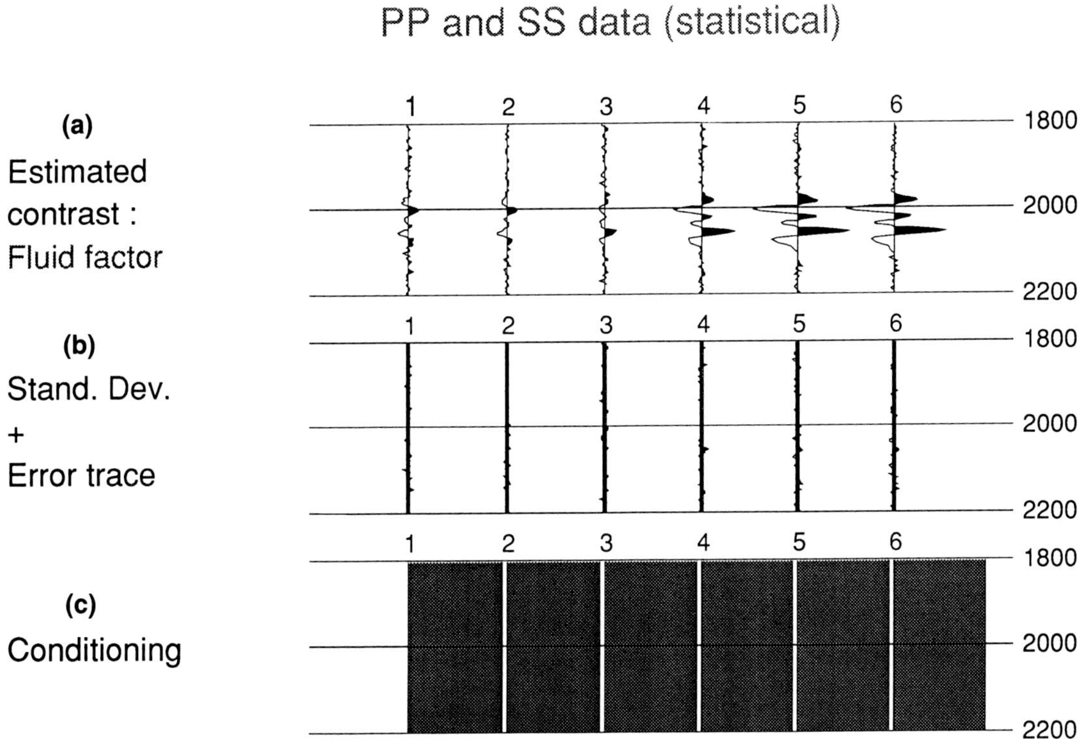

Fig. 16. Fluid factor results obtained with *statistically* weighted stacking of P-P and S-S data combined. (a) The estimates look significantly less noisy than the results obtained with P-P data only. (b) The actual error in the estimated contrasts, together with the standard deviation much smaller than those for P-P data only. (c) The conditioning of the estimated contrasts is much better ($K_i \approx 1$) than obtained with P-P data alone.

Figure 16 shows the fluid factor contrast results obtained by statistically weighted stacking of both *P-P* and *S-S* data. Comparison with Figure 15 shows that the incorporation of *S-S* data resulted in more accurate estimates for ΔF, due to a more accurate relative *S*-wave velocity contrast estimate. We also see that the standard deviation of the estimates has decreased, to indicate the higher accuracy. On the average the actual error is well below this one-standard deviation limit. Also the conditioning has improved.

For the sake of completeness, Figure 17 shows the fluid factor contrast results obtained from inversion of *S-S* data alone. Since, the *S-S* data contain absolutely no information on the *P*-wave velocity, the fluid factor contrast result will rely completely on the empirical input as indicated by the conditioning curve with zero amplitude. The empirical input consists, in part, of the mudrock line relation between $\Delta V/V$ and $\Delta W/W$ for *water* filled sandstones. Results satisfying the mudrock relation will show a negligible fluid factor contrast since the factor indicates *deviations* from this relation. This deviation is exactly, what we observe in Figure 18 in the gas filled part of the section: negligible contrasts where there should have been significant contrasts. The importance of correct prior (empirical) information is shown for those cases where it is needed. In the case of *P-P* data alone or *P-P* and *S-S* data, incorporation of the erroneous empirical information caused no harm, since the answer was determined (mainly) by the seismic input. However, in the case of *S-S* data alone, it becomes of key importance. Note that the necessity of caution is clearly indicated by the conditioning curves (Figure 17). However, even in this case we see that the (very much larger) standard deviations in Figure 17 are still indicative of the errors one might expect in practice, where the true contrasts are not available.

CONCLUSIONS

In conclusion we can say that (deterministically or statistically) weighted stacking shows an improvement

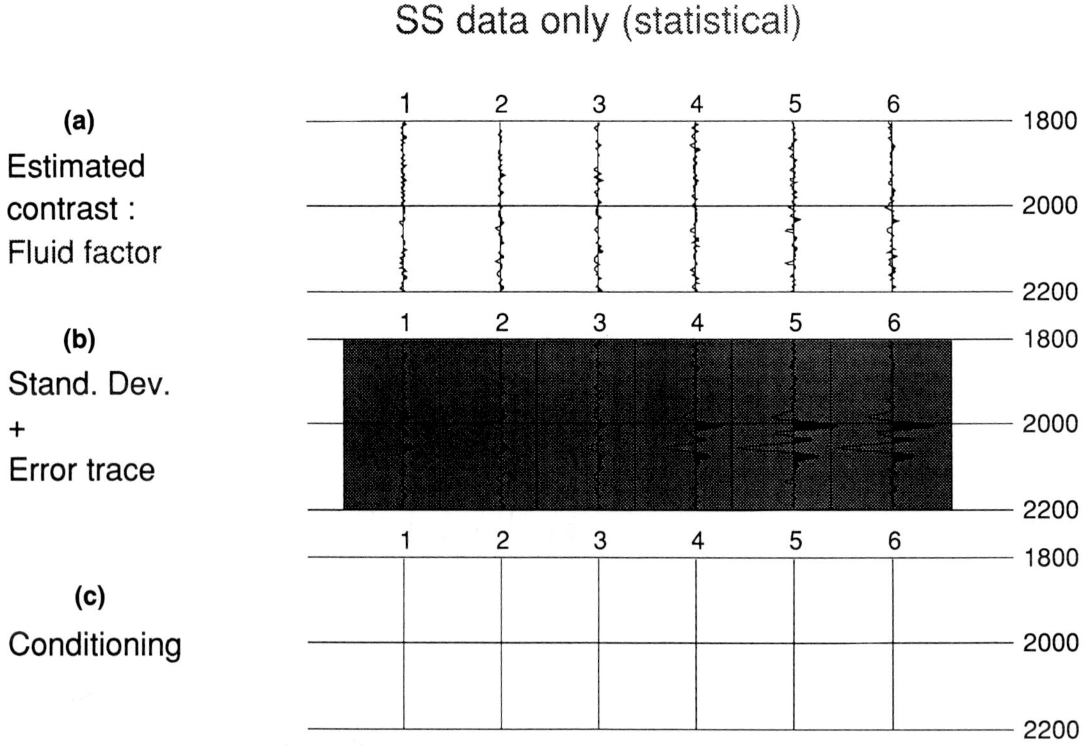

Fig. 17. Fluid factor results obtained with *statistically* weighted stacking of *S-S* data alone. (a) The estimates show almost no contrasts. (b) The actual error in the estimated contrasts together with the standard deviation; even for these large errors they are (on the average) well below the one-standard deviation limit. (c) The conditioning of the estimated contrasts indicates that the result is completely determined by the empirical information.

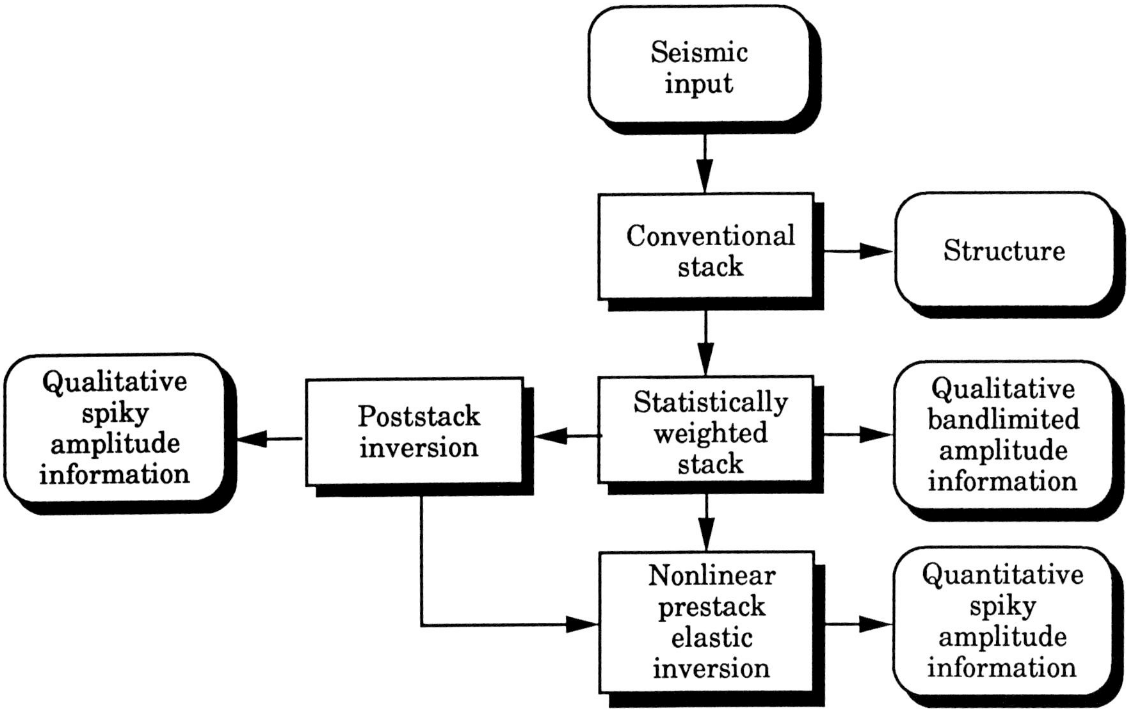

Fig. 18. Processing sequence for AVO inversion.

over normal stacking for a qualitative analysis of multioffset amplitudes in the case of significant AVO (e.g., strong increasing AVO or polarity flip). The statistically Bayesian based extension allows a more sensible stabilization than plain hard substitution, since the solution is allowed to deviate from the empirical relations provided the seismic input is strong enough. The effect of unstabilized, ill-resolved, parameters on the results is recognized and taken into account. Furthermore, determination of quality measures are allowed for accuracy and dependence on seismic input. Although the method involves the linear approximation (small offsets) the weighted stacking method is fast and robust. Therefore, the method can be used to identify significant anomalies (i.e., potentially interesting areas requiring further processing) and may supply good start values for the nonlinear methods (Figure 18). In principle, the combined inversion of *P-P* and *S-S* data yields significantly more accurate and better determined (by the seismic input) estimates for the relative contrasts. Finally, the availability of an accurate macro model is a prerequisite; for accurate offset to angle conversion and for alignment of corresponding reflectivity information on *P-P* and *S-S* sections.

ACKNOWLEDGMENT

This research was carried out within the DELPHI-project at the Delft University of Technology. We thank the sponsors of the DELPHI-project for their support. Furthermore, we thank the reviewers for their constructive suggestions.

REFERENCES

Aki, K., and Richards, P. G., 1980, Quantitative seismology: Freeman and Co.

Ball, V., 1987, Depth inversion of impedance and Poisson's ratio using the Insight interactive modeling system: Presented at the 57th Ann. Internat. Mtg., Soc. Expl. Geophys.

Balogh, D., Snyder, G., and Barney, W., 1986, Examples of a new approach to offset amplitude analysis: Presented at the 56th Ann. Internat. Mtg., Soc. Expl. Geophys.

Berkhout, A. J., and Wapenaar, C. P. A., 1990, Delphi: Delft philosophy on acoustic and elastic inversion: The Leading Edge, **9**, No. 2, 30–33.

Brac, J., Dequirez, P. Y., Herve, F., Jacques, C., Lailly, P.,

Richard, V., and Tran Van Nieuh, D., 1988, Inversion with a priori information: An approach to integrated stratigraphic interpretation: Presented at the 58th Ann. Internat. Mtg., Soc. Expl. Geophys.

Castagna, J. P., Batzle, M. L., and Eastwood, R. L., 1985, Relationships between compressional-wave and shear-wave velocities in clastic silicate rocks: Geophysics, **50**, 571–581.

Chiburis, E. F., 1987, Studies of amplitude versus offset in Saudi Arabia: Presented at the 57th Ann. Internat. Mtg., Soc. Expl. Geophys.

Crase, E., Pica, A., Noble, M., McDonald, J., and Tarantola, A., 1990, Robust elastic nonlinear waveform inversion: Application to real data: Geophysics, **55**, 527–538.

Duijndam, A. J. W., 1987, Detailed Bayesian inversion of seismic data: PhD thesis, Delft Geophys.

Duijndam, A. J. W., 1988a, Bayesian estimation in seismic inversion: Part I Principles: Geophys. Prosp., **36**, 878–898.

Duijndam, A. J. W., 1988b, Bayesian estimation in seismic inversion: Part II Uncertainty analysis: Geophys. Prosp., **36**, 899–918.

Gardner, G. H. F., Gardner, L. W., and Gregory, A. R., 1974, Formation velocity and density-the diagnostic basics for stratigraphic traps: Geophysics, **39**, 770–780.

Gelfand, V., and Larner, K., 1984, Seismic lithologic modeling: The Leading Edge, **3**, No. 11, 30–35.

Gelfand, V., Ng, P., Nguyen, H., and Lamer, K., 1986, Seismic lithologic modeling of amplitude-versus-offset data: Presented at the 56th Ann. Internat. Mtg., Soc. Expl. Geophys.

Haas, J. C. de, and Berkhout, A. J., 1988, On the information content of P-P, P-SV, SV-SV and SV-P reflections: Presented at the 58th Ann. Internat. Mtg., Soc. Expl. Geophys.

Haas, J. C. de, and Berkhout, A. J., 1989, The information content of elastic reflections: Presented at the 51st Mtg., Eur. Assn. Expl. Geophys.

Jackson, D. D., and Matsu'ura, M.,1985, A Bayesian approach to nonlinear inversion: J. Geophys. Res., **90**, 581–591.

Lörtzer, G. J. M., and Berkhout, A. J., 1989, Statistical extensions for weighted stacking with result analysis: Presented at the 59th Ann. Internat. Mtg., Soc. Expl. Geophys.

Lörtzer, G. J. M., and Berkhout, A. J., 1988, Evaluation of weighted stacking techniques for AVO inversion: Presented at the 50th Mtg., Eur. Assn. Expl. Geophys.

Menke, W., 1984, Geophysical data analysis: Discrete inverse theory: Academic Press Inc.

Rijssen, E. P. F. van, and Herman, G. C., 1988, Resolution of elastic inversion: Presented at the 50th Mtg., Eur. Assn. Expl. Geophys.

SEG, 1988, Workshop on amplitude versus offset: Presented at the 57th Ann. Internat. Mtg., Soc. Expl. Geophys.

Smith, G. C., and Gidlow, P. M., 1986, Weighted stacking for rock property estimation and detection of gas: Presented at the 48th Mtg., Eur. Assn. Expl. Geophys.

Smith, G. C., and Gidlow, P. M., 1987, Weighted stacking for rock property estimation and detection of gas: Geophys. Prosp., **35**, 993–1014.

Spratt, S., 1987, Effect of Normal Move Out errors on Amplitude Versus Offset-derived shear reflectivity: Presented at the 57th Ann. Internat. Mtg., Soc. Expl. Geophys.

Tarantola, A., 1987, Inverse problem theory: Elseviers Science Publ.

Correlation and Interpretation of *P-P* and *P-SV* Data Zamora Gas Field, Yolo County, California

J. F. Zaengle and C. W. Frasier‡*

SUMMARY

A three-component seismic line acquired over the Zamora gas field, northern Sacramento Valley, California, displays conventional (*P-P*) and converted-wave (*P-SV*) seismic anomalies. Well log and seismic modeling results indicate that "bright-spot" reflection strengths are related to variations in fluid saturation and lithology. *P-P* sections, which measure *P* impedance reflectivities, are complemented by *P-SV* sections, which measure *S* impedance reflectivities in much the same way that independently recorded *SH-SH* data do. Joint interpretation of these data sets may improve risk evaluation and well site selection in areas where rapid changes in reservoir quality are known to exist.

A statistical wavelet extraction without well control from the *P-P* data produced a clean zero-phase wavelet, which when convolved with the reflectivity series from a central well in the field produced an excellent match to the stacked *P* section.

However, the events on the stacked *P-SV* section were so different in character and amplitude from the *P-P* events that an initial correlation between the two sections was not possible. To overcome this the *P-SV* data were smoothly compressed to *P-P* traveltimes during migration using the ARCO mudrock function, which empirically relates V_s to V_p for brine saturated shale sequences.

This time compression, although approximate, allows an interpreted correlation of events between the two data sets with interesting amplitude relations. A low-porosity, carbonate-cemented sandstone shallow in the section displays high reflectivity events on both the *P-P* and *P-SV* sections, as expected because of the presence of "hard streaks" in the formation. The top of the Capay shale has a stronger *P-SV* than *P-P* amplitude, due to a larger decrease in *S* than *P* impedance in the shale. The Starkey producing sand has a strong *P-P* amplitude anomaly with no visible *P-SV* seismic reflection. The absence of a bright spot on the *P-SV* section is due to the presence of gas which results in a larger decrease in V_p than in V_s in the Starkey sand. Amplitude with offset models using rock physics programs support these conclusions.

Using the interpreted correlations the interval times of *P-P* and *P-SV* events were used to calculate V_p/V_s ratios for the principal lithologic units in the Zamora area. Well log correlations to the seismic events and V_p/V_s ratio data indicate that changes in V_p/V_s can be directly related to changes in the sand/clay ratio and the consolidation state of the sediments both laterally and vertically in the section. These results indicate that a joint interpretation of *P-P* and *P-SV* sections can be a potentially powerful tool to interpret lithologic variability in stratigraphically complex areas, in much the same way as *P-P* and *SH-SH* data have been used, and without the need to shoot each line twice, once for each source.

The V_p/V_s ratios determined from seismic event correlations match those derived empirically using sonic logs and the ARCO mudrock line function. This result indicates that the ARCO mudrock line function provides a good "first-order" approximation of V_s when other *S*-velocity data are not available. In the Zamora area the V_p/V_s ratios vary from 2.2 for consolidated sands to 3.5 for unconsolidated shallow shales.

INTRODUCTION AND BACKGROUND

Seismic detection of hydrocarbon reservoirs in many basins is complicated because bright-spot anomalies can result from a variety of lithologic and fluid associations. This report highlights the results of a seismic field trial experiment in the Zamora gas field, northern Sacramento Valley, California, that used three-component seismic data, well logs, rock physics

*Department of Geological Science, University of Michigan, Ann Arbor, MI 48109-1063
‡Chevron Oil Field Research Company, P.O. Box 446, La Habra, CA 90633-0446.

data, and amplitude variation-with-offset (AVO) modeling strategies to differentiate gas reservoir seismic anomalies from those formed by cement-induced impedance contrasts.

The Zamora gas field is located approximately 60 miles northeast of the San Francisco Bay area, in Yolo County, California (Figure 1). Gas production in the Zamora area is from late Cretaceous and Tertiary age deltaic sandstones that occur at depths of approximately 3500 to 4000 ft (Figure 2). Structure maps and cross-sections indicate gas accumulation is controlled in large part by structural trapping mechanisms where closure is provided by subtle, low-relief, anticlinal structures. Well site selection and risk evaluation in this area can be complicated by variable stratigraphy, cementation patterns, and fluid saturation which cause reservoir quality to vary significantly both laterally and vertically in the section.

One difficult problem faced by interpreters in the Zamora area is differentiating low porosity, carbonate-cemented sands from porous, gas-bearing reservoirs, each of which produces high-amplitude reflections on conventional *P*-wave seismic data. An experimental, three-component line was shot with *P* vibrators from north to south across several bright spot anomalies that have been tested by wells along the line of section (Figure 2). The availability of good quality geologic control made this an ideal area to test the potential of using *P-SV*-wave data as a prospecting tool.

The primary purpose of the field trial was to process and jointly interpret the *P-P* section (using the vertical component data) and the *P-SV* section (from the inline horizontal component only). The motivation for this is that the *P*-wave reflectivity is sensitive to changes in *P*-wave impedance, whereas the reflectivity of *P-SV*-wave reflection data is sensitive to changes in *S*-wave impedance (Figure 3). Although not obvious converted-wave reflection coefficients at non-zero offset scale approximately like *S*-wave reflection coefficients at normal incidence. This is shown in the Appendix using

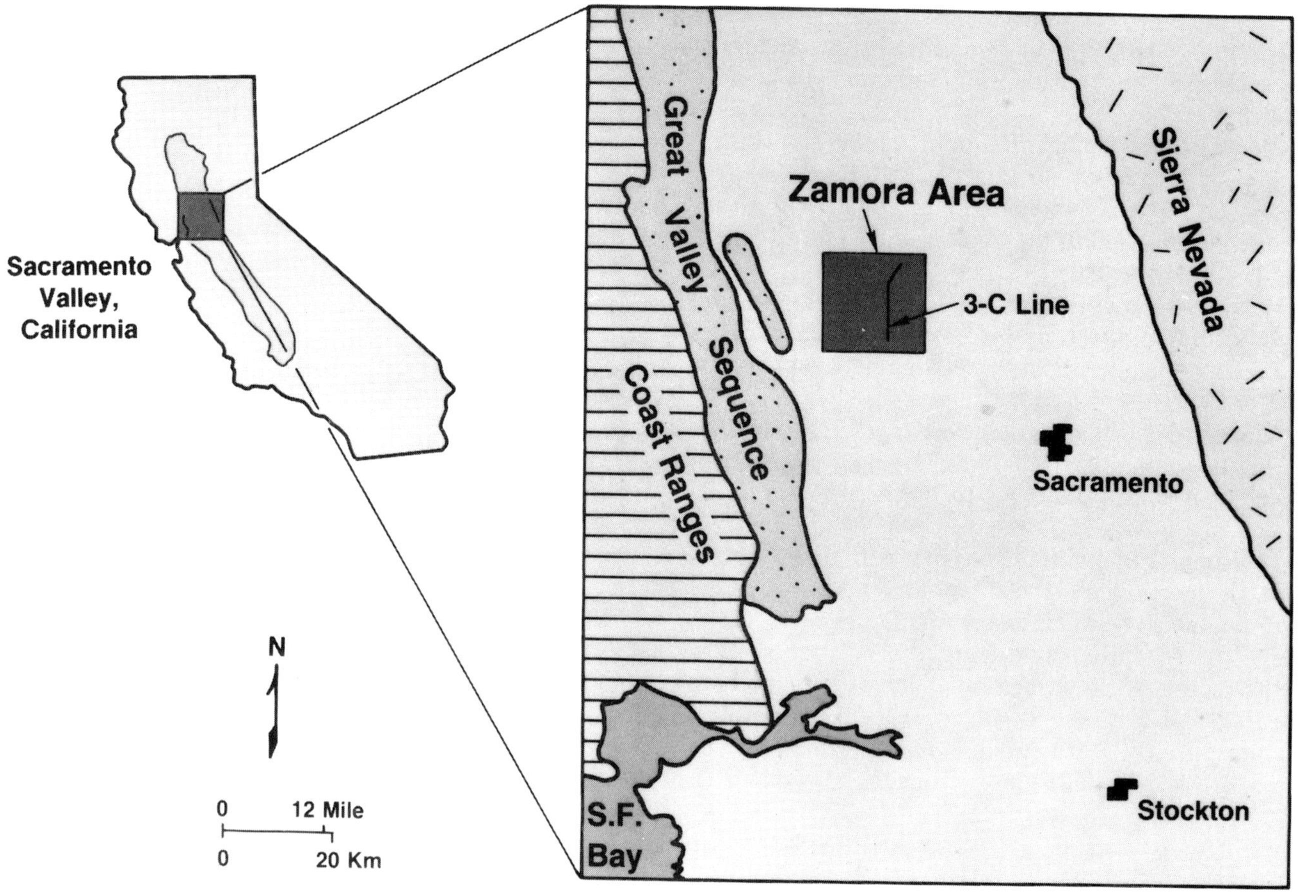

Fig. 1. Zamora Gas Field Location Map.

the modified Bortfeld equations for *P-SV* and *SH-SH* reflection coefficients given in Aki and Richards (1980). As a result, a converted wave stack will show variations in amplitudes with changing *S* impedance that mimic amplitude changes on an *SH-SH* section.

This variation of the direct hydrocarbon detection methods using separately recorded *P* and *SH* data was described in Ensley (1984) and Robertson and Pritchett (1985) concerning their work in the Sacramento Valley.

Differences in the sensitivity of *P-P* and *P-SV* data potentially allow an interpreter to detect a classic bright spot on the *P*-wave section and to verify the bright spot as being gas-related by noting its absence on the converted-wave section. This is because *S*-wave reflectivity is insensitive to variations in the gas/fluid mixture saturating a potential reservoir. The subsurface model assumes that a reservoir sand is capped by a shale with about the same *P*-wave velocity as a brine saturated sand. It is important to note that there are many other cases where the simple subsurface model is not appropriate.

Alternately, an amplitude anomaly that appears on both the *P-P* and *P-SV* sections is most likely related to changes in the rigidity of the formation, which would reflect changes in lithology, porosity distribution, and variations in the consolidation state or "hardness" of the prospective section.

When interpretations of *P-SV* data are combined with rock physics modeling and AVO modeling strategies, there is an improved framework for evaluating the relative impact of lithologic and fluid effects on bright spot development.

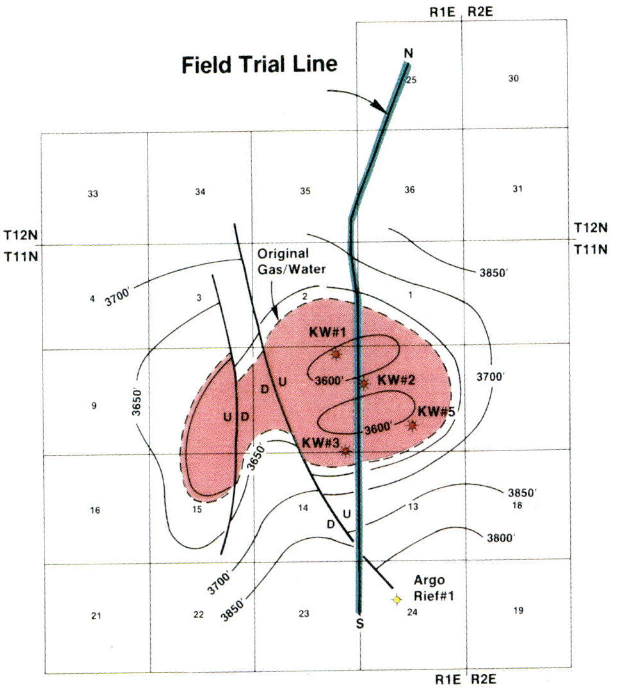

Fig. 2. Zamora field structure map drawn on the top of the gas-producing Starkey sandstone that occurs at depths of approximately 3600 ft. The position of the multicomponent seismic line and available subsurface well control is indicated.

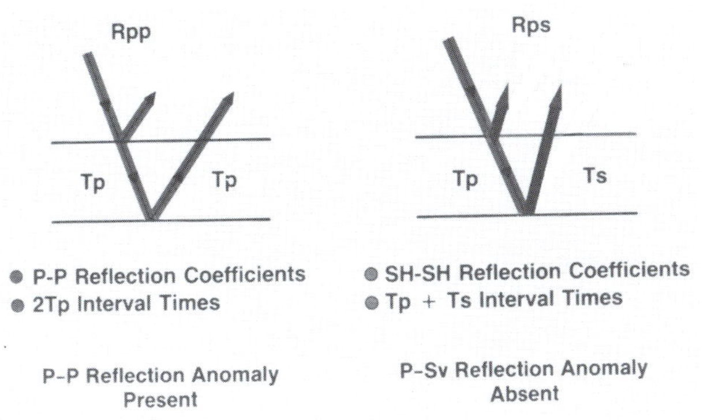

Fig. 3. Schematic diagram illustrating *P-P* and *P-SV* anomalies from a gas-sand. See the Appendix for the relation between *P-SV* and *SH-SH* reflection strength.

SEISMIC DATA ACQUISITION AND PROCESSING

The 5.7 mile-long Zamora line is 60-fold, split-spread data acquired by using four *P*-wave vibrators and 110 ft groups of three-component Mark Product L-15B geophones, with a natural frequency of 5 Hz. Checkshot surveys were recorded along the line of section to a depth of 600 ft. These surveys provided valuable near-surface velocity information for making statics corrections and for modeling.

The *P-SV* section was stacked by common-reflection point (CRP) using a scheme similar to that described in Frasier and Winterstein (1990). The major effect of that processing is to eliminate the severe depth point smearing caused by gathering the asymmetric rays by common midpoint (CMP).

WAVELET EXTRACTION AND SYNTHETIC SEISMOGRAM MODELS

The processed *P*-wave seismic data are very high quality and display two pronounced bright-spots, one at approximately 0.6 s and another at approximately 1.1 s, both of which have been tested by wells that have good quality sonic and density logs (Figure 4).

Wavelet extraction and zero offset models were prepared to test the reliability of well log ties with the field trial data.

Wavelets were calculated from the *P-P* section using a sparse reflectivity statistical model without sonic log control. These wavelets are zero-phase wavelets with little side lobe energy (Figure 5).

The extracted wavelet was convolved with the reflectivity series calculated from the Knaggs-Wallace #1 sonic/density logs and other wells along the line of section to produce a synthetic seismic trace. An exceptionally good match between the filtered synthetic and the field trial *P-P* data (Figure 5) was obtained. The results of this modeling provided confidence that reliable correlations could be made between available lithologic control and at least the *P-P* seismic data.

Posted along with the synthetic trace in Figure 5 is the sonic log from the Knaggs-Wallace No. 1 Well. This data indicates that the bright-spot shallow in the section has very fast sonic traveltimes on the order of 105 ms/ft which translates to *P*-wave velocities in excess of 9500 ft/s. Log analysis and well cuttings indicate that this interval represents a tight, carbonate-cemented sand bed with porosities on the order of 10 percent.

In contrast, the deep bright spot anomaly has relatively slow sonic traveltimes of approximately 160 ms/ft. *P*-wave velocities for this interval are on the order of 5500–6000 ft/s. Well log analysis and formation tests indicate that this interval is a porous, gas-producing sand with saturations of greater than 80 percent.

Other significant reflections to note on the *P*-wave section (Figure 6) are those from the top of the Capay and Sacramento shales. The Capay shale displays a relatively continuous, low-amplitude reflection at approximately 0.920 s. The Sacramento shale consists of a series of high continuity and amplitude reflections at 1.480 s to approximately 1.600 s. The reflections generated from each of these formations will be key to correlating and interpreting the *P-P* and *P-SV* data.

CONVERTED-WAVE EVENT CORRELATIONS

Perhaps the most important part of the interpretation procedure involved the correlation of *P*-wave events with their equivalent reflections on the converted-wave data (Figure 7). It seemed apparent early on that the high reflectivity converted (*P-SV*) events at 2.800 s tied with Sacramento shale reflections on the *P*-wave data (Figures 6 and 7). As a first approximation, the high reflectivity *P-SV* events at 1.900 s

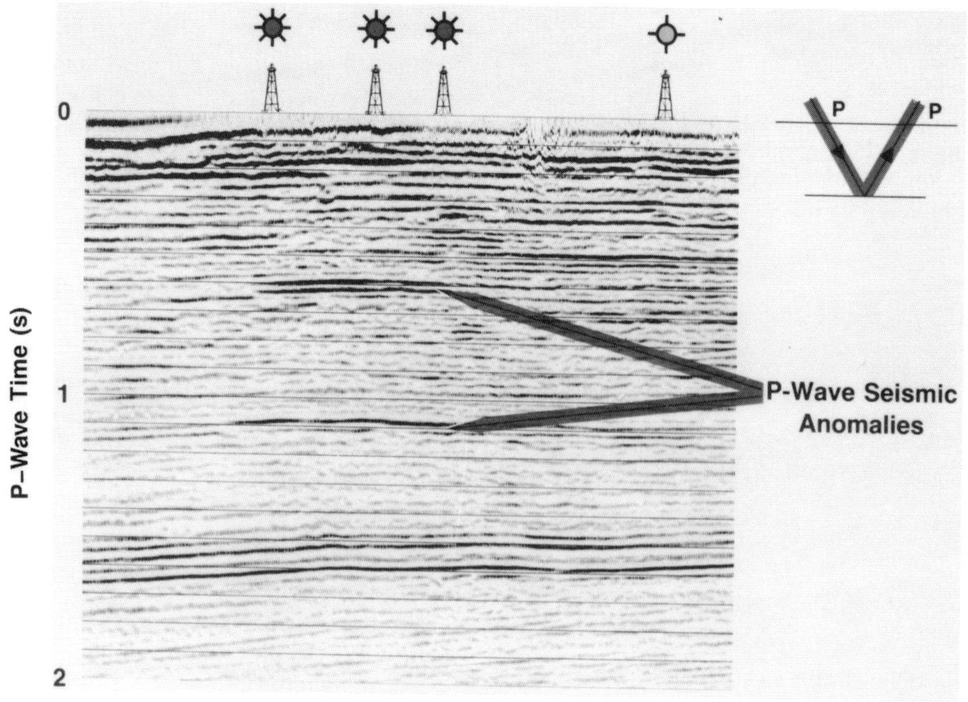

Fig. 4. Processed *P*-wave section from the Zamora field area (vertical geophone components).

appeared to tie with Starkey gas sand *P-P* anomalies. However, when *P-P* and *P-SV* traveltimes were used to estimate V_p/V_s ratios, the values obtained were less than 2.0, which is extremely low for sand/shale lithologies at this depth in the Sacramento Valley.

In order to correlate the events between the two sections a "low-frequency compression" of *P-SV* time to *P-P* time was made using the ARCO mudrock line. The logic for using this strategy is that the ARCO mudrock line generally provides a good approximation of V_p/V_s ratios, especially in brine-saturated, relatively unconsolidated sand and shale sequences like those in the Sacramento Valley. In addition, rock physics measurements of V_p and V_s for sands and shales collected at Chevron display a very good correspondence with this empirical function (Figure 8).

COMPARISON OF *P-P* AND "COMPRESSED" *P-SV* SECTIONS

When the *P*-wave section is compared with the compressed *P-SV* section (Figures 9 and 10), it is evident that deep reflections associated with the top of the Sacramento shale occur at roughly the same *P*-wave times (1.5 s). Confident that this correlation was a realistic one, it became clearer how other reflections correlated from section to section.

The low-porosity, cemented sandstone shallow in the section (approximately 0.6 s) displays high reflectivity events on both the *P-P* and *P-SV* sections. The presence of well developed events on both the *P-P* and *P-SV* sections is interpreted to be due to "hard streaks" that result in a large increase in both V_p and V_s velocity.

The top of the Capay shale has a much sharper *P-SV* than *P-P* reflection due to a much larger decrease in V_s than V_p velocity. This is one of the few cases where we've seen event reflectivities and continuity much stronger on the *P-SV* data than on the *P*-wave data.

The Starkey producing sand is represented by a strong *P-P* anomaly and a barely visible *P-SV* seismic reflection. The absence of a bright-spot on the *P-SV* section indicates that the *S* impedances are nearly balanced across the interface at the top of the sand.

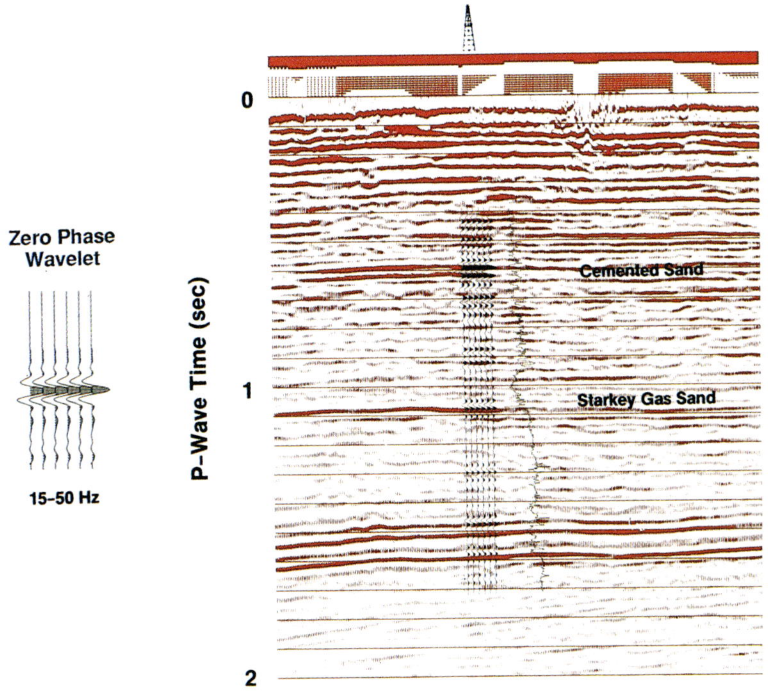

Fig. 5. (a) Zero-phase wavelet extracted from the Zamora field trial data by a statistical method without well control. (b) Synthetic seismogram and sonic log trace from the Knaggs-Wallace No. well.

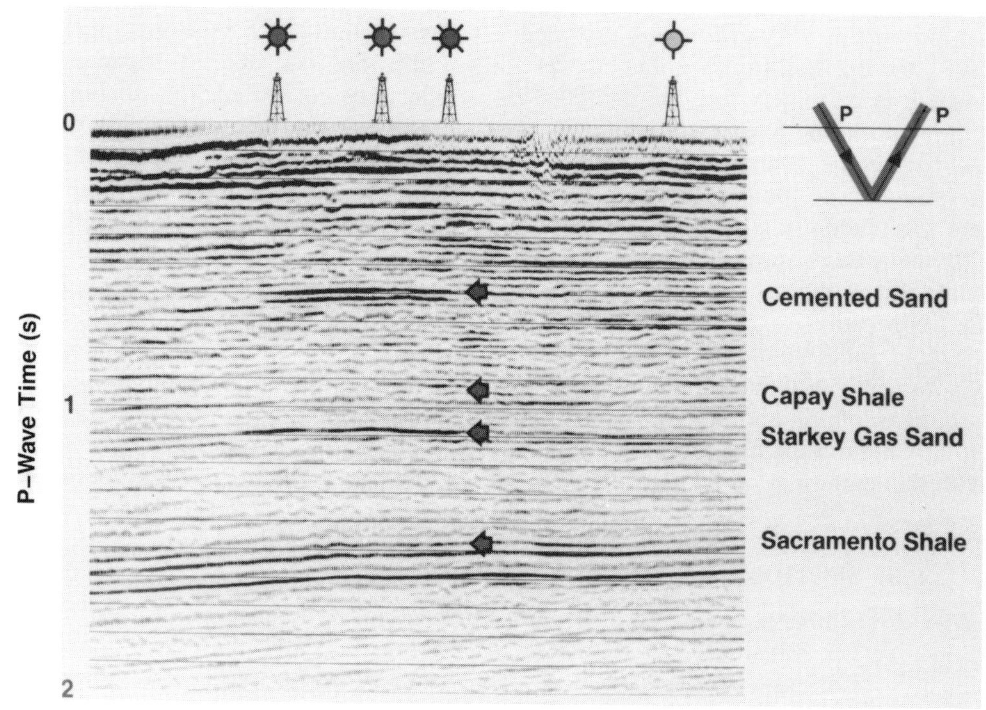

Fig. 6. Processed *P*-wave seismic section from the Zamora field area. Correlations between seismic events and the top of important stratigraphic intervals are indicated.

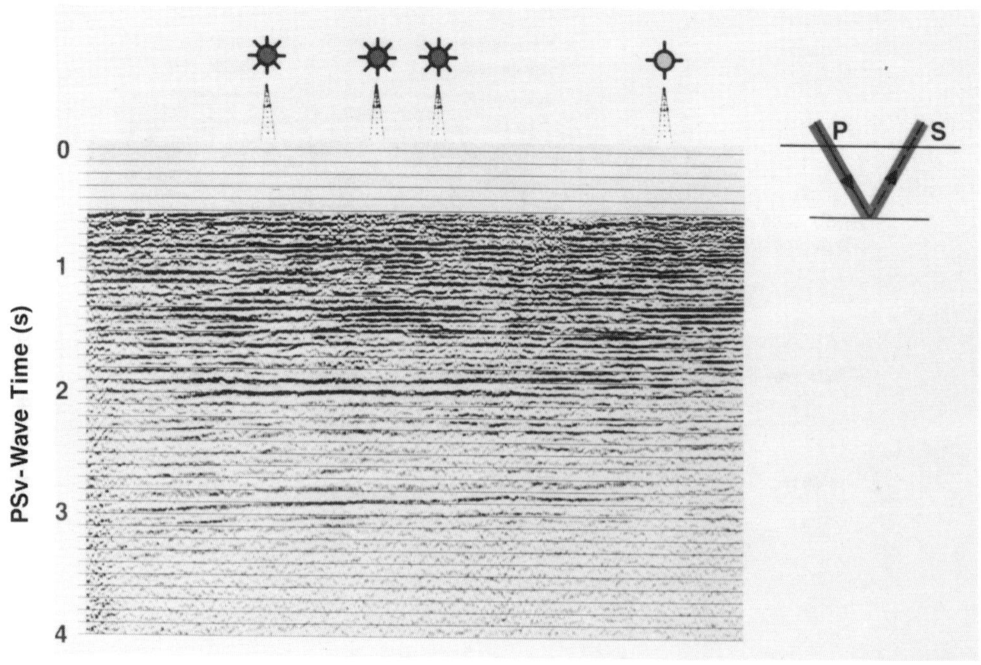

Fig. 7. Processed converted-wave (*P-SV*) section from the Zamora field area (inline horizontal geophone components). Note that this section is not obviously correlated to the (*P-P*) section of the previous figure.

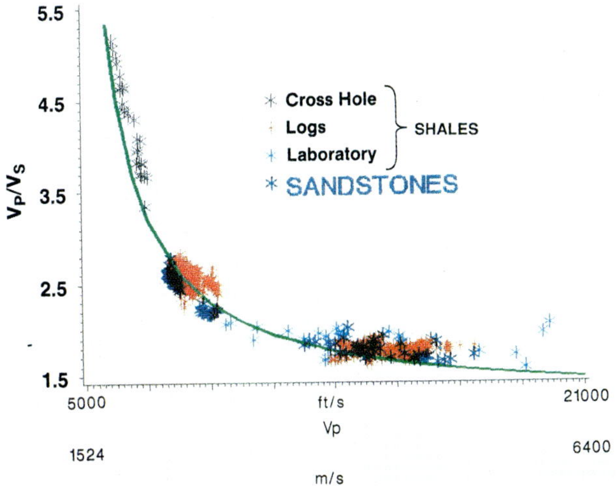

Fig. 8. Assorted rock physics data for brine-saturated sediments compared with the ARCO mudrock line. All data are shales except for indicated sandstones.

V_P/V_S RATIOS CALCULATED FROM *P-P* AND *P-SV* TRAVELTIMES

Having done an approximate correlation of events on each section using the smoothly compressed *P-SV* time section, we returned to the uncompressed time sections to look in detail at the interval V_p/V_s ratios. The *P-P* and *P-SV* seismic events were timed and used to calculate V_p/V_s ratios for the principle lithologic units in the Zamora area. This was accomplished using a simple equation that relates interval V_p/V_s ratios to *P-P* and *P-SV* traveltimes:

$$V_p/V_s = \text{Tss/Tpp} = (2\text{Tps}-\text{Tpp})/\text{Tpp}$$

where Tss, Tps, and Tpp are one-way interval traveltimes for *S*, *P-SV* and *P*-waves, respectively.

Interval V_p/V_s ratios for the major lithostratigraphic units are displayed in Figure 11. Comparison of the V_p/V_s ratio data with log data from wells along the line of section indicates that relative changes in V_p/V_s ratios can be used as an indicator of gross lithologic variation (sand/clay ratios). For instance, V_p/V_s ratios calculated for the Capay shale interval (essentially 100 percent clay) range from 3.50 to 3.55 with a mean of 3.53. The Starkey/Peterson interval (which is greater than 90 percent brine-saturated sandstone) has V_p/V_s ratios that range from 2.51 to 2.56 with a mean of 2.53.

In addition, within any one lithostratigraphic unit there is a general decrease in V_p/V_s ratio with increasing depth of burial. This observation is apparently related to an increase in the consolidation state of the

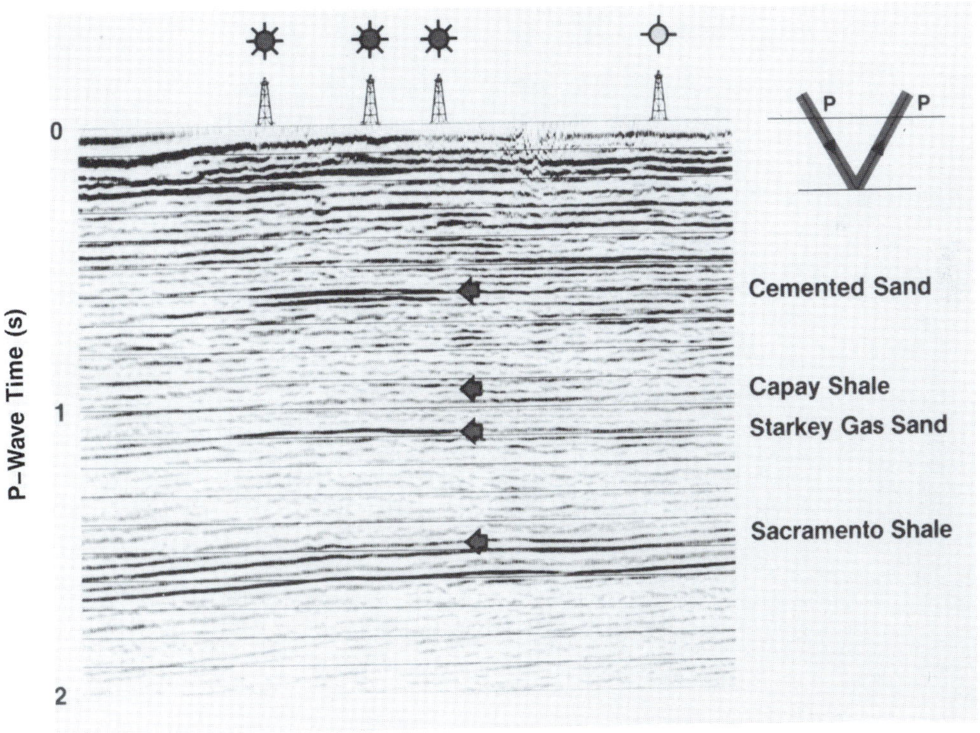

Fig. 9. Processed *P*-wave seismic section from the Zamora field area. Correlations between seismic events and the top of important stratigraphic intervals are indicated.

sediments as they are more deeply buried and become better lithified. This conclusion is supported by the dramatic differences in V_p/V_s ratio between the Capay and Sacramento shales, both of which are marine shales with similar bulk compositions. Presumably the major difference between the two units is their degree of lithification.

The V_p/V_s ratios determined independently from seismic event correlations provide a "higher frequency" profile of V_p/V_s with depth, whereas the low frequency compression of *P-SV* time to *P-P* time was a first correction to approximately align the entire sections in time so that correlations could be interpreted. Granted that there is artistic license in this correlation, it is interesting to compare the V_p/V_s ratios from the picked events with expected ratios obtained from applying the ARCO mudrock line to the Knaggs-Wallace No. 1 well sonic log as shown in Figure 12.

The sonic log is converted to an approximate *S*-wave sonic log by applying the ARCO mudrock line, from which a synthetic V_p/V_s log is derived. Superposed on this last log are measured V_p/V_s ratios from correlated events picked near the well on Figure 11. The good agreement between synthetic V_p/V_s logs and picked events suggests that the ARCO mudrock line is a good "first-order" approximation of V_s for brine-saturated sand-shale sequences in the absence of any real in-situ *S* velocities.

AMPLITUDE VARIATION WITH OFFSET MODELS

In order to test the validity of event correlations and the interpretations made from the stacked data, amplitude variation with offset (AVO) reflectivity models were prepared. Sonic and density logs from the

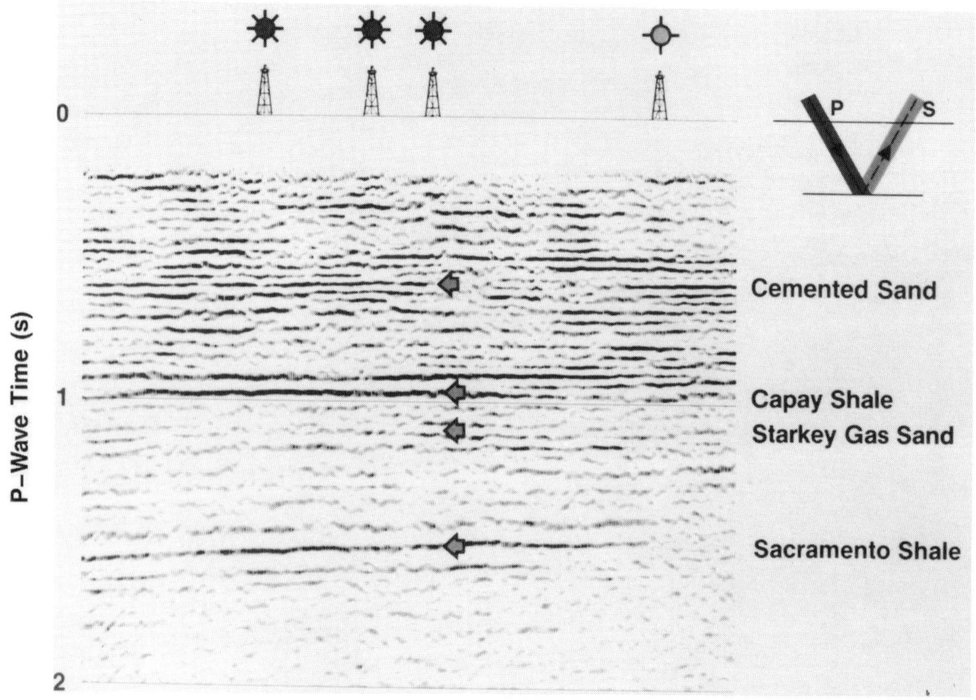

Fig. 10. Processed converted-wave (*P-SV*) section compressed to *P-P* time using the ARCO mudrock line. Events that correlate with the *P*-wave section are indicated.

Knaggs-Wallace Well No. 1 well provided the information necessary for calculating *P*-wave velocities and subsurface reflectivity. Rock physics programs were used to estimate V_s for sand beds, while V_s for shaly, brine-saturated units were estimated using the ARCO mudrock line (Castagna, et al., 1985).

STARKEY GAS SAND AVO MODELS

AVO models for the Starkey gas sand were prepared to evaluate the effects of fluid saturation on AVO response for *P-P* reflections.

Well log analysis and formation tests indicate that productive Starkey reservoirs have porosities on the order of 32 percent with gas saturations of over 80 percent. The first AVO example (with Poisson's ratios on the order of .30 to .32; Figure 13) does a good job of matching events in time, however, it falls short in matching the increase in AVO that is seen in the actual seismic data.

A second model was prepared to increase the *S* impedance contrast between overlying shale and Starkey sand. This was done by increasing the V_s velocities for the sands (+10 percent) and lowering it for the shales (−10 percent). The modeled result (Figure 14) does give a detectable increase in AVO with Poisson's ratios on the order of .25 for the sand, but again falls short of that observed in the real data.

A third model was prepared to see if a better correspondence could be achieved by varying the V_s for the sands (+20 percent) and shales (−20 percent). The modeled result (Figure 15) gives a very detectable increase in AVO that comes very close to the field trial data. The Poisson's ratio for gas sands in this model are on the order of .15–.18, which is within the scatter of rock physics predictions. However, uncertainties in offset gain corrections applied to data make more precise conclusions unwarranted. In this data 1 s AEC functions were used to trace equalize the data before offset modeling was done.

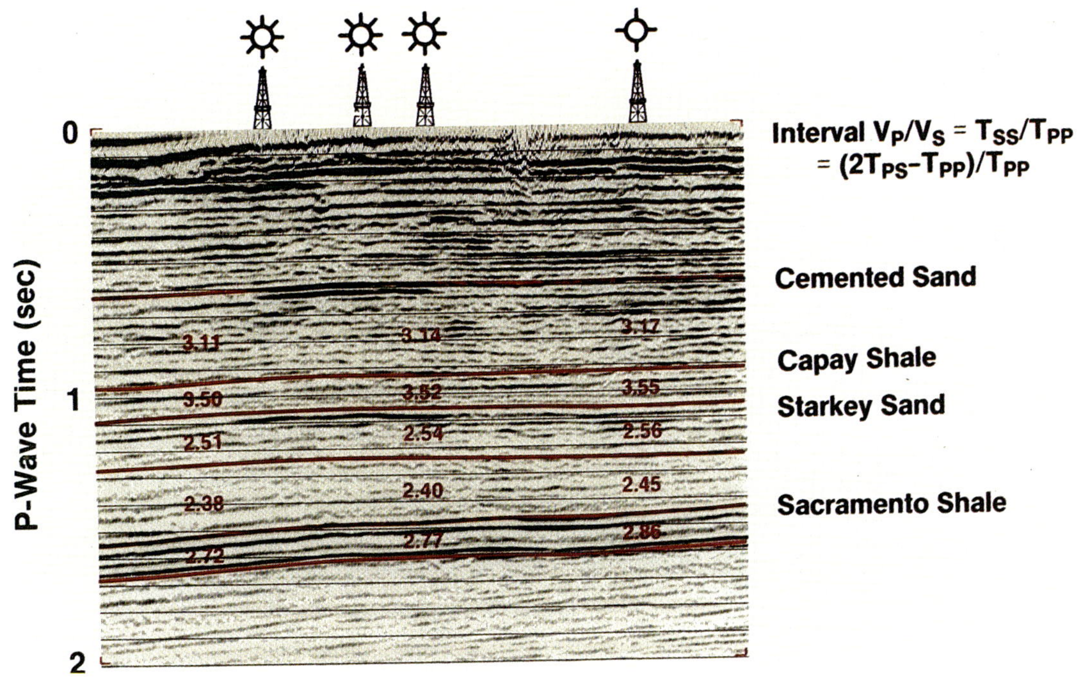

Fig. 11. V_p/V_s ratios for important lithostratigraphic intervals calculated using *P-P* and *P-SV* traveltimes.

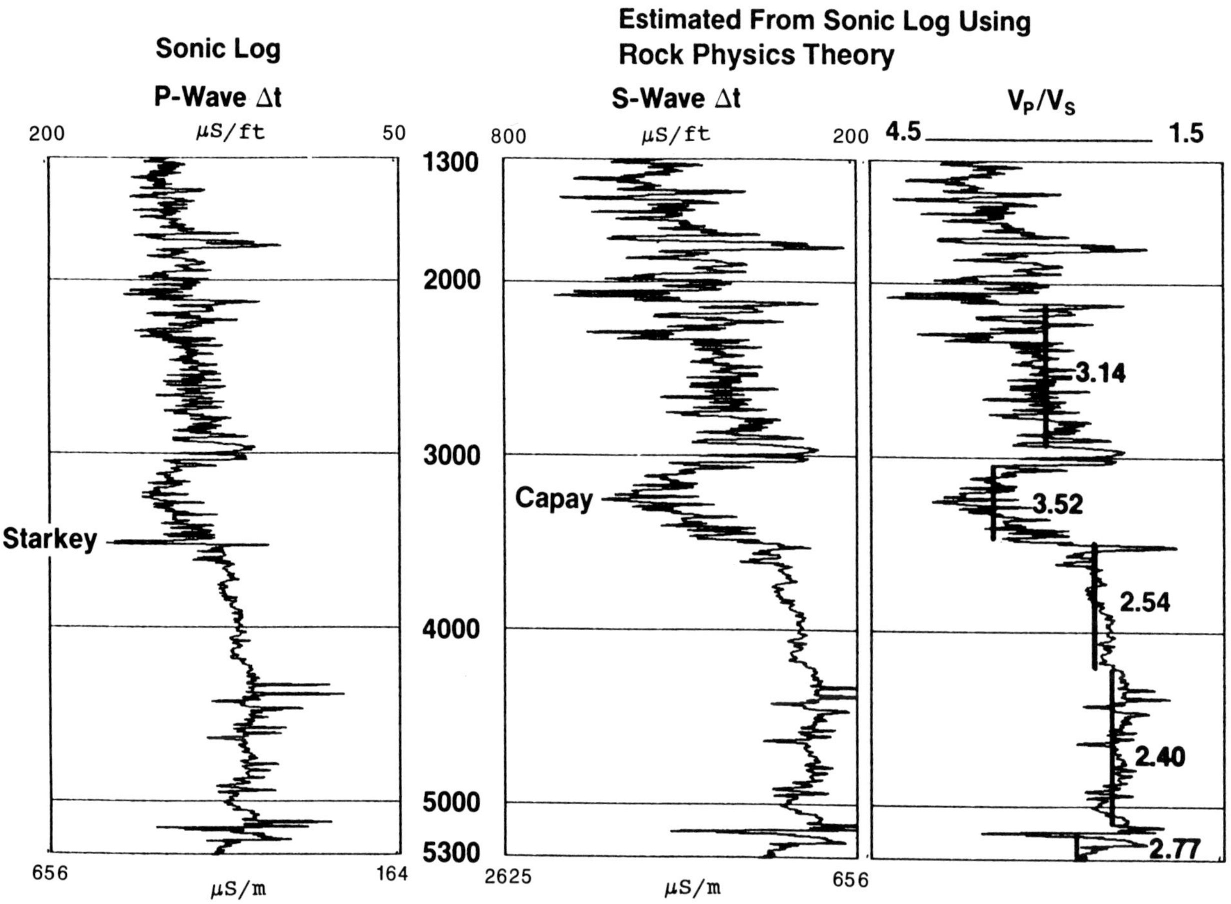

Fig. 12. *S*-wave traveltime and V_p/V_s ratio logs estimated from the Knaggs-Wallace No. 1 *P*-wave sonic log using the ARCO mudrock line and rock physics programs. V_p/V_s ratios estimated using Knaggs-Wallace No. 1 *P*-wave sonic log using the ARCO mudrock line compare favorably with those calculated independently from 3-C seismic event correlations.

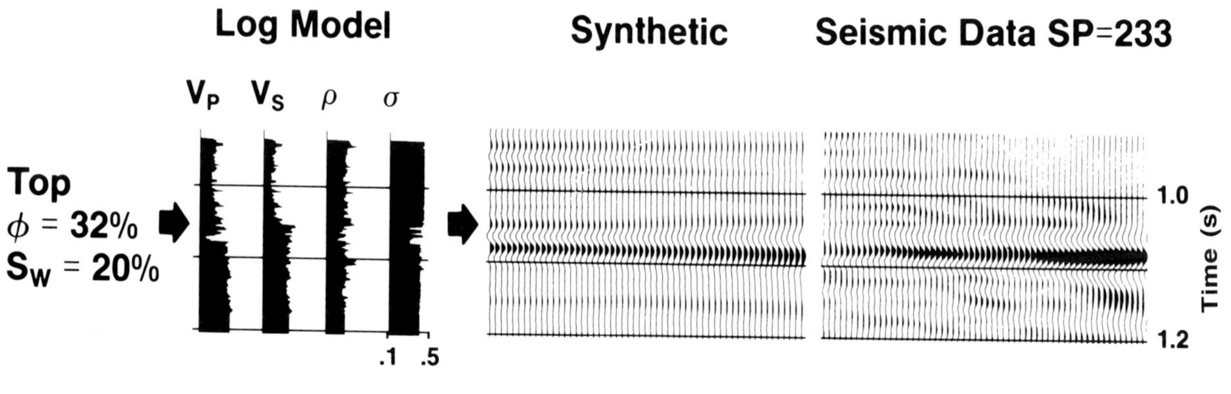

Fig. 13. Starkey gas sand AVO model for gas-saturated case (porosity = 32 percent; gas saturation = 80 percent).

A fourth model was prepared for a brine-saturated sand case with the same porosity of 32 percent. In the example (Figure 16), the absence of gas results in very low reflectivity events that mimic those observed on the stacked data where the reservoir sand is increasingly shaly and water wet.

SHALLOW BRIGHT-SPOT AVO MODELS

AVO models were prepared for the shallow carbonate-cemented sand that displays a high reflectivity *P*-wave event (Figure 9). Well log analysis and well cuttings indicate that this interval has porosities on the order of 10 percent, with the remaining pore space occluded by calcite cement. The synthetic (Figure 17) does a good job of matching the timing and reflectivity of events that are seen in the field trial data. These data support the conclusion that variations in carbonate cementation are controlling the amplitude of seismic events in this interval.

CONCLUSIONS

An integrated interpretation strategy that utilizes *P-P* and *P-SV* seismic reflection data, rock physics relationships, and AVO modeling strategies can provide extra evidence to differentiate between high-amplitude events from gas sands and low-porosity, well-cemented sands. This seems to be useful in the Sacramento Valley in areas where no shear log data are available.

In brine saturated sand-shale sequences the ARCO mudrock line provides a good scaling function to compress *P-SV* reflection times to *P-P* times so that events on the two sections can

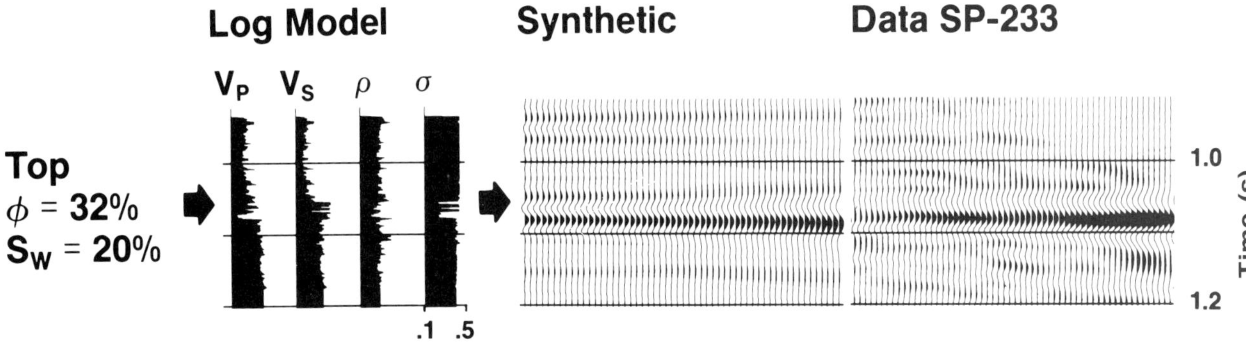

Fig. 14. Starkey gas sand AVO model for gas-saturated case (porosity = 32 percent; gas saturation = 80 percent) using alternate V_s estimates for sands (+10 percent) and shales (−10 percent).

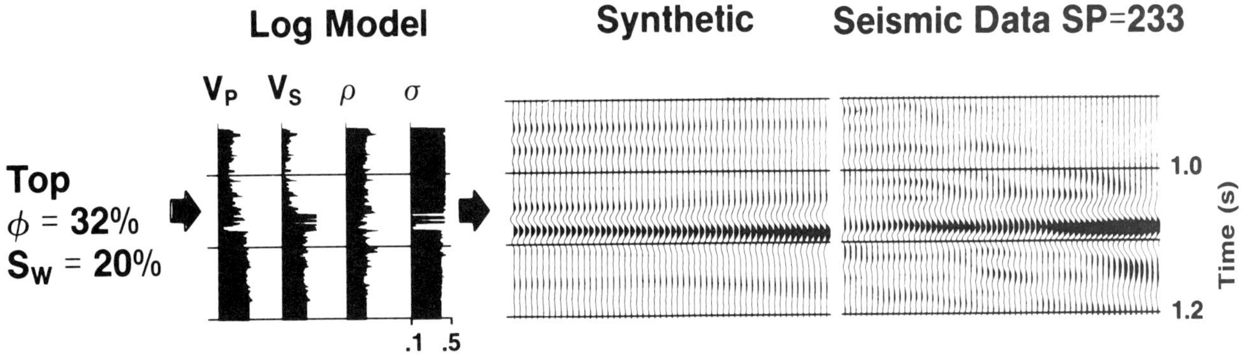

Fig. 15. Starkey gas sand AVO model for gas-saturated case (porosity = 32 percent; gas saturation = 80 percent) using alternate V_s estimates for sands (+20 percent) and shales (−20 percent).

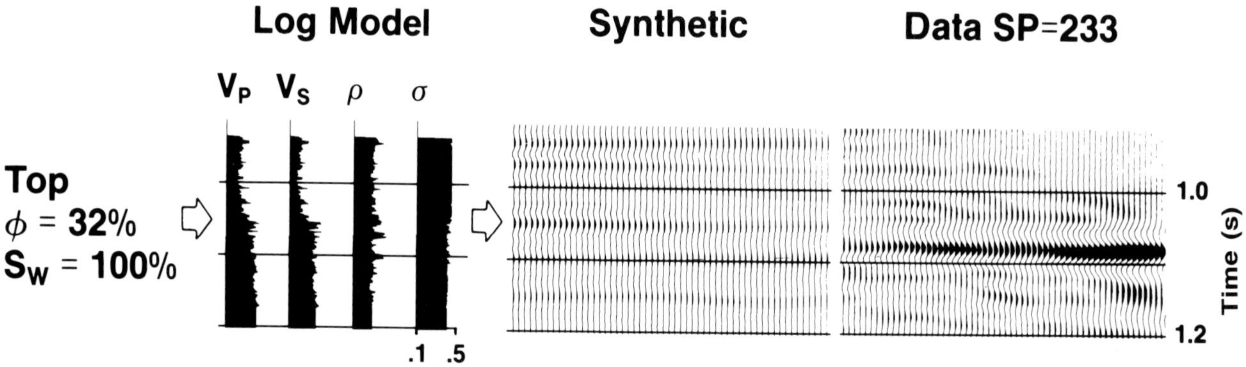

Fig. 16. Starkey gas sand AVO model for brine-saturated case (porosity = 32 percent; gas saturation 0 percent).

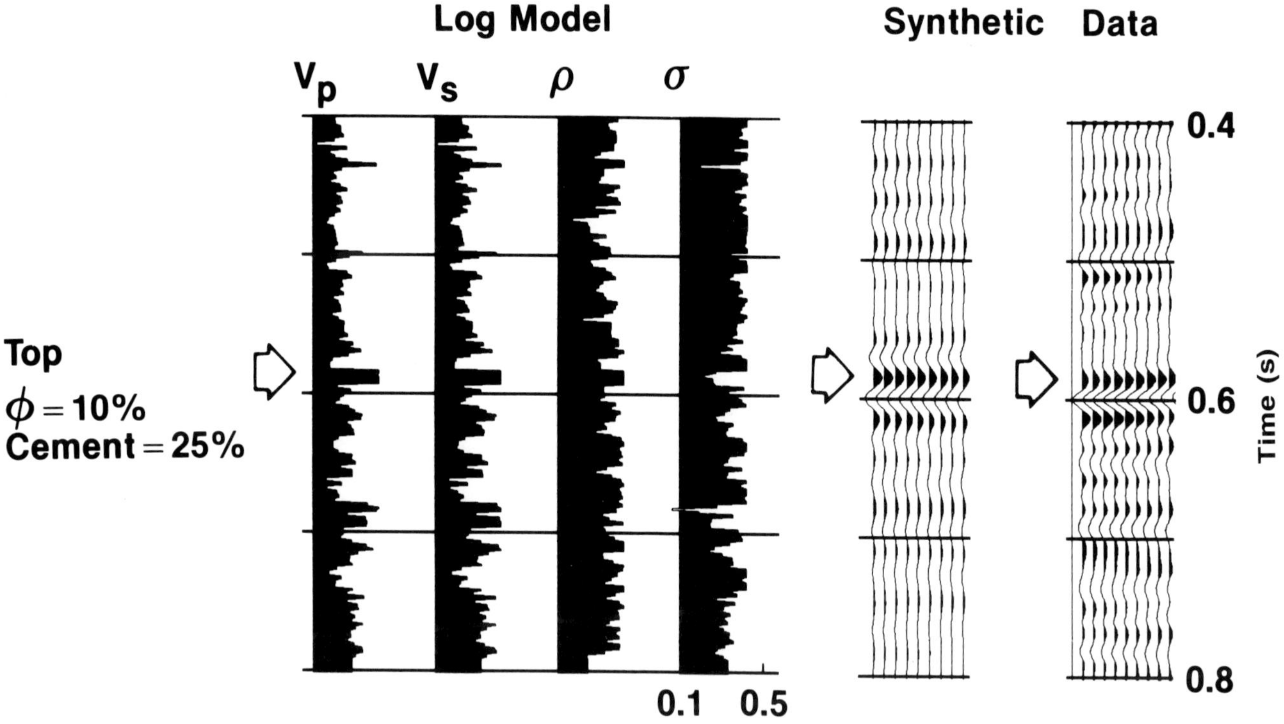

Fig. 17. Shallow bright-spot AVO model for carbonate cemented cased (porosity = 10 percent; carbonate cement = 25 percent).

be correlated in a gross sense. Once event correlations are made, the finer interval V_p/V_s ratios determined from seismic event times provide higher resolution information which agrees with sonic log predictions using the ARCO mudrock line.

Interval values of V_p/V_s ratios obtained this way may indicate gross lithologic variations (i.e., sand/clay ratio, consolidation state) both laterally and vertically in the section.

AVO modeling of both the Starkey gas sand and a well-cemented sand in the section provide extra evidence to reinforce the interpretations obtained from the stacked sections, in spite of processing uncertainties in the seismic data.

In the final analysis the interpreter still has to make a judgment based on the available data. Although any one lithology indicator may be risky, an integrated study with several seismic indicators will significantly improve an explorationist's ability to evaluate exploration risks before drilling.

ACKNOWLEDGMENTS

The authors thank Chevron for permission to publish this paper. In particular, we thank Fred Herkenhoff of Chevron's Western Region for support of multi-component recording, and Al Hardin and Don Winterstein of Chevron Oil Field Research Company for valuable recording and processing help.

APPENDIX—RELATION BETWEEN CONVERTED AND SHEAR-WAVE REFLECTIVITIES

This appendix shows why the converted *P-SV* reflectivity at non-zero offset mimics *SH-SH* reflectivity at zero offset, within a scale factor, and is insensitive to *P* impedance changes across interfaces. The simplest way to show this is with the modified Bortfeld reflection coefficients given in Aki and Richards (1980). These coefficients are good approximations to the exact reflection formulas in cases of small impedance contrasts across an interface and at small incident angles less than critical. The formulas are much simpler than the exact reflection coefficients and are ideal for interpreters to use for modeling and analysis of data.

The reflection coefficients are expressed in terms of reflectivity components $\delta V_p/V_p$, $\delta V_s/V_s$ and $\delta\rho/\rho$ at an interface. These terms are the fractional changes in *P*-wave velocity, *S*-wave velocity, and density across the reflecting interface. These fractional changes equal the actual contrasts from above to below the interface divided by the average parameter value of the two layers.

If we take the formula for *P-SV* reflectivity from Aki and Richards (1980) and take the near offset approximation, keeping only linear terms in sin θ, the incident angle, we obtain, after some algebra,

$$R_{ps}(\theta) = -4\,\frac{\sin\theta}{r}\left[\frac{(2+r)}{8}\frac{\delta\rho}{\rho} + \frac{1}{2}\frac{\delta V_s}{V_s}\right] \quad (1)$$

where $r = V_p/V_s$.

Note that contrasts in V_p are not in the formula. We can rewrite this expression as

$$R_{ps}(\theta) = +4\,\frac{\sin\theta}{r}\left[R_{ss} - \frac{(r-2)}{8}\frac{\delta\rho}{\rho}\right] \quad (2)$$

where R_{ss} is the zero offset approximate reflectivity for *SH* waves given by

$$R_{ss} = -\frac{1}{2}\left[\frac{\delta\rho}{\rho} + \frac{\delta V_s}{V_s}\right].$$

In the case of older, consolidated sediments where $r = 2$ we see that the second term inside the brackets in equation (2) vanishes. Then the R_{ps} reflectivity at a given offset will scale up or down proportional to the R_{ss} reflectivity. As shown in this paper and in previous studies (Ensley, 1984; Robertson and Pritchett, 1985) r in the north Sacramento Valley can be 3 or higher. If $r = 3$ then R_{ps} doesn't quite scale proportionally to R_{ss} but is close to it since the extra density contrast effect inside the brackets of equation (2) is small, i.e., $(1/8)\delta\rho/\rho$.

For seismic data this simple relation between R_{ps} and R_{ss} can be useful for interpreters looking at amplitude changes or anomalies in reflections without detailed processing or modeling analysis.

REFERENCES

Aki, K., and Richards, P. G., 1980, Quantitative seismology, Theory and methods, Volume I, W. H. Freeman and Company.

Castagna, J., Batzle, M., Eastwood, R., 1985, Relationships between compressional-wave and shear-wave velocities in clastic silicate rocks: Geophysics, **50**, 571–581.

Ensley, R. A., 1984, Comparison of *P*- and *S*- wave seismic data: A new method for detecting gas reservoirs: Geophysics, **49**, 1420–1431.

Frasier, C. W., and Winterstein, D. F., 1990, Analysis of conventional and converted mode reflections at Putah Sink, California, using three-component data: Geophysics, **55**, 646–659.

Robertson, J. D., and Pritchett, W. C., 1985, Direct hydrocarbon detection using comparative *P*-wave and *S*-wave seismic sections: Geophysics, **50**, 383–393.

Index